Principles of
Engineering Economic Analysis

Principles of
Engineering Economic Analysis

third edition

John A. White
School of Industrial and Systems Engineering
Georgia Institute of Technology
Atlanta, Georgia

Marvin H. Agee
Department of Industrial Engineering and Operations Research
Virginia Polytechnic Institute and State University
Blacksburg, Virginia

Kenneth E. Case
School of Industrial Engineering and Management
Oklahoma State University
Stillwater, Oklahoma

John Wiley & Sons
WILEY New York • Chichester • Brisbane • Toronto • Singapore

Library of Congress Cataloging in Publication Data:

White, John A., 1939–
    Principles of engineering economic analysis / John A. White,
Marvin H. Agee, Kenneth E. Case.—3rd ed.
      p.   cm.
    Bibliography: p.
    Includes index.
    ISBN 0-471-61320-7
    1. Engineering economy.   I. Agee, Marvin H., 1931–   II. Case,
Kenneth E.   III. Title.
TA177.4.W48 1989
658.1'55—dc19

88-26094
CIP

Printed in the United States of America

10 9 8

Printed and bound by Courier Companies, Inc.

About the Authors

JOHN A. WHITE, Ph.D., P.E.

Dr. John A. White is Regents' Professor of Industrial and Systems Engineering and Gwaltney Professor of Manufacturing at the Georgia Institute of Technology. Currently, he is serving as Assistant Director for Engineering at the National Science Foundation. His more than twenty years teaching experience has been gained at Georgia Tech, Virginia Tech, and The Ohio State University. He is a graduate of the University of Arkansas (BSIE), Virginia Tech (MSIE), and The Ohio State University (Ph.D.); additionally, he received an honorary doctorate in engineering from the Katholieke Universitiet of Leuven in Belgium.

He is the co-author of five texts: *Facility Layout and Location: An Analytical Approach,* Prentice-Hall, 1974; *Analysis of Queueing Systems,* Academic Press, 1975; *Principles of Engineering Economic Analysis,* Second Edition, John Wiley & Sons, 1984; *Capital Investment Decision Analysis for Management and Engineering,* Prentice-Hall, 1980; and *Facilities Planning,* John Wiley & Sons, 1984. He also is Editor of the *Production Handbook,* Fourth Edition, John Wiley & Sons.

Dr. White is a member of the National Academy of Engineering. He is past-president of the Institute of Industrial Engineers and past-chairman of the American Association of Engineering Societies.

MARVIN H. AGEE, Ph.D.

Dr. Marvin H. Agee is Professor of Industrial Engineering and Operations Research at Virginia Polytechnic Institute and State University. He is a graduate of Virginia Tech (BSIE, MSIE) and The Ohio State University

(Ph.D.), and has over thirty years teaching experience at these two universities.

Dr. Agee is a Fellow of the Institute of Industrial Engineers (IIE) and a senior member of the Society of Manufacturing Engineers (SME). He has held numerous offices in IIE and Alpha Pi Mu. In IIE, he has served as the Director of Continuing Education, Chairman of the Gardner-Gilbreath Scholarship Fund Trustees, and Director of the Facilities Planning and Design Division. He is also a Fellow of IIE and a recipient of this organization's Outstanding Educator Award. He has held the positions of Executive Secretary, Vice President, President, and Advisory Board Member in Alpha Pi Mu. He has also held local or regional offices in SME and the American Society of Engineering Education (ASEE), and has served on the College–Industry Council on Material Handling Education.

Dr. Agee is also the co-author of three books: *Quantitative Analysis for Management Decisions,* Prentice–Hall, 1976; *Study Guide for Professional Engineers' Examination in Industrial Engineering,* IIE, 1982; and *Principles of Engineering Economic Analysis,* Second Edition, John Wiley & Sons, 1984.

KENNETH E. CASE, Ph.D., P.E.

Dr. Kenneth E. Case is Regents Professor of Industrial Engineering and Management at Oklahoma State University. He is a graduate of OSU (BSEE, MSIE, and Ph.D.) and has taught at the university level at both O.S.U. and Virginia Tech. He is a licensed professional engineer and was named the Outstanding Engineer in Oklahoma in 1987.

Dr. Case is a Fellow of the Institute of Industrial Engineers which he served as President in 1986–87. Other IIE positions held include Vice President of Systems and Technical Divisions, Director of Professional Registration, Director of Conference Program Coordination, and Editor of the Applied Probability and Statistics Department of *IIE Transactions.*

Dr. Case is a Senior Member of the American Society for Quality Control which he served as Tulsa Section Chairman in 1977–78 when that section won the Berg Award as the most outstanding in its class. He is a member of the Editorial Board of Journal of Quality Technology. He has presented numerous seminars for the Education and Training Institute of the ASQC, on topics including statistical process quality control, quality engineering, basic statistical quality control, and quality costs. He is an ASQC Certified Quality Engineer and an ASQC Certified Reliability Engineer.

Preface

This textbook is for those who are interested in learning the principles involved in analyzing economic investment alternatives. Although the material is useful for accounting, economics, finance, and management students, its primary audience will be engineering and engineering technology students. Courses for which the book is intended are variously titled engineering economy, economic analysis, industrial economy, economic decision analysis, engineering decision analysis, and engineering economic analysis.

Over the past 30 years we have taught engineering economic analysis to more than 4000 persons, including undergraduate and graduate students from engineering and nonengineering academic programs. Additionally, we have taught refresher courses for professional engineering registration examinations and continuing education short courses on the subject. However, in addition to teaching others to perform economic analyses, we have also put into practice the principles and techniques we have taught. Specifically, our consulting and industrial experience required the application of the material presented here. Based on that experience, we perceived a need for a textbook that provided a unified treatment of economic analysis principles. We also felt the need to address the subject from a *cash flow viewpoint*.

The following are some of the important features of this textbook.

1. A cash flow approach is taken throughout.
2. A systematic seven-step approach is given for performing a comparison of investment alternatives.
3. The study period must be specified explicitly instead of unknowingly employing a least common multiple of lives study period.

4. Comprehensive treatments are provided of:
   a. Cost concepts, including cost terminology, cost estimation, accounting principles, nonmonetary considerations, and cost of capital determinations.
   b. Inflation, as it affects the economic comparison of investment alternatives.
   c. Both post-1986 ACRS and pre-1987 methods of depreciation.
   d. Income taxes, including capital gains and losses, depreciation recapture, and depletion.
   e. Economic analyses in the public sector, emphasizing benefit-cost analysis procedures.
   f. Break-even, sensitivity, and risk analyses, with the latter including analytical and simulation approaches.
   g. Decision models, including decisions under risk and uncertainty and decision tree analyses.

5. Approximately 200 worked-out examples and 400 problems are provided, with answers to even-numbered problems given at the end of the text.

6. Section numbers are provided for material covered by each end-of-chapter problem to facilitate a self-study program.

7. Modeling uniform, gradient, and geometric series, modeling changing interest rates, and modeling equity amounts in loan payments are presented to emphasize the importance of *modeling* cash flow series instead of using more conventional cookbook approaches.

8. New and expanded tables of compound interest factors are provided, including both discrete and continuous compounding, discrete and continuous flows, and gradient and geometric cash flow series.

9. Separate chapters on measuring investment worth and the comparison of alternatives are provided to permit a detailed understanding of the measures of investment worth prior to their use in comparing alternatives.

10. Methods of measuring investment worth include present worth, annual worth, future worth, capitalized worth, payback period, savings/investment ratio, internal rate of return, and external rate of return methods.

11. Methods of comparing investment alternatives include both ranking and incremental cash flow approaches.

12. Depreciation is covered in conjunction with the discussion of income taxes, since depreciation is not a cash flow and it influences cash flows through taxation.

13. Replacement analysis is covered in the discussion of comparison of alternatives, since a cash flow analysis of replacement problems is essentially no different than any other type of investment alternative.

14. The data have been packaged in a nine-chapter format in the sequence most natural to the coverage of the material.

15. The text is practice-oriented; only material that is required in the performance of engineering economic analyses is presented, since we believe that including tangential and purely theoretical material has a deleterious effect on the student.

16. A detailed Solutions Manual has been prepared that includes detailed solutions to all exercises.

The manuscript has been tested in the classroom. The orientation of the material and its organization have been influenced by the recommendations of numerous colleagues and undergraduate and graduate students.

The first seven chapters have been developed from our collective experience of teaching the subject to engineering sophomores from all disciplines; the material in the remaining two chapters has been covered in a senior-level elective course. Although elementary calculus is a sufficient prerequisite for the first seven chapters, an introductory course in probability theory is suggested as a prerequisite for the final two chapters.

Many people influenced the development of this textbook. In particular, we have benefited from either taking coursework from or teaching with John R. Canada, Wolter J. Fabrycky, Richard S. Leavenworth, William T. Morris, G. T. Stevens, Jr., and Gerald J. Thuesen, who have authored or coauthored one or more texts on the subject. Our approach to the subject has also been influenced by our associations with Ronald G. Askin, Richard H. Bernhard, William E. Biles, Leland T. Blank, James R. Buck, Lynn E. Bussey, Claus Christiansen, Thomas P. Cullinane, Stuart E. Dreyfus, Carl B. Estes, Gerald A. Fleischer, David R. Freeman, Jorge Haddock, Lynwood A. Johnson, Marilyn S. Jones, Raymond P. Lutz, Leon F. McGinnis, Chan S. Park, Robert S. Kaplan, W. J. Kennedy, Jr., Jack R. Lohmann, Paul A. Nelson, M. Wayne Parker, Gunter P. Sharp, Donald R. Smith, William G. Sullivan, J. M. A. Tanchoco, Suleyman Tufekci, and Donovan B. Young. Therefore, the influence of others will be evident in various places in the book. Because it is impossible to credit the many individual contributions, we simply acknowledge gratefully their influence on our understanding of the subject.

Particular emphasis is given to the contribution of R. Dale Atkins. He provided the camera-ready, laser printed interest tables, which substantially improved the Third Edition.

Finally, we thank our wives Mary Lib, Barbara, and Lynn, for their patience and understanding during the project. They constantly demonstrated considerable insight on the subject by the gentle reminder, "If you know so much about investments, why aren't we rich?"

John A. White
Marvin H. Agee
Kenneth E. Case

Contents

CHAPTER

Introduction

1.1 BACKGROUND

The subject matter discussed in this book is variously referred to as *economic analysis, engineering economy,* and *economic decision analysis,* among other names. Traditionally, the application of economic analysis techniques in the comparison of engineering design alternatives has been referred to as *engineering economy.* However, the emergence of a widespread interest in economic analysis in public sector decision making has brought about greater use of the more general term, *economic analysis.*

Our approach to the subject will be a *cash flow approach.* A cash flow occurs when money actually changes hands from one individual to another or, for that matter, from one organization to another. Thus, money received and money dispersed (spent) constitute cash flows. We also will emphasize the fundamentals of economic analysis without dwelling on some of the philosophical issues associated with the subject; for those interested in pursuing the latter, references are provided.

1.2 THE PROBLEM-SOLVING PROCESS

Economic analyses are typically performed as a part of the overall problem-solving process. In designating a new or improved product, a manufacturing process, or a system to provide a desired service, the "problem solver" is involved in performing the following steps.

1. Formulation of the problem.
2. Analysis of the problem.

1

3. Search for alternative solutions to the problem.

4. Selection of the preferred solution.

5. Specification of the preferred solution.

The problem-solving process begins when a decision maker is dissatisfied with something, or recognizes a need, and then decides to do something about it. The process ends with plans for correcting the dissatisfaction or satisfying the need. The term *problem* should therefore be considered in a general and unlimited context. Examples range from easily defined problems such as high material scrap rates in production, unsafe working conditions, and poor utilization of machinery to somewhat hazy problems such as "something must be done to maintain business survival."

The *formulation of the problem* involves the establishment of its boundaries, and it is aided by taking a black box approach. As Krick [11] describes the black box, an originating state of affairs (state A) and a desired state of affairs (state B) exist. Also, a transformation must take place in going from state A to state B, as depicted in Figure 1.1. More than one method of performing the transformation from state A to state B exists, and there is unequal preferability of these methods. The solution to the problem is visualized as a black box of unknown, unspecified contents having input A and output B.

The *analysis of the problem* consists of a relatively detailed phrasing of the characteristics of the problem, including restrictions and the criteria to be used in evaluating alternatives. Considerable fact gathering is involved, including the real restrictions on the problem, which must be satisfied. Consequently, any budget, quality, safety, personnel, environmental, and service level constraints that may exist are identified.

In this step of the problem-solving process, the problem definition is judged to be critical since, in most cases, the solution of a problem is directly determined by the problem definition. According to Stoll [19], engineering design is accomplished through the problem-solving process. In the context of designing new products for manufacture, engineering design changes are inherent in the process as more complete design information is obtained (from prototype tests, analysis, customer feedback, etc.). Stoll emphasizes that time and effort spent "up front" on careful and thorough definition of the problem is necessary to achieve a design solution having acceptable quality, cost and delivery schedule.

The *search for alternative solutions to the problem* involves the use of the engineer's creativity in developing feasible solutions to the design problem.

FIGURE 1.1. Black box approach.

A number of aids to stimulate one's creativity are provided by Krick [11]. In particular, he suggests that one should:

1. Exert the necessary effort.
2. Not get bogged down in details too soon.
3. Make liberal use of the questioning attitude.
4. Seek many alternatives.
5. Avoid conservatism.
6. Avoid premature rejection.
7. Avoid premature acceptance.
8. Refer to analogous problems for ideas.
9. Consult others.
10. Attempt to divorce one's thoughts from the existing solution.
11. Try the group approach.
12. Remain conscious of the limitations of the mind in the process of idea generation.

Unfortunately, the natural tendency is to mix the *evaluation* of alternatives with the *generation* of alternative solutions. As a result, the search process usually terminates with the development of the first alternative that can be justified economically. By separating the search process from the selection process, we enhance the chance of generating a number of alternatives that can be justified economically.

The *selection of the preferred solution* consists of the measurement of the alternatives, using the appropriate criteria. The alternatives are compared with the constraints, and infeasible alternatives are eliminated. The benefits produced by the feasible alternatives are then compared. Among the criteria considered for choosing among alternatives is the economic performance of each alternative.

The *specification of the preferred solution* consists of a detailed description of the solution to be implemented. Predictions of the performance characteristics of the solution to the problem are included in the specification.

The problem is not solved, however, until the preferred solution is successfully implemented. Upon completion of the problem-solving process discussed before, someone must approve the recommended solution and authorize its implementation. (This is discussed later in Section 1.3.2). Implementation may itself be considered another problem (or project) whereby similar steps are followed to accomplish the solution implementation. Once a preferred solution is accepted and authorized by the appropriate decision maker, implementation must be planned and then managed through to completion. Typically, a project team is assigned to accomplish implementation, with a project manager assigned to be accountable for completion. Project management is a subject outside the scope of this book but a considerable literature exists on the subject. A sample of the literature is given in the bibliography at the end of this chapter [7, 9, 12, 14].

Example 1.1

As an illustration of the application of the problem-solving approach, consider the case of a locally owned electrical power generating company faced with an air pollution problem that had to be solved. The power plant was using low-grade coal with sulfur content of 1.9% by weight, and the air pollution control board for the state directed the power plant to comply with newly adopted standards for air quality. A consulting engineer was hired to study the problem and recommend the best course of action.

The engineer met with representatives of the air pollution control board, the power company, county and city officials, and suppliers of various fuel alternatives and associated power-generating equipment. An analysis of the problem yielded an agreement to use a 20-year planning horizon, forecasts of power requirements over the planning horizon, and a specification of the criteria to be used in evaluating the alternatives. Specifically, a benefit-cost analysis was to be performed; a cost of $150/ton of sulfur pollutant released into the air was to be employed in the analysis.

A search for alternative solutions to the problem resulted in a consideration of bunker C fuel oil, low-sulfur furnace fuel oil, and natural gas. The methods described in Chapter Seven were employed to compare the alternatives using a benefit-cost analysis; bunker C fuel oil was the recommended energy source.

Based on the report provided by the consulting engineer, the city contracted for the installation of storage tanks for the fuel oil, piping, pumps, and the other equipment necessary for converting the furnaces from coal to fuel oil.[1]

Example 1.2

As a second illustration of the problem-solving procedure, a leading manufacturer of automotive bearings was faced with the need to expand its distribution operations. A study team was formed to analyze the problem and develop a number of feasible alternative solutions. After analyzing the problem and projecting future distribution requirements, the following feasible alternatives were determined:

[1] For additional discussion of the case study on which the above was based, see [5].

1. Consolidate all distribution activities and expand the existing distribution center, located in Michigan.
2. Consolidate all distribution activities and construct a new distribution center, location to be determined.
3. Decentralize the distribution function and build several new distribution centers geographically dispersed in the United States.

After considering the pros and cons of each alternative, the officers of the company directed the study team to pursue the second alternative.

An extensive plant location study was performed. The location study resulted in five candidate locations being selected for final consideration. The criteria used to make the final selection included:

1. Land cost and availability.
2. Labor availability and cost.
3. Proximity to supply and distribution points (present and future).
4. Taxes (property and income) and insurance rates.
5. Transportation (access to rail and interstate highways).
6. Community attitudes.
7. Building costs.

Based on site visits to each location, the officers of the company selected a site in Alabama and directed the study team to develop design alternatives for the material-handling system to be used in the distribution center.

Applying the problem-solving procedure, four alternatives were obtained for evaluation. The first alternative involved the use of pallet racks, lift trucks, and flow racks; the second alternative included the use of an automated stacker crane system, lift trucks, conveyors, and flow racks; the third alternative suggested narrow-aisle, guided picking machines, high-rise shelving, conveyors, driverless tractor trains, and lift trucks; the fourth alternative consisted of an automated stacker crane system, a rail-guided picking vehicle system, a sortation and accumulation conveyor system, and high-rise, narrow-aisle lift trucks.

A planning horizon of 10 years was used in performing an economic analysis of each design alternative. The fourth alternative was the most economical and was recommended to top management. Based on the detailed presentation and the economics involved, management approved a budget of $9 million to implement the recommendations of the study team.

As demonstrated by the preceding illustrations, engineers and systems analysts must be prepared to defend their solutions to problems. Economic performance is among the several criteria used to evaluate each alternative. Monetary considerations seldom can be ignored. If economics are not considered in the criteria used in the final evaluation of the alternatives, they are usually involved in an initial screening of the alternatives. In fact, money can be a constraint on the alternatives that can be considered, as well as the basis for the final selection.

This textbook treats in detail the step in the problem-solving process that involves the selection of the preferred solution. The process of measuring cash flows and benefits and the consideration of multiple objectives in selecting the preferred design are also treated.

In comparing alternatives, the *differences* in the alternatives will be emphasized. Consequently, the aspects of the alternatives that are the same normally will not be included in the analysis.

1.3 NONMONETARY CONSIDERATIONS

1.3.1 Multiple Criteria in the Decision Process

The principal emphasis of this textbook is on the use of logical methodology to choose an investment project from among several investment projects available to the decision maker, where the criterion for choice will be some single economic measure of effectiveness. The investment project chosen, however, may be different from the project recommended by the engineer, who based the recommendation solely on an economic criterion. Managers typically have multiple criteria to be considered in reaching a final decision about the alternative to be adopted. Among the factors to be considered are quality, safety, environmental impact, community attitudes, labor-management relationships, cash flow position, risks, system reliability, system availability, system maintainability, system operability, system flexibility, impact on personnel levels, training requirements, comparisons with competitors, impact on the different units within the organization, ego, customers' preferences, capital requirements, and economic justification.

To illustrate a situation in which both economic factors and noneconomic, or nonmonetary, factors are involved, suppose that the management of a firm, because of a significant increase in market demand, has the choice between (1) going to a second shift of production using existing machinery and (2) installing more highly mechanized machinery in order to meet the new production schedules. Clearly, a comparison of the costs for each of these alternatives over a planning horizon is in order, and one objective of management would no doubt be to meet production schedules at the lowest possible annual cost of production.

Although considerable investigation and study are required, many of the factors involved in a decision that may at first seem noneconomic can be expressed in (or reduced to) monetary values. For instance, in the above example, maintenance personnel may have to be trained in the installation and repair of the new machinery. Most likely, these training costs can be determined and charged to the "mechanization" alternative. However, other factors involved are not so easily reduced to monetary values and, indeed, some factors cannot be so reduced. For example, the installation of the mechanized machinery would probably reduce the number of personnel required with the present production system, whereas the second shift alternative would result in additional employees. If persons are laid off or transferred to other jobs within the firm, their salaries could be considered annual savings accruing to the mechanized machinery alternative and, conversely, the salaries of additional employees integral with the second shift alternative would be annual costs. However, layoffs or transfers could have a deleterious effect on the morale of the remaining employees. On the other hand, adding new personnel could have a beneficial effect on morale. Assessing the cost or gain of these expected changes in morale is virtually impossible. Nevertheless, the potential effect on production that morale changes may have must be considered by management.

Factors that affect a decision but cannot be expressed in monetary terms are often called *intangibles* or *irreducibles*. Practically all real-world business decisions involve both monetary and intangible factors. In the above production example, assume that the alternative of a second shift results in a total annual cost of $100,000 for both shifts over a 5-year planning horizon with no persons laid off or transferred or new employees hired. For the mechanized machinery alternative, there would be an estimated annual cost of $85,000 over the 5-year planning horizon with six persons laid off. Thus, this oversimplified decision is reduced to two measures: a monetary annual cost and the single intangible factor of employee morale. If management chooses the second shift alternative, this would imply that management places a higher subjective utility value on "good employee morale" than on the $15,000 difference in annual costs between the two alternatives. That is, the objective of good employee morale is more important than the objective of lowest annual cost (at least for this difference of $15,000). Instances of decision situations in which intangible considerations outweigh monetary ones are frequent, and the engineer should not become distraught over this fact. Within the hierarchy of management responsibilities for a firm, the higher the level of management, the more likely it is that intangible considerations will be given greater subjective weight. Engineering project proposals should therefore reflect a knowledge of intangibles that management will wish to consider.

Miller and Starr have made interesting observations in regard to the decision-making process.[2]

1. Being unable to satisfactorily describe goals in terms of one objective, *people customarily maintain various objectives.*

2. Multiple objectives are frequently in conflict with each other, and when they are, a *suboptimization* problem exists.

[2] David W. Miller and M. K. Starr, *Executive Decisions and Operations Research*, Second Edition, Prentice-Hall, 1969, pp. 52–53.

3. At best, we can only optimize as of *that time* when the decision is made. This will frequently produce a suboptimization when viewed in subsequent times.

4. Typically, decision problems are so complex that any attempt to discover *the* set of optimal actions is useless. Instead, people set their goals in terms of outcomes that are *good enough.*

5. Granted all the difficulties, human beings make every effort to be *rational* in resolving their decision problems.

Today's business world is characterized by continual and rapid change. Managers of these businesses are faced with an ever-increasing amount of new information and technology, coupled with ever-increasing global competition. These managers must establish long-term (strategic) goals and objectives for their firms to survive and progress. Example goals might be to (1) shorten the time between the design of a new product or service and customer delivery, (2) improve the quality of the products or services, and (3) reduce the amount of materials inventoried. Associated with each of the goals could be several objectives. By the same token, a single objective may affect each of the goals. For example, the objective of automating a particular area of a manufacturing plant could contribute toward shortening the lead time from design to customer delivery, improving product quality, and lowering the amount of work-in-process inventory. An engineering project to automate the particular area of the plant could generate alternative hardware/software systems to accomplish the objective. The evaluation of each alternative system should therefore include factors that would measure progress towards each of the three goals. These factors would most likely be both economic (initial investment costs and annual operating expenses) and noneconomic (expanded capacity, machine utilization, etc.) in nature.

The decision problem to select a particular hardware/software system in the previous paragraph would be greatly simplified if the various measures could be transformed to a single measure, such as values on some "utility" scale. Then each measure (dollars, machine utilization percentages, etc.) for a given alternative could be converted to a utility value, the new utility values could be weighted by their relative importance, and the weighted utility values could be aggregated by an appropriate mathematical function. A single utility value would thereby result for each alternative, and a selection would be made by choosing the alternative having the maximum utility value. This area of study is that of multi-attribute or multi-criteria decision making. This is an important area of study but is outside the scope of this book. Some selected references are given in the bibliography of this chapter [1, 4, 6, 8, 16, 17, 18, 20] and the bibliography of Chapter Nine.

There will be some additional discussion on multiple objectives in Chapter Nine. Otherwise, it will be assumed throughout the book that only a single economic measure of effectiveness is relevant in comparing alternative projects.

1.3.2 Selling the Project

One of the more important ingredients in gaining approval from management for an economically and technically justified investment alternative is the sales technique of the individual who presents the alternative to the manager. It is not unusual for a manager to adopt a weaker alternative because of the persuasive powers of the person who presented the alternative.

As Lutz [13] points out, *what is said* to support a project justification and *how it is said* influence the ultimate disposition of the funding request. As he puts it, "Communication content may outweigh the technical aspects of the proposal. The decision as to whether or not a project is funded often hinges on how successfully you communicate with the decision makers using their rules, words and decision criteria. Structuring your proposal for ease of acceptance is imperative and should be closely examined to try and increase the likelihood of obtaining capital budgeting dollars." Lutz argues that managers will often accept and implement a weak alternative because they

1. Recognize the existence of a problem.
2. Understand the recommended solution.
3. Know how to implement the recommendation.
4. Anticipate immediate improvements with positive financial results.

A tendency exists for problem solvers to spend too much time trying to develop the recommended solution to the problem and too little time in determining the best way to sell the solution to management. Effective communication is essential; in particular, it is important to communicate with management in a way that can be understood and accepted [10].

Paraphrasing Stoll [19], the business goals of most manufacturing companies are twofold: (1) make as good a product as possible, in as short a time as possible, and for as little cost as possible; and (2) sell as many as possible, as fast as possible, for as much as possible. A similar statement would apply to firms that sell services rather than products.

The engineer should realize that his or her proposed investment project is just one of several which will compete for limited funds available. Managers will be concerned about how the proposed investment fits into their long-range goals and the risk or uncertainty involved in the proposed project. They must consider how the project, if implemented, will affect the various organizational units within the firm [13]. Automating a production department will affect the maintenance department, which may not have the skills to maintain new automated equipment.

A proposal requiring capital expenditures may require approval at several managerial levels. The level(s) and number of levels required is a function of multiple variables, but particularly the magnitude of the capital request. For example, a departmental manager may have the authority to approve requests up to $10,000, a plant manager may have the authority to approve requests up to $100,000, but corporate management must approve requests exceeding $100,000. The decision criteria at each management level may be, and probably are, different. The engineer is therefore reminded again to practice effective communication by using management's rules, words, and decision criteria. In this regard, it should be noted that many senior managers do not have technical degrees or backgrounds. The analyst or engineer making the proposal should therefore explain complex technical topics in a manner which an intelligent nontechnical manager can understand.

Given that a proposed investment project is technically sound and economically justified, the proposal should also clearly specify and emphasize any noneconomic benefits expected to accrue from project implementation. The nature of the noneconomic benefits depends of course on the type of project proposed. However, some example noneconomic benefits might be better utilization of floor space, reduction in computer data entry errors, reduction in material flow congestion and backtracking, safer working conditions, improvements in production scheduling, standardization of spare parts inventory, and so forth. Although managers necessarily have to focus attention on economic benefits, they can and do appreciate that noneconomic benefits directly or indirectly contribute to the long-range progress and success of the business.

It is important to write a proposal that includes an *executive summary*. The executive summary normally reduces the performance characteristics of the recommendation to a few pages. Technical aspects of the recommended solution and the technical details of the economic analysis are usually provided in the main report for the manager who wishes to obtain additional understanding of the information in the executive summary. A face-to-face presentation is also quite important in achieving effective communication.

Communication techniques such as the use of visual aids, voice control, dress, structure, and clarity of expression are necessary to sell the proposed solution. However, it is probably more important to know your audience well and to prepare your proposal using their language. Since the language of managers is often the language of finance, Chapter Two introduces some fundamental financial terminology used by managers.

1.4 PRESENT ECONOMY EXAMPLES

Most of the methodology presented in this book is based on the fundamental concept that money has a time value. Investment projects that have a life cycle of several years will almost invariably have associated revenues (or savings) and expenses that occur periodically over the life cycle. This series of cash flows constitute a particular cash flow pattern and alternative projects can have different cash flow patterns over their respective life cycles. In order to compare such alternative projects at a common point in time by an economic measure of effectiveness, the time value of money must be considered.

The time value of money concept simply recognizes that a given dollar amount has different values at different times. In order to adjust these different values to a common reference point on the time scale, an interest rate is used to make the adjustment. This time value of money concept will be developed and illustrated in Chapter Three. Economic measures of effectiveness will also be defined and illustrated in Chapter Three.

There are other situations in which time is not a significant economic factor and thus, the time value of money does not have to be considered. Degarmo [3], one of the early authors on the subject of engineering economy, has labeled these types of economic problems as *present economy studies*. He states that the time value of money does not have to be considered when one or more of the following conditions hold: (1) there is no investment of capital and only out-of-pocket costs are involved, (2) after any first cost (capital investment) is paid, the long-term costs will be the same for all alternatives or will be proportional to the first cost, and (3) the alternatives will have essentially identical results regardless of the capital investment involved.

In the business and industrial world, present economy studies typically involve the choice among alternative designs, materials, or methods. Four examples are subsequently presented for illustrative purposes and the end-of-chapter problems will provide additional insight into the nature of such problems. Two of the illustrative examples do involve capital investments but the decision question posed can be answered without considering the time value of money.

Example 1.3

A particular metal component part can be machined on either an engine lathe or a turret lathe. In either case, the machines are not used exclusively for this particular part. Rather, parts are produced in batches according to a customer's order. If a customer's order is produced on the engine lathe, the following information is relevant:

Set-up time required per order	= negligible
Direct labor and machine time per part	= 10 minutes
Direct labor rate	= $7.50/hour
Machine rate ·	= $12.00/hour
Cutting tool cost per part	= $0.20/unit

If the part is produced on the turret lathe, a set-up of the machine is required each time a customer's order is processed. The set-up is required only once per order if the order size is less than or equal to 300 units. If the order size is greater than 300 units, then a set-up is required every 300 units. Relevant time and cost data is given below:

Set-up time required per order (both the operator and
   machine time is charged to the order)   = 2 hours
Direct labor and machine time per part   = 5 minutes
Direct labor rate   = \$9.00/hour
Machine rate   = \$15.00/hour
Cutting tool cost per part   = \$0.10/unit

The material cost per unit is the same regardless of the machine used. Which is the economical machine to use for order sizes of 25, 100, and 500 units?

Let x = the number of units produced per customer order. Then, it will be convenient to develop a cost-per-order formula for each type of machine. That is,

Engine lathe cost per order:

Direct labor cost/minute   = \$7.50/60 = \$0.125/minute
Direct labor cost/unit   = (\$0.125/minute)(10 minute/unit)
                                 = \$1.25/unit
Machine cost/minute   = \$12.00/60 = \$0.20/minute
Machine cost/unit   = (\$0.20/minute)(10 minute/unit)
                                 = \$2.00
Cutting tool cost/unit   = \$0.20/unit
Thus, the cost per order is
   \$(1.25 + 2.00 + 0.20)/
      unit (x units/order)   = \$3.45x

Turret lathe cost per order:

Set-up cost/order   = (n)(2 hour/set-up)
                              × (\$9.00/hour + \$15.00/hour)
                          = (\$48/set-up)(n)
where n = the number of set-ups
                               required per order
Direct labor cost/minute   = \$9.00/60 ≐ \$0.15/minute
Direct labor cost/unit   = (\$0.15/minute)(5 minute/unit)
                                 = \$0.75/unit
Machine cost/minute   = \$15.00/60 = \$0.25/minute
Machine cost/unit   = (\$0.25/minute)(5 minute/unit)
                                 = \$1.25/minute
Cutting tool cost/unit   = \$0.10/unit
The cost per order is then
   \$48 n + \$(0.75 + 1.25 + 0.10)/unit
   (x)   = \$48 n + \$2.10 x

The comparative costs per order size of 25, 100, and 500 units are calculated as follows:

For an order size of 25 units

       Engine lathe cost = \$3.45(25)   = \$ 86.25
       Turret lathe cost = \$48(1) + \$2.10(25)   = \$100.50

For an order size of 100 units

Engine lathe cost = $3.45(100)	= $345.00
Turret lathe cost = $48(1) + $2.10(100)	= $258.00

For an order size of 500 units

Engine lathe cost = $3.45(500)	= $1725
Turret lathe cost = $48(2) + $2.10(500)	= $1146

Example 1.4

The SMA Corporation, a food processing firm, ships cartons of canned food to wholesale distributors via motor freight (tractor and trailer). The cartons are 8 inches wide × 12 inches long × 6 inches high and are stacked in layers on wooden pallets, with the largest surface area of the cartons (8 inches × 12 inches) resting on the pallet. The pallets are 40 inches long × 48 inches wide × 6 inches high and the cartons are arranged on the pallet in the pattern sketched in Figure 1.2. Three layers of cartons are stacked on each pallet and then "strapped" with metal bands to stabilize the load. Each pallet with 60 cartons is then termed a unit load. The total height of the unit load is 24 inches, as shown in Figure 1.3.

The inside dimensions of the trailer (closed van) are 34 feet long × 7 feet 6 inches wide × 6 feet 10 inches high. The freight cost to ship to a particular wholesale distributor is $1000/trailer load. The SMA Corporation therefore wishes to ship the maximum number of unit loads per trip. The company has a choice of two types of wooden pallets. The types are two-way entry and four-way entry, that is, a fork-lift truck (used to place and retrieve the unit loads from the trailer) can insert the forks into the pallet from two directions

FIGURE 1.2. Pallet pattern.

FIGURE 1.3. Unit load.

or from four directions, as shown in Figure 1.4. The implication of this choice is that, for the two-way entry pallet, the fork-lift truck can only address the pallet perpendicular to the 48-inch dimension of the pallet. The two-way pallets cost $6 each and the four-way pallets cost $8 each. If the unit loads can be stacked only three tiers high inside the trailer (three unit loads per stack), which pallet style will result in the lowest freight cost per unit load? (Assume negligible clearance required between each adjacent stack and between a stack and the walls of the trailer).

The trailer is 7 feet 6 inches (or 90 inches) wide. Since the 40-inch × 48-inch, two-way entry pallet is 48 inches wide, two rows of pallets cannot be stacked in the trailer (in the width direction). The trailer is 34 feet (or 408 inches) long and the number of unit load stacks that can be placed in the trailer length-direction is 408 inches/40 inches = 10.2, or 10 stacks. With 3 unit loads/stack, 10 stacks/row, and 1 row/trailer, 30 unit loads can be shipped per trailer. This results in a unit load cost per load of $1000/30 = $33.33/unit load. The unit cost of the pallets is $6, and the total cost per unit load shipped is $(33.33 + 6) = $39.33.

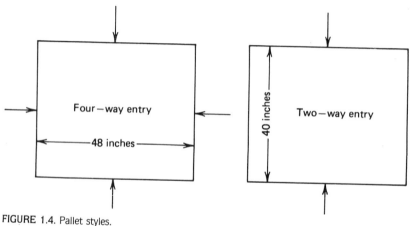

FIGURE 1.4. Pallet styles.

If the four-way entry pallet is used, two rows of pallets can be stacked in the trailer (in the width direction), with 10 inches of clearance between the rows. That is, 90-inch trailer width − 2(40-inch unit load width) = 10-inch clearance. With a trailer length of 408 inches, the number of load stacks per row is 408 inches/48 inches = 8.5, or 8 stacks/row. Thus, two rows/trailer × 8 stacks/row × 3 unit loads/stack = 48 unit loads/trailer. The freight cost per unit load shipped is then $1000/48 = $20.83. With a unit cost of $8/pallet, the total cost per unit load shipped is $28.33.

The four-way entry pallets should be purchased.

Example 1.5

The Hoakie Hahn Company manufactures a single product but in six different styles. The market area for the product is intrastate only, and the company also owns six retail outlet stores within the state. Because of customer preferences, only certain styles are inventoried at each outlet store. The decision at the present time concerns the warehouse space at each outlet store.

Based upon an analysis of past sales records and other current marketing information, an increase in future sales for each store is predicted. An increase in warehouse space at each store location is therefore required. If the company only has $100,000 available for warehouse expansion, how much capital must be borrowed to finance the construction of the needed space?

Relevant data are given in Tables 1.1 and 1.2. Although some of the warehouses have a small amount of unused space currently available, assume they are at capacity. Further, as a gross estimate, assume that the space required is proportional to the volume of product (style) manufactured per month.

TABLE 1.1
Sales Volume Data for Example 1.5

	Sales Volume	
Style	Present Sales Volume (units/month)	Predicted Sales Volume (percent of present volume)
1	860	120
2	5300	140
3	370	200
4	900	100
5	2200	80
6	230	150

TABLE 1.2

Space Requirement Data for Example 1.5

		Space Requirements	
Outlet Store	Styles Inventoried	Present Warehouse Space (square feet)	Cost of Additional Space (dollars/square foot)
A	1,2,5	4000	25
B	2,4	2000	35
C	3,5,6	3600	30
D	1,2,4	7400	40
E	1,3,6	700	25
F	3,5,6	4200	30

The predicted sales for each style are

$$\text{Style 1} = (\ 860)(1.20) = 1032 \text{ units/month}$$
$$\text{Style 2} = (5300)(1.40) = 7420 \text{ units/month}$$
$$\text{Style 3} = (\ 370)(2.00) = \ \ 740 \text{ units/month}$$
$$\text{Style 4} = (\ 900)(1.00) = \ \ 900 \text{ units/month}$$
$$\text{Style 5} = (2200)(0.80) = 1760 \text{ units/month}$$
$$\text{Style 6} = (\ 230)(1.50) = \ \ 345 \text{ units/month}$$

The present sales indices for each outlet store are

$$\text{Store A} = (\ 860 + 5300 + 2200) = 8360 \text{ units/month}$$
$$\text{Store B} = (5300 + \ \ 900) \qquad\quad = 6200 \text{ units/month}$$
$$\text{Store C} = (\ 370 + 2200 + \ 230) = 2800 \text{ units/month}$$
$$\text{Store D} = (\ 860 + 5300 + \ 900) = 7060 \text{ units/month}$$
$$\text{Store E} = (\ 860 + \ 370 + \ 230) = 1460 \text{ units/month}$$
$$\text{Store F} = (\ 370 + 2200 + \ 230) = 2800 \text{ units/month}$$

The indices for each outlet store, based on forecast sales, are

$$\text{Store A} = (1032 + 7420 + 1760) = 10212 \text{ units/month}$$
$$\text{Store B} = (7420 + \ \ 900) \qquad\quad = \ 8320 \text{ units/month}$$
$$\text{Store C} = (\ 740 + 1760 + \ 345) = \ 2845 \text{ units/month}$$
$$\text{Store D} = (1032 + 7420 + \ 900) = \ 9352 \text{ units/month}$$
$$\text{Store E} = (1032 + \ 740 + \ 345) = \ 2117 \text{ units/month}$$
$$\text{Store F} = (\ 740 + 1760 + \ 345) = \ 2845 \text{ units/month}$$

An estimate of the total warehouse space needed at each store is

$$\text{Store A} = (10212/8360)(4000 \text{ square feet}) = 4886 \text{ square feet}$$
$$\text{Store B} = (\ 8320/6200)(2000 \text{ square feet}) = 2684 \text{ square feet}$$
$$\text{Store C} = (\ 2845/2800)(3600 \text{ square feet}) = 3658 \text{ square feet}$$
$$\text{Store D} = (\ 9352/7060)(7400 \text{ square feet}) = 9802 \text{ square feet}$$

Store E = (2117/1460)(700 square feet) = 1015 square feet
Store F = (2845/2800)(4200 square feet) = 4268 square feet

The cost of the additional warehouse space needed at each store is

Store A = (4886 − 4000)($25) = $ 22,150
Store B = (2684 − 2000)($35) = 23,940
Store C = (3658 − 3600)($30) = 1,740
Store D = (9802 − 7400)($40) = 96,080
Store E = (1015 − 700)($25) = 7,875
Store F = (4268 − 4200)($30) = 2,040
TOTAL $153,825

Since the $153,825 needed is greater than the $100,000 available, the company would have to borrow $53,825 to finance all the estimated warehouse expansion required.

Example 1.6

For one product that a company manufactures, the purchased raw material is 2-inch diameter × 10-feet long extruded aluminum bar stock. This material costs $1.05/pound. Since the company purchases an average of 100,000 pounds/month, it is considering manufacturing the raw material, which would require the purchase of an extrusion machine and a cutoff saw. Raw material for the extruder can be purchased in 4-inch × 4-inch × 57-inch long cast aluminum ingots for $0.50/pound. The company's process engineer estimates the extruding operation would cost $0.30/pound, and the subsequent cutoff operation would cost $0.05/pound. The engineer also estimates there would be a 10% scrap loss of material at the extrusion stage and a 5.7% loss of material at the cutoff operations. (1) If the company manufactures the extruded rod, what are the estimated annual savings? (2) If the company manufactures the extruded rod, what is the production cost for each rod? (Aluminum density = 0.1 pound/cubic inch).

(1) The present annual cost of the raw material is

(100,000 pounds/month)($1.05/pound)(12 months/year) = $1,260,000

If the company manufactures the extruded rod from cast ingots, the scrap losses must be considered to arrive at the average quantity of ingots to purchase each month. Thus,

let x = the pounds of ingots purchased per month

Then, since there is a 10% loss at the extrusion stage, 90% or $0.90x$ will output from this stage. The quantity, $0.9x$, then enters the cutoff stage. Since there is a 5.7% loss at this stage, then 94.3% of the input quantity will output from the cutoff stage, or $(0.90x)(0.943) = 0.8487x$ is the quantity output from the cutoff operation. This quantity must equal 100,000 pounds/month. The monthly purchase quantity should therefore be

$$0.8487x = 100,000 \text{ pounds/month}$$

$$x = 117,827.3 \text{ pounds/month}$$

The monthly input and output quantities for each manufacturing stage are

Extrusion input:	117,827.3 pounds
Extrusion output:	$(117,827.3)(.90) = 106,044.6$ pounds
Cutoff input:	106,044.6 pounds
Cutoff output:	$(106,044.6)(.943) = 100,000$ pounds

The monthly production costs are

Raw material costs:	$(117,827.3 \text{ pounds/month})(\$0.50/\text{pound})$ $= \$58,913.65/\text{month}$
Extrusion costs:	$(117,827.3 \text{ pounds/month})(\$0.30/\text{pound})$ $= \$35,348.19/\text{month}$
Cutoff costs:	$(106,044.6 \text{ pounds/month})(\$0.05/\text{pound})$ $= \$5302.23/\text{month}$

The total costs are

$$(\$58,913.65 + \$35,348.19 + \$5302.23)(12)$$
$$= \$1,194,768.80/\text{year}$$

Thus, the annual savings from manufacturing the extruded rod are

$$(\$1,260,000) - (\$1,194,768.80) = \$65,231.20$$

(2) The volume of each extruded rod is

$$(\pi)(10)(12) = 377.00 \text{ cubic inch}$$

and the weight of each rod is

$$(377.00 \text{ cubic inch})(0.1 \text{ pound/cubic inch}) = 37.70 \text{ pounds}$$

The number of rods extruded per month is

$$(100,000 \text{ pounds/month}) \div (37.70 \text{ pounds/unit}) = 2652.52 \text{ units/month}$$

The cost of an extruded rod is

$$\frac{(\$58,913.65 + \$35,348.19 + \$5,302.23)/\text{month}}{2652.52 \text{ units/month}}$$

or $37.54/unit.

1.5 OVERVIEW OF THE TEXT

The material in the book has been organized in a manner consistent with the logical sequence of steps followed in performing an economic evaluation of investment alternatives. Chapter Two provides a discussion of cost concepts and includes discussions of elements of *costs:* future costs, sunk costs, fixed costs, variable costs, average costs, and an introduction to project cost estimation.

After acquainting the reader with the types of data that are germane to investment projects, Chapter Three presents some important fundamental concepts involving the *time value of money*. In fact, Chapter Three provides the foundation for the remainder of the book; therefore the reader must understand the time value of money operations presented in this chapter.

Chapter Four presents various *economic measures for determining a project's worth*. Included are the measures of annual worth, future worth, present worth, rates of return, savings/investment ratios, and payback period. The impact of inflation and deflation on a project's economic feasibility is discussed.

The economic measures for determining project worth are used in Chapter Five for *comparing investment alternatives*. The subject of replacing capital assets is also emphasized in this chapter.

Chapter Six addresses the issues of *depreciation allowances* and *income taxes* and their incorporation in economic analyses; Chapter Seven presents *benefit-cost analysis* which is often used for the economic analysis of projects in the public sector.

In Chapter Eight, supplementary analysis techniques are presented, particularly the effects of *risk and uncertainty* on the analysis of economic investment alternatives. In modeling the effects of risk and uncertainty mathematically, two approaches are taken: a prescriptive (normative) approach and a descriptive approach. A prescriptive model prescribes the action that, in some sense, is optimal; a descriptive model describes the behavior of the situation modeled. Chapter Eight provides a descriptive treatment of the effects of risk and uncertainty by presenting the subjects of *break-even, sensitivity,* and *risk analysis*. Then, prescriptive models are presented in Chapter Nine and are classified as models for *decision making under risk and uncertainty*.

BIBLIOGRAPHY

1. Chernoff, Herman, and Moses, Lincoln E., *Elementary Decision Theory*, Wiley, 1959.

2. Cox, John L. and Dimsdale, Parks B., "Selling New Ideas Properly Aids Their Acceptance Within an Organization," *Industrial Engineering*, December, 1986, pp. 30–41.

3. Degarmo, E. Paul, *Engineering Economy*, Macmillan, 1960. (Also, later editions in 1967, 1973—with John R. Canada; 1979, 1984—with John R. Canada and William Sullivan).

4. Fishburn, Peter C., *Decisions and Value Theory*, Wiley, 1964.

5. Fleischcher, G. A., *Benefit-Cost Analysis Applied to Air Pollution Control: A Case Study Based on the Dixon, Tiller County/U.S.A. Reference Community*,

National Communicable Disease Center, Bureau of Disease Prevention and Environmental Control, Public Health Service, U.S. Department of Health, Education, and Welfare, Atlanta, GA, 1968.

6. Frazelle, Ed, "Suggested Techniques Enable Multi-Criteria Evaluation of Material Handling Alternatives," *Industrial Engineering*, February, 1985, pp. 42–48.

7. Gessner, Robert A., *Manufacturing Information Systems: Implementation Planning*, Wiley, 1984.

8. Kepner, Charles H., and Tregoe, Benjamin B., *The Rational Manager*, McGraw-Hill, 1965.

9. Kerzner, Harold, *Project Management: A Systems Approach to Planning, Scheduling, and Controlling*, 2 ed., Van Nostrand Reinhold Company, 1984.

10. Klausner, R. F., "Communicating Investment Proposals to Corporate Decision-Makers," *The Engineering Economist, 17*(1), 1971, pp. 45–55.

11. Krick, E. V., *An Introduction to Engineering and Engineering Design*, Wiley, 1965.

12. Levine, Harvey A., *Project Management Using Microcomputers*, Osborne McGraw-Hill, 1986.

13. Lutz, R. P., "How to Justify the Purchase of Material Handling Equipment to Management," *Proceedings of the 1976 MHI Material Handling Seminar*, The Material Handling Institute, Pittsburgh, PA, 1976.

14. Moder, Joseph J., Phillips, Cecil R., and Davis, Edward W., *Project Management With CPM, PERT, and Precedence Programming*, 3 ed., Van Nostrand Reinhold Company, 1983.

15. O'Guin, Michael C., "Information Age Calls For New Methods of Financial Analysis in Implementing Manufacturing Technologies," *Industrial Engineering*, November, 1987, pp. 36–40.

16. Raiffa, Howard, *Decision Analysis—Introductory Lectures on Choice Under Uncertainty*, Addison-Wesley, 1968.

17. Schlaifer, Robert, *Probability and Statistics for Business Decisions*, McGraw-Hill, 1959.

18. Starr, Martin Kenneth, *Product Design and Decision Theory*, Prentice-Hall, 1963, Chapters 1 and 6.

19. Stoll, Henry W., "Design For Manufacture," Chapter 13 in *Tool and Manufacturing Engineers Handbook*, Vol 5—*Manufacturing Management*, Society of Manufacturing Engineers, 1988.

20. Sullivan, William G., "Models IEs Can Use To Include Strategic, Non-Monetary Factors in Automation Design," *Industrial Engineering*, March, 1986, pp. 42–50.

PROBLEMS

1. A particular product can be manufactured by either of two machining sequences, S_1 or S_2. Sequence S_1 consists of processing through five machines: M1, M2, M3, M4, and M5. For Sequence S_2, processing is by three machines: MA, MB, and MC.

The scrap rate for each of these machines is as follows: M1 = 2%, M2 = 5%, M3 = 3%, M4 = 8%, M5 = 4%, MA = 5%, MB = 3%, and MC = 2%. Assume a batch size of 1000 units is to be processed through each sequence. The raw material cost is $1.00 per unit. The total processing cost through sequence S_1 is $5000 and through sequence S_2 is $6000. Which sequence of processing results in the lowest unit cost of final product? (1.4)*

2. A frequently occurring maintenance task is currently performed by a work crew of three persons. This crew typically accomplishes the task in 5 hours. However, when the three-person crew performs the task, only one person is busy 100% of the time. The second person is busy 70% of the time and the third person is busy only 30% of the time. The task can be performed by either one, two, or three persons.

   The regular wage rate for each of the three persons is $8.00/hour. The overtime (greater than 8 hours) wage rate is $12.00/hour. Whenever the task is performed, tools and equipment are required, and these are valued at $5.00/hour.

   Based on the total cost (labor and equipment) for the task, what is the most economical crew size? *Hint:* First determine the total work content (hours) of the task. Then determine the total cost for each crew size. (1.4)

3. A study made in the assembly department of a small electronics manufacturing firm revealed that workers typically produce 100–150 assemblies in an 8-hour working day. For simplicity, assume the production rates are 100, 125, or 150 units per day. If 100 units are produced, an average of 3% are rejected. When 125 units are produced, an average of 8% are rejected, and 20% are rejected at the 150 units/day rate. The total material cost per unit is $5.50.

   If workers are paid $0.65 per acceptable assembly, what is the preferred production rate from the worker's point of view? Which production rate results in the lowest unit cost for the company? (1.4)

4. Mr. Joe Smith is considering the purchase of a new car for business purposes. He has narrowed his choice to three cars; the cost details are given below:

Car Type	A	B	C
Fixed cost/month (dollars)	$350	$275	$600
Miles per gallon of fuel	30	25	35
Estimated cost of maintenance (dollars/mile)	$ 0.12	$ 0.17	$ 0.10

Mr. Smith estimates he will travel about 2000 miles/month. The cost of fuel is $1.00/gallon.

   a. Based on a total cost/month comparison, determine the most economical car to purchase.

   b. If Car Type C were purchased, what would be the average cost per mile of travel? (1.4)

---

* Indicates the relevant section in the chapter.

5. A small grocery store is considering the purchase of a personal computer (PC) system to be able to keep more complete inventory records. There is a choice of three alternative local suppliers. Specifications and cost details for each PC system are given below:

PC Type	A	B	C
Time required to perform all the desired activities (hours/day)	4	6	3
Hourly wage of the operator (dollars/hour)	$6	$6	$ 6
Fixed cost for system (dollars/day)	$8	$5	$15
Cost of power (dollars/hour)	$0.30	$0.30	$ 0.60

Assume that the operator of the PC system can do other productive work at the store when not using the microcomputer. Based on a total cost per day comparison,

a. Determine the most economical system.

b. Would your purchase recommendation change if the hourly wage were $10? *Note:* Total cost per day equals the sum of the fixed costs, labor cost, and power costs. (1.4)

6. The relationship between cutting tool life (T) in minutes, and cutting speed (V) in feet/minute, is expressed by Taylor's equation:

$$VT^n = K$$

where K is a constant. From the above equation, it is possible to mathematically derive an equation to find the tool life resulting in minimum unit cost. The equation is:

$$T_c = \left(t_c + \frac{c_c}{c_o}\right)\left(\frac{1}{n} - 1\right)$$

where T_c = tool life for minimum unit cost, minutes
           t_c = tool change time, minutes
           c_c = cost per cutting edge, dollars/edge
           c_o = cost of labor and overhead, dollars/minute
           n = Taylor's tool-life exponent

A metal machining company is evaluating three different types of cutting tool inserts for one of their high-volume NC turning operations. The specifications and cost information available for the inserts are as follows:

	Insert Type		
	Tungsten Carbide	Coated Carbide	Ceramic
n	0.22	0.27	0.38
t_c (minutes)	2	2	2
c_c (dollars/edge)	0.20	0.90	1.50
c_o (dollars/minute)	0.5	0.5	0.5
K	150	250	550

For each insert type, determine the tool life for minimum unit cost and the corresponding cutting speed. Which insert type is preferred to maximize the metal removal rate (proportional to cutting speed) and hence the production rate? (1.4)

7. A group of 10 university students is planning a 7-day trip to Daytona Beach, Florida. This destination is 950 miles away from their campus. The group has decided to share automobile expenses equally and is considering three alternative car rentals. For simplicity, assume only the following cost data is relevant. The cost for Car A is $18.50/day plus $0.10/mile traveled, and the capacity of Car A is four people. The cost for Car B is $27.50/day plus $0.08 per mile traveled, with a capacity of four people. The cost for Car C is $32.00/day plus $0.12 per mile traveled, with a capacity of six people. The group estimates that local travel in Daytona will be about 350 miles. (1.4)

    a. What is the most economic alternative travel plan for the group of students?

    b. Given the answer from (*a*), determine the per person cost for travel.

8. The process engineer in a medium-sized machine shop has the choice of machining a particular part on either of two machines. Orders for this part are received regularly, but on a recurring basis. The order size varies. When the order is processed on Machine A, four setup operations of 1 hour each are required. Once set up, the per unit machining time on Machine A is 0.5 hours.

    When the order is processed on Machine B, eight setup operations of 0.75 hour each are required. Once set up, the per unit machining time on Machine B is 0.55 hours. The hourly wage rates for the machine operators are $5 and $4.50 for Machine A and Machine B, respectively. The hourly overhead rates (including setup time) are $10 and $9 for Machine A and Machine B, respectively. (1.4)

(a) Which machine is preferred if the order size is 100 units? 500 units?
(b) What is the break-even order size?

9. The American Grinding Company is faced with a decision to select one of the following two types of grinding wheels for a rough grinding operation. The criterion of selection is to minimize the unit cost per cubic inch of metal removed. The unit cost is given by the following expression:

$$\text{Unit cost} = \text{cost of machining} + \text{cost of wheel}$$

$$= \frac{c_o}{R} + \frac{c_w}{G_r}$$

where c_o = Cost of labor and overhead, dollars/hour
           R = Material removal rate, cubic inch/minute
           c_w = Cost of wheel, dollars/cubic inch
           G_r = Grinding ratio, cubic inch of metal/cubic inch of wheel

Also, the grinding ratio and the material removal rate are related by

$$G_r R^n = K$$

where n and K = constants.

The alternatives being considered are:

	Grinding Wheel Type	
	A	B
c_w	0.1	0.09
c_o	10	10
n	1	1
K	4400	2600

For each wheel type, develop a unit cost equation as a function of the variable R. Compare the equations and recommend which grinding wheel to purchase. (1.4)

10. In Example 1.6 of Chapter 1, assume that the company has decided to manufacture extruded rods and has two alternative sources of aluminum ingots. One source provides 4-inch × 4-inch × 57-inch ingots for $0.50/pound and the other source provides 5-inch × 6-inch × 100-inch ingots for $0.40/pound. The process engineer estimates a 10% scrap loss for the 57-inch-long ingot and a 15% scrap loss in the 100-inch-long ingot at the extrusion stage. The cutoff loss is estimated at 5% for each of the ingot types. Assume the monthly consumption of the extruded bar stock is 100,000 pounds. Which size ingot should the company purchase? (1.4)

11. For Example 1.3 in Chapter 1, determine the order size at which it will be economical to switch over from the engine lathe to the turret lathe. (1.4) *Hint:* Define turret lathe cost per order equations over the production ranges, $1 \leq x < 300$ and $300 \leq x < 600$. Then compare engine lathe and turret lathe production costs using the boundary values for each range. Draw conclusions from these comparisons.

12. In Example 1.4 of Chapter 1, assume that the alternative pallets available to the company are 32 inches long × 32 inches wide × 6 inches high, and 40 inches long × 48 inches wide × 6 inches high, both of two-way entry type. The pallets cost $4 and $6, respectively. Find which pallet style will result in the lowest freight cost per carton. (1.4)

13. In Example 1.6 of Chapter 1, let there be a fixed overhead cost of $500/month connected with the extruder machine. For simplicity, it is assumed there is a zero fixed cost associated with the cutoff saw. Using a total monthly production cost comparison (purchase extruded stock vs. produce extruded stock), find the volume of production at which it will be economical for the company to start manufacturing the extruded rods. *Hint:* Define y = the required production volume in pounds/month, which is the output quantity from the cutoff operation, and define x = the required purchase volume in pounds/month, which is the quantity required as input to the extruder. Express total monthly production costs as a function of the variable y; set the two comparison equations equal to each other and solve for y. (1.4)

CHAPTER

Cost Concepts

2.1 INTRODUCTION

Engineering economic analysis is primarily concerned with comparing alternative projects on the basis of an economic measure of effectiveness. This comparison process utilizes a variety of cost terminologies and cost concepts, and it will be helpful to present these prior to the discussion in Chapter Four of the economic measures of effectiveness for comparing alternative projects. To implement the discussion of cost terminology, a typical production situation will now be described.

Let us assume that the business of a small manufacturing firm is job-shop machining. That is, the firm produces a variety of products and component parts according to customer order. Any given order may be for quantities of as few as five parts or as many as several hundred parts. The firm has periodically received orders to manufacture a part, which we will identify as Part No. 163H, for the Deetco Corporation. The part has been manufactured in a four-stage production sequence consisting of (1) sawing bar stock to length, (2) machining on an engine lathe, (3) machining on an upright drill press, and (4) packaging. The unit cost to produce Part No. 163H by this sequence has been $25, where the unit cost consists of the major cost elements of direct labor, direct materials, and overhead (prorated costs for insurance, taxes, electric power, marketing expenses, etc.). The firm is now in the process of negotiations with the Deetco Corporation to obtain a contract for producing 10,000 of these parts over a period of 4 years, or an average of 2500 units/year. A contract for this volume of parts is highly desirable but, in order to obtain the contract, the firm must lower the unit cost.

An engineer for the firm has been assigned to determine production methods to lower the unit cost. After study, the engineer recommends the purchase of a small turret lathe. With the turret lathe, the processing sequence of Part No. 163H would consist essentially of (1) machining bar stock on the turret lathe and (2) packaging. The estimated unit cost for Part No. 163H by this production method would be $15. Furthermore, the production rate by the new method would be increased over the old method because the turret lathe would replace the sawing, engine lathe, and drill press operations.

If the turret lathe is purchased, the saw, engine lathe, and drill press would not be sold but would be kept for other jobs for which the firm may receive orders. The turret lathe would be reserved for the production of Part No. 163H, but about 25% excess production capacity could be devoted to other jobs.

The incremental investment required to purchase the turret lathe and the new tooling required, as well as installing the machine, is $50,000. The physical life of the turret lathe is judged to be about 25 years, but assume that federal tax laws permit the investment capital to be recovered through annual depreciation charges in 5 years. At the end of 5 years, the firm estimates the market, or salvage, value of the turret lathe would be $25,000. If the maximum unit price that the Deetco Corporation will pay for Part No. 163H is $22, should the firm accept the contract for 10,000 parts and then purchase the turret lathe in order to execute the contract?

This particular question will not be answered here or in a later chapter of the book. Rather, the example situation has been cited to illustrate one type of decision with which this book is concerned. The reader can also appreciate that considerable research and investigation is required to determine or estimate the cost figures used in the turret lathe example situation. This cost information is typically obtained from a variety of sources, such as company production records, accounting records, manufacturer's catalogs, publications from the U.S. Government Printing Office, and so forth. The engineer should therefore be familiar with cost terminology, cost factors, and cost concepts as used by different specialists if effective comparisons and intelligent recommendations are to be made.

2.2 COST TERMINOLOGY

Both cost definitions and cost concepts are included in this section and will be discussed under six categories: (1) life-cycle costs, (2) past and sunk costs, (3) future and opportunity costs, (4) direct, indirect, and overhead costs, (5) fixed and variable costs, and (6) average and marginal costs.

2.2.1 Life-Cycle Costs

The *life-cycle cost* for an item is the sum of all expenditures associated with the item during its entire service life. The term *item* should be interpreted in the general sense as a machine, a unit of equipment, a product line, a project, a building, a system, and so forth. Life-cycle costs may include engineering design and development costs, fabrication and testing costs, operating and maintenance costs, and disposal costs. Life-cycle costs may also be expressed as the summation of acquisition, operation, maintenance, and disposal costs. Thus, life-cycle cost terminology may vary from author to author, but the basic meaning of the term is clear.

This textbook is predominantly concerned with the economic justification of engineering projects, the replacement of existing projects or capital assets, and the economic comparison of alternative projects. For the purpose of these types of analyses, we will define life-cycle costs to consist of (1) first cost (or initial investment), (2) operating and maintenance costs, and (3) disposal costs. There is obviously a time period involved for the life cycle and, upon disposal, the item may have a salvage value.

The *first cost* of an item is considered the total initial investment required to get the item ready for service; such costs are usually nonrecurring during the life of the item. For the purchase of a numerically controlled machine tool, Steffy et al. [17] state that the first cost of the machine tool may consist of the following major elements: (1) the basic machine cost, (2) costs for training personnel, (3) shipping and installation costs, (4) initial tooling costs, and (5) supporting equipment costs. The installation costs may include, for example, the costs of preparing a foundation; vibration and noise insulation; providing heat, light, and power supply; and cost of testing. Supporting equipment costs may include computer hardware and software, and a spare parts inventory.

For some other item, a different set of first-cost elements may be appropriate. Some projects may include *working capital* for inventories, accounts receivable, and cash for wages, materials, and so forth. In any case, it is emphasized that the first cost of an item normally involves many more cost elements than just the basic purchase price. Whether the first-cost elements are aggregated or maintained separately depends on income tax considerations and whether or not a before-tax or after-tax economic analysis is desired. Certain income tax laws and depreciation methods are presented in Chapter Six, and further discussion on this particular point is therefore deferred.

Operating and *maintenance* costs are recurring costs that are necessary to operate and maintain an item during its useful life. Operating costs usually consist of labor, material, and overhead items. Depending upon the accounting system used by a firm, a wide range of cost factors may be included in the major cost classification of overhead. Typical overhead items are fuel or electric power, insurance premiums, inventory charges, indirect labor (as opposed to direct labor), administrative and management expenses, and so forth. It is usually assumed that *operating and maintenance costs* are annual costs, but maintenance costs may not be on a recurring, annual basis. That is, a regular, annual schedule of minor or preventive maintenance may be followed, or it could be policy that maintenance is performed only when necessary, such as when a major overhaul is required. In most cases, the maintenance policy would consist of both preventive maintenance and maintenance on an "as needed" basis. In any case, repair and upkeep result in costs that must be recognized in the economic analysis of engineering projects.

When the life cycle of an item has ended, *disposal costs* usually result. Disposal costs may include labor and material costs for removal of the item, shipping costs, or special costs; an example of special costs being the cost of disposing of hazardous materials. Although disposal costs may be incurred at the end of the life cycle, most items have some monetary value at the time of disposal. This value is the *market* or *trade-in value* (i.e., the actual dollar worth for which the item may be sold at the time of disposal). Then, after deducting the cost of disposal, the net dollar worth at the time of disposal is termed the *salvage value*.

The market value, the disposal costs, and the salvage value are usually not known with certainty and therefore must be estimated. For an item that satisfies the United States Internal Revenue Service (IRS) definition of a capital asset and that decreases in value over time through physical deterioration, the IRS has approved various depreciation methods that can serve to estimate the rate of deterioration and consequent decrease in value of the asset. The value of the capital asset at the end of a given accounting period during the asset's life is termed the *book value*. The book value is thus an estimate of the market value; further discussion on depreciation accounting is deferred until Chapter Six. *Scrap value,* on the other hand, refers only to the value of the material of which the item is made. For example, a 4-year-old automobile may have a scrap value of $500 but a market value of $5000. A distinction between these terms is not particularly important for evaluating potential investment projects—and salvage value will be used as the general term to denote the end-of-life value. For example, a trade-in value of $3000 minus disposal costs of $500 equals a net salvage value of $2500.

The life cycle obviously involves a time horizon, and the end of an item's life may be judged from either a functional or an economic point of view. The economic life of an item is generally shorter than the functional life. For example, an engine lathe may remain functionally useful for 40 years or more, but, because of periodic advancements in machine design technology, newer engine lathes have higher production rates; the economically useful life of an engine lathe may be only 10 years. The economic life of an item is usually a matter of company policy that is greatly influenced by income tax considerations. In this textbook, the end of life for an item will mean the end of its economic life, rather than its functional life.

2.2.2 Past and Sunk Costs

Past costs are historical costs that have occurred for the item under consideration. *Sunk costs* are past costs that are unrecoverable. The distinction is perhaps best made through examples. Assume that an investor purchases 100 shares of common stock in the ABC Corporation through a broker at

$25/share. In addition, the investor pays $85 in brokerage fees and other charges. Just 2 months later, and before receiving any dividend payments, the purchaser resells the 100 shares of common stock through the same broker at $35/share minus $105 for selling expenses. The purchaser realizes a net profit of $810 ($3500 − $2500 − $85 − $105) on these transactions. At the time of sale, the $2500 and $85 are past costs, but because these are recovered after the sales transaction, sunk costs are not incurred. If, on the other hand, the investor were to sell the 100 shares 2 months after purchase and the market price were $20/share, with a $70 charge for selling fees, the investor would incur a capital loss of $655 ($2000 − $2500 − $85 − $70). In this instance, some of the past costs would be recovered, but the $655 capital loss would be a sunk cost. If the investor reasons that the market price will decline further or if he or she simply needs the money, the $655 sunk cost should be ignored if the shares are to be sold for $20 each. However, sunk costs are not totally irrelevant to a present decision. They may qualify as capital losses and serve to offset capital gains or other taxable income and thus reduce income taxes paid. Examples in Chapter Six will illustrate this point. Also, past costs and sunk costs provide information that can improve the accuracy of estimating future costs for similar items.

Another example of sunk costs is the purchase and sale of an item of equipment. Assume the equipment is purchased for $10,000 and the salvage value at the end of 5 years of service is estimated to be $5000. For illustrative purposes, we will further assume that the annual decrease in value for the equipment through physical deterioration, or depreciation, is $1000. The $1000 annual cost of depreciation is a cost of production that, in theory, should be allocated to the output of the equipment. After allocating this and other manufacturing costs, general and administrative costs, and marketing costs to each unit of production, the total unit cost is determined. A profit is then added to each unit of production in order to arrive at the unit selling price. Thus, when a unit is sold, a portion of each sales dollar returns a portion of the depreciation expense. In this illustration, it is assumed that sales will return, or recover, the total estimated depreciation expense of $5000 (first cost minus estimated salvage value) for the 5-year period. However, if the equipment has a market value of only $2000 at the end of 5 years, there is a $3000 ($5000 − $2000) sunk cost. The $3000 capital loss represents an error in estimating the rate of depreciation, and the owner cannot insist that the equipment is worth $5000 when the market value for the 5-year-old equipment is, in fact, only $2000. If the equipment is kept, it is argued that the true value being kept is thus only $2000.

2.2.3 Future and Opportunity Costs

All costs that may occur in the future are termed *future costs*. These future costs may be operating costs for labor and materials, maintenance costs, overhaul costs, and disposal costs. In any case, by virtue of occurring in the future, these costs are rarely known with certainty and must therefore be estimated. This is also true for future revenues or savings if these are involved in a given project. Estimates of future costs or revenues are uncertain and subject to error. Thus, the economic analysis is simplified if certainty of future costs, revenues, or savings is assumed. This assumption is made until Chapter Eight, where concepts of probability are introduced.

The cost of forgoing the opportunity to earn interest, or a return, on investment funds is termed an *opportunity cost*. This concept is best explained by means of illustrations. For example, if a person has $1000 and stores this cash in a home safe, the person is forgoing the opportunity to earn interest on the money by establishing a savings account in a local bank that pays, for example, 9% annual compound interest. (Of course, investments other than savings accounts are possible). For a 1-year period, the person is forgoing the opportunity to earn (0.09) ($1000) = $90. The $90 amount is thus termed the opportunity cost associated with storing the $1000 in the home safe.

A similar illustration of an opportunity cost is to assume that a person has $5000 cash on hand. This amount is considered *equity capital* if the $5000 was not borrowed (i.e., there is no debt obligation involved). The person has available secure investment opportunities such as establishing a personal savings account in a commercial bank or purchasing other financial instruments. From the available investment opportunities, suppose the optimum combination of risk (security level) and interest yield on the investment results in a 10% annual interest. Thus, the investment of $5000 would yield (0.10) ($5000) = $500 each year. If the person, instead of investing the $5000, purchases an automobile for the same amount for personal use, the person will forgo the opportunity to earn $500 interest/year. The $500 amount is again termed an annual opportunity cost associated with purchasing the automobile.

The same logic applies in defining an annual opportunity cost for investments in business and engineering projects. The purchase of an item of production machinery with $20,000 of equity funds prevents this money from being invested elsewhere with greater security and/or higher profit potential. This concept of *opportunity cost* is fundamental to the study of engineering economy and is a cost element that is included in virtually all methodologies for comparing alternative projects. In later chapters, the concept of opportunity costs will be discussed under the heading of *minimum attractive rate of return (MARR)*.

Some individuals define *MARR* as the *cost of capital*. As used in this text, the term *cost of capital* refers to the cost of obtaining funds for financing projects through debt obligations. These funds are usually obtained from external sources by (1) borrowing money from banks or other financial organizations (e.g., insurance companies and pension funds) and (2) issuing bonds. These debt obligations are normally long-term, as opposed to short-term, obligations for the purchase of supplies and raw materials. The debt obligations result in interest payments on, say, a monthly, quarterly, semiannual, or annual basis. The interest payments are thus a cost of borrowed capital. Financing projects through issuing bonds is a method of obtaining capital funds that is probably less known to the reader than borrowing money from a bank at a stated interest rate. Some elaboration of bonds is therefore appropriate.

Bonds are issued by various organizational units—partnerships, corporations (profit or nonprofit), governmental units (municipal, state, and federal), or other legal entities. The sale of bonds represents a legal debt of the issuing organization; bonds are generally secured by its assets. Examples are mortgage bonds or collateral bonds. Debenture bonds, on the other hand, are promissory notes or just a promise to pay. In any case, the purchaser of a bond has legal claim to the assets of the issuing unit but has no ownership privileges in the issuing unit. Purchasers of the common or preferred stock of an organizational unit do hold ownership status but may or may not have voting privileges, depending on the stipulations of the particular stock issue. In the sense that bonds are debt obligations and not ownership shares, bonds are considered a more secure investment than either common or preferred stock. This statement should not be taken as a universal truth, however, since the security level for a bond or a stock depends on many factors, economic and otherwise; the principal factor is the financial soundness of the issuing unit. Further details on interest payments on bank loans and interest payments on issued bonds are covered in Chapters Three and Four, respectively.

Another method of financing engineering projects is through the use of *equity funds* (i.e., through ownership capital or cash that is debt free). The use of such funds incurs an opportunity cost, as mentioned previously, and will again be discussed in Chapter Four as the minimum attractive rate of return (*MARR*) for an investment project. However, the point is that the cost of capital is the cost of borrowed funds. Chapter Seven further discusses the cost of capital for financing public utility projects.

2.2.4 Direct, Indirect, and Overhead Costs

It will be helpful to provide definitions of direct, indirect, and overhead costs in the context of a manufacturing environment. A typical cost structure for manufacturing, adapted from Ostwald,[1] is provided in Figure 2.1.

The *cost of goods sold,* as shown in Figure 2.1, is the total cost of manufacturing a product. An amount of profit is then added to this total cost to arrive at a selling price. Such a cost structure is helpful in arriving at a unit cost, which is a primary objective of cost accounting. The term *cost of goods sold,* as used here, has a different meaning from the term *cost of goods sold* used in general accounting practice, particularly for retail businesses. As will be mentioned later in the section on accounting principles, general accounting defines the cost of goods sold to be beginning-of-the-year inventory plus

---

[1] Phillip F. Ostwald, *Cost Estimating for Engineering and Management,* Prentice-Hall, 1974, p. 55.

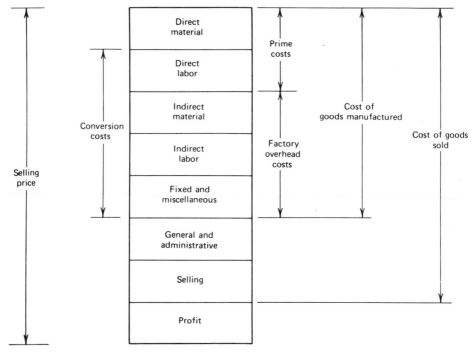

FIGURE 2.1. A cost structure for manufacturing.

purchases minus end-of-the-year inventory. Different meanings for the same terminology are unfortunate, but they do occur in the literature on accounting, and the reader is cautioned on this point. To simplify the treatment of the total cost of goods sold (as defined by Figure 2.1), the major cost elements can be defined as direct material, direct labor, and overhead costs.

Direct material and labor costs are the costs of material and labor that are easily measured and can be conveniently allocated to a specific operation, product, or project.

Indirect costs for both labor and material, on the other hand, are either very difficult or impossible to assign directly to a specific operation, product, or project. The expense of directly assigning such costs is prohibitive, and actual direct costs are therefore considered to be indirect for accounting purposes.

As an example of these different cost elements, suppose the raw material for a given part is a rectangular gray iron casting. The casting is milled on five sides, the unmachined surface is painted and air-dried, then four through holes are drilled and tapped. The finished parts are stacked in wooden boxes, 30/box, and are delivered to a customer.

In this example, the direct labor required per part to machine, paint, and package is probably readily determined. The labor required to receive the raw materials, handle parts between work stations, load boxes onto a truck, and deliver material to the customer is less easily identified and assigned to each part. This labor would be classified as indirect labor, especially if the labor in receiving, handling, shipping, and delivery is responsible for dealing with many different parts during the normal workday. The unit purchase price of the gray iron casting is an identifiable direct material cost. The cost of paint used per part may or may not be easily determined; if it is not, it is an example of indirect material cost. Also, any lubricating oils used during the machining processes would be an indirect material cost, not readily assigned on a cost per part basis.

Overhead costs consist of all costs of manufacturing other than direct material and direct labor. A given firm may identify different overhead categories such as factory overhead, general and administrative overhead, and marketing expenses. Furthermore, overhead amounts may be allocated to a total plant, departments within a plant, or even to a given item of equipment. Typical specific items of cost included in the general category of overhead are: indirect materials, indirect labor, taxes, insurance premiums, rent, maintenance and repairs, supervisory and administrative (technical, sales, and management) personnel, and utilities (water, electric power, etc.). Depreciation expenses are also usually included in the general overhead, but may occasionally be considered a part of direct costs. It is the task of cost accounting to assign a proportionate amount of these costs to various products manufactured or to services provided by a business organization.

2.2.5 Fixed and Variable Costs

Fixed costs do not vary in proportion to the quantity of output. General administrative expenses, taxes and insurance, rent, building and equipment depreciation, and utilities are examples of cost items that are usually invariant with production volume and hence are termed *fixed costs*. Such costs may be fixed only over a given range of production; they may then change and be fixed for another range of production. *Variable* costs vary in proportion to quantity of output. These costs are usually for direct material and direct labor.

Many cost items have both fixed and variable components. For example, a plant maintenance department may have a constant number of maintenance personnel at fixed salaries over a wide range of production output. However, the amount of maintenance work done and replacement parts required on equipment may vary in proportion to production output. Thus, total annual maintenance costs for a plant over several years would consist of both fixed and variable components. Indirect labor, equipment depreciation, and electrical power are other cost items that may consist of fixed and variable components. Determining the fixed and variable portion of such a cost item may not be possible; if it is possible, the expense of establishing detailed measurement techniques and accounting records may be prohibitive. A comprehensive discussion on this issue is outside the scope of this book, and the reader is referred to books on general cost accounting for further reading.

Certain total costs (TC), then, can be expressed as the sum of fixed (FC) and variable costs (VC). As an example, the total annual cost for operating a personal automobile for a given year might be expressed as

$$TC(x) = FC + VC(x)$$

where x = miles per year. Costs for insurance, license tags, depreciation, certain maintenance, and interest on borrowed money if the automobile were financed are essentially fixed costs, independent of the miles traveled per year. Expenses for gasoline, oil, tire replacements, and certain maintenance are proportional to, or functional with, the mileage per year. One could argue, however, that depreciation expenses are comprised of both fixed and variable components. Arbitrarily assigning numerical values to the total cost function, assume that

$$TC(x) = \$950 + \$0.15x$$

is a valid relationship for a given year in question (the expression is restricted to a given year, since actual depreciation expenses, and hence the fixed expenses, may vary from year to year). This relationship is linear in terms of x, however, the variable cost component is often a nonlinear function. Figure 2.2 graphically illustrates the total cost function.

Now let us consider Figure 2.2 as a total cost function for a production line in a manufacturing firm where the output from the line is a single product. Furthermore, let it be assumed that each unit of production can be sold for $\$R$ and that the total revenue (TR) is a linear function of the production quantity:

$$TR(x) = \$Rx$$

Adding this functional relationship to Figure 2.2 and modifying the terminology for this example yields Figure 2.3.

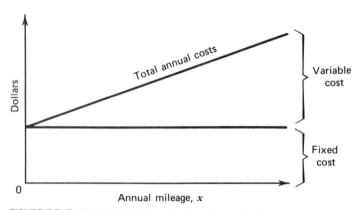

FIGURE 2.2. Total annual costs as a function of annual mileage.

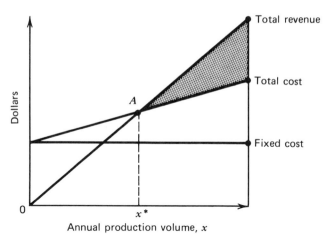

FIGURE 2.3. Relationship between total revenue and total cost as a function of annual production volume.

It is noted from Figure 2.3 that the total annual revenue equals the total annual cost at point A or at an annual production volume of x^* units. Thus, at x^*,

$$TR(x^*) = TC(x^*)$$
$$= FC + VC(x^*)$$

and x^* is termed the annual production volume required in order to *break even*. Certain important observations can now be made. If the production volume is less than x^*, an annual net loss will occur, the amount of which is equal to $TC(x) - TR(x)$, evaluated for a particular value of x. By the same token, if the production volume is greater than x^*, then an annual net revenue or profit will result (the shaded region of Figure 2.3). The amount of annual profit is equal to $TR(x) - TC(x)$, evaluated for a particular value of x.

It is generally desirable to have a "low" break-even value. For the general example of Figure 2.3, this can be accomplished in three independent ways: (1) increasing the slope of the total revenue line, (2) decreasing the slope of the variable cost line, and (3) decreasing the magnitude of the fixed cost line. Increasing the slope of the total revenue line means increasing the selling price of the product, which may be a poor marketing strategy in a competitive market environment where sales would be lost. Fixed costs, although not literally "fixed" in all cases, are difficult to reduce. Thus, reducing variable costs for direct material and labor usually offers the greatest opportunity to the engineer or analyst for profit improvement.

The concept of break-even analysis is general. Assuming that a break-even point exists, then, for two relationships $y = g(\cdot)$ and $y = h(\cdot)$ that are functions of a single variable x, the value of x for break-even, say x^*, may be determined from equating $g(x) = h(x)$ and solving for x^*. The concept can be

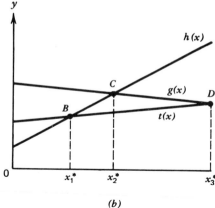

(a)   (b)

FIGURE 2.4. Graphical representation of break-even point. (a) A single break-even value. (b) Three break-even values.

extended to more than two functions of a single variable, say $y = h(x)$, $y = g(x)$, and $y = t(x)$. If these are all linear functions, then Figure 2.4 depicts two of the possible results.

In Figure 2.4b, there is no unique break-even value of x involving all three functional relationships. The linear equations $y = h(x)$ and $y = t(x)$ intersect at point B or $x = x_1^*$, which is then the break-even value for these two relationships. Point C, or $x = x_2^*$, is the break-even value for $y = h(x)$ and $y = g(x)$. Point D, or $x = x_3^*$, is the break-even value for $y = g(x)$ and $y = t(x)$.

The concept of break-even analysis also extends to nonlinear functions, with one or more break-even values, and functions of more than a single variable, which may be of linear or nonlinear form. However, examples and problems dealing only with functions of a single variable will be presented in this chapter.

Example 2.1

The cost of tooling and direct labor required to set up for a machining job on a turret lathe is $300. Once set up, the variable cost to produce one finished unit consists of $2.50 for material and $1 for labor to operate the lathe. For simplicity, it is assumed these are the only relevant fixed and variable costs. If each finished unit can be sold for $5, determine the production quantity required to break even and the net profit (or loss) if the lot size is 1000 units. Letting x = the production volume in units, then

$$TR(x) = TC(x) = FC + VC(x)$$

and

$$\$5x = \$300 + (\$2.50 + \$1.00)x$$

Solving yields $x = x^*$ (the break-even value), or

$$x^* = \$300/\$1.50 = 200 \text{ units}$$

For a production output of 1000 units, the net profit, P, is calculated to be

$$P = \$5(1000 \text{ units}) - (\$2.50 + \$1.00)(1000 \text{ units}) - \$300$$
$$= \$1200$$

Example 2.2

This example concerns the decision of selecting between two alternative methods of processing crude oil in a producing oil field, where the basis for the decision is the number of barrels of crude oil processed per year. The two methods of processing the crude oil are (1) a manually operated tank battery or (2) an automated tank battery. The tank batteries consist of heaters, treaters, storage tanks, and so forth, that remove salt water and sediment from crude oil prior to its entrance to pipelines for transport to an oil refinery.[2]

For each alternative, fixed costs and variable costs are involved. Fixed costs include items such as pumper labor, maintenance (fixed over the production quantity of interest), taxes, certain energy costs (power to operate control panels and motors in continuous operation), and, for manual tank batteries, a cost for oil "shrinkage." Variable costs for chemical additives, heating, and noncontinuous operating motors are proportionate with the volume of oil being processed. This relationship is assumed to be linear over the production quantity of interest, since the data given in Tables 2.1 and 2.2 are based on a production quantity of 500 barrels/day.

In addition to the fixed and variable costs given in Table 2.1 for the automatic tank battery operation, other annual fixed costs are

$$D_1 = \text{annual cost of depreciation and interest}$$
$$= \$3082$$

$$M_1 = \text{annual cost of maintenance, taxes, and labor}$$
$$= \$5485$$

Letting x = the number of barrels of oil processed per year, the total annual cost, $TC_1(x)$, for the automatic tank battery operations is given by

$$TC_1(x) = FC_1 + VC_1(x)$$
$$= (\$982 + \$3082 + \$5485)$$
$$+ \$0.01136x$$
$$= \$9549 + \$0.01136x$$

---

[2] The example is taken, with slight modification, from Ferguson and Shamblin [6] by permission of the publisher.

TABLE 2.1

Cost Data for Automatic Tank Battery Operations

Fixed cost	
Control panel power	$0.15/day
Circulating pump power (3 horsepower)	0.82/day
Maintenance	1.00/day
Meter calibration	0.40/day
Chemical pump power ($\frac{1}{4}$ horsepower)	0.32/day
Total	$2.69/day
or $982/year ($2.69/day × 365 days)	
Variable cost	
Pipeline pump (5 horsepower @ 50% utilization)	$0.63/day
Chemical additives (7.5 quarts/day)	3.75/day
Inhibitor (2 quarts/day)	1.00/day
Gas (10.8 MCF/day × $0.0275/MCF)	0.30/day
Total	$5.68/day
or $0.01136/barrel based on 500 barrels/day	

In addition to the fixed and variable costs given in Table 2.2 for the manual tank battery operations, other annual fixed costs are

$$D_2 = \text{annual cost of depreciation and interest}$$
$$= \$2017$$

$$M_2 = \text{annual cost of maintenance, taxes, and labor}$$
$$= \$7921$$

Then, the total annual cost, $TC_2(x)$, for the manual tank battery operation is given by

$$TC_2(x) = FC_2 + VC_2(x)$$
$$= (\$358 + \$2017 + \$7921)$$
$$+\$0.00810x$$
$$= \$10,296 + \$0.00810x$$

By equating the two total cost functions, the break-even production volume can be determined as

$$TC_1(x) = TC_2(x)$$

$$\$9549 + \$0.01136x = \$10,296 + \$0.00810x$$

$$x^* = 229,141 \text{ barrels/year}$$

The interpretation of the break-even point in this example is that $x^* = 229,141$ barrels/year is the *point of indifference* between the choice of the two alternatives. If production volume is less than x^*, then the first alternative, or the automatic tank battery operation, would be preferred. For in-

---

TABLE 2.2

Cost Data for Manual Tank Battery Operation

Fixed cost	
Chemical pump power	$0.16/day
Circulating pump power	0.82/day
Total	$0.98/day
or $358/year ($0.98/day × 365 days)	
Variable cost	
Chemical additives (7.5 quarts/day)	$375/day
Gas	0.30/day
Total	$4.05/day
or $0.00810/barrel based on 500 barrels/day	

---

stance, if $x = 200,000$ barrels/year, then $TC_1(x) = \$11,821$ and $TC_2(x) = \$11,916$. Similarly, if production volume is greater than x^*, the manual tank battery operation would be preferred.

2.2.6 Average and Marginal Cost

The *average cost* of one unit of output (unit cost) is the ratio of total cost and quantity of output (miles traveled, production volume, etc.). That is,

$$AC(x) = \frac{TC(x)}{x}$$

where $AC(x)$ = average cost per unit of x
   $TC(x)$ = total cost for x units of output
   x = output quantity

The average cost is usually a variable function of the output quantity, and normally decreases with an increasing quantity of output. Using the automobile example from Section 2.2.5, which had a total cost function of $\$(950 + 0.15x)$, the average cost, in dollars per mile, is given by

$$AC(x) = \frac{950 + 0.15x}{x}$$

$$= \frac{950}{x} + 0.15$$

If the automobile travels 10,000 miles/year, then the average operating cost is $\$(950/10000 + 0.15) = \$0.245/\text{mile}$. For a total annual travel distance of 20,000 miles, the unit operating cost decreases to $\$0.1975/\text{mile}$.

It should be noted from the average cost function of this example, $\$(950/x + 0.15)$, that as the output quantity x increases, the proportion of the fixed cost allocated to each unit of output decreases. This relationship is a fundamental economic principle often referred to as the *economies of scale,* a principle underlying the economic benefits of mass production. Such a relationship assumes, however, that the variable cost coefficient remains constant over the range of the output variable x. In a production environment, it is very likely that the variable cost coefficient will increase as the production volume increases (due to increased maintenance expenses, defective product, etc.).

For a total cost function that is continuous in the output (or independent) variable x, marginal cost is defined as the derivative of the total cost function (dependent variable) with respect to x, or $dTC(x)/dx$. This is true for continuous functions that are linear or nonlinear in the output variable x. In the special case of a continuous total cost function that is linear in x, such as $TC(x) = (\$950 + 0.15x)$, then $dTC(x)/dx = \$0.15$. In this case, marginal cost is the constant value $\$0.15$ and is the cost required to increase the output quantity x by one unit.

If the total cost function is discontinuous and defined only for discrete values of x (for example, $x = 1, 2, 3, \ldots$), then difference equations must be used to determine marginal costs. For example, $TC(6) - TC(5)$ is the marginal cost of increasing the output quantity from $x = 5$ to $x = 6$. Thus, in the discrete case, marginal cost is always the cost required to increase the output quantity x by one unit at a specified level of output. This is true for a discrete total cost function which has either a linear or nonlinear trend.

The concept of marginalism is general and applies to other mathematical functions as well. For example, marginal revenues can be determined from total revenue functions, marginal profit values can be determined from total profit functions, and so forth. If these functions are defined for discrete values of x or are continuous functions that are linear in x, then marginal revenue (profit) is the additional revenue (profit) received from selling one more unit of the output quantity x, at a specified level of output.

Marginal and average values corresponding to a specified output quantity are generally different. If the marginal cost is smaller than the average cost, an increase in output will result in a reduction of unit cost. This can be seen by recalling the familiar automobile problem, when $TC(x) = \$950 + \$0.15x$. The average cost is $AC(x) = \$950/x + \$0.15$ and the marginal cost is $MC(x) = \$0.15$. Thus, for all nonnegative finite values of x, marginal cost is always smaller than the average cost, and the unit cost will continue to decrease as x is increased. Such a relationship is not true in general for nonlinear total cost functions. The following example will illustrate some of the cost, revenue, and profit relationships.

Example 2.3

A small firm blends and bags chemicals, primarily for home gardening purposes. The market area for the firm is local, and all sales are to wholesale distributors. For one pesticide dust product, sales and production cost records over the past 10 seasons have been reviewed and analyzed. The following equations *approximate* the relationships among selling price, sales volume, production costs, and profit before income taxes. (The functional form for selling price implicitly reflects a fundamental relationship between price and demand. Namely, as the selling price is decreased, demand for the item increases. Alternatively, in order to increase the demand, the selling price must be reduced).

Let t = number of tons per season

$SP(t)$ = selling price in order to sell t tons
   = $\$(800 - 0.8t)$

$TR(t)$ = total revenue when t tons are sold at a particular selling price
   = Selling price $\times$ Demand
   = $\$(800 - 0.8t)t$
   = $\$(800t - 0.8t^2)$

$MR(t)$ = the marginal revenue at a sales volume of t tons

   $= \dfrac{dTR(t)}{dt} = \$(800 - 1.6t)$

$TC(t)$ = the total production cost for t tons
   = $\$(10,000 + 400t)$

$TP(t)$ = total profit when t tons are sold
   = $TR(t) - TC(t)$
   = $\$(800t - 0.8t^2) - \$(10,000 + 400t)$
   = $\$(-0.8t^2 + 400t - 10,000)$

$AP(t)$ = average profit per ton when t tons are sold
   = $TP(t)/t$
   = $\$(-0.8t + 400 - 10,000/t)$

The equations apply for the range, $0 \le t < 1000$.

We first note that the total revenue will be maximized when 500 tons are produced and sold per season. That is, by calculus,

$$\frac{dTR(t)}{dt} = 800 - 2(0.8)t = 0 \qquad \text{and} \qquad t = 500 \text{ tons}$$

The total revenue with a sales volume of 500 tons is

$$TR(500) = \$800(500) - \$0.8(500)^2$$
$$= \$200,000$$

The marginal revenue at the output level of 500 tons is

$$MR(500) = \$800 - \$1.6(500) = 0$$

Thus, the rate of change in the $TR(t)$ function with respect to t is zero when $TR(t)$ is evaluated at $t = 500$. The $TR(t)$ function is a strictly concave function with a unique maximum value at $TR(500)$. For sales from $t = 1$ to 500, the total revenue function is increasing at a decreasing rate. For sales volumes from $t = 500$ to 1000, total revenues are decreasing at an increasing rate.

Maximizing total revenues is not the issue in this example however. We wish to maximize profits. Again, by calculus,

$$\frac{dTP(t)}{dt} = 2(-0.8)t + 400 = 0 \quad \text{and} \quad t = 250 \text{ tons}$$

Thus, total profit will be maximized for a sales volume of 250 tons, and the maximum profit per season would be

$$TP(250) = -\$0.8(250)^2 + \$400(250) - \$10,000$$
$$= \$40,000$$

The average profit per ton when 250 tons are sold is

$$AP(250) = -0.8(250) + 400 - 10,000/250$$
$$= \$160/\text{ton}$$

Finally, we note that there are two break-even points in this example. By equating $TR(t) = TC(t)$, or

$$800t - 0.8t^2 = 10,000 + 400t$$

we obtain

$$-0.8t^2 + 400t - 10,000 = 0$$

Solving for the positive roots of this quadratic equation yields

$$t = 26.39, 473.61$$

For a sales volume in the range, $26.39 \le t \le 473.61$, the firm will make a profit. Sales volumes outside this range will result in total costs exceeding total revenues and a net loss to the firm.

2.3 ESTIMATION

The estimation of future events or the outcomes of present actions taken is obviously a fact of life for every individual, group, and organization. Family budgeting, weather forecasts, market forecasts of demand for consumer products, and predicting the annual national revenues from income taxation are only a few examples of the almost limitless number and variety of estimates that are made in our personal and business lives. In this textbook, we are concerned with estimation in the specific context of factors relevant to comparing alternative engineering/investment projects and making a selection from these projects. The annual revenues or savings, the initial and annual recurring costs, the life of a project, and the future salvage value of capital assets such as buildings and equipment that may be associated with a given project are rarely, if ever, known with certainty.

Many different terms pertain to the general subject of estimation. No attempt will be made in this text to enumerate and explain all the terms exhaustively; selected terminology will be given throughout the book as needed to explain the topics contained in a given chapter or section. Furthermore, an in-depth study of estimation procedures and the accuracy of estimated values is the study of mathematical statistics and probability theory, about which a vast literature exists.

Chapters Three to Seven are concerned with comparing alternative projects when the estimated values for relevant factors are single-valued or point estimates. In such cases, the single-valued estimates are considered certain to occur. Moving a step toward realism, an interval estimate could be made for the value of a given factor such as annual costs. That is, a high and low value or range might be estimated for the factor. Chapter Eight discusses factors that contribute to uncertainty in estimated values and considers deviations from single-valued or point estimates. Chapter Nine continues the discussion on decision making with the topics of risk and uncertainty, defining these terms and posing methodology for selecting a feasible alternative.

Although it is difficult to state precisely in quantitative terms, there is a relationship between the accuracy of an estimate and the cost of making the estimate. Intuitively, as more detailed information is obtained to provide the basis for an estimate and as more mathematical preciseness is exercised in calculating the estimate, the more accurate the estimate should be. However, as the level of detail increases, the greater the cost involved in making the estimate. Ostwald[3] has conceptualized this notion by the function

$$C_T = C(M) + C(E)$$

where C_T = the total cost of making the estimate, dollars
          $C(M)$ = the functional cost of making the estimate, dollars
          $C(E)$ = the functional cost of errors in the estimate, dollars

As depicted in Figure 2.5, the total cost of making an estimate reaches a minimum value when the amount of detail reaches a value D_1. Quantitatively defining the amount of detail is at best difficult and may be a practical impossibility. However, this concept of the total cost of an estimate varying with the amount of detail involved in making the estimate is realistic and is important to the general subject of estimation. In the abbreviated discussion on cost estimation techniques that follows, it will be noted that the individual techniques are based on varying amounts of detail with implied differences in the cost of making the estimate.

The texts by Ostwald [13, 14] are primary references on the subject of cost estimating. He defines four categories of estimated items: operations, products, projects, and systems. The categories are based on the scope of activity involved, but the distinction among them is not sharp and unequivocal. An *operation* is considered a basic or lowest-level activity. Machining, assembly, and painting are examples of operations for cost estimation pur-

---

[3] Op. cit., p. 455.

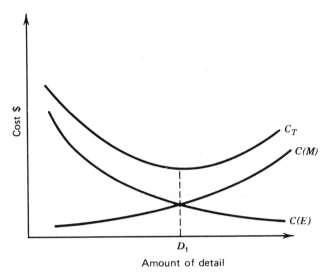

FIGURE 2.5. Cost of increasing detail.

poses. Therefore, operations are subelements of manufacturing a product or providing a service. Producing a given product or providing a given service may be a subelement of a project, and a system may consist of several projects. A system may, for example, be defined as a total manufacturing plant (or several plants), a regional health care program of facilities and personnel, or a total program of national defense.

Generally, the larger the scope of the categorical item being estimated, the more difficult it is to make accurate estimates of the costs and revenues involved. Ostwald presents a comprehensive discussion of estimating techniques for all categories, and Vinton [19] provides procedural detail for estimating the cost of certain manufacturing operations and manufactured products.

2.3.1 Project Estimation

For all categories of estimated items previously mentioned, three principal classes of estimates, based on accuracy and degree of detail, may be defined: (1) order-of-magnitude estimates, (2) preliminary estimates, and (3) detailed estimates. *Order-of-magnitude estimates* are usually gross estimates based on experience and judgment and made without formal examination of the details involved. Accuracies of ±50%, relative to actual cost, can be expected.

Preliminary estimates are also gross estimates, but more consideration is given to detail in making the estimate than for order-of-magnitude estimates. Certain subelements of the overall task are individually estimated, engineering specifications are considered, and so forth. Estimating the cost for manufacturing new components or products before designs and production plans are complete is an example of preliminary cost estimation. This type of estimate enables process engineering and product engineering groups to compare alternative designs or manufacturing methods and assess the economic impact of them. An accuracy level of about ±20% of actual cost can be expected with preliminary estimates.

Detailed estimates are expected to result in an accuracy level of ±5% of actual cost. In preparing the estimate, each subelement of the overall task is considered, and an attempt is made to assign a realistic cost to the subelement. Pricing a product or contract bidding usually involves detailed estimates of the costs involved.

Estimating the cost of a project may involve all of the above types of estimates. The time available to make the estimate, the information available, the experience and knowledge of the persons making the estimate, and the type and scope of the project are all factors that affect the desired accuracy of the project's cost estimate and the cost of making the estimate. When a comparison of alternative projects is based on future cash flows, the cost and revenue figures are necessarily estimates. These estimates will normally vary from actual cost and revenue figures.

In subsequent chapters, examples illustrating the principles of engineering economic analysis will primarily involve the following estimated items: first cost, operating costs, maintenance costs, recurring revenues or savings, useful life, salvage value, income taxes, and cost of capital. These cost items have been discussed previously in this chapter under the section on cost terminology.

Estimates of the functionally useful physical life of an item of equipment may be obtained from manufacturers and suppliers. Alternatively, if a company repeatedly buys a particular item of equipment and keeps accurate maintenance records, these records may be used to obtain an estimate of the functional life of the item. For example, suppose records reveal that 100% of the items survive the first 3 years of service. Then, 10% of the items fail in the fourth year, 20% of the items fail in the fifth year, 50% in the sixth year, 15% in the seventh year, and the remaining 5% fail in the eighth year. A weighted-average time of failure for any equipment item is then [(0.10)(4) + (0.20)(5) + (0.50)(6) + (0.15)(7) + (0.05)(8)] or 5.85 years. This is an estimate of the functional life of this particular equipment item.

In making an economic comparison of alternative equipment items, machines, or projects, the analyst may be interested only in a particular period of time or planning horizon. This planning horizon may be (and often is) different from the functionally useful life of one or more of the items, machines, or projects under consideration. For example, a job contract may cover a 5-year period. To execute the contract, machines and equipment items must be purchased. There are alternative machine and equipment items which can accomplish the job. These may have functional lives of 10, 15, or 20 years. An economic comparison of the alternative machines and equipment items would probably adopt a 5-year planning horizon (the length of the contract), and only the relevant cost and revenue figures for the 5-year period would be used in the comparison. Further discussion on this point is deferred until Chapter Five.

2.3.2 General Sources of Data

There are many sources for providing data to make the various estimates required in comparing alternative investment projects. Morris [10] devotes a chapter to the general subject of gathering data for economic analysis purposes and cites both internal and external sources of information. Examples of sources within a firm are sales records, production control records, inventory records, quality control records, purchasing department records, work measurement and other industrial engineering studies, maintenance records, and personnel records. Some, all, or additional records are input into the accounting function of the firm, which then compiles various financial reports for management. A principal objective of the accounting system is to determine per-unit costs for direct materials, direct labor, and overhead involved in manufacturing a product or providing a service. Since overhead costs, by definition, cannot be allocated as direct charges to a given product or service, they are prorated among the various products or services by somewhat arbitrary methods, some of which are presented later in the chapter. Thus, a caution is raised concerning the use of cost accounting data for estimating the cost of overhead items associated with a project. Nevertheless, the accounting system can and usually does serve as an important, if not primary, internal source of detailed estimates on operating costs, maintenance costs, and material costs, among others.

Sources of data external to the firm may be grouped into two general classes: published information that is generally available and information (published or otherwise) available on request. Available published information includes the vast literature of trade journals, professional society journals, U.S. government publications, reference handbooks, other books, and technical directories. Information not generally available except by request includes many sources listed in the previous category. For instance, many professional societies and trade associations publish handbooks, other books, special reports, and research bulletins that are available on request. Manufacturers of equipment and distributors of equipment are excellent sources of technical data, and most will readily supply this information without charge. Additionally, various government agencies, commercial banks (particularly holding companies involved in leasing buildings and equipment), and research organizations (commercial, governmental, industrial, and educational) may be sources of data to aid the estimating process.

2.4 GENERAL ACCOUNTING PRINCIPLES

As already mentioned, the engineer should have some understanding of basic accounting practice and cost accounting techniques in order to obtain data from the firm's accounting system. If accounting is classified into general accounting and cost accounting, then cost accounting is judged more important to the engineer as a source of data for making cost estimates pertinent to engineering projects. Cost accounting will therefore receive the greater emphasis in this text; in either case, the treatment of accounting is cursory and is directed toward fundamental accounting concepts instead of comprehensive accounting detail.

According to Niven and Ohman [12], an accounting system is concerned with the four primary functions of *recording, classifying, summarizing,* and *interpreting* the financial data of an organization, whether nonprofit or a business organization. The discussion to follow assumes a business organization, and, in virtually all such organizations, general accounting information is summarized in at least two basic financial reports:

1. A *balance sheet,* or statement of financial conditions, provides a summary listing of the assets, liabilities, and net worth accounts of the firm as of a particular date.

2. A *profit and loss statement* (or *income statement*) shows the revenues and expenses incurred by the firm during a stated period of time—a month, quarter, or year.

2.4.1 Balance Sheet

Records of financial transactions and a variety of internal reports serve as inputs to the accounting system, providing information on sales and other revenues and the expenses incurred in obtaining the revenues. Revenues and expenses for a specified period are then summarized on the profit and loss statement. The net profit or loss resulting for this period is then transferred to the net worth section of the balance sheet. Thus, although the two basic financial reports provide different financial pictures of the firm, they are directly related. Before illustrating this fact, discussion of certain terminology is necessary.

The items listed on a balance sheet are usually classified into three main groups: assets, liabilities, and net worth items. Subgroups may also be identified, such as current and fixed assets or liabilities. *Assets* are properties owned by the firm, and *liabilities* are debts owed by the firm against these assets. The dollar difference between assets and liabilities is the *net worth* of the business, which measures the investment made by the owners or stockholders of the business plus any accumulated profits left in the business by the owners or stockholders. Another term for net worth is *owners' equity.* A fundamental accounting equation is thus defined:

$$\text{Assets} - \text{Liabilities} = \text{Net Worth}$$

Rewriting, we have

$$\text{Assets} = \text{Liabilities} + \text{Net Worth}$$

and the usual format of a balance sheet follows the equation in this form.

Current assets include cash and other assets that can be readily converted into cash; an arbitrary period of 1 year is usually assumed as a criterion for conversion. Similarly, *current liabilities* are the debts that are due and payable within 1 year from the date of the balance sheet in question. *Fixed assets,* then, are the properties owned by the firm that are not readily converted into cash within a 1-year period, and *fixed liabilities* are long-term debts due and payable after 1 year from the date of the balance sheet. Typical current-asset items are cash, accounts receivable, notes receivable, raw material inventory, work in process, finished goods inventory, and prepaid expenses. Fixed-asset items are land, buildings, equipment, furniture, and fixtures. Items that are typically listed under current liabilities are accounts payable, notes payable, interest payable, taxes payable, prepaid income, and dividends payable. Fixed liabilities may also be notes payable, bonds payable, mortgages payable, and so forth. Net worth items appearing on a balance sheet are less standard and, to a degree, depend on whether the business is a sole proprietorship, a partnership, or a corporation. The size of the corporation is also an influencing factor on item designation. However, items such as capital stock, retained earnings, capital surplus, or earned surplus appear under the net worth group. An example of a balance sheet for a hypothesized firm is exhibited in Table 2.3.

The sample balance sheet in Table 2.3 is "balanced" since assets = liabilities + net worth. It gives a statement of the financial condition of the organization as of a specific date—the close of an accounting period. Although depreciation will be covered in Chapter Six, note the fixed asset portion of the balance sheet. For example, the building originally cost $200,000, and depreciation expenses have been charged annually, so that the total depreciation charges (as of the date of the balance sheet) have been $50,000, the amount entered as *depreciation reserve* for the building. In theory, the depreciation reserve is an accumulated amount of funds held to repurchase the asset when its functionally useful life terminates. The first cost of the depreciable asset (building, in this case) minus the amount in the depreciation reserve equals the *book value.* If the book value were a true estimate of the salvage, or market, value, then the sum of the amount in the depreciation reserve plus the book value provides the firm with an amount of funds equal to the original purchase price. This sum can be applied toward the purchase of a replacement asset. It is the usual case, however, that the market value of the asset when sold differs from the book value. The difference results in either a capital gain or loss and affects the firm's income taxes. (This point will be covered more fully in Chapter Six.)

A similar explanation applies to the fixed asset of equipment. In this particular balance sheet, the equipment account is an aggregate for all the equipment owned by the company instead of an individual listing of each equipment item. There could be separate equipment accounts, grouped according to equipment class. In any case, a company normally keeps individual records on equipment items, which are then summarized on the balance sheet.

TABLE 2.3

Sample Balance Sheet

Jax Tool and Engineering Company, Inc.
Balance Sheet
December 31, 19—

Assets

Current assets			
Cash		$ 25,000	
Accounts receivable		115,000	
Raw materials		8,500	
Work in process		7,000	
Finished goods inventory		3,000	
Small tool inventory		12,500	
Total current assets			$171,000
Fixed assets			
Land		$ 30,000	
Building	$200,000		
Less: Depreciation Reserve	50,000	150,000	
Equipment	$750,000		
Less: Depreciation Reserve	150,000	600,000	
Office equipment		10,000	
Total fixed assets			$790,000
Total assets			$961,000

Liabilities and Net Worth

Current liabilities			
Accounts payable		$ 32,000	
Taxes payable		15,000	
Total current liabilities			$ 47,000
Fixed liabilities			
Mortgage loan payable		$130,000	
Equipment loan payable		350,000	
Total fixed liabilities			$480,000
Total liabilities			$527,000
Common stock		$325,000	
Retained earnings		80,000	
Earned surplus (for current year)		29,000	
Total equity			$434,000
Total liabilities and equity			$961,000

2.4.2 Income Statement

The second basic financial report compiled by the accounting system is the *income statement* or *profit and loss* statement. For the current accounting period, the income statement provides management with (1) a summary of the revenues received, (2) a summary of the expenses incurred to obtain the revenues, and (3) the profit or loss resulting from business operations. The format of the income statement varies, and the revenue and expense items depend on the type of business involved. Thus, the income statement given in Table 2.4 is illustrative only and still concerns our hypothesized Jax Tool and Engineering Company. Let us further assume that the period of time covered is 1 year, which has ended as of the date of the balance sheet given in Table 2.3.

The format for the income statement in Table 2.4 is an oversimplification of such a statement for even a small manufacturing firm. For example, the *cost of goods sold* entry may be considerably more detailed to reflect multiple product lines, the depreciation items may be detailed to a greater extent to reflect multiple classes of assets, and several common expense items—such as employee benefits contributions, insurance premiums, and advertising—are not included.

TABLE 2.4

Sample Income Statement

Jax Tool and Engineering Co., Inc.
Income Statement
Year Ended December 31, 19__

Sales		$1,200,000
Less cost of goods manufactured:		
Direct labor	$420,000	
Direct materials	302,000	
Indirect labor	112,000	
Depreciation	98,000	
Repairs and maintenance	41,500	
Utilities	11,500	985,000
Gross profit		$ 215,000
Less other expenses:		
Administration	$ 76,000	
Marketing	49,000	
Interest payments	35,000	160,000
Net profit before income taxes		$ 55,000
Less income taxes		26,000
Net profit (posted to earned surplus)		$ 29,000

Another version of the income statement is shown in Table 2.5. This format is typical for a retail business, and the major difference in form concerns the method of determining the cost of goods sold item.

2.5 COST ACCOUNTING PRINCIPLES

2.5.1 Cost Allocation Methods

The balance sheet and the income statement in Section 2.4 are considerably removed both in time and in detail from decisions at the usual engineering project level. More important to the engineer as a source of cost information is the cost accounting system within a particular firm. The firm may be involved in manufacturing or providing services, and if it is involved in manufacturing, production may be on a job-shop or process basis. There are some fundamental differences in cost accounting procedures for determining manufacturing costs versus determining the cost of providing a service; also, there are differences in accounting procedures if manufacturing is on a job-shop or process basis. In order to concentrate on basic principles instead of details, the cost accounting system assumed will be that of a job-shop manu-

TABLE 2.5
Sample Income Statement

Jax Tool and Engineering Co., Inc.
Income Statement
Year Ended December 31, 19—

Sales		$1,200,000
Less cost of goods sold:		
Inventory, January 1, 19—	$ 26,000	
Plus purchases	432,000	
	$458,000	
Less inventory, December 31, 19—	44,000	414,000
Gross profit		$ 786,000
Less expenses:		
Direct labor	$420,000	
Depreciation—building	10,000	
Depreciation—equipment	30,000	
Repairs and maintenance	41,500	
Indirect labor	218,000	
Utilities	9,800	
Supplies, tooling	1,700	$ 731,000
Net profit before income taxes		$ 55,000
Less income taxes		26,000
Net profit (posted to earned surplus)		$ 29,000

facturing firm. Thus, the emphasis will be on determining the per-order costs for a job order.

The total cost of producing any job order consists of direct material, direct labor, and overhead costs. An additional item of cost could be special tooling or equipment purchases strictly for the job order in question. This definition of total cost does not detail the overhead cost into factory overhead, general overhead, and marketing expenses in order to simplify the presentation. Materials for a given job order may include purchased parts and in-house fabricated parts, and the cost for direct materials is determined primarily from purchase invoices. Questions of scrap allowances and averaging material costs, which fluctuate over time, present problems in obtaining accurate direct material costs, but determining such costs is reasonably straightforward. Direct labor time spent on a job order is normally recorded by operators on labor time cards, and the direct labor cost is determined by applying the appropriate labor cost rates. The labor rates, as determined by the accounting system, will normally include the cost of employee fringe benefits in addition to the basic hourly rate. Although accurately determining the direct labor cost for a given job is a major accounting problem, it is more readily determined than the overhead cost.

Overhead costs typically cannot be allocated as direct charges to any single job order and must, therefore, be prorated among all the job orders on some rational basis. Common methods for distribution are

1. The rate per direct labor hour.
2. A percentage of direct labor cost.
3. A percentage of prime cost (direct material plus direct labor cost).

If a single overhead rate for the entire manufacturing firm is to be used, this would, of course, be an average overhead rate for the entire factory, assuming the expense of providing and using an hour of factory facilities is about the same throughout the factory. To illustrate the three methods listed, it is assumed that a company experienced the following costs in the previous year 19__.

Total direct labor hours	48,000
Total direct labor cost	$480,000
Total direct material cost	$600,000
Total Overhead cost	$360,000

Then, the *overhead rate per direct labor hour* would be

$$\text{Rate} = \frac{\text{Overhead cost}}{\text{Direct labor hours}}$$

$$= \frac{\$360,000}{48,000}$$

$$= \$7.50/\text{direct labor hour}$$

If a particular job order requires 40 hours of direct labor with an average rate of $12.50/hour and $850 of direct materials, then the total cost of the job would be computed as

Direct material cost	= $ 850
Direct labor cost = 40 hours × $12.50/hour	= 500
Overhead cost = 40 hours × $ 7.50/hour	= 300
TOTAL COST	$1650

Determining the overhead rate as a *percentage of direct labor cost* for this company will yield

$$\% = \frac{\text{Overhead cost}}{\text{Direct labor cost}} (100\%)$$

$$= \frac{\$360,000}{\$480,000} (100\%)$$

$$= 75\%$$

For the same job above, the total cost would be computed as

Direct material cost	= $ 850
Direct labor cost	= 500
Overhead cost = (0.75)($500)	= 375
TOTAL COST	$1725

Determining the overhead rate as a *percentage of prime cost* for this company will yield

$$\% = \frac{\text{Overhead cost}}{\text{Direct labor cost} + \text{Direct material cost}} (100\%)$$

$$= \frac{\$360,000}{\$1,080,000} (100\%)$$

$$= 33\tfrac{1}{3}\%$$

For the same job above, the total cost would be computed as

Direct material cost	= $ 850
Direct labor cost	= 500
Overhead cost = (0.333)($1350)	= 450
TOTAL COST	$1800

Determining the overhead cost for a job order by the *rate per direct labor hour* method will yield the same result as the *percentage of direct labor cost* method, provided that the rate per direct labor hour used on the job in question is equal to the average factory labor rate. The "percentage of prime cost" method will necessarily yield a different assignment of overhead to a job order than the other two methods. The choice among these three methods is arbitrary; indeed, other methods are used by cost accountants in distributing overhead costs to a given job order. The *rate per direct labor hour* method is perhaps most commonly used.

Whatever method is chosen from the above for distributing overhead costs to job orders in a current year, the rates or percentages are based on the previous year's cost figures. Thus, overhead rates may change from year to year within a particular firm.

Since an average overhead rate for the entire factory may very well be too gross an estimate when actual overhead costs differ among departments within the factory, cost accounting may determine individual overhead rates for departments or cost centers. Furthermore, the hourly rates for direct labor may vary among these cost centers. A further refinement is to determine overhead rates for individual machines within cost centers. Then, as particular job orders progress through cost centers (departments and/or machines), the direct labor time (or machine time) spent on the job order in the various cost centers is recorded, the appropriate labor or machine rates and overhead rates are applied, and the total cost for the job is calculated.

The following example illustrates the variety of methods used to distribute overhead to a given cost center; the total overhead for the cost center will then be distributed to particular products by yet another method.

Example 2.4

The following information has been accumulated for the Deetco Company's two departments during the past year. (See Table 2.6.)

The Deetco Company distributes depreciation overhead based on (1) the first cost of equipment in each department, (2) a zero salvage value of the equipment in 10 years, and (3) a constant annual (or straight-line) rate of depreciation. All overhead other than depreciation is first distributed to each department according to the number of employees in each department, and then an overhead rate per direct labor hour is computed for each department.

What selling price should the company quote on Job Order D if raw material costs are estimated as $900, estimated direct labor hours required in

TABLE 2.6

Data for Example 2.4

	Department A	Department B	Total
Direct materials cost	$720,000	$240,000	$960,000
Direct labor cost	$260,000	$140,000	$400,000
Direct labor hours	25,200	16,200	41,400
Number of employees	14	9	23
First cost of equipment	$250,000	$200,000	$450,000
Annual depreciation	$ 25,000	$ 20,000	$ 45,000
Other factory overhead			$150,000
General overhead			$350,000

Departments A and B are 30 hours and 100 hours respectively, and profit is to be calculated as 25% of selling price?

For Department A, the total overhead allocation is determined as follows:

Annual depreciation		= $ 25,000
Other factory overhead	= (14/23)($150,000)	= 91,304
General overhead	= (14/23)($350,000)	= 213,043
TOTAL OVERHEAD COSTS		$329,347

Thus, the overhead rate for Department A per direct labor hour is

$$\text{Rate} = \frac{\$329,347}{25,200 \text{ hours}}$$

$$= \$13.07/\text{direct labor hour}$$

For Department B, the total overhead allocation is calculated as follows:

Annual depreciation		= $ 20,000
Other factory overhead	= (9/23)($150,000)	= 58,696
General overhead	= (9/23)($350,000)	= 136,957
TOTAL OVERHEAD COSTS		$215,653

Thus, the overhead rate for Department B per direct labor hour is

$$\text{Rate} = \frac{\$215,653}{16,200 \text{ hours}}$$

$$= \$13.31/\text{direct labor hour}$$

The estimated total cost for Job Order D is then computed as

Direct material cost		= $ 900.00
Direct labor cost for Dept. A	= ($260,000/25,200)(30 hours)	= 309.52
Overhead cost for Dept. A	= ($13.07)(30 hours)	= 392.10
Direct labor cost for Dept. B	= ($140,000/16,200)(100 hours)	= 864.20
Overhead cost for Dept. B	= ($13.31)(100 hours)	= 1331.00
Total cost		$3796.82

If x = the selling price of Job Order D, then

$$x = \text{Total cost} + \text{Profit}$$
$$= \$3796.82 + (0.25)x$$

and

$$x = \frac{3796.82}{0.75} = \$5062.43$$

2.5.2 Standard Costs

Although the first task of cost accounting is to determine per-item or per-order costs, another major purpose of cost accounting is to interpret financial data so that management can (1) measure changes in production efficiency and (2) judge the adequacy of production performance. Establishing cost standards can be of great assistance in achieving these objectives. A standard-cost system involves, in advance of manufacture, (1) the preparation of standard rates for material, labor, and overhead; and (2) the application of these rates to the standard quantities of material and labor required for a job order, or for each production operation required to complete the job order.[4]

Since a process-type manufacturing firm such as an oil refinery outputs the same product (or a few products) over a long time period, cost standards are more readily determined for process firms than for job-shop firms where the variety of output is large and varies with customer order. However, the number and type of production operations required to complete various job orders are finite for a given manufacturing firm. Each job order is, of course, made up of single units. Thus, a standard amount of material can be determined for each unit, and standard labor times and machine times can be determined for each unit. It is usually the responsibility of the work measurement function within the firm to determine these standard quantities. Then, by applying standard unit material costs and standard labor rates, standard unit costs for material, labor, and overhead can be determined. The standard costs then serve as a basis for measuring production efficiency and performance over time. Deviations from standard costs may be caused by several factors, especially (1) raw material price variations and (2) actual quantities of material and labor used versus the standard amounts of these items. This latter factor is the one of primary concern in determining production efficiency and performance, measures of which provide information to management to aid in cost control.

2.6 SUMMARY

In this chapter we provided an introduction to the language of accountants, financial analysts, and managers. To be successful in selling engineering designs, one must learn to communicate; to communicate effectively, all parties must speak the same language. Additionally, the data sources for economic analyses are often to be found in the accounting systems; hence, it is essential that the engineer be familiar with accounting principles. To complete the coverage of cost concepts, a brief introduction to cost estimation was presented.

BIBLIOGRAPHY

1. Apple, James M., *Material Handling Systems Design*, Ronald Press, 1972, Chapters 13 to 15.

---

[4] Frank W. Wilson and Philip D. Harvey (Editors), *Tool Engineers Handbook*, Second Edition, McGraw-Hill, 1959, pp. 2–23.

2. Blank, Leland T., and Tarquin, Anthony J., *Engineering Economy*, Second Edition, McGraw-Hill, 1983.
3. Clark, Forest D., and Lorenzoni, A. B., *Applied Cost Engineering*, Second Edition, Revised and Expanded, Marcel Dekker, 1985.
4. De La Mare, R. F., *Manufacturing Systems Economics*, Holt, Rinehart & Winston, 1982.
5. DeGarmo, E. Paul, Sullivan, William G., and Canada, John R., *Engineering Economy*, 7 ed., Macmillan, 1984.
6. Ferguson, Earl J., and Shamblin, James E., "Break-Even Analysis," *The Journal of Industrial Engineering*, *18* (8), August 1967.
7. Hartley, Ronald V., *Cost and Managerial Accounting*, Allyn & Bacon, 1983.
8. Kepner, Charles H., and Tregoe, Benjamin B., *The Rational Manager*, McGraw-Hill, 1965.
9. Morris, William T., *Engineering Economic Analysis*, Reston, 1976.
10. Morris, William T., *The Analysis of Management Decisions*, Revised Edition, Irwin, 1964, Chapter 5.
11. Niswonger, C. Rollin, and Fess, Philip E., *Accounting Principles*, 12 ed., South-Western, 1977.
12. Niven, Williams, and Ohman, Anka, *Basic Accounting Procedures*, Prentice-Hall, 1964, p. 2.
13. Ostwald, Phillip F., *Cost Estimating for Engineering and Management*, Prentice-Hall, 1974.
14. Ostwald, Phillip F., *Cost Estimating*, Second Edition, Prentice-Hall, 1984.
15. Polimeni, Ralph S., Fabozzi, Frank J., and Adelberg, Arthur H., *Cost Accounting: Concepts and Applications for Managerial Decision Making*, 2 ed., McGraw-Hill, 1986.
16. Riggs, James L., and West, Thomas M., *Engineering Economics*, Third Edition, McGraw-Hill, 1986.
17. Steffy, Wilbert, Smith, Donald H., and Souter, Donald, *Economic Guidelines for Justifying Capital Purchases*, Industrial Development Division, University of Michigan, Ann Arbor, Mich., 1973, p. 89.
18. Thuesen, G. J. and Fabrycky, W. J., *Engineering Economy*, 7 ed., Prentice-Hall, Inc., 1989.
19. Vinton, Ivan R. (Editor), *Realistic Cost Estimating for Manufacturing*, Society of Manufacturing Engineers, Dearborn, Mich., 1968.

PROBLEMS

1. A small manufacturing firm is converting from a manual inventory system to a microcomputer system. The incremental investment required would be $16000. Due to the rapid increase in technology, it is assumed that the salvage value will be zero any time after the system is purchased. It is estimated that a net annual savings of $2500 in operating costs will result when the new computer is purchased. Assume that the time value of money is zero. (2.2.5)

a. How many years of savings are required to recover the incremental investment?

b. What annual savings are required to recover the incremental investment in 4 years?

2. A small metal-machining company produces parts according to customer order. One particular part is frequently ordered in batch sizes of 15–150 units. The company has determined that the part can be produced on three different machine tools: M1, M2, M3. An economic analysis reveals the following data:

Machine	Fixed Cost/Order	Variable Cost/Unit
M1	$300	$9
M2	$500	$5
M3	$750	$3

a. By either a graphical or analytical approach, determine the most economical machine(s) to use for order sizes from 0–150 units.

b. For an order size of 75 units, what is the minimum total cost for machining the order? (2.2.5)

3. An engine lathe or a turret lathe can be used to produce a job order of Part 173. If the job order is produced on the engine lathe, Machinist B performs all required operations (i.e., the tooling setup and the operation of the lathe during the machining cycle). On the engine lathe, a tooling setup time of 15 minutes and machining time of 30 minutes are required to produce each part. Machinist B earns $9/hour, and the overhead rate for each machining hour of engine lathe operation is $14.00. The tooling costs per lot for production on the engine lathe are estimated to be $500. If the job order is produced on the turret lathe, Machinist A must do the initial tooling setup for the job, and 4 hours are required for this. Once the setup is done, no further tooling setup per part is required, and Machinist B will operate the turret lathe for the machining cycles. Machinist A earns $12/hour. The machining time for Part No. 173 is 10 minutes on the turret lathe, and the overhead rate for each machining hour is $25. The tooling costs per lot for production on the turret lathe is $750. For what job-order size would the turret lathe be economically preferred to the engine lathe? (2.2.5)

4. In a stable economic environment, the A. B. Jax Specialty Foundry Co. can produce a maximum of 1500 gray iron railroad car wheels per month and sell these for $300 each. Now, after prolonged labor problems in the coal industry, the foundry can only sell an average of 500 wheels/month at $250 each. The coal industry strike is temporary, but the future over the next 6 months to a year appears very uncertain.

The foundry has a depreciable investment in buildings and equipment of $300,000, and the value decreases at a rate of 6⅔% each year.

At a production rate of 500 wheels/month, direct labor costs will increase from $55/wheel at maximum production rate to $75/wheel. Direct material costs per wheel will remain at $50. Other annual overhead costs theoretically vary in linear fashion from $60,000 at zero output to $105,000 at maximum production capacity. Would you recommend that the A. B. Jax Specialty Foundry Co. shut down temporarily or continue to operate at this reduced capacity? (2.2.5)

5. A commercial machine shop regularly produces a stainless steel component for a major electronics manufacturer. The machine shop purchases the component from a nearby specialty steel company in semifinished condition, performs drilling and milling operations on the part, and ships it to the electronics firm.

The machining operations are those that can readily be performed on a tape-controlled drill press with a turret head. Thus, the management of the machine shop feels the purchase of such a machine to produce only this part is economically justified. An engineer is then assigned the task of determining the production quantity for break-even, assuming a time value of money (minimum attractive rate of return) equal to zero.

The engineer compiles the following information and cost estimates. The tape-controlled machine will have an installed first cost of $60,000, which includes the necessary electronic software, cutting tools and holders, and work-holding devices. Training of the operator for the machine is included in the purchase price. The economic life of the machine is assumed to be 10 years with a salvage value of $20,000 at that time. The decrease in asset value is estimated at $4000/year and judged to be an annual fixed cost. Other fixed costs are $1500/year. The steel parts are sold to the electronics firm for $15.30/unit. The variable unit costs are estimated as $1.70 for direct labor, $7.00 for direct material, and $3.50 for overhead (excluding depreciation of the machine—the $4000/year fixed cost mentioned previously).

What annual sales volume (number of parts) is required in order to break even on the machine purchase if linearity is assumed? (2.2.5)

6. A subsidiary plant of a major furniture company manufactures wooden pallets primarily for the parent company but also sells pallets to other industrial customers. Although the plant produces multiple pallet styles and sizes, approximately 95% of sales is due to a single pallet style and size. The subsidiary plant can therefore be considered a single-product plant. The plant has the capacity to produce 300,000 pallets per year. Presently, the plant is operating at 70% of capacity. The selling price of the pallet is $18.25 and the variable cost per pallet is $15.75. (The selling price and variable cost per pallet are linearly related to production quantity.) At zero output, the subsidiary plant's annual fixed costs are about $550,000 and are approximately constant up to the maximum production quantity per year. (2.2.5)

   a. With the present 70% of capacity production, what is the expected annual profit or loss for the subsidiary plant?

   b. What annual volume of sales (units) is required in order for the plant to break even?

   c. What would be the annual profit or loss if the plant were operating at 90% of capacity?

   d. If fixed costs could be reduced by 40%, what would be the effect on the break-even sales volume?

7. A first production run of a new product is made with the following data compiled:

Direct material cost	= $ 5.00/unit
Direct labor cost	= $ 2.90/unit
Overhead cost	= $ 4.10/unit
Selling price	= $20.00/unit

The above costs are based on all good units produced and no rejects or scrap. Parts are inspected only once, after all the manufacturing operations are performed but prior to shipping. During the production run, a scrap rate of 30% has occurred. What is the maximum scrap rate permissible in order to break even (i.e., total costs equal to total revenues)? (2.2.5)

8. A typical gasoline-powered farm tractor with PTO rating of 35 horsepower has an operating cost of about \$3.50/hour of use (for fuel, lubricants, oil filters, repairs, and maintenance), excluding labor cost, and an annual fixed cost of about \$1200 (for depreciation, insurance and taxes, housing, and opportunity costs).

Similar data for a diesel-powered tractor of the same size are \$2.75/hour of use for operating cost and \$1425 annual fixed cost.[5] (2.2.5)

   a. What is the number of operating hours per year for break-even between the two tractors?

   b. If the estimated number of operating hours per year is 600, what annual savings are estimated if the diesel-powered tractor is purchased instead of the gasoline-powered tractor?

   c. For the diesel-powered tractor operated at 600 hours per year, what revenues/hour must be generated in order to break-even?

9. A large coal mining company owns and operates a fleet of four-wheel drive vehicles. Maintenance records over a period of five years reveal that about 20% of the time, tires are replaced after 10,000 miles of use. Fifteen percent of the tires are replaced after 15,000 miles, 25% are replaced after 20,000 miles, 30% are replaced after 25,000 miles, and 10% are replaced after 30,000 miles of use. Determine a weighted-average estimate of tire life. (2.3.1)

10. On January 3, 19__, a manufacturing firm purchased a drill press for \$15,000, a fork-lift truck for \$25,000, and a microcomputer system for \$10,000. For depreciation purposes, the estimated lives of these items are 10, 7, and 5 years, respectively. Annual depreciation expenses for the items are \$1300, \$3000, and \$2000, respectively. Assume it is now three years later and December 31, 19__. How would these items appear under the Fixed Assets portion of the Balance Sheet? What is the total Book Value and the total amount in the Depreciation Reserve for the items? (2.4.1)

11. K. Z. Moley purchased a small retail restaurant business and opened on July 1, 19__. At the date of opening, he had invested \$20,000 of equity funds with the following breakdown: \$10,000 in equipment, \$7000 in inventory items, and \$3000 in operating cash.

   a. Prepare a balance sheet for the business as of July 1, 19__. (2.4.1)
   The data below summarize the gross sales and expenses for the restaurant during the first 3-month period.

Gross sales	\$32,000
Purchases	12,500

[5] Example problem based on Publication 510 (revised February 1974) entitled "Farm Machinery Performance and Costs," Extension Division, Virginia Polytechnic Institute and State University, Blacksburg, Virginia.

Salaries	6,000
Advertising expense	1,000
Rent expense	1,800
Expense for utilities	900
Expense for misc. supplies	600

For the purchases, $10,000 was paid with cash and $2500 is still owed. At the close of the 3-month period, the end-of-period inventory is worth $5500.

b. Prepare an income statement for the 3-month period covered. (2.4.2)

c. If, at the end of the 3-month period on October 1, 19___, the net worth (or ownership) account is $22,000, determine the amount of the cash account in order to balance the accounting equation as of October 1, 19___. (2.4.1)

12. A successful building contractor purchased a 150-acre farm to operate on a part-time basis and raise beef cattle. The purchase price of the farm was $150,000. The contractor paid 30% of the purchase price in cash and financed the remainder over 15 years at a 10% annual simple interest rate with a mortgage, payable to a local bank. Soon after the purchase of the farm, the contractor purchased cattle for $25,000, paid $10,000 in cash, and gave a promissory note to the seller for the balance. The note carried an 8% annual simple interest rate and was to be paid off within 5 years. During the first full year of operations, the farm resulted in the following revenues and expenses.

Calves sold (income)	$12,600
Timber sold (income)	5,000
Labor expenses	1,500
Expenses for machinery hired	2,750
Veterinarian fees	350
Fertilizer expenses	1,800
Property taxes	1,625
Feed purchased	275
Expenses for repairs	950
Interest expenses	11,700
Insurance expenses	1,000
Legal fees	350
Expenses for miscellaneous supplies	425

Prepare an income statement for the farming operation and determine the net profit (loss) before income taxes. (2.4.2)

13. An income statement for the WAC Company covering a calendar year ending December 31 is as follows.

Income Statement

Gross income from sales	$247,000
Less: cost of goods sold	138,800
Net income from sales	$108,200

Operating expenses		
Rent	$ 9,700	
Salaries	30,200	
Depreciation	5,800	
Advertising	4,300	
Insurance	1,500	$ 51,500
Net profit before income taxes		56,700
Less: income taxes		23,973
Net profit after income taxes		$ 32,727

The Balance Sheet Accounts as of December 31, are

Cash	$94,227
Accounts receivable	8,000
Notes payable	25,000
Raw material inventory	10,000
Work-in-process inventory	15,000
Accounts payable	6,000
Declared dividends	20,000
Finished goods inventory	18,500
Land	30,000
Original cost of building	80,000
Building—Reserve for depreciation	8,000
Original cost of equipment	40,000
Equipment—Reserve for depreciation	4,000

a. Determine the Earned Surplus for the year in question.
b. Prepare a Balance Sheet as of December 31. (2.4.1, 2.4.2)

14. An oil refinery produces one base type of crude oil for the southeastern U.S. market. The two equations below give the relationships that approximate the total cost and the total profit per week in dollars. (*Note:* This problem is for illustration purposes only; the total cost function and the cost coefficients are not necessarily realistic.)

$$\text{Total cost, } TC(x) = 50,000 + 20.2x + 0.0001x^2, \text{ for } x \geq 0$$
$$= 0 \text{ otherwise}$$

$$\text{Total profit, } TP(x) = Sx - TC(x)$$

where x = Amount of crude oil produced, barrels/week
S = sales price of crude oil, dollars/barrel
= $35/barrel.

a. At what level of production in barrels/week is the cost/barrel minimum? What is the minimum cost/barrel?

b. What is the maximum weekly profit that the company can make? At what level of production is the maximum weekly profit attainable?

c. Over what range of production is profit possible?

d. Suppose the refinery capacity is limited to 50,000 barrels/week, what is the range of production over which profit is possible? (2.2.6)

15. Production of a particular type of annual crop is a function of fertilizer used and is given by the following relationship:

$$P(t) = \text{crop production, barrels/acre}$$
$$= 0.417t - 0.00125t^2$$

where t = amount of fertilizer used, pounds/acre.

The crop can be sold at a price of \$15/barrel. Given that one barrel of crop weighs about 120 pounds, the total cost of crop production is

$$TC(t) = \$(220 - 2t) \text{ per acre}$$

a. How much fertilizer should the farmer use for this crop per acre of land to maximize his annual profit?

b. What is the maximum annual profit per acre of land for this crop?

c. Between what range of fertilizer application is a profit possible? (2.2.6)

16. Assume the total annual inventory cost for a particular item carried in inventory is

$$TC(Q) = \text{annual cost of ordering} + \text{annual cost}$$
$$\text{of carrying inventory, dollars/year}$$
$$= P_c \left(\frac{A}{Q}\right) + (0.60\ Q + 150),$$

where P_c = cost of preparing a purchase order (a constant value, regardless of the quantity ordered)
$\quad\quad$ = \$15 per purchase order.
$\quad Q$ = the number of items ordered each time a purchase order is placed.
$\quad A$ = total annual demand for this item
$\quad\quad$ = 2000 units.
A/Q = the number of orders placed each year.

Note that the total annual inventory cost $TC(Q)$ is a function of the order size Q. (2.2.6)

a. Determine the value of Q that will minimize the total annual inventory cost (i.e., determine the economic order quantity Q^*).

b. What will be the corresponding total annual inventory cost, $TC(Q^*)$?

c. For the economic order quantity Q^*, determine the number of orders that will be placed each year.

d. If the cost of placing a purchase order is decreased by 50%, what is the effect on the economic order quantity?

17. A worker in a company is on an incentive system. The worker earns \$1.50 for each part produced. In *addition*, the worker earns a guaranteed wage of \$5/hour regardless of production output. The product is sold at a price of $(5 - 0.4t)$, dollars/unit, where t is the number of units produced per hour. (2.2.6)

a. At what rate of production will the company earn the most profit from this worker's output? (Assume all units produced can be sold.)

b. What is the amount of maximum profit per hour?

c. At what rate of production will the revenue be maximized? What is the corresponding profit per hour?

d. Within what range of production can the company make profit?

18. When electrical current flows in a conductor there is a power loss due to electrical resistance. Resistance is inversely proportional to the size of the conductor and thus, the cost of power loss will decrease with increased conductor size. However, as the size of the conductor increases, the investment in the conductor also increases. Define A to be the cross-sectional area (square inches) of a conductor and assume the total cost per year (TC) of transmitting a given quantity of current over a fixed distance by a copper conductor is given by:

$$TC = \$20 + \$30A + \frac{\$270}{A}$$

a. Determine the cross-sectional area of the copper conductor which will result in minimum total cost/year.

b. Using the cross-sectional area determined in (a), what is the minimum total cost per year to operate the conductor system?

c. For 2-inch and 4-inch diameter copper conductors, determine the total cost per year.

19. An order for 5000 units of Part D142 is received by the J. T. Kling Engineering Company, a small machine shop. The raw material for this part is SAE 1020 steel rectangular bar stock. Each unit of bar stock costs $30 and yields 20 D142 parts. The basic manufacturing sequence, with standard operation times per part and machine overhead rates (dollars/hour), is given in the following table.

Manufacturing Operation	Standard Time Per Part	Machine Overhead Rate
Cutoff on power hacksaw	2 minutes	$1.20/hour
Mill two sides; deburr	5 minutes	2.00/hour
Drill three ⅜-inch diam. holes	3 minutes	1.60/hour
Surface grind one side; deburr	3 minutes	1.40/hour
Package	0.25 minutes	—

The direct labor time per part is the same as the machine time per part (packaging is a manual operation only). The tooling cost for this job order is estimated to be $500. In addition to tooling costs and machine overhead costs, factory overhead costs (for indirect labor, indirect materials, utilities, etc.) are $15 per direct labor hour. The average direct labor hour rate, including fringe benefits, is $10 per hour. (2.5.1)

a. Assuming a zero scrap rate during production, determine the total estimated costs for the job order of 5000 units.

b. Determine the *unit* selling price if profit is to be 25% of the total cost.

20. The welding department of a mining equipment manufacturing plant consists of four cost centers: manual arc welding (A), semiautomatic welding (B), furnace brazing and heat treating (C), and finishing (D). Some oxyacetylene cutting is also done in Center B. Assume that is is possible to allocate departmental overhead expenses directly to each cost center, and that the following data for the welding department were compiled last year by the accounting system.

Cost Center	Departmental Expenses	Direct Labor Hours	Direct Labor Cost	Direct Material
A	$21,000	10,000	$75,000	$12,000
B	10,000	4,000	18,000	12,000
C	5,400	1,500	9,000	4,500
D	4,800	2,800	11,200	3,000

Compute the overhead rate (or rates) applicable by the following methods. (2.5.2)

a. Percentage of direct labor cost for each cost center.

b. Percentage of prime cost for each cost center.

c. Rate per direct labor hour for each cost center.

21. Given the welding department and four cost centers found in Problem 20, the direct labor hours and costs for each cost center remain the same, but new and additional data are given below (assume cost data are for the previous year).

Cost Center	Square Feet Occupied	Cost of Machinery	Number of Direct Labor Employees
A	900	$10,000	5
B	400	9,000	2
C	600	11,700	1
D	500	4,800	1
Total	2,400	$35,500	9

Expenses other than for direct labor and materials chargeable to the welding department last year were

Maintenance	$ 6,400
Gas and electricity	20,000
Supervision and other indirect labor	30,000

Miscellaneous supplies	7,500
Equipment depreciation	4,600
Building depreciation	8,000

Determine an overhead rate per direct labor hour for each cost center if the welding department expenses above are first allocated to each cost center as follows. (2.5.1)

a. Maintenance expenses and equipment depreciation expenses are allocated according to the value of equipment (percent of total) in each cost center.

b. Building depreciation expenses are allocated according to the floor space occupied (percent of total) by each cost center.

c. Supervision and other indirect labor expenses are allocated according to the number of direct labor employees (percent of total) of each cost center.

d. Supplies and gas and electricity expenses are allocated according to the number of direct labor hours (percent of total) for each cost center.

CHAPTER

Time Value of Money Operations

3.1 INTRODUCTION

Design alternatives are normally compared by using a host of different criteria, including system performance and economic performance. Among the system performance characteristics that are of concern, quality, safety, and customer service considerations are of primary importance. Among the economic performance characteristics normally considered are initial investment requirements, return on investment, and the cash flow (CF) profile. Since the cash flow profiles are usually quite different among the several design alternatives, in order to compare the economic performances of the alternatives, one must compensate for the differences in the timing of cash flows. A fundamental concept underlies much of the material covered in the text: *money has a time value*. By this, we mean that the *value* of a given sum of money depends on *when* the money is received.

Example 3.1

To illustrate the concept of the time value of money, suppose a wealthy individual approaches you and says, "Because of your outstanding ability to manage money, I am prepared to present you with a tax-free gift of $1000. However, if you prefer, I will postpone the presentation for 1 year, at which time I will guarantee that you will receive a tax-free gift of $X." Would you choose to receive the $1000 now or the $X 1 year from now if $X equaled (1) $1000, (2) $1100, (3) $2000, (4) $10,000?

In presenting this situation to numerous students, no students preferred receiving $X in case (1), very few students preferred $X in case (2), most students preferred receiving $2000 a year from now, and all students indicated a preference for $10,000 a year from now. The point is that the value of $1000 1 year from now was perceived to be less than the value of $1000 at present. For most students, the value of $1100 1 year from now was believed to be less than the value of $1000 at present. Only a few students felt that $2000 a year from now was less valuable than $1000 at present. All students believed $10,000 a year from now was more valuable than $1000 at present. Thus, for each individual student, some value (or range of values) of $X exists for which one would be indifferent between receiving $1000 now versus receiving $X a year from now. If, for example, one is indifferent for $X equal to $1250, then we would conclude that $1250 occurring one year from now has a *present value of* $1000 *for that particular individual.*

Some would argue that the reason the students prefer receiving $X today rather than receiving $X 1 year later is due to the effects of *inflation.* They would contend that the purchasing power of $X decreases over time during periods of inflation. In fact, most people have experienced the impact of inflation and cost of living increases.

While it is true that the *value, economic worth,* or *purchasing power of money* changes over time during periods of inflation and deflation, it is also true that people will prefer to receive $X today rather than 1 year later even when there is no inflation or deflation. One reason for this preference is the *opportunity cost of money.*

As noted in Chapter Two, "the cost of forgoing the opportunity to earn interest, or a return, on investment funds is termed an *opportunity cost.*" The $X received today can be invested at some interest rate so that 1 year later the investment will be worth more than $X.

Because people favor current consumption to postponed consumption, it is necessary that postponed consumption be compensated or rewarded with "interest" or a "return on investment." Without such compensation, consumption would not be postponed and capital would not be available for investment.

The concept of inflation and its consideration in evaluating investment alternatives is extremely important. However, before addressing the subject of inflation, it is essential that a number of fundamental concepts be addressed related to the time value of money in the comparison of the economic performances of design alternatives. In this chapter we consider inflation effects after we examine a number of mathematical operations that are based on the time value of money, with an emphasis on modeling cash flow profiles.

The following examples are intended to illustrate the time value of money in the absence of inflation or deflation.

Example 3.2

To continue our consideration of different cash flow situations, examine closely the two cash flow profiles given in Table 3.1. Both alternatives involve an investment of $100,000 in ventures that last for 4 years. Alternative A involves an investment in a minicomputer by a consulting engineer who is planning on providing computerized design capability for clients. Since the engineer anticipates that competition will develop very quickly if the plan proves to be successfu, a declining revenue profile is anticipated.

Alternative B involves an investment in a land development venture by a group of individuals. Different parcels of land are to be sold over a 4-year period. The land is anticipated to increase in value. There will be differences in the sizes of parcels to be sold. Consequently, an increasing revenue profile is anticipated.

The consulting engineer has available funds sufficient to undertake either investment, but not both. The cash flows shown are cash flows after taxes and other expenses have been deducted. Both investments result in $160,000 being received over the 4-year period; hence, a net cash flow of $60,000 occurs in both cases.

Which would you prefer? If you prefer Alternative B, then you are not acting in a manner consistent with the concept that money has a time value. The $60,000 difference at the end of the first year is worth more than the $60,000 difference at the end of the fourth year. Likewise, the $20,000 difference at the end of the second year is worth more than the $20,000 difference at the end of the third year.

Example 3.3

As another illustration of the impact of the time value of money on the preference between investment alternatives, consider investment alterna-

TABLE 3.1

Cash Flow Profiles for Two Investment Alternatives

End of Year (EOY)	CF		
	A	B	(A − B) Difference
0	−$100,000	−$100,000	$0
1	+ 70,000	+ 10,000	+60,000
2	+ 50,000	+ 30,000	+20,000
3	+ 30,000	+ 50,000	−20,000
4	+ 10,000	+ 70,000	−60,000

tives C and D, having the cash flow profiles depicted in Figure 3.1. The cash flow diagrams indicate that the positive cash flows for Alternative C are identical to those for Alternative D, except that the former occur 1 year sooner; both alternatives require an investment of $6000. *If exactly one of the alternatives must be selected,* then Alternative C would be preferred to Alternative D, based on the time value of money.

Example 3.4

A third illustration of the effect of the time value of money on the selection of the preferred investment alternative is presented in Figure 3.2. Either Alternative E or Alternative F must be selected; the only differences in the performance characteristics of the two alternatives are economic differ-

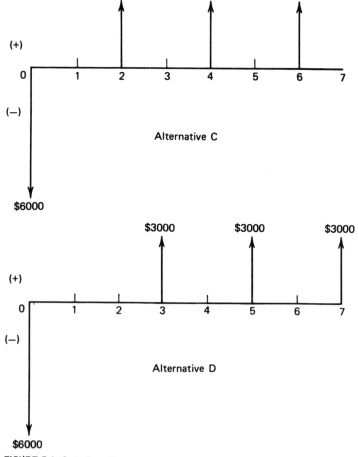

FIGURE 3.1. Cash flow diagrams for Alternatives C and D.

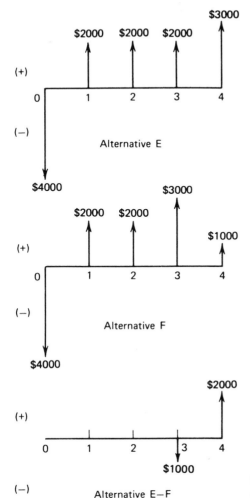

FIGURE 3.2. Cash flow diagrams for Alternatives E and F.

ences. As shown in Figure 3.2, the economic differences reduce to a situation in which the receipt of $1000 is delayed in order to receive $2000 a year later. For this illustration, we would conclude that most of the students polled earlier would prefer Alternative E to Alternative F, since most of the students preferred $X when it equaled $2000.

In order to provide motivation and to provide a familiar scenario, the treatment of time value of money operations will be presented frequently in the context of personal finance. However, each of the concepts examined in this chapter can and does occur in the business world.

In this chapter we emphasize end-of-period cash flows and end-of-period compounding.[1] Depending on the financial institution involved, in personal finance transactions, savings accounts might not pay interest on deposits made in "the middle of a compounding period." Consequently, answers obtained using the methods we describe should not be expected to be exactly the same as those provided by the financial institution.

The beginning-of-period cash flows can be handled very easily by noting that the end of period t is the beginning of period $t + 1$. To illustrate, rental payments might be made at the beginning of each month. However, one can think of the payment made at the beginning of, say, March as having been made at the end of February.

In Chapter Four, end-of-year cash flows are assumed unless otherwise noted. There, it is realized that monetary transactions take place during a calendar year, but it is convenient to ignore any compounding effects within a year and deal directly with end-of-year cash flows.

3.2 INTEREST CALCULATIONS

In considering the time value of money, it is convenient to represent mathematically the relationship between the current or *present* value of a single sum of money and its *future* value. Letting time be measured in years, if a single sum of money has a current or *present* value of P, its value in n years would be equal to

$$F_n = P + I_n$$

where F_n is the accumulated value of P over n years, or the *future* value of P, and I_n is the increase in the value of P over n years. I_n is referred to as the accumulated *interest* in borrowing and lending transactions and is a function of P, n, and the *annual interest rate, i*. The annual interest rate is defined as the change in value for $1 over a 1-year period.

Over the years, two approaches have emerged for computing the value of I_n. The first approach considers I_n to be a linear function of time. Since i is the rate of change over a 1-year period, it is argued that P changes in value by an amount Pi each year. Hence, it is concluded that I_n is the product of P, i, and n, or,

$$I_n = Pin$$

and

$$F_n = P(1 + in)$$

This is called the *simple interest* approach.

The second approach used to compute the value of I_n is to interpret i as *the rate of change in the accumulated value of money*. Hence, it is argued that the following relation holds:

---

[1] The exception to this will be the treatment of continuous cash flows.

$$I_n = iF_{n-1}$$

and

$$F_n = F_{n-1}(1 + i)$$

This approach is referred to as the *compound interest* approach.

The approach to be used in any particular situation depends on how the interest rate is defined. Since practically all monetary transactions are based currently on compound interest rates instead of on simple interest rates, we will assume compounding occurs unless otherwise stated.

A convenient method of representing the time value of money is to visualize positive and negative cash flows as though they were generated by a borrower and a lender. In particular, suppose you loaned $1000 for 1 year to an individual who agreed to pay you interest at a rate of 10%/year. The $1000 is referred to as the *principal* amount. At the end of 1 year, you would receive $1100 from the borrower. Thus, we might say that $1100 1 year from now is worth $1000 today based on a 10% interest rate or, conversely, we might say that $1000 today has a value of $1100 1 year from now based on a 10% interest rate.

If the individual borrowed the $1100 for an additional year, then you would be owed $1210, since the interest on $1100 for 1 year equals $(0.10) \times ($1100)$, or $110. Equivalently, borrowing $1000 for 2 years at 10% interest yields $1210 owed if interest is *compounded* annually. Compound interest involves the computation of interest charges during a time period based on the unpaid principal amount plus any accumulated interest charges up to the beginning of the time period. The interest amount due for a given interest period converts to principal for the purpose of calculating the interest amount due in the subsequent interest period. The relationship between the principal amount and the compounding of interest is given in Table 3.2.

TABLE 3.2

Illustrating the Effect of Compound Interest

End of Period	(A) Amount Owed	(B) Interest for Next Period	(C) = (A) + (B) Amount Owed for Next Period*	
0	P	Pi	$P + Pi$	$= P(1 + i)$
1	$P(1 + i)$	$P(1 + i)i$	$P(1 + i) + P(1 + i)i$	$= P(1 + i)^2$
2	$P(1 + i)^2$	$P(1 + i)^2 i$	$P(1 + i)^2 + P(1 + i)^2 i$	$= P(1 + i)^3$
3	$P(1 + i)^3$	$P(1 + i)^3 i$	$P(1 + i)^3 + P(1 + i)^3 i$	$= P(1 + i)^4$
⋮	⋮	⋮	⋮	
$n - 1$	$P(1 + i)^{n-1}$	$P(1 + i)^{n-1}i$	$P(1 + i)^{n-1} + P(1 + i)^{n-1}i = P(1 + i)^n$	
n	$P(1 + i)^n$			

* Notice, the value in column (C) for the end of period $(n - 1)$ provides the value in column (A) for the end of period n.

Recall that whenever the interest charge for any time period is based only on the unpaid principal amount and not on any accumulated interest charges, simple interest calculations apply. The interest due for a given interest period does not convert to principal for the purpose of calculating the interest amount due in the subsequent interest period. In the preceding illustration, the amount owed at the end of 2 years would be the $1000 principal plus the interest charges of $(0.10)(\$1000) = \$100$ each year or $1200 total. Thus, in this example, the $1210 for the compounding case compares with the $1200 for the simple interest case. (In some cases, the difference is much more dramatic!)

Example 3.5

Person A borrows $4000 from Person B and agrees to pay $1000 plus accrued interest at the end of the first year and $3000 plus the accrued interest at the end of the fourth year. What are the amounts for the two payments if 8% annual simple interest applies? For the first year, the payment is $1000 + $(0.08)(\$4000) = \$1320$. For the fourth year, the payment is $3000 + $(0.08)(\$4000 - \$1000)(3) = \$3720$.

3.3 SINGLE SUMS OF MONEY

To illustrate the mathematical operations involved in modeling cash flow profiles using compound interest, first consider the investment of a single sum of money, P, in a savings account for n interest periods. Let the interest rate per interest period be denoted by i, and let the accumulated total in the fund n periods in the future be denoted by F. As shown in Table 3.2, assuming no monies are withdrawn during the interim, the amount in the fund after n periods equals $P(1 + i)^n$. As a convenience in computing values of F (the future worth) when given values of P (the present worth), the quantity $(1 + i)^n$ is tabulated in Appendix A for various values of i and n. The quantity $(1 + i)^n$ is referred to as the *single sum, future worth factor* and is denoted $(F|P\ i,n)$. The expression $(F|P\ i,n)$ is read as "the F, given P factor at $i\%$ for n periods." The above discussion is summarized as follows.

Let P = the equivalent value of an amount of money at time zero, or present worth
F = the equivalent value of an amount of money at time n, or future worth
i = the interest rate per interest period
n = the number of interest periods

Thus, the future worth is related to the present worth as follows:

$$F = P(1 + i)^n \qquad\qquad (3.1)$$

where i is expressed as a decimal amount. Equivalently,

$$F = P(F|P\ i,n) \qquad\qquad (3.2)$$

where i is expressed as a percentage amount. A cash flow diagram depicting the relationship between F and P is given in Figure 3.3. Remember that F occurs n periods after P.

Example 3.6

An individual borrows $1000 at 12% compounded annually. The loan is paid back after 5 years. How much should be repaid?

Using the compound interest tables in Appendix A for 12% and five periods, the value of the $(F|P\ 12,5)$ factor is found to be 1.7623. Thus,

$$
\begin{aligned}
F &= P(F|P\ 12,5) \\
&= \$1000(1 + 0.12)^5 \\
&= \$1000(1.7623) \\
&= \$1762.30
\end{aligned}
$$

The amount to be repaid equals $1762.30.

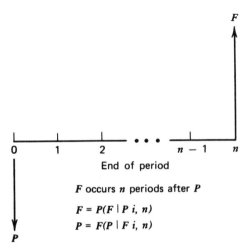

F occurs n periods after P

$F = P(F \mid P\ i, n)$

$P = F(P \mid F\ i, n)$

FIGURE 3.3. Cash flow diagram of the time relationship between P and F.

Since we are able to determine conveniently values of F when given values of P, i, and n, it is a simple matter to determine values of P when given values of F, i, and n. In particular, since

$$F = P(1 + i)^n$$

on dividing both sides by $(1 + i)^n$, we find that the present worth and future worth have the relation

$$P = F(1 + i)^{-n} \qquad (3.3)$$

or

$$P = F(P|F\ i,n) \qquad (3.4)$$

where $(1 + i)^{-n}$ and $(P|F\ i,n)$ are referred to as the *single sum, present worth factor*.

Example 3.7

To illustrate the computation of P given F, i, and n, suppose you wish to accumulate \$10,000 in a savings account 4 years from now and the account pays interest at a rate of 9% compounded annually. How much must be deposited today?

$$\begin{aligned} P &= F(P|F\ 9,4) \\ &= \$10{,}000(0.7084) \\ &= \$7084 \end{aligned}$$

3.4 SERIES OF CASH FLOWS

Having considered the transformation of a single sum of money to a future worth equivalent when given a present worth amount and vice versa, we generalize that discussion to consider the conversion of a series of cash flows to present worth and future worth equivalents. In particular, let A_t denote the magnitude of a cash flow (receipt or disbursement) at the end of time period t. Using discrete compounding, the present worth equivalent for the cash flow series is equal to the sum of the present worth equivalents for the individual cash flows. Consequently,

$$P = A_1(1 + i)^{-1} + A_2(1 + i)^{-2} + \cdots + A_{n-1}(1 + i)^{-(n-1)} + A_n(1 + i)^{-n} \quad (3.5)$$

or, using the summation notation,

$$P = \sum_{t=1}^{n} A_t(1 + i)^{-t} \qquad (3.6)$$

or, equivalently,

$$P = \sum_{t=1}^{n} A_t(P|F\ i,t) \qquad (3.7)$$

Example 3.8

Consider the series of cash flows depicted by the cash flow diagram given in Figure 3.4. Using an interest rate of 6%/interest period, the present worth equivalent is given by

$$\begin{aligned}
P &= \$300(P|F\ 6,1) - \$300(P|F\ 6,3) + \$200(P|F\ 6,4) \\
&\quad + \$400(P|F\ 6,6) + \$200(P|F\ 6,8) \\
&= \$300(0.9434) - \$300(0.8396) + \$200(0.7921) \\
&\quad + \$400(0.7050) + \$200(0.6274) \\
&= \$597.04
\end{aligned}$$

The future worth equivalent is equal to the sum of the future worth equivalents for the individual cash flows. Thus,

$$F = A_1(1 + i)^{n-1} + A_2(1 + i)^{n-2} + \cdots + A_{n-1}(1 + i) + A_n \qquad (3.8)$$

or, using the summation notation,

$$F = \sum_{t=1}^{n} A_t(1 + i)^{n-t} \qquad (3.9)$$

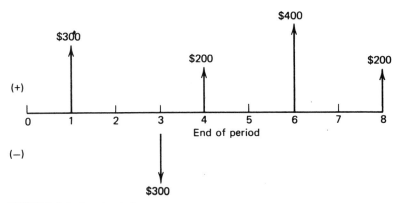

FIGURE 3.4. Series of cash flows.

$$F = \sum_{t=1}^{n} A_t(F|P\ i, n - t) \qquad (3.10)$$

Alternately, since we know the value of future worth is given by

$$F = P(1 + i)^n \qquad (3.11)$$

substituting Equation 3.6 into Equation 3.11 yields

$$F = (1 + i)^n \sum_{t=1}^{n} A_t(1 + i)^{-t}$$

Hence,

$$F = \sum_{t=1}^{n} A_t(1 + i)^{n-t} \qquad (3.12)$$

Example 3.9

Given the series of cash flows in Figure 3.4, determine the future worth at the end of the eighth period using an interest rate of 6%/interest period.

$$
\begin{aligned}
F &= \$300(F|P\ 6,7) - \$300(F|P\ 6,5) + \$200(F|P\ 6,4) \\
&\quad + \$400(F|P\ 6,2) + \$200 \\
&= \$300(1.5036) - \$300(1.3382) + \$200(1.2625) \\
&\quad + \$400(1.1236) + \$200 \\
&= \$951.56
\end{aligned}
$$

Alternately, since we know the present worth is equal to $597.04,

$$
\begin{aligned}
F &= P(F|P\ 6,8) \\
&= \$597.04(1.5938) \\
&= \$951.56
\end{aligned}
$$

Note that in computing the future worth of a series of cash flows, *the future worth amount obtained occurs at the end of time period n.* Thus, if a cash flow occurs at the end of time period n, it earns no interest in computing the future worth amount at the end of the nth period.

Obtaining the present worth and future worth equivalents of cash flow series by summing the individual present worths and future worths respectively, can be quite time-consuming if many cash flows are included in the series. However, with the development and widespread use of pocket calculators (some of which have the capability of performing compound interest calculations automatically), microcomputers, minicomputers, and time-sharing systems, it is not uncommon to treat all series in the manner described above.

When such hardware is not available, it is possible to use more efficient solution procedures if the cash flow series have one of the following forms.

Uniform Series of Cash Flows

$$A_t = A \qquad\qquad t = 1, \ldots, n$$

Gradient Series of Cash Flows

$$A_t = \begin{cases} 0 & t = 1 \\ A_{t-1} + G & t = 2, \ldots, n \end{cases}$$

Geometric Series of Cash Flows

$$A_t = \begin{cases} A & t = 1 \\ A_{t-1}(1 + j) & t = 2, \ldots, n \end{cases}$$

3.4.1   Uniform Series of Cash Flows

A uniform series of cash flows exists when all of the cash flows in a series are equal. In the case of a uniform series the present worth equivalent is given by

$$P = \sum_{t=1}^{n} A(1 + i)^{-t} \tag{3.13}$$

where A is the magnitude of an individual cash flow in the series.

Letting $X = (1 + i)^{-1}$ and bringing A outside the summation yields

$$P = A \sum_{t=1}^{n} X^t$$

$$= AX \sum_{t=1}^{n} X^{t-1}$$

Letting $h = t - 1$ gives the geometric series

$$P = AX \sum_{h=0}^{n-1} X^h \tag{3.14}$$

Since the summation in Equation 3.14 represents the first n terms of a geometric series, the closed form value for the summation is given by

$$\sum_{h=0}^{n-1} X^h = \frac{1 - X^n}{1 - X} \tag{3.15}$$

Hence, on substituting Equation 3.15 into Equation 3.14, we obtain

$$P = AX \left(\frac{1 - X^n}{1 - X} \right)$$

Replacing X with $(1 + i)^{-1}$ yields the following relationship between P and A.

$$P = A\left[\frac{(1 + i)^n - 1}{i(1 + i)^n}\right]$$
(3.16)

more commonly expressed as

$$P = A(P|A\ i,n)$$
(3.17)

where $(P|A\ i,n)$ is referred to as the *uniform series, present worth factor* and is tabulated in Appendix A for various values of i and n.

Example 3.10

An individual wishes to deposit a single sum of money in a savings account so that five equal annual withdrawals of $2000 can be made before depleting the fund. If the first withdrawal is to occur 1 year after the deposit and the fund pays interest at a rate of 12% compounded annually, how much should be deposited?

Because of the relationship of P and A, as depicted in Figure 3.5, in which P occurs one period before the first A, we see that

$$P = A(P|A\ 12,5)$$
$$= \$2000(3.6048)$$
$$= \$7209.60$$

Thus, if $7209.60 is deposited in a fund paying 12% compounded annually, then five equal annual withdrawals of $2000 can be made.

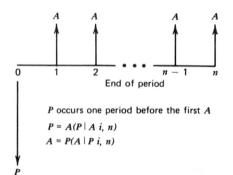

P occurs one period before the first A

$P = A(P|A\ i, n)$

$A = P(A|P\ i, n)$

FIGURE 3.5. Cash flow diagram of the relationship between P and A.

Example 3.11

In Example 3.10, suppose that the first withdrawal will not occur until 3 years after the deposit.

As depicted in Figure 3.6, the value of P to be determined occurs at $t = 0$, whereas a straightforward application of the $(P|A\ 12,5)$ factor will yield a single sum equivalent at $t = 2$. Consequently, the value obtained at $t = 2$ must be moved backward in time to $t = 0$. The latter operation is easily performed using the $(P|F\ 12,2)$ factor. Therefore,

$$
\begin{aligned}
P &= A(P|A\ 12,5)(P|F\ 12,2) \\
&= \$2000(3.6048)(0.7972) \\
&= \$5747.49
\end{aligned}
$$

Deferring the first withdrawal for 2 years reduces the amount of the deposit by $\$7209.60 - \$5747.49 = \$1462.11$.

The reciprocal relationship between P and A can be expressed as

$$
A = P\left[\frac{i(1 + i)^n}{(1 + i)^n - 1}\right]
\tag{3.18}
$$

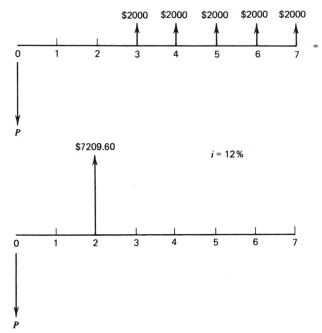

FIGURE 3.6. Equivalent cash flow diagrams.

or as

$$A = P(A|P\ i,n)$$

(3.19)

The expression $(A|P\ i,n)$ is called the *capital recovery factor* for reasons that will become clear in Chapter Four. The $(A|P\ i,n)$ factor is used frequently in both personal financing and in comparing economic investment alternatives.

Example 3.12

Suppose $10,000 is deposited into an account that pays interest at a rate of 15% compounded annually. If 10 equal, annual withdrawals are made from the account, with the first withdrawal occurring 1 year after the deposit, how much can be withdrawn each year in order to deplete the fund with the last withdrawal?

Since we know that A and P are related by

$$A = P(A|P\ i,n)$$

then

$$
\begin{aligned}
A &= P(A|P\ 15,10) \\
  &= \$10,000(0.1993) \\
  &= \$1993
\end{aligned}
$$

Example 3.13

Suppose that in Example 3.12 the first withdrawal is delayed for 2 years, as depicted in Figure 3.7. How much can be withdrawn each of the 10 years?

The amount in the fund at $t = 2$ equals

$$
\begin{aligned}
V_2 &= P(F|P\ 15,2) \\
    &= \$10,000(1.3225) \\
    &= \$13,225
\end{aligned}
$$

Therefore, the size of the equal annual withdrawals will be

$$
\begin{aligned}
A &= V_2(A|P\ 15,10) \\
  &= \$13,225(0.1993) \\
  &= \$2635.74
\end{aligned}
$$

Thus, delaying the first withdrawal for 2 years increases the size of each withdrawal by $642.74.

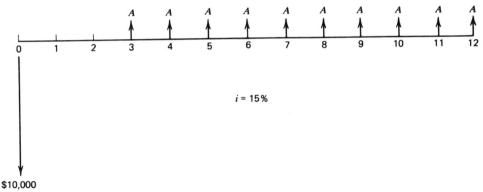

FIGURE 3.7. Cash flow diagram of deferred payment example.

The future worth of a uniform series is obtained by recalling that

$$F = P(1 + i)^n \tag{3.20}$$

Substituting Equation 3.16 into Equation 3.20 for P and reducing yields

$$F = A \left[\frac{(1 + i)^n - 1}{i} \right] \tag{3.21}$$

or, equivalently,

$$F = A(F|A\ i,n) \tag{3.22}$$

where $(F|A\ i,n)$ is referred to as the *uniform series, future worth factor*.

Example 3.14

If annual deposits of $1000 are made into a savings account for 30 years, how much will be in the fund immediately after the last deposit if the fund pays interest at a rate of 8% compounded annually?

$$F = A(F|A\ 8,30)$$
$$= \$1000(113.2831)$$
$$= \$113,283.10$$

The reciprocal relationship between A and F is easily obtained from Equation 3.21. Specifically, we find that

$$A = F\left[\frac{i}{(1 + i)^n - 1}\right] \tag{3.23}$$

or, equivalently,

$$A = F(A|F\ i,n) \tag{3.24}$$

The expression $(A|F\ i,n)$ is referred to as the *sinking fund factor,* since the factor is used to determine the size of a deposit one should place (sink) in a fund in order to accumulate a desired future amount. As depicted in Figure 3.8, *F occurs at the same time as the last A.* Thus, the last A or deposit earns no interest.

Example 3.15

If $150,000 is to be accumulated in 35 years, how much must be deposited annually in a fund paying 8% compounded annually in order to accumulate the desired amount immediately after the last deposit?

$$A = F(A|F\ 8,35)$$
$$= \$150,000(0.0058)$$
$$= \$870$$

3.4.2 Gradient Series of Cash Flows

A gradient series of cash flows occurs when the value of a given cash flow is greater than the value of the previous cash flow by a constant amount, G.

F occurs at the same time as the last A

$F = A(F\,|\,A\ i,\ n)$

$A = F(A\,|\,F\ i,\ n)$

End of period

FIGURE 3.8. Cash flow diagram of the relationship between A and F.

Consider the series of cash flows depicted in Figure 3.9. The series can be represented by the sum of a uniform series and a gradient series. By convention, the gradient series is defined to have the first positive cash flow occur at the end of the second time period. The size of the cash flow in the gradient series occurring at the end of period t is given by

$$A_t = (t - 1)G \quad t = 1, \ldots, n \qquad (3.25)$$

The gradient series arises when the value of an individual cash flow differs by a constant, G, from the preceding cash flow. As an illustration, if an individual receives an annual bonus and the size of the bonus increases by \$100 each year, then the series is a gradient series. Also, operating and maintenance costs tend to increase over time because of both inflation effects and a gradual deterioration of equipment; such costs are often approximated by a gradient series.

The present worth equivalent of a gradient series is obtained by recalling

$$P = \sum_{t=1}^{n} A_t(1 + i)^{-t} \qquad (3.26)$$

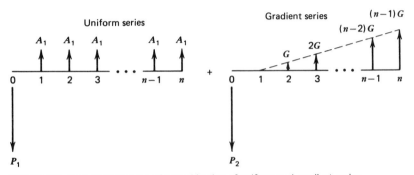

FIGURE 3.9. Cash flow diagram of a combination of uniform and gradient series.

Substituting Equation 3.25 into Equation 3.26 gives

$$P = \sum_{t=1}^{n} (t - 1)G(1 + i)^{-t} \qquad (3.27)$$

or, equivalently,

$$P = G \sum_{t=1}^{n} (t - 1)(1 + i)^{-t} \qquad (3.28)$$

As an exercise you may wish to show that the summation reduces to

$$P = G \left[\frac{1 - (1 + ni)(1 + i)^{-n}}{i^2} \right] \qquad (3.29)$$

Expressing the term in brackets in terms of interest factors already treated yields

$$P = G \left[\frac{(P|A\ i,n) - n(P|F\ i,n)}{i} \right] \qquad (3.30)$$

or, equivalently,

$$P = G(P|G\ i,n) \qquad (3.31)$$

where $(P|G\ i,n)$ is the *gradient series, present worth factor* and is tabulated in Appendix A.

A uniform series equivalent to the gradient series is obtained by multiplying the value of the gradient series present worth factor by the value of the $(A|P\ i,n)$ factor to obtain

$$A = G \left[\frac{1}{i} - \frac{n}{i} (A|F\ i,n) \right]$$

or, equivalently,

$$A = G(A|G\ i,n) \qquad (3.32)$$

where the factor $(A|G\ i,n)$ is referred to as the *gradient-to-uniform series conversion factor* and is tabulated in Appendix A. To obtain the future worth equivalent of a gradient series at time n, multiply the value of the $(A|G\ i,n)$ factor by the value of the $(F|A\ i,n)$ factor.

It is not uncommon to encounter a cash flow series that is the sum or difference of a uniform series and a gradient series. To determine present worth and future worth equivalents of such a composite, one can deal with each special type of series separately.

Example 3.16

Maintenance costs for a particular production machine increase by $1000/year over the 5-year life of the equipment. The initial maintenance cost is $3000. Using an interest rate of 8% compounded annually, determine the present worth equivalent for the maintenance costs.

A cash flow diagram for this example is given in Figure 3.10. Note that the cash flow series consists of the sum of a uniform series of $3000 and a gradient series with G equal to $1000. Converting the uniform series to a present worth amount gives

$$P_1 = \$3000(P|A\ 8,5)$$
$$= \$3000(3.9927)$$
$$= \$11{,}978.10$$

Converting the gradient series to a present worth amount gives

$$P_2 = G(P|G\ 8,5)$$
$$= \$1000(7.3724)$$
$$= \$7372.40$$

(*Notice that n equals 5 even though only four positive cash flows are present in the gradient series.*)

Hence, the present worth equivalent of the maintenance costs will be

$$P = P_1 + P_2$$
$$= \$11{,}978.10 + \$7372.40$$
$$= \$19{,}350.50$$

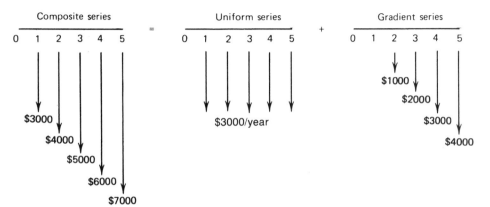

FIGURE 3.10. Cash flow diagram for a gradient series example.

The uniform series equivalent of the cash flow series given in Figure 3.10 can be determined by summing the base series of $3000 and the uniform series equivalent of the gradient series. Hence,

$$A = \$3000 + G(A|G\ 8,5)$$
$$= \$3000 + \$1000(1.8465)$$
$$= \$4846.50$$

Converting the maintenance costs to a future worth amount can be performed either by converting the present worth amount to a future worth amount,

$$F = \$19,350.50(F|P\ 8,5)$$
$$= \$19,350.50(1.4693)$$
$$= \$28,431.69,$$

or by converting the uniform series amount to a future worth amount,

$$F = \$4846.50(F|A\ 8,5)$$
$$= \$4846.50(5.8666)$$
$$= \$28,432.48,$$

with the $0.79 difference due to round-off error in the interest tables.

Example 3.17

Five annual deposits are made into a fund that pays interest at a rate of 8% compounded annually. The first deposit equals $800; the second deposit equals $700; the third deposit equals $600; the fourth deposit equals $500; and the fifth deposit equals $400. Determine the amount in the fund immediately after the fifth deposit.

As depicted by the cash flow diagrams in Figure 3.11, the cash flow series can be represented by the difference in a uniform series of $800 and a gradient series of $100. The uniform series equivalent of the gradient series is given by

$$A = G(A|G\ 8,5)$$
$$= \$100(1.8465)$$
$$= \$184.65$$

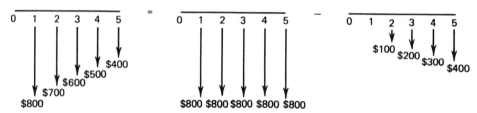

FIGURE 3.11. Cash flow diagrams for the decreasing gradient series example.

Therefore, a uniform series having cash flows equal to $800 − $184.65, or $615.35, is equivalent to the original cash flow series. The future worth equivalent is found to be

$$F = A(F|A\ 8,5)$$
$$= \$615.35(5.8666)$$
$$= \$3610.01$$

3.4.3 Geometric Series of Cash Flows

The geometric cash flow series, as depicted in Figure 3.12, occurs when the size of a cash flow increases (decreases) by a fixed percent from one time period to the next. If j denotes the percent change in the size of a cash flow from one period to the next, the size of the tth flow can be given by

$$A_t = A_{t-1}(1 + j) \qquad t = 2, \ldots, n$$

or, more conveniently,

$$A_t = A_1(1 + j)^{t-1} \qquad t = 1, \ldots, n \qquad (3.33)$$

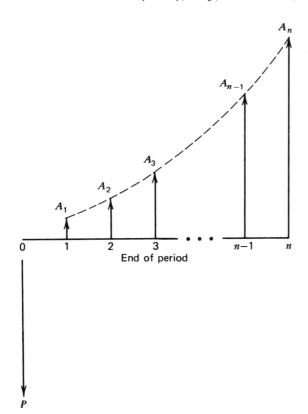

FIGURE 3.12. Cash flow diagram of the geometric series.

The geometric series is used to represent the growth (positive j) or decay (negative j) of costs and revenues undergoing annual percentage changes. As an illustration, if labor costs increase by 10% a year, then the resulting series representation of labor costs will be a geometric series.

The present worth equivalent of the cash flow series is obtained by substituting Equation 3.33 into Equation 3.6 to obtain

$$P = \sum_{t=1}^{n} A_1(1 + j)^{t-1}(1 + i)^{-t} \tag{3.34}$$

or

$$P = A_1(1 + j)^{-1} \sum_{t=1}^{n} \left(\frac{1 + j}{1 + i}\right)^{t} \tag{3.35}$$

As an exercise, the reader may wish to show that the following relationship results:

$$P = \begin{cases} A_1 \left[\dfrac{1 - (1 + j)^n(1 + i)^{-n}}{i - j} \right] & i \neq j \\[4mm] \dfrac{nA_1}{1 + i} & i = j \end{cases} \tag{3.36}$$

or

$$P = A_1(P|A_1\ i,j,n) \tag{3.37}$$

where $(P|A_1\ i,j,n)$ is the *geometric series, present worth factor* and is tabulated in Appendix A for various values of i, j, and n.

For the case of $j \geq 0$ and $i \neq j$, the relationship between P and A can be conveniently expressed in terms of compound interest factors previously considered

$$P = A_1 \left[\frac{1 - (F|P\ j,n)(P|F\ i,n)}{i - j} \right] \qquad i \neq j, \qquad j \geq 0 \tag{3.38}$$

Example 3.18

A professional sports team has negotiated a 5-year contract with one of its "super stars." The contract calls for a salary of $1,125,000 the first year, followed by 10% annual increases. The salary will be paid out over a 20-year period, with the amount paid each year being 10% greater than the previous year. The general manager of the team and the athlete's agent agree that the present worth of the contract will be equated to the present worth of the payments using an annual compound rate of 10%.

a. The general manager wishes to compute the present worth equivalent cost to the team. They have been able to invest money and earn a 15% annual return.

b. The athlete wants to know how much will be paid each of the first 5 years, how much will be paid the twentieth year, and the present worth equivalent of the contract if "money is worth" (i) 5%, (ii) 8%, and (iii) 10% over the 20-year period.

For this example, the present worth equivalent of the contract can be determined for the general manager as follows. First, $A_1 = \$1,125,000$, $j = 10\%$, $i = 15\%$, and $n = 5$. The present worth equivalent will be

$$P = A_1(P|A_1\ 15,10,5)$$
$$= \$1,125,000(8.0169)$$
$$= \$9,019,012.50$$

The general manager and the agent agreed that the amount to be paid the athlete the first year can be determined by setting the present worth equivalent of the payments equal to the present worth equivalent of the contract at 10% interest. Hence,

$$A_1(P|A_1\ 10,10,20) = \$1,125,000(P|A_1\ 10,10,5)$$
$$A_1(18.1818) = \$1,250,000(4.5455)$$
$$A_1 = \$312,503.44$$

The payment to be made in year k can be determined as follows,

$$A_k = \$312,503.44(F|P\ 10,k-1)$$

Hence,

$$A_2 = \$312,503.44(F|P\ 10,1)$$
$$= \$343,753.78$$

$$A_3 = \$312,503.44(F|P\ 10,2)$$
$$= \$378,129.16$$

$$A_4 = \$312,503.44(F|P\ 10,3)$$
$$= \$415,942.08$$

$$A_5 = \$312,503.44(F|P\ 10,4)$$
$$= \$457,536.29$$

and the final payment will be

$$A_{20} = \$312,503.44(F|P\ 10,19)$$
$$= \$1,911,239.79$$

In order to determine the present worth equivalent of the contract for the athlete, note that the size of the annual payments increases at an annual rate of $j = 10$. Hence, the present worth equivalent of the contract, given an interest rate of $i\%$, is

$$P = \$312{,}503.44(P|A_1\ i\%,10,20)$$

For $i = 5\%$,

$$P = \$312{,}503.44(30.7105)$$
$$= \$9{,}597{,}137$$

For $i = 8\%$,

$$P = \$312{,}503.44(22.1686)$$
$$= \$6{,}927{,}764$$

For $i = 10\%$,

$$P = \$312{,}503.44(18.1818)$$
$$= \$5{,}681{,}875$$

Since the future payment amounts are guaranteed, the present worth of the contract from the athlete's perspective will increase with a decrease in the time value of money.

The future worth equivalent of the geometric series is obtained by multiplying the value of the geometric series present worth factor and the $(F|P\ i,n)$ factor to obtain

$$F = \begin{cases} A_1\left[\dfrac{(1+i)^n - (1+j)^n}{i-j} \right] & i \neq j \\[2ex] nA_1(1+i)^{n-1} & i = j \end{cases} \tag{3.39}$$

or

$$F = A_1(F|A_1\ i,j,n) \tag{3.40}$$

where $(F|A_1\ i,n)$ is the *geometric series, future worth factor* and is tabulated in Appendix A. From Equation 3.39, notice that $(F|A_1\ i, j, n) = (F|A_1\ j,i,n)$.

Example 3.19

An individual receives an annual bonus and deposits it in a savings account that pays 8% compounded annually. The size of the bonus increases by 10%

each year; the initial deposit was \$500. Determine how much will be in the fund immediately after the tenth deposit.

In this case, $A_1 = \$500$, $i = 8\%$, $j = 10\%$, and $n = 10$. Thus, the value of F is given by

$$F = A_1(F|A_1\ 8,10,10) = A_1(F|A_1\ 10,8,10)$$
$$= \$500(21.7409)$$
$$= \$10,870.45$$

The interest factors developed to this point are summarized in Table 3.3. Values of the factors given in Table 3.3 are provided in Appendix A.

3.5 MULTIPLE COMPOUNDING PERIODS IN A YEAR

Not all interest rates are stipulated as annual compounding rates. For example, if the interest rate is stipulated as 12% compounded quarterly, then the interest period is a 3-month period with 3% interest rate/interest period.

TABLE 3.3

Summary of Discrete Compounding Interest Factors

To Find	Given	Factor	Symbol
P	F	$(1 + i)^{-n}$	$(P|F\ i,n)$
F	P	$(1 + i)^n$	$(F|P\ i,n)$
P	A	$\dfrac{(1 + i)^n - 1}{i(1 + i)^n}$	$(P|A\ i,n)$
A	P	$\dfrac{i(1 + i)^n}{(1 + i)^n - 1}$	$(A|P\ i,n)$
F	A	$\dfrac{(1 + i)^n - 1}{i}$	$(F|A\ i,n)$
A	F	$\dfrac{i}{(1 + i)^n - 1}$	$(A|F\ i,n)$
P	G	$\dfrac{1 - (1 + ni)(1 + i)^{-n}}{i^2}$	$(P|G\ i,n)$
A	G	$\dfrac{(1 + i)^n - (1 + ni)}{i[(1 + i)^n - 1]}$	$(A|G\ i,n)$
P	$A_{1,j}$	$\dfrac{1 - (1 + j)^n(1 + i)^{-n}}{i - j}$	$(P|A_1\ i,j,n)^*$
F	$A_{i,j}$	$\dfrac{(1 + i)^n - (1 + j)^n}{i - j}$	$(F|A_1\ i,j,n)^*$

$* i \neq j.$

Thus, if $1000 is borrowed at an interest rate of 12% compounded quarterly, then the amount owed at the end of 5 years, or 20 interest periods, is obtained as follows:

$$
\begin{aligned}
F &= P(F|P\ 3,20) \\
&= \$1000(1 + 0.03)^{20} \\
&= \$1000(1.8061) \\
&= \$1806.10
\end{aligned}
$$

With quarterly compounding, $1806.10 is to be repaid.

Example 3.20

As another illustration of multiple compounding periods within a year, suppose interest is stated as 8% compounded quarterly. If you borrow $1000 for 1 year, how much must be repaid?

Using an interest rate of 2%/3-month period, after four interest periods the amount owed is given by:

$$
\begin{aligned}
F &= P(F|P\ 2,4) \\
&= \$1000(1.0824) \\
&= \$1082.40
\end{aligned}
$$

Thus, $1082.40 would be owed. Note that if interest were stated as 8.24% compounded annually,

$$
\begin{aligned}
F &= P(F|P\ 8.24,1) \\
&= \$1000(1.0824) \\
&= \$1082.40
\end{aligned}
$$

and the same amount would be owed.

From the results of Example 3.20 it can be concluded that 8% compounded quarterly is equivalent to 8.24% compounded annually. The rate of 8% is referred to as the *nominal annual interest rate;* the rate of 8.24% is referred to as the *effective annual interest rate*. The effective interest rate is defined as the annual compounding rate that is equivalent to the stated interest rate.

Letting r denote the nominal annual interest rate, m denote the number of compounding periods per year, i denote the interest rate per interest period, and i_{eff} denote the effective annual interest rate,

$$
\begin{aligned}
i_{eff} &= \left(1 + \frac{r}{m}\right)^{m} - 1 \\
&= (1 + i)^{m} - 1
\end{aligned}
$$

$$i_{\text{eff}} = \left(F|P \frac{r}{m}, m \right) - 1 \qquad\qquad (3.41)$$

To illustrate with 8% compounded semiannually, r equals 0.08, m equals 2, and

$$i_{\text{eff}} = (1 + 0.04)^2 - 1$$
$$= (F|P\ 4,2) - 1$$
$$= 0.0816$$

Thus, 8% compounded semiannually is equivalent to 8.16% compounded annually (i.e., 8.16% is the effective annual interest rate when interest is 8% compounded semiannually).

Suppose interest is stated as 18% compounded monthly. The effective annual interest rate is given by

$$i_{\text{eff}} = (1 + 0.015)^{12} - 1$$
$$= (F|P\ 1.5,12) - 1$$
$$= 0.1956$$

Thus, 18% compounded monthly has an effective annual interest rate of 19.56%.

Example 3.21

An individual borrowed $1000 and paid off the loan with interest after 4.5 years. The amount paid was $1500. What was the effective annual interest rate for this transaction? Letting the interest period be a 6-month period, it is seen that the payment of $1500 and the debt of $1000 are related by the expression

$$F = P(F|P\ i,n)$$

Thus,

$$\$1500 = \$1000(F|P\ i,9)$$

or

$$\$1500 = \$1000(1 + i)^9$$

Dividing both sides by $1000 gives

$$1.50 = (1 + i)^9$$

Taking the logarithm of both sides yields

$$\log 1.50 = 9 \log (1 + i)$$

Dividing both sides by 9 and taking the antilog of the result provides the relation

$$(1 + i) = 1.046$$
$$i = 0.046$$

Thus, the 6-month interest rate is approximately 4.6%. Computing the effective annual interest rate yields

$$i_{\text{eff}} = (1 + i)^2 - 1$$
$$= (1 + 0.046)^2 - 1$$
$$= 0.0943$$

The effective annual interest rate for the loan transaction was approximately 9.43%.

Instead of using logarithms, we could have searched the interest tables for a value of i that yielded a value of 1.50 for the $(F|P\ i,9)$ factor. With interpolation, i would be found to equal approximately 4.6%. The computation of the effective annual interest rate would follow the same procedure used to determine the effective annual interest rate for 9.2% compounded semiannually.

In the United States, the federal government requires that anyone borrowing money be informed of the total amount of interest paid and the *annual percentage rate* associated with the transaction. Depending on how the lender computes the annual percentage rate, the percentage quoted can be either the nominal annual interest rate, the effective annual interest rate, or something else entirely. To illustrate, many revolving charge accounts charge an interest rate of 1.5%/month, which is equivalent to 18% compounded monthly or 19.56% compounded annually. In many cases, the annual percentage rate is stated to be 18%.

Since, in personal financing, the annual percentage rate might be computed differently among the alternative sources of funds, the effective annual interest rate is a very good basis for comparing alternative financing plans. The effective annual interest rate is also useful when the timing of cash flows does not coincide with the end of interest periods.

A number of situations involve cash flows occurring at intervals that are not the same as the compounding intervals. Consider the following two possibilities: money is compounded *more* frequently than the occurrence of cash flows and money is compounded *less* frequently than the occurrence of cash flows.

Example 3.22

Suppose an individual makes monthly deposits into a fund that pays interest at a rate of 8% compounded quarterly. Depending on how the financial institution interprets the rate of "8% compounded quarterly," money deposited during a quarter might not earn any interest. We interpret 8% compounded quarterly to mean that deposits earn interest equivalent to 8% compounded quarterly. Hence, the monthly interest rate to be applied to the monthly deposits should be equivalent to 8% compounded quarterly. Letting i be the monthly rate, we note that the effective interest rate for $i\%$/month should be the same as for 8% compounded quarterly. Therefore

$$i_{\text{eff}} = (1 + i)^{12} - 1$$
$$= \left(1 + \frac{0.08}{4}\right)^4 - 1$$

or

$$(1 + i)^{12} = (1.02)^4$$

or

$$(1 + i)^3 = 1.02$$

Thus,

$$1 + i = (1.02)^{1/3}$$
$$i = (1.02)^{1/3} - 1$$
$$i = 0.0066227$$

Consequently, a monthly interest rate of 0.66227% can be used to determine the balance in the fund at the end of any month.

Example 3.23

Suppose quarterly deposits are made into a fund that pays interest at a rate of 6% compounded monthly. By the same arguments used in the previous example, we know that

$$i_{\text{eff}} = (1 + i)^4 - 1$$
$$= \left(1 + \frac{0.06}{12}\right)^{12} - 1$$

or

$$(1 + i)^4 = (1.005)^{12}$$
$$i = (1.005)^3 - 1$$
$$= 0.015075$$

Thus, a quarterly interest rate of 1.5075% can be used to determine the balance in the fund at the end of any quarter.

3.6 CONTINUOUS COMPOUNDING

In the discussion of the effective interest rate in the previous section, it was noted that as the frequency of compounding in a year increases, the effective interest rate increases. Since monetary transactions occur daily or hourly in most businesses, and money is normally "put to work" for the business as soon as it is received, compounding is occurring quite frequently. If one wishes to account explicitly for such rapid compounding, then *continuous compounding relations should be used.* Continuous compounding means that each year is divided into an infinite number of interest periods. Mathematically, the single payment compound amount factor under continuous compounding is given by

$$\lim_{m \to \infty} \left(1 + \frac{r}{m} \right)^{mn} = e^{rn}$$

where n is the number of years, m is the number of interest periods per year, and r is the nominal annual interest rate. Given P, r, and n, the value of F can be computed using continuous compounding as follows.

$$F = P\, e^{rn}$$

(3.42)

or

$$F = P(F|P\ r,n)_\infty$$

(3.43)

where $(F|P\ r,n)_\infty$ denotes the *continuous compounding, single sum, future worth factor.* The subscript ∞ is provided to denote that continuous compounding is being used. The interest tables for continuous compounding are given in Appendix B.

Example 3.24

If $2000 is invested in a fund that pays interest at a rate of 12% compounded continuously, after 5 years the cumulative amount in the fund will total

$$F = P(F|P\ 12\%,5)_\infty$$
$$= \$2000(1.8221)$$
$$= \$3644.20$$

Thus, a withdrawal of $3644.20 will deplete the fund after 5 years.

The effective interest rate under continuous compounding is easily obtained using the relation

$$i_{\text{eff}} = e^r - 1$$

(3.44)

or

$$i_{\text{eff}} = (F|P\ r,1)_\infty - 1$$

(3.45)

To illustrate, if interest is 12% compounded continuously, then the effective interest rate is given by

$$i_{\text{eff}} = (F|P\ 12,1)_\infty - 1$$
$$= 0.1275$$

Thus, 12.75% compounded annually is equivalent to 12% compounded continuously.

The inverse relationship between F and P indicates that

$$P = F\ e^{-rn}$$

(3.46)

or

$$P = F(P|F\ r,n)_\infty$$

(3.47)

where $(P|F\ r,n)_\infty$ is called the *continuous compounding, single sum, present worth factor*.

3.6.1 Discrete Flows

If it is assumed that cash flows are discretely spaced over time, then the continuous compound relations for the uniform, gradient, and geometric series can be obtained. Substituting e^{-rn} for $(1 + i)^{-n}$, $e^r - 1$ for i, and e^{rn} for $(1 + i)^n$ in the remaining discrete compounding formulas yields the continuous compound interest factors summarized in Table 3.4. Values for these factors are provided in Appendix B.

Example 3.25

To illustrate the use of the continuous compound interest factors, suppose $1000 is deposited each year into an account that pays interest at a rate of 12% compounded continuously. Determine both the amount in the account immediately after the tenth deposit and the present worth equivalent for 10 deposits.

TABLE 3.4

Summary of Continuous Compounding Interest Factors for Discrete Flows

To Find	Given	Factor	Symbol
P	F	e^{-rn}	$(P\vert F\ r,n)_\infty$
F	P	e^{rn}	$(F\vert P\ r,n)_\infty$
F	A	$\dfrac{e^{rn}-1}{e^r-1}$	$(F\vert A\ r,n)_\infty$
A	F	$\dfrac{e^r-1}{e^{rn}-1}$	$(A\vert F\ r,n)_\infty$
P	A	$\dfrac{e^{rn}-1}{e^{rn}(e^r-1)}$	$(P\vert A\ r,n)_\infty$
A	P	$\dfrac{e^{rn}(e^r-1)}{e^{rn}-1}$	$(A\vert P\ r,n)_\infty$
P	G	$\dfrac{e^{rn}-1-n(e^r-1)}{e^{rn}(e^r-1)^2}$	$(P\vert G\ r,n)_\infty$
A	G	$\dfrac{1}{e^r-1}-\dfrac{n}{e^{rn}-1}$	$(A\vert G\ r,n)_\infty$
P	A_1,c	$\dfrac{1-e^{(c-r)n}}{e^r-e^c}$	$(P\vert A_1\ r,c,n)_\infty^*$
F	A_i,c	$\dfrac{e^{rn}-e^{cn}}{e^r-e^c}$	$(F\vert A_1\ r,c,n)_\infty^*$

* $r \neq c$.

The amount in the fund immediately after the tenth deposit is given by the relation

$$F = \$1000(F\vert A\ 12,10)_\infty$$
$$= \$1000(18.1974)$$
$$= \$18,197.40$$

The present worth equivalent for 10 deposits is obtained using the relation

$$P = \$1000(P\vert A\ 12,10)_\infty$$
$$= \$1000(5.4810)$$
$$= \$5481$$

In the case of the geometric series the size of the tth cash flow will be assumed to be given by

$$A_t = A_{t-1}e^c \qquad t = 2, \ldots, n \qquad (3.48)$$

or, equivalently,

$$A_t = A_1 e^{(t-1)c} \qquad t = 1, \ldots, n \qquad (3.49)$$

where c is the nominal compound rate of increase in the size of the cash flow. The resulting expressions for the *continuous compounding, geometric series present worth factor* and the *continuous compounding, geometric series future worth factor* are given, respectively, by $(P|A_1\ r,c,n)_\infty$ and $(F|A_1\ r,c,n)_\infty$, as given in Table 3.4 for the case of $r \neq c$. As an exercise you may wish to derive the appropriate expressions when $r = c$.

Example 3.26

An individual receives an annual bonus and deposits it in a savings account that pays 8% compounded continuously. The size of the bonus increases each year at a rate of 10% compounded continuously; the initial deposit was $500. Determine how much will be in the fund immediately after the tenth deposit.

In this case, $A_1 = \$500$, $r = 8\%$, $c = 10\%$, and $n = 10$. Thus, the value of F is given by

$$
\begin{aligned}
F &= A_1(F|A_1\ 8,10,10)_\infty \\
&= \$500(22.5162) \\
&= \$11,258.10
\end{aligned}
$$

Comparing the result with that obtained in Example 3.19, the effect of continuous compounding increases the amount in the fund by $11,258.10 − $10,870.45 = $387.65.

If discrete flows occur during a year, it is necessary to define r consistent with the spacing of cash flows. To illustrate, suppose semiannual deposits are made into an account paying 12% compounded continuously. In this case the nominal semiannual rate should be determined such that its effective interest rate is 12.75%, or $r = 6.18\%$.

Since the differences in discrete and continuous compounding are not great in most cases, it is not uncommon to see discrete compounding used when continuous compounding is more appropriate. The arguments given for this are that errors in estimating the cash flows will probably offset any attempts to be very precise by using continuous compounding and that the interest rate used in discrete compounding is actually the effective interest rate resulting from continuous compounding.

3.6.2 Continuous Flow

Thus far only discrete cash flows have been assumed. It was assumed that cash flows occurred at, say, the end of the year. In some cases money is expended throughout the year on a somewhat uniform basis. Costs for labor, carrying inventory, and operating and maintaining equipment are typical examples. Others include capital improvement projects that conserve energy, water, or process steam.

Consequently, as a mathematical convenience, instead of assuming that money flows in discrete increments at the end of monthly, weekly, daily, or hourly time periods, it is assumed that money flows continuously during the time period at a uniform rate. Instead of having a uniform series of discrete cash flows of magnitude A, it assumed that a *total* of $\bar{A}$ dollars flows uniformly and continuously throughout a given time period. Such an approach to modeling cash flows is referred to as the *continuous flow* approach.

To illustrate the continuous flow concept, suppose you are to divide $1000 into k equal amounts to be deposited at equally spaced points in time during a year. The interest rate per period is defined to be r/k, where r is the nominal rate. Thus, the present worth of the series of k equal amounts is

$$P = \frac{\$1000}{k} \left(P|A \, \frac{r}{k}, k \right)$$

or

$$P = \frac{\$1000}{k} \left[\frac{(1 + (r/k))^k - 1}{(r/k)(1 + (r/k))^k} \right]$$

which reduces to

$$P = \$1000 \left[\frac{1}{r} - \frac{1}{r(1 + (r/k))^k} \right]$$

Taking the limit of P as k approaches infinity gives

$$\lim_{k \to \infty} P = \$1000 \left(\frac{1}{r} - \frac{1}{re^r} \right)$$

or

$$P = \$1000 \left(\frac{e^r - 1}{re^r} \right)$$

In general, for n years,

$$P = \bar{A} \left(\frac{e^{rn} - 1}{re^{rn}} \right) \tag{3.50}$$

or

$$P = \bar{A}(P|\bar{A}\ r,n)$$

$$(3.51)$$

where $(P|\bar{A}\ r,n)$ is referred to as the *continuous flow, continuous compounding uniform series present worth factor* and is tabulated in Appendix B.

The remaining continuous flow, continuous compound interest factors are summarized in Table 3.5. Values of the factors are given in Appendix B for various values of r and n. With continuous flow and continuous compounding, the continuous annual cash flow, $\bar{A}$, is equivalent to $Ar/(e^r - 1)$, when discrete flow and continuous compounding is used. Hence, the discrete flow equivalent of $\bar{A}$ is given by

$$A = \frac{\bar{A}(e^r - 1)}{r}$$

or

$$A = \bar{A}(F|\bar{A}\ r,1)$$

TABLE 3.5
Summary of Continuous Compounding Interest Factors for Continuous Flows

To Find	Given	Factor	Symbol
P	$\bar{A}$	$\dfrac{e^{rn} - 1}{re^{rn}}$	$(P|\bar{A}\ r,n)$
$\bar{A}$	P	$\dfrac{re^{rn}}{e^{rn} - 1}$	$(\bar{A}|P\ r,n)$
F	$\bar{A}$	$\dfrac{e^{rn} - 1}{r}$	$(F|\bar{A}\ r,n)$
$\bar{A}$	F	$\dfrac{r}{e^{rn} - 1}$	$(\bar{A}|F\ r,n)$

Example 3.27

What are the present worth and future worth equivalents of a uniform series of continuous cash flows totalling $10,000/year for 10 years when interest is compounded continuously at a rate of 20%/year?

The present worth equivalent is given by

$$P = \$10,000(P|\bar{A}\ 20,10)$$
$$= \$10,000(4.3233)$$
$$= \$43,233$$

The future worth equivalent is given by

$$F = \$10,000(F|\bar{A}\ 20,10)$$
$$= \$10,000(31.9453)$$
$$= \$319,453$$

Continuous flow and continuous compounding are concepts that are used to represent more closely the realities of business transaction. In fact, several of the computer programs developed by industries and governmental agencies and used to perform economic analyses incorporate both concepts. We have presented both concepts since each has experienced increased usage during the past decade.

3.7 EQUIVALENCE

Throughout the preceding discussion we have used the term *equivalence* without defining what was meant by the term. This was done intentionally in order to introduce subtly the notion that two cash flow series or profiles are *equivalent* at some specified interest rate, $k\%$, if their present worths are *equal* using an interest rate of $k\%$.

Example 3.28

The two cash flow profiles shown in Figure 3.13 are equivalent at 15%, since each has a present worth of $424.80. You may wish to verify this claim.

Of course, if the present worths of two cash flow profiles are equal at $k\%$, their "worths" at any given point in time will be equal when using a discount (interest) rate of $k\%$. Likewise, if two cash flow profiles are equivalent at $k\%$, their respective uniform series equivalents will be equal when expressed over the same time period. Hence we could state that the two cash flow profiles shown in Figure 3.13 are equivalent at 15%, since each has a worth of $1299.46 at the end of period eight; alternately, we could assert their equivalence at 15% on the basis that each can be represented by a uniform series of $126.72/period over the interval [1, 8].

Example 3.29

What single sum of money at $t = 6$ is equivalent to the cash flow profile shown in Figure 3.14 if $i = 10\%$?

The present worth of the cash flow profile is given by

$$P = -\$400(P|F\ 10,1) + \$100(P|A\ 10,3)(P|F\ 10,1)$$
$$+\$100(P|A\ 10,3)(P|F\ 10,5)$$
$$= -\$400(0.9091) + \$100(2.4869)(0.9091) + \$100(2.4869)(0.6209)$$
$$= \$16.85$$

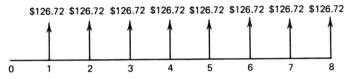

FIGURE 3.13. Cash flow diagrams for the equivalence example.

Moving $16.85 forward in time to $t = 6$ gives

$$F = \$16.85(F|P\ 10,6)$$
$$= \$16.85(1.7716)$$
$$= \$29.86$$

Thus, at 10% a positive cash flow of $29.86 at $t = 6$ is equivalent to the cash flow profile shown in Figure 3.14.

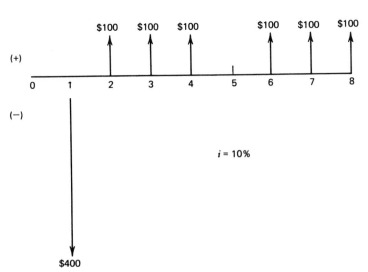

FIGURE 3.14. Cash flow diagram for the equivalence example.

Example 3.30

Using an 8% discount rate, what uniform series over five periods, [1, 5], is equivalent to the cash flow profile given in Figure 3.15?

The cash flow profile in Figure 3.15a consists of the difference in a uniform series of $500 and a gradient series, with $G = \$100$. A uniform series equivalent of the cash flow profile can be obtained for the interval [2, 6] as follows:

$$A = \$500 - \$100(A|G\ 8,5)$$
$$= \$500 - \$100(1.8465)$$
$$= \$315.35$$

The uniform series of $315.35 over the interval [2, 6] must be converted to a uniform series over the interval [1, 5]. Thus, *each* of the five cash flows must be moved back in time one period. The discounted value of $315.35 over one time period using an 8% discount rate is

$$P = \$315.35(P|F\ 8,1)$$
$$= \$315.35(0.9259)$$
$$= \$291.98$$

(a)

(b)

(c)

FIGURE 3.15. Cash flow diagrams for the equivalence example.

Consequently, a uniform series of $291.98 over the interval [1, 5] is equivalent to the cash flow profile given in Figure 3.15a. If you have doubts concerning the equivalence, compare their present worths using an 8% interest rate.

Example 3.31

Determine the value of X that makes the two cash flows, given in Figure 3.16, equivalent when a discount rate of 15% is used.

Equating the future worths of the two cash flow profiles at $t = 4$ gives

$$\$200(F|A\ 15,4) + \$100(F|A\ 15,3) + \$100$$
$$= [\$200 + X(A|G\ 15,4)](F|A\ 15,4)$$

Canceling 200(F|A\ 15,4)$ on both sides yields

$$\$100(3.4725) + \$100 = X(1.3263)(4.9934)$$

Solving for X gives a value of $67.53.

Example 3.32

For what interest (discount) rate are the two cash flow profiles shown in Figure 3.17 equivalent?

Converting each cash flow profile to a uniform series over the interval [1, 5], gives

$$-\$4000(A|P\ i,5) + \$1500 = -\$7000(A|P\ i,5) + \$1500 + \$500(A|G\ i,5)$$

or

$$\$3000(A|P\ i,5) = \$500(A|G\ i,5)$$

$i = 15\%$

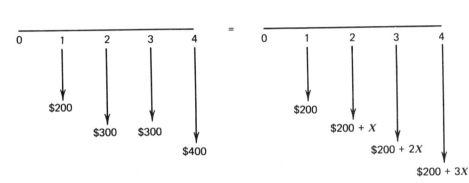

FIGURE 3.16. Cash flow diagrams for the equivalence example.

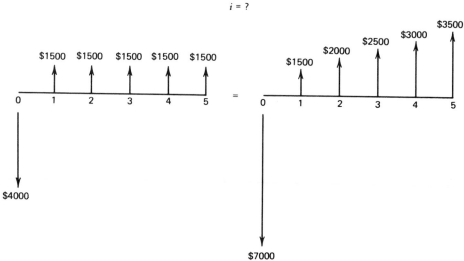

FIGURE 3.17. Cash flow diagrams for the equivalence example.

which reduces to

$$(A|G\ i,5) = 6(A|P\ i,5)$$

On searching through the interest tables at $n = 5$, it is found that the $(A|G\ i,5)$ factor is six times the value of the $(A|P\ i,5)$ factor for an interest rate between 12% and 15%. Specifically, with a 12% interest rate,

$$(A|G\ 12,5) - 6(A|P\ 12,5) = 1.7746 - 6(0.2774) = 0.1102$$

and, using a 15% interest rate,

$$(A|G\ 15,5) - 6(A|P\ 15,5) = 1.7228 - 6(0.2983) = -0.0670$$

Interpolating for i gives

$$i = 0.12 + \frac{(0.15 - 0.12)(0.1102)}{(0.1102 + 0.0670)}$$

or

$$i = 0.1386$$

Therefore, using a discount rate of approximately 13.86% will establish an equivalence relationship between the cash flow profiles given in Figure 3.17.

Example 3.33

A firm purchases a machine for $30,000, keeps it for 5 years, and sells it for $6000. During the time the machine was owned by the company, operating and maintenance costs totaled $8000 the first year, $9000 the second year,

$10,000 the third year, $11,000 the fourth year, and $12,000 the fifth year. The firm uses a 15% interest rate in performing economic analyses. Determine the single sum of money occurring at (1) $t = 0$ and (2) $t = 5$, which is equivalent to the cash flow history for the machine. Also, determine the uniform series occurring over the interval [1, 5] that is equivalent to the cash flow profile for the machine.

In the economic analysis literature, the single sum equivalent at time zero for a cash flow profile is called the *present worth* or *present value* for the cash flow profile. We will use the term present worth and denote it as PW. For the example,

$$PW = -\$30,000 - \$8000(P|A\ 15,5) - \$1000(P|G\ 15,5)$$
$$+\$6000(P|F\ 15,5)$$
$$= -\$59,609.57$$

Hence, a single expenditure of $59,609.57 at time zero would have been equivalent to the cash flows experienced during the ownership of the machine.

The single sum equivalent at the end of the life of a project is termed the *future worth* or *future value* for the project. We denote the future worth by FW and compute its value as follows:

$$FW = -\$30,000(F|P\ 15,5) - [\$8000 + 1000(A|G\ 15,5)](F|A\ 15,5)$$
$$+\$6000$$
$$= -\$119,897$$

Alternately, the future worth can be obtained from the present worth:

$$FW = PW(F|P\ i,n)$$
$$= -\$59,609.57(F|P\ 15,5)$$
$$= -\$119,898.69$$

(The difference of $1.69 is due to round-off error in the tables). Hence, a single expenditure of $119,897 at time 5 would have been equivalent to the cash flows associated with the machine.

A uniform series equivalent for a series of yearly cash flows is referred to as the *annual worth* or *equivalent uniform annual cost* for the project. The latter expression is most appropriate for the type of example under consideration, since the resulting uniform series is a cost, not an income, series. The annual worth designation is AW; $EUAC$ denotes the equivalent uniform annual cost. We will use both designations throughout the text.

For the example problem, the $EUAC$ determination is performed as follows:

$$EUAC = \$30,000(A|P\ 15,5) + [\$8000 + \$1000(A|G\ 15,5)]$$
$$-\$6000(A|F\ 15,5)$$
$$= \$17,782$$

Hence an annual expenditure of $17,782/year for 5 years is equivalent to the cash flow profile associated with the machine investment. Alternately, the equivalent uniform cost can be obtained from either the present worth or the future worth.

$$EUAC = -PW(A|P \ i,n)$$

$$EUAC = -FW(A|F \ i,n)$$

Present worth, future worth, and annual worth computations are used often in comparing economic investment alternatives having different cash flow profiles. Consequently, we will have need for PW, FW, and AW calculations for the discussion of alternative comparisons in Chapter Four.

3.8 CHANGING INTEREST RATES

The preceding discussion assumed that the interest rate did not change during the time period of concern. Recent experience indicates that such a situation is not likely if the time period of interest extends over several years (i.e., more than one interest rate may be applicable). Considering a single sum of money and discrete compounding, if i_t denotes the interest rate appropriate during time period t, the future worth equivalent for a single sum of money can be expressed as

$$F = P(1 + i_1)(1 + i_2) \cdots (1 + i_{n-1})(1 + i_n) \tag{3.52}$$

and the inverse relation

$$P = F(1 + i_n)^{-1}(1 + i_{n-1})^{-1} \cdots (1 + i_2)^{-1}(1 + i_1)^{-1} \tag{3.53}$$

Example 3.34

Consider the situation depicted in Figure 3.18 in which an individual deposited $1000 in a savings account that paid interest at an annual compounding rate of 8% for the first 3 years, 10% for the next 4 years, and 12% for the next 2 years. How much was in the fund at the end of the ninth year?

Letting V_t denote the value of the account at the end of time period t, we see that

$$V_3 = \$1000(F|P \ 8,3)$$
$$= \$1000(1.2597)$$
$$= \$1259.70$$

Likewise,

$$V_7 = \$1259.7(F|P \ 10,4)$$
$$= \$1259.7(1.4641)$$
$$= \$1844.33$$

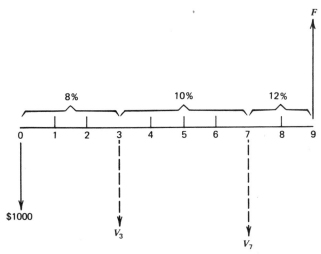

FIGURE 3.18. Cash flow diagram for a changing interest rate example.

Similarly,

$$F = V_9 = \$1844.33(F|P\ 12,2)$$
$$= \$1844.33(1.2544)$$
$$= \$2313.53$$

Alternately, the amount in the account at the end of 9 years, is given by

$$F = \$1000(1.08)(1.08)(1.08)(1.10)(1.10)(1.10)(1.10)(1.12)(1.12)$$
$$= \$2313.55$$

(The difference in answers is due to cumulative round-off errors in the interest tables.)

Extending the consideration of changing interest rates to series of cash flows, the present worth of a series of cash flows can be represented as

$$P = A_1(1 + i_1)^{-1} + A_2(1 + i_1)^{-1}(1 + i_2)^{-1} + \cdots$$
$$+ A_n(1 + i_1)^{-1}(1 + i_2)^{-1} \cdots (1 + i_n)^{-1} \qquad (3.54)$$

The future worth of a series of cash flows can be given by

$$F = A_n + A_{n-1}(1 + i_n) + A_{n-2}(1 + i_{n-1})(1 + i_n) + \cdots$$
$$+ A_1(1 + i_2)(1 + i_3) \cdots (1 + i_{n-1})(1 + i_n) \qquad (3.55)$$

Example 3.35

Consider the cash flow diagram given in Figure 3.19 with the appropriate interest rates indicated. Determine the present worth, future worth, and uniform series equivalents for the cash flow series.

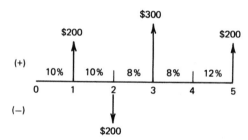

FIGURE 3.19. Cash flow diagram for a changing interest rate example.

Computing the present worth gives

$$P = \$200(P|F\ 10,1) - \$200(P|F\ 10,1)(P|F\ 10,1)$$
$$+ \$300(P|F\ 8,1)(P|F\ 10,1)(P|F\ 10,1)$$
$$+ \$200(P|F\ 12,1)(P|F\ 8,1)(P|F\ 8,1)(P|F\ 10,1)(P|F\ 10,1)$$
$$= \$200(P|F\ 10,1) - \$200(P|F\ 10,2)$$
$$+ \$300(P|F\ 10,2)(P|F\ 8,1) + \$200(P|F\ 10,2)(P|F\ 8,2)(P|F\ 12,1)$$
$$= \$372.62$$

The future worth is given by

$$F = \$200 + \$300(F|P\ 8,1)(F|P\ 12,1) - \$200(F|P\ 8,2)(F|P\ 12,1)$$
$$+ \$200(F|P\ 10,1)(F|P\ 8,2)(F|P\ 12,1)$$
$$= \$589.01$$

The uniform series equivalent is obtained as follows.

$$P = A(P|F\ 10,1) + A(P|F\ 10,2) + A(P|F\ 8,1)(P|F\ 10,2)$$
$$+ A(P|F\ 8,2)(P|F\ 10,2) + A(P|F\ 12,1)(P|F\ 8,2)(P|F\ 10,2)$$

$$\$372.62 = A[(0.9091) + (0.8264) + (0.9259)(0.8264) + (0.8573)(0.8264)$$
$$+ (0.8929)(0.8573)(0.8264)]$$

$$\$372.62 = 3.842\ A$$

$$A = \$96.99$$

Thus, $96.99/time period for five time periods is equivalent to the original cash flow series.

3.9 CONSIDERATION OF INFLATION

As noted previously, any consideration of cash flows in today's economy must include a consideration of inflation. Although it is important to consider the impact of inflation on investments made within one country, it is especially important to do so in multinational investment situations. The inflation rates in Argentina, Brazil, Canada, France, Great Britain, Italy, Japan, Mexico, and the United States, for example, can be dramatically different. Hence, a firm that is faced with making decisions concerning the investment of capital in various nations must give strong consideration to the inflationary economies in the countries in question.

Inflation can be defined as an increase in the amount of currency in circulation in a given country, resulting in a relatively sharp and sudden fall in its value and, therefore, a rise in prices. Hence, inflation affects adversely the purchasing power of money.

To illustrate what we mean by the term, *purchasing power of money,* suppose a firm purchases 1 million pounds of material each year, and the price of the material increases by 10% per year. Obviously, the quantity of material the firm can purchase with a fixed amount of money, that is, the purchasing power of the firm's money, decreases over time. The only way the firm can afford to continue purchasing the material is to decrease its usage rate or to increase its own source of funds. In the latter case, the firm might increase the price of the products it produces; if so, then the purchasing power of its customers' money will be decreased. In these situations, the continuing spiral of price increases does not contribute real increases in the firm's profits; instead, it results in an inflated representation of the firm's profits. The overall process is referred to as *inflation.*

Over the past two decades, we have experienced price level increases at both single-digit and double-digit rates. In some countries, triple-digit rates describe the price increases they have experienced. It has become widely accepted to refer to price level increases as inflation and to refer to price level decreases as deflation. (Under deflationary conditions, the deflation rate can be incorporated in economic analyses as a negative inflation rate.)

Although we have become accustomed to living in inflationary times, we are often very uncertain about how inflation should be treated, if at all, in an economic analysis. This section will present some background material on inflation, followed by two equivalent approaches for performing economic analyses under inflation.

3.9.1 Background on Inflation

One of the most difficult aspects of inflation is measuring it. To illustrate our claim, consider the following types of rates: *inflation rates, real interest rates, combined interest rates,* and *commodity escalation rates.*

As noted above, in recent years it has been apparent that the costs of goods and services are often affected by inflation. As a result, it is appropriate for interest rates and rates of return to reflect both the time value of money and the effects of inflation. For example, your decision to invest in a fund paying 12% interest would be quite different under negligible inflation conditions and 10% inflation conditions. We call the time value of money (or desired return) *in the absence of inflation* the *real interest rate* and denote it by d. The *combined interest rate* is denoted by i and the *inflation rate* is denoted by j. The three rates are related as follows:

$$1 + i = (1 + d)(1 + j)$$

or

$$i = d + j + dj \tag{3.56}$$

To complicate matters, it is not uncommon to find that prices and expenses for various goods and services increase (and decrease) at different rates, due to different external forces. For example, the price paid for a dozen eggs has scarcely changed over the past decade or more; whereas, the cost of new housing has increased dramatically in some parts of the United States. Oil prices change at different rates than the price of bread; changes in aluminum and steel prices are not closely correlated with changes in textile prices or the prices for microcomputers. In this case, we would consider the rate at which unit prices changed for each commodity to be an individual *commodity escalation rate*. (It is important to note that the commodity escalation rate for an individual commodity can be greater or less than the overall inflation rate.)

In practice, when it comes time to measure the inflation rate, what we observe are the values of commodity escalation rates and the combined interest rates. The value of the inflation rate is not as evident. As a result, in estimating the value of the underlying inflation rate, a surrogate measure is generally used. One popular surrogate measure of the inflation rate in the United States is the yearly change in the *Consumer Price Index* (CPI); another is the *Producer Price Index* (PPI), also known as the *Wholesale Price Index* (WPI). A number of composite indexes are used by firms to forecast the inflation rate; for example both *Business Week* and *Forbes* publish their own indexes.

The challenge is not in obtaining values of indexes. Rather, it is in knowing how to convert the values into forecasts of future inflation rates. As an example, *Industry Week* regularly publishes monthly updates on a dozen indexes; it also provides monthly updates on approximately twenty "business barometers," including aluminum output (in thousands of metric tons), appliance shipments (in millions of units), auto output (in thousands of units), electric power output (in billions of kilowatt-hours), machine tool orders (in millions of dollars), new construction (in billions of dollars), prime interest rate, and retail sales (in billions of dollars).

Table 3.6 presents CPI values for recent years. From annual changes in the CPI, a value can be assigned to the overall annual inflation rate.

Some do not agree with the use of changes in the CPI as estimates of inflation. They argue that the inflation rate cannot be an exact representation of the purchasing power of *your* money without *your* purchases being accurately reflected by the makeup of commodities included in the computation of the CPI.

TABLE 3.6

An Indicator of the General Inflation Rate in the United States

Year	Consumer Price Index[a] (Base is 100 in 1967)	CPI Rate of Inflation in Percent
1975	161.2	9.1
1976	170.5	5.8
1977	181.5	6.5
1978	195.4	7.7
1979	217.4	11.3
1980	246.8	13.5
1981	272.4	10.4
1982	289.1	6.1
1983	298.4	3.2
1984	311.1	4.3
1985	322.2	3.6
1986	328.4	1.9
1987	340.4	3.7
1988[b]	354.3	4.1

[a] Source: 1975–1985 statistics from [17]; 1986–1988 statistics from [2].

[b] Forecast by *Industry Week* [2].

As an example, if you never intend to buy or sell a house, then it would seem that the change of prices in houses will not have an impact on the purchasing power of your money. Likewise, if you are a vegetarian, then the change in prices for beef, pork, and chicken would not seem to affect you directly. However, although changes in some prices might not affect you directly, they will affect others—and those people, in turn, will increase the prices of their goods and services in order to recover the increased amount of money spent on items you do not buy.

Because the overall economy is so interdependent, it is difficult to determine the inflation rate for a person or an individual firm. Hence, changes in composite indexes, such as the CPI and PPI (or WPI), are used by individuals and organizations to estimate the impact of inflation on their economic investments. Since the rates of change in CPI and PPI (or WPI) reflect the changes in prices of a mixture of commodities, then any estimate of inflation derived from these rates will be a weighted average of the annual rates of change in the costs of a wide range of commodities. For this reason, the resulting estimate of the inflation rate is referred to as a "market basket" rate.

In addition to consumer prices, producer prices, and wholesale prices, the Commerce Department and the Labor Department, among others, publish a wide range of economic data for a variety of industry groups. Depending on the industry, inflation estimates might be more accurately developed using one or more such sources. To illustrate the difficulty in estimating the impact inflation will have on a particular industry, consider the following comments made by Cooper and Madigan in *Business Week:*

> Although the GNP (Gross National Product) report raises key questions about economic growth this year, the outlook for inflation is coming into clearer focus. Basically, the prices and costs at yearend points to no significant pickup in inflation from last year's pace.
>
> In the fourth quarter, the fixed-weight price index in the GNP report rose at a 3.7% annual rate, compared with 3.4% in the third. For the full year, this broad measure of prices advanced only 4%.
>
> In the consumer sector only, the slim 0.1% increase in December's consumer price index left last year's prices 4.4% above December, 1986. That rate of gain was the fastest since 1981. However, just as the miniscule 1.2% inflation rate in 1986 reflected almost solely the collapse in oil prices, 1987's faster pace was largely the result of the 18% rebound in prices of energy goods.
>
> A better reading on the economy's underlying inflation potential comes by excluding the volatile energy and food sectors. By that measure prices rose 4.2% last year, about the same as in both 1985 and 1986. That indicates little, if any, inflationary pressure at the start of 1988. [5]

A further difficulty in estimating the inflation rate is the need to forecast future inflation rates, since our economic analyses deal with future cash flows. Historical data in the form of CPI and PPI indexes might not provide a good basis for estimating future rates of inflation. Because of the difficulty in forecasting inflation, some prefer to estimate all future cash flow values in terms of "today's dollars" or "constant dollars." When constant dollars are used, the appropriate discount rate to use is the real interest rate, d.

Because of the need to distinguish between cash flows that incorporate explicitly the impact of inflation and those that do not. The former are called *then-current* cash flows, whereas the latter are called *constant worth* cash flows.

When constant worth cash flows can be represented as a uniform series, the equivalent then-current cash flows will be represented by a geometric series. Let year zero be used to establish the base for constant worth cash flows and C_k denote the constant worth amount of a cash flow occurring at the end of year k. If

$$C_k = T_0 \quad \text{for } k = 1, \ldots, n \quad (3.57)$$

then the then-current equivalent cash flow will be

$$\begin{aligned} T_k &= T_0(1 + j)^k \\ &= T_0(F|P\ j,k) \end{aligned} \quad \text{for } k = 1, \ldots, n \quad (3.58)$$

Likewise, if the then-current cash flows increase at a commodity escalation rate of g, and

$$T_k = T_0(1 + g)^k$$
$$= T_0(F|P\ g,k) \qquad \text{for } k = 1, \ldots, n \qquad (3.59)$$

then

$$C_k = T_k(1 + j)^{-k}$$
$$= T_k(P|F\ j,k) \qquad \text{for } k = 1, \ldots, n \qquad (3.60)$$

or

$$C_k = T_0(1 + g)^k(1 + j)^{-k}$$
$$= T_0(F|P\ g,k)(P|F\ j,k) \qquad \text{for } k = 1, \ldots, n \qquad (3.61)$$

Since the commodity escalation rate is a combined rate, consisting of inflation and real escalation, it is possible to ascertain the *real commodity escalation rate, v,* as follows:

$$v = (1 + g)/(1 + j) - 1 \qquad (3.62)$$

Thus, if the constant worth cash flows increase at a real escalation rate of v such that

$$C_k = T_0(1 + v)^k$$
$$= T_0(F|P\ v,k) \qquad (3.63)$$

then

$$T_k = T_0(1 + g)^k$$
$$= T_0(F|P\ g,k) \qquad (3.64)$$

Similarly, the present worth equivalent of a constant worth cash flow is given by

$$P = C_k(P|F\ d,k) \qquad (3.65)$$

and the present worth equivalent of a then-current cash flow is given by

$$P = T_k(P|F\ i,k) \qquad (3.66)$$

where i, j and d are related by Equation 3.56.

Example 3.36

The need for maintenance occurs at an increasing rate of 8%/year; the maintenance cost this year totals $1000. Furthermore, the labor and parts required to maintain the equipment increase due to inflation at a rate of 10%/year. The real interest rate for the firm is 12%/year. It is desired to determine the present worth equivalent of the then-current and constant worth maintenance cost in the fifth year. In this case, $v = 0.08$, $j = 0.10$, $T_0 = \$1000$, and $d = 0.12$.

The commodity escalation rate for maintenance costs is made up of an 8% real escalation rate and a 10% inflation rate. Hence, the commodity escalation rate will be equal to the sum of 8%, 10%, and their product, or 18.8%.

The *then-current maintenance cost* will total $1000(F|P\ 18.8,5)$ or $2366.31 in the fifth year; the *constant worth maintenance cost* in the fifth year will be $1000(F|P\ 8,5)$ or $1469.30. To obtain the present worth equivalent of the maintenance cost, the real interest rate of 12% is applied to the constant worth amount, whereas a combined interest rate of $0.12 + 0.10 + 0.12(0.10)$ or 0.232 is applied to the then-current amount. Thus,

$$P = \$1469.30(P|F\ 12,5)$$
$$= \$833.68$$

or

$$P = \$2366.31(P|F\ 23.2,5)$$
$$= \$2366.31(1.232)^{-5}$$
$$= \$833.72$$

(The $0.04 difference in the two answers is due to round-off error in the value of the $(P|F\ 12,5)$ interest factor.)

Example 3.37

The commodity escalation rate for the cost of raw materials during a 3-year project is expected to increase at an annual rate of 10%. The general inflation rate is also expected to be 10%/year. A unit of raw material costs $500 today.

The then-current and constant worth costs per unit during each of the next 3 years are given in Table 3.7. Values of $T_0 = \$500$, $g = 0.10$, and $j = 0.10$ are used in Equations 3.59 and 3.61 to generate the table.

Example 3.38

Energy on the 3-year project of Example 3.37 currently costs $55.00/thousand kilowatt-hours. Energy costs will escalate at a rate of 22%, and inflation will remain constant at 10%/year.

TABLE 3.7
Then-Current and Constant Worth Cost of Raw Material

Year	Then-Current Dollars	Constant Worth Dollars
1	$550.00	$500.00
2	605.00	500.00
3	665.50	500.00

TABLE 3.8

Then-Current and Constant Worth Cost of
Energy (per 1000 kilowatt-hours)

Year	Then-Current Dollars	Constant Worth Dollars
1	$67.10	$61.00
2	81.86	67.65
3	99.87	75.04

The then-current and constant worth energy rates are presented in Table
3.8. $T_0 = \$55.00/1000$ kilowatt-hours, $g = 0.22$, and $j = 0.10$ are used in
Equations 3.63 and 3.64 to generate the values in the table.

3.9.2 Economic Analysis Under Inflation

Either of the following procedures should be used to perform economic
analyses in conditions of inflation.

1. Express all cash flows in terms of their then-current amounts, and
use an interest rate that combines the real discount rate and the inflation rate, $i = d + j + dj$.

2. Express all cash flows in terms of their constant worth amounts and
use the real discount rate alone without an inflation rate component.

Using the first approach to account for inflation, the present worth equivalent of a series of then-current cash flows is computed as follows:

$$PW = \sum_{t=0}^{n} T_t(1 + d)^{-t}(1 + j)^{-t}$$

$$= \sum_{t=0}^{n} T_t(1 + d + j + dj)^{-t}$$

$$= \sum_{t=0}^{n} T_t(1 + i)^{-t} \tag{3.67}$$

The second approach computes the present worth equivalent using constant
worth dollar amounts as follows:

$$PW = \sum_{t=0}^{n} C_t(1 + d)^{-t} \tag{3.68}$$

*In subsequent sections of the book we implicitly assume that either the
discount rate used with then-current dollars includes a component for inflation, or the cash flows are in the form of constant worth dollars.*

Example 3.39

To illustrate the two approaches for dealing with inflation, consider two cash flow series for a project over a 3-year period. Expenses are $10,000 initially, with operating and maintenance costs of $2000 in the first year, increasing at a rate of 15%/year. Incomes begin in year 1 at $8000 and decrease at a rate of 20%/year; the general inflation rate is 10%, and the real discount rate is 7%.

Tables 3.9 and 3.10 give the present worth of the project under each approach. Note that the commodity escalation rate for operating and maintenance costs is $g_1 = 0.15$, the commodity escalation rate for income is $g_2 = -0.20$, $d = 0.07$, and $j = 0.10$. Had the then-current net cash flows been discounted using only the real discount rate of 7%, the present worth would have been $1208.91 instead of the $-$424.79 obtained considering inflation. Thus, what would have seemed to be a profitable investment without considering inflation actually has a negative present worth when the effects of inflation are included in the analysis.

As a further complication in dealing with inflation, the inflation rate tends to change from one time period to the next. If j_t denotes the inflation rate for period t and d_t denotes the real discount rate for period t, the present worth can be expressed in then-current dollars as

$$PW(d_t, j_t) = T_0 + \sum_{k=1}^{n} T_k \prod_{t=1}^{k} (1 + j_t)^{-1}(1 + d_t)^{-1} \tag{3.69}$$

where $\prod_{t=1}^{k}$ represents the product of the following terms for t ranging from 1 through k.

Inflation is a much-discussed subject in the area of economic investment analysis. Some argue that inflation effects can be ignored, since inflation will affect all investments in roughly the same way. Thus, it is argued, the relative differences in the alternatives will be approximately the same with or without inflation considered. Others argue that the inflation rate during the past decade has been so dynamic that an accurate prediction of the true

TABLE 3.9

Then-Current Project Analysis

End of Year, t	Then-Current Expenses, $g_1 = .15$	Then-Current Incomes, $g_2 = -.20$	Net Cash Flow, T_t	Discount Factor, $(1 + d + j + dj)^{-t}$ $= (1 + .177)^{-t}$	Discounted Net Cash Flow, $T_t(1 + .177)^{-t}$
0	$-$10,000	—	$-$10,000	1.000	$-$10,000.00
1	$-$ 2,000	$8,000	6,000	.850	5,097.71
2	$-$ 2,300	6,400	4,100	.722	2,959.59
3	$-$ 2,645	5,120	2,475	.613	1,517.91
				Total	$-$$ 424.79

TABLE 3.10

Constant Worth Project Analysis

End of Year, t	Constant Worth Expenses	Constant Worth Incomes	Net Cash Flow, C_t	Discount Factor $(1 + d)^{-t}$ $= (1 + .07)^{-t}$	Discounted Net Cash Flow, $C_t(1 + .07)^{-t}$
0	−$10,000.00	—	−$10,000.00	1.000	−$10,000.00
1	− 1,818.18	$7,272.73	5,454.55	.935	5,097.71
2	− 1,900.83	5,289.26	3,388.43	.873	2,959.59
3	− 1,987.23	3,846.73	1,859.50	.816	1,517.92
				Total	−$ 424.79

inflation rate and its impact on future cash flows is not possible. Another argument for ignoring explicitly the effects of inflation is that it is accounted for implicitly, since cash flow estimates for the future are made by individuals conditioned by an inflationary economy. Thus, it is argued, any estimates of future cash flows probably incorporate implicitly inflationary effects. A final argument for ignoring inflation in comparing investment alternatives involving only negative cash flows is that an alternative that is preferred by ignoring inflation effects will be even more attractive when effects of inflation are incorporated in the analysis.

The above arguments are certainly valid in some instances. However, counterarguments can be given for each. Actually, it is not difficult to explicitly account for inflation. Use of either of the two approaches illustrated is recommended.

3.10 PRINCIPAL AMOUNT AND INTEREST AMOUNT IN LOAN PAYMENTS

With both personal and corporate investments that are financed from borrowed funds, income taxes might be affected by the amount of interest paid. Hence, it is quite important to know how much of each payment is interest and how much is reducing the principal amount borrowed initially. To illustrate this situation, suppose you borrowed $10,000 and paid it back using four equal annual installments, with interest computed at 10% compounded annually. The payment size is computed to be

$$A = P(A|P\ 10,4)$$
$$= \$10,000(0.3155)$$
$$= \$3155$$

The interest accumulation the first year is 0.10(10,000), or $1000. Therefore, the first payment consists of a $1000 interest payment and a $2155 principal payment. The unpaid balance at the beginning of the second year is

$10,000 − $2155, or $7845; consequently, the interest charge the second year is 0.10($7845), or $784.50. Thus, the second payment consists of a $784.50 interest payment and a principal payment of $3155 − $784.50, or $2370.50. The unpaid balance at the beginning of the third year is $7845 − $2370.50, or $5474.50, so the interest charge the third year is 0.10($5474.50), or $547.45. Thus, the third payment consists of a $547.45 interest payment and a principal payment of $3155 − $547.45, or $2607.55. The unpaid balance at the beginning of the fourth year is $5474.50 − $2607.55, or $2866.95; the interest charge the fourth year is 0.10($2866.95), or $286.70. Thus, the fourth payment consists of a $286.70 interest payment and a principal payment of $3155 − $286.70, or $2868.30. (The principal payment in the fourth payment should equal the unpaid balance of $2866.95 at the beginning of the fourth year. The difference of $1.35 is due to round-off errors in computing the payment size.)

The amount of principal remaining to be repaid immediately after making payment $(t − 1)$ can be found by stripping off the interest on the remaining $n − t + 1$ payments; letting U_{t-1} denote the unpaid principal after making payment $t − 1$, we know that

$$U_{t-1} = A(P|A\ i, n − t + 1) \tag{3.70}$$

or

$$U_{t-1} = A \sum_{j=1}^{n-t+1} (1 + i)^{-j}$$

where A denotes the size of the individual payments and i represents the interest rate used to compute A.

The amount by which payment t reduces the unpaid principal will be designated E_t and is given by the relation $E_t = U_{t-1} − U_t$. Hence,

$$E_t = A \sum_{j=1}^{n-t+1} (1 + i)^{-j} − A \sum_{j=1}^{n-t} (1 + i)^{-j} \tag{3.71}$$

or

$$E_t = A \left[\sum_{j=1}^{n-t} (1 + i)^{-j} + (1 + i)^{-(n-t+1)} \right] − A \sum_{j=1}^{n-t} (1 + i)^{-j}$$

Thus,

$$E_t = A(1 + i)^{-(n-t+1)} \tag{3.72}$$

or

$$\boxed{E_t = A(P|F\ i, n − t + 1)} \tag{3.73}$$

Recalling how the payment size is determined, we express E_t as

$$E_t = P(A|P\ i,n)(P|F\ i,n - t + 1) \qquad (3.74)$$

where P is the original principal amount borrowed, and E_t is the amount of payment t, which is an *equity* payment against principal). Letting I_t be the amount of payment t, which is an interest payment, it is seen that

$$I_t = A - E_t.$$

$$I_t = A[1 - (P|F\ i,n - t + 1)] \qquad (3.75)$$

Example 3.40

To illustrate the use of the formulas for computing the values of E_t and I_t, suppose \$10,000 is borrowed at 12% annual interest and repaid with five equal annual payments. The payment size is found to be

$$A = \$10,000(A|P\ 12,5)$$

$$A = \$10,000(0.2774)$$
$$= 2774$$

Thus

$$E_1 = \$2774(P|F\ 12,5) = \$1574, \qquad I_1 = \$1200$$

$$E_2 = \$2774(P|F\ 12,4) = \$1763, \qquad I_2 = \$1011$$

$$E_3 = \$2774(P|F\ 12,3) = \$1975, \qquad I_3 = \$799$$

$$E_4 = \$2774(P|F\ 12,2) = \$2211, \qquad I_4 = \$563$$

$$E_5 = \$2774(P|F\ 12,1) = \$2477, \qquad I_5 = \$297$$

Example 3.41

An individual purchases a \$130,000 house and makes a down payment of \$30,000. The remaining \$100,000 is financed over a 25-year period at 12% compounded monthly. The monthly house payment is computed to be

$$A = \$100,000(A|P\ 1,300)$$
$$= \$1053$$

The individual keeps the house for 5 years and decides to sell it. How much equity is there in the house?

The amount of principal remaining to be paid can be determined by computing the present worth of the remaining 240 monthly payments using a 1% interest rate, or

$$P = \$1053(P|A\ 1,240)$$
$$= \$95,633$$

Thus, the individual's equity equals ($100,000 − $95,633) or $4367, plus the value of the down payment, or $34,367. It is interesting (or perhaps depressing if you are a home owner) to determine that over the 5-year period the individual made payments totaling $63,180, of which $58,813 was for interest; that is, $58,813 = $63,180 − $4367.

3.11 BOND PROBLEMS

Although bonds are important financial instruments in the business world as investment opportunities, there are additional reasons for considering bond problems in an economic analysis text, as well as at this particular point in the text.

1. Bond problems illustrate the notion of equivalence covered previously in this chapter. That is, the purchase price of a bond is equivalent to the returns from the bond at an appropriate compound interest rate. Hence, they provide a useful mechanism for demonstrating a number of the time value of money calculations developed in this chapter.

2. Bond problems are convenient investment opportunities and will be used to motivate our presentation in Chapter Four of various measures of the economic worths of investment alternatives.

3. The issuance and sale of bonds is a mechanism by which capital may be raised to finance engineering projects; as such, they contribute to the firm's overall cost of capital, as described in Chapter Five.

An organizational unit desiring to raise capital may issue bonds totaling, say, $1 million, $5 million, $25 million, or more. A financial brokerage firm usually handles the issue on a commission basis and sells smaller amounts to other organizational units or individual investors. Individual bonds are normally issued in even denominations such as $500, $1000, or $5000. The stated value on the individual bond is termed the *face* or *par value*. The par value is to be repaid by the issuing organization at the end of a specified period of time, say 5, 10, 15, 20, or even 50 years. Thus, the issuing unit is obligated to *redeem* the bond at par value at *maturity*. Furthermore, the issuing unit is obligated to pay a stipulated *bond rate* on the face value during the interim between date of issuance and date of redemption. This might be 10% payable quarterly, 9½% payable semiannually, 11% payable annually, and so forth. For the purpose of the problems to follow, it is emphasized that the bond rate applies to the par value of the bond.

Example 3.42

A person purchases a $5000, 5-year bond on the date of issuance for $5000. The stated bond rate is 10% semiannually, and the interest payments are received on schedule until the bond is redeemed at maturity for $5000. The bond rate per interest period is 10%/2 = 5%. Thus, the bond holder receives (0.10/2)($5000) = $250 payments every 6 months. A cash flow diagram for the duration of the investment is given in Figure 3.20, where time periods are 6-month intervals.

It is noted from Figure 3.20 that the $5000 expenditure at $t = 0$ yields $250 each interest period for 10 periods and a $5000 redemption value at $t = 10$. Thus, the $5000 investment at $t = 0$ yielded the revenues from $t = 1$ through $t = 10$. What annual interest rate or *return on investment* did the $5000 investment yield?

One might intuitively answer the above question by stating that the $5000 investment was exactly returned (no loss or gain in capital) at redemption and, since during the interim an interest rate of 10% semiannual was received, then the yield *must* be 10% semiannual. Accepting this argument for the moment, one might further pose certain hypotheses concerning the transaction. For instance, it is hypothesized that the $5000 at $t = 0$ is equivalent (has the same present worth) to the revenue cash flows if the time value of money is 10% compounded semiannually. That this hypothesis is true is shown by the relation

$$P = A(P|A\ 5,10) + F(P|\ 5,10)$$

or

$$\$5000 = (0.05)(\$5{,}000)(7.7217) + \$5000(0.6139),$$

and

$$\$5000 = \$5000$$

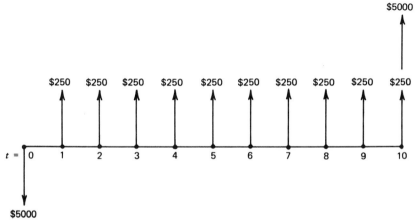

FIGURE 3.20. Cash flow for a $5000 bond.

A second hypothesis concerning the transaction is that if more than the par value is paid for the bond at $t = 0$ and all revenue figures remain the same, then a return on investment less than the bond rate of 10% semiannual will be received on the investment. For example, if $5500 were paid for the bond at $t = 0$, the present worth of the $5500 outgo is not equal (equivalent) to the present worth of the revenues at a 10% semiannual yield. This is shown by

$$\$5500 \neq (0.05)(\$5000)(P|A\ 5,10) + \$5000(P|F\ 5,10)$$

$$\$5500 \neq \$250(7.7217) + \$5000(0.6139)$$

$$\$5500 \neq \$5000$$

This result now raises the pertinent question, what *is* the rate of return on the investment if $5500 is paid for the bond and the revenues remain the same? Intuition may suggest that the yield will be less than 10% semiannual because the purchase value of $5500 decreases (a loss of investment capital) to $5000 on redemption. Furthermore, the semiannual payments of $250 are not equal to 5% of the $5500 purchase price. In order to answer the question more precisely, let us answer the alternate question of "what interest rate per period (or yield per period) will make the future worth of all cash flows equal to zero?" That is, what interest rate satisfies the following equation?

$$-\$5500(F|P\ i,10) + (.05)(\$5000)(F|A\ i,10) + \$5000 = \$0$$

The solution to the above equation gives the answer to the original question, and the task is to solve for the positive roots of the polynomial. However, an approximate solution to avoid such tedium is by trial and error. An iterative procedure follows.

For $i = 3\%$ (6% semiannual)

$$-\$5500(F|P\ 3,10) + (.05)(\$5000)(F|A\ 3,10) + \$5000 \neq \$0$$

$$-\$5500(1.3439) + (\$250)(11.4639) + \$5000 \neq \$0$$

$$\$474.53 \neq \$0$$

For $i = 4\%$ (8% semiannual)

$$-\$5500(F|P\ 4,10) + (.05)(\$5000)(F|A\ 4,10) + \$5000 \neq \$0$$

$$-\$5500(1.4802) + (\$250)(12.0061) + \$5000 \neq \$0$$

$$-\$139.58 \neq \$0$$

From these two trials (for $i = 3\%$ and $i = 4\%$), the future worth of $0 is bracketed, as shown in Table 3.11. Using the data of Table 3.11, we can solve for X by linear interpolation, or

$$\frac{0.04 - 0.03}{-\$139.58 - \$474.53} = \frac{0.04 - X}{-\$139.58 - \$0.00}$$

or

TABLE 3.11			
Bond Yield Interpolation			
For i =	0.03	X	0.04
FW of revenues =	$474.53	$0.00	−$139.58

$$X = 0.3772 \quad \text{or} \quad 3.772\%$$

Thus, the equivalent yield on the $5500 investment is approximately 3.772%/period, or (2)(3.772) = 7.544% semiannually, or an effective annual yield of $[(1 + 0.03772)^2 − 1](100\%)$ = 7.686%. The figures support the *a priori* intuition of a return on investment less than 10% semiannual, and the second hypothesis is accepted.

Bond problems arise in economic analysis because many bonds trade daily through financial markets such as the New York Stock Exchange. Thus, bonds may be purchased for less than, greater than, or equal to par value, depending on the economic environment. They may also be sold for less than, greater than, or equal to par value. Furthermore, once purchased, bonds may be kept for a variable number of interest periods before being sold. A variety of situations can occur, but only three basic types of bond problems can occur. These will be presented after formalizing the discussion thus far. We now employ the following notation.

P = the purchase price of a bond
F = the sales price (or redemption value) of a bond
V = the par or face value of a bond
r = the bond rate per interest period
i = the yield rate (return on investment or rate of return) per interest period
n = the number of interest payments received by the bondholder
$A = Vr$ = the interest or coupon payment received

The general expression relating these is

$$P = Vr(P|A\ i,n) + F(P|F\ i,n) \tag{3.76}$$

Now, the three types of bond problems follow:

1. Given P,r,n,V, and a desired i, find the sales price F.
2. Given F,r,n,V, and a desired i, find the purchase price P.
3. Given P,F,r,n, and V, find the yield rate i that has been earned on the investment.

Each of these cases is illustrated in the following examples.

Example 3.43

A $1000, 12% semiannual bond is purchased for $1050 at an arbitrary $t = 0$. If the bond is sold at the end of 3 years and six interest payments, what must be the selling price if this bond is to be a good investment compared to investing at 5% per 6-month period? The cash flow for this example is given by Figure 3.21.

The present worth of the cash flow depicted in Figure 3.21 follows.

$$P = Vr(P|A\ 5,6) + F(P|F\ 5,6)$$

or

$$\$1050 = (\$1000)(0.06)(5.0757) + F(0.7462).$$

Solving the above equation for F yields a value of $999.01. As long as the selling price at the end of three years is at least $999.01 the bond will yield a return of at least 10% compounded semiannually.

Example 3.44

If a $1000, 12% semiannual bond is purchased at an arbitrary $t = 0$, held for 3 years and six interest payments, and redeemed at par value, what must the purchase price have been in order for the bond to be preferred over investing at 14% compounded semiannually?

From the cash flow given in Figure 3.22, the present worth of future cash flows is

$$P = Vr(P|A\ 7,6) + F(P|F\ 7,6)$$

or

$$P = (\$1000)(0.06)(4.7665) + \$1000(0.6663)$$

and

$$P = \$952.29.$$

FIGURE 3.21. Cash flow for a $1000 bond—determine sales price.

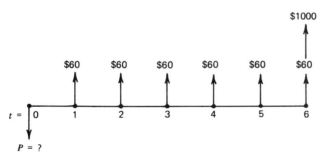

FIGURE 3.22. Cash flow for a $1000 bond—determine purchase price.

As long as the purchase price was less than $952.29, the bond was better than investing at 14% compounded semiannually.

Example 3.45

If a $1000, 12% quarterly bond is purchased at $t = 0$ for $1020 and sold 3 years later for $950, (a) what was the quarterly yield on the investment, and (b) what was the effective annual return? From the cash flow diagram in Figure 3.23, and setting the present worth equal to zero, we have

$$-\$1020 + (\$1000)(0.03)(P|A\ i,12) + \$950(P|F\ i,12) = \$0$$

It is then necessary to solve for the unknown i by trial and error as follows: For $i = 2\%$,

$$-\$1020 + \$30(10.5753) + \$950(0.7885) \neq \$0$$

$$-\$46.33 \neq \$0$$

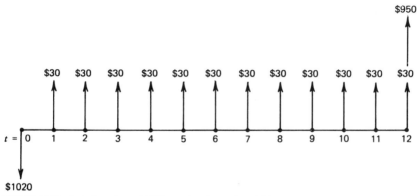

FIGURE 3.23. Cash flow for a $1000 bond—determine yield.

For $i = 2\frac{1}{2}\%$,

$$-\$1020 + \$30(10.2578) + \$950(0.7436) \neq \$0$$

$$\$5.85 \neq \$0$$

Then, interpolating gives

$$\frac{0.025 - 0.020}{\$5.85 - (-\$46.33)} = \frac{0.025 - X}{\$5.85 - \$0}$$

$$X = 0.02444 \qquad \text{or} \qquad 2.444\% \text{ per quarter}$$

The effective annual return is

$$[(1 + 0.02444)^4 - 1]100\% = 10.140\%$$

3.12 SUMMARY

In this chapter we developed the time value of money concept and defined a number of mathematical operations consistent with that concept. To motivate the discussion, we emphasized the subject from the viewpoint of personal financing. In subsequent chapters we apply the concepts developed in this chapter to the study of investment alternatives from the viewpoint of an ongoing enterprise.

BIBLIOGRAPHY

1. Altany, D. R., "Running Out of Room to Grow," *Industry Week*, Vol. 236, No. 6, March 21, 1988, pp. 19–20.

2. Altany, D. R., "GNP Correction in Store," *Industry Week*, Vol. 236, No. 6, March 21, 1988, pp. 101–102.

3. Blank, L. T., and Tarquin, A. T. *Engineering Economy*, 2 ed., McGraw-Hill, 1983.

4. Canada, J. R., and White, J. A. *Capital Investment Decision Analysis for Management and Engineering*, Prentice-Hall, 1980.

5. Cooper, J. C., and Madigan, K. "Business Outlook," *Business Week*, No. 3037, February 8, 1988, p. 17.

6. DeGarmo, E. P., Canada, J. R., and Sullivan, W. G., *Engineering Economy*, 7 ed., Macmillan, 1984.

7. Grant, E. L., Ireson, W. G. and Leavenworth, R. S. *Principles of Engineering Economy*, 7 ed, Wiley, 1982.

8. Jones, B. W., *Inflation in Engineering Economy Analysis*, Wiley, 1982.

9. Lutz, R. P., "Discounted Cash Flow Techniques," Chapter 9.3, *Handbook of Industrial Engineering*, (G. Salvendy, editor), Wiley, 1982.

10. Newman, D. G., *Engineering Economic Analysis*, 2 ed., Engineering Press, 1983.

11. Riggs, J. L., and West, T. M., *Engineering Economics,* 3 ed., McGraw-Hill, 1986.

12. Smith, G. W., *Engineering Economy,* 3 ed., Iowa State University Press, 1979.

13. Taylor, G. A., *Managerial and Engineering Economy,* 3 ed., Van Nostrand Reinhold, 1980.

14. Thuesen, G. J. and Fabrycky, W. J., *Engineering Economy,* 6th ed., Prentice-Hall, 1984.

15. Thuesen, G. J., "Economic Analysis," Chapter 6.2, *Production Handbook,* 4th ed., (J. A. White, editor), Wiley, 1987.

16. White, J. A., "Engineering Economy," *Handbook of Engineering Fundamentals,* 4th ed., (M. Souders and O. W. Eshbach, editors), Wiley, in-press.

17. *Statistical Abstract of the United States 1987,* 107th edition, U.S. Department of Commerce, Bureau of the Census, 1987, pp. 464–465.

PROBLEMS

1. Pat sold Mike a used automobile for $3000. Mike paid $500 as a down payment and gave Pat three personal notes for the remainder due. The principal of each note was $1000, $750, and $750. The $1000 note was due 1 year after the purchase, the second note was due 2 years after the purchase, and the third note was due 3 years after the purchase. The annual simple interest agreed to by Pat and Mike was 10% for the first note, 12% for the second note, and 15% for the third note. How much total interest was paid to Pat by Mike? (3.2)

2. A debt of $2500 is incurred at $t = 2$. An annual simple interest rate of 8% on the unpaid balance is agreed on. Three equal payments of $1000 are made at $t = 4$, 5, and 6 to pay off the debt, including interest payments. Determine the size of each payment. (3.2)

3. How long does it take a deposit in an 8% compound interest fund to double in value? (3.3)

4. How much money today is equivalent to $10,000 in 10 years, with interest at 8% compounded annually? (3.3)

5. If $5000 is deposited into a fund paying 6% compounded annually, what sum will be accumulated at the end of 10 years? What would be the sum accumulated at the end of 5 years if the fund paid 12% compounded annually? What is suggested regarding doubling the interest rate and halving the length of the time period? If you had $5000 available for investment and the two options were available, which would you choose if you had to choose one of them? Justify your choice. If you chose the shorter duration investment, what will you do with your accumulated monies over the next 5-year period? Should the answer to this question influence your choice? (3.3)

6. If a fund pays 10% compounded annually, what single deposit now will accumulate $12,000 at the end of the tenth year? If the fund pays 5% compounded annually, what single deposit is required now in order to accumulate $6000 at the end of the tenth year? (3.3)

7. Maria deposits $500, $1200, and $2000 at $t = 1$, 2, and 3, respectively. If the fund pays 8% compounded per period, what sum will be accumulated in the fund at (a) $t = 3$ and (b) $t = 6$? (3.4)

8. Suppose you wanted to become a millionaire at retirement. If an annual compound interest rate of 8% could be sustained over a 40-year period, how much would have to be deposited yearly in the fund in order to accumulate $1 million? (3.4)

9. Juan deposits $2000 in a savings account that pays 8% compounded annually. Exactly 2 years later he deposits $3000; 2 years later he deposits $4000; and 4 years later he withdraws all of the interest earned to date and transfers it to a fund that pays 10% compounded annually. How much money will be in each fund 4 years after the transfer? (3.4)

10. A debt of $1000 is incurred at $t = 0$. What is the amount of three equal payments at $t = 1$, 2, and 3 that will repay the debt if money is worth 10% compounded per period? (3.4.1)

11. Five deposits of $500 each are made at $t = 1, 2, 3, 4$, and 5 into a fund paying 8% compounded per period. How much will be accumulated in the fund at (a) $t = 5$, and (b) $t = 10$? (3.4.1)

12. What equal annual deposits must be made at $t = 2, 3, 4, 5$, and 6 in order to accumulate $15,000 at $t = 8$ if money is worth 10% compounded annually? (3.4.1)

13. If annual deposits of $800 are made in a fund paying 12% interest compounded annually, how much money will be in the fund immediately after the fifth deposit? (3.4.1)

14. Kim deposits $1000 in a savings account; 4 years after the deposit, half of the account balance is withdrawn. $2000 is deposited annually for an 8-year period, with the first deposit occurring 2 years after the withdrawal. The total balance is withdrawn 15 years after the initial deposit. If the account earned interest of 8% compounded annually over the 15-year period, how much was withdrawn at each withdrawal point? (3.4.1)

15. John borrows $15,000 at 15% compounded annually; he pays off the loan over a 5-year period with annual payments. Each successive payment is $500 greater than the previous payment. How much was the first payment? (3.4.2)

16. Solve Problem 15 for the case in which each successive payment is $500 less than the previous payment. (3.4.2)

17. Land is purchased for $25,000. It is agreed that land will be paid for over a 5-year period with annual payments and using a 12% annual compound interest rate. Each payment is to be $2000 greater than the previous payment. Determine the size of the last payment. (3.4.2)

18. Solve Problem 17 for the case in which each successive payment is $2000 less than the previous payment. (3.4.2)

19. Solve Problem 15 for the case in which each successive payment is to be 10% greater than the previous payment. (3.4.3)

20. Solve Problem 15 for the case in which each successive payment is to be 10% less than the previous payment. (3.4.3)

21. Solve Problem 17 for the case in which each successive payment is to be 12% greater than the previous payment. (3.4.3)

22. Solve Problem 17 for the case in which each successive payment is to be 20% less than the previous payment. (3.4.3)

23. In a new, highly automated factory, labor costs are expected to decrease at an annual compound rate of 5%; material costs are expected to increase at an annual compound rate of 6%; and energy costs are expected to increase at an annual compound rate of 3%. The labor, maintenance, and energy costs the first year are $2 million, $2 million, and $1,500,000. (3.4.3)
(a) What will be the value of each cost during the fifth year?
(b) Using an interest rate of 6% compounded annually, what uniform annual costs over a 5-year period would be equivalent to the cumulative labor, maintenance, and energy costs?

24. Mary works for a company that pays an annual bonus, the size of which is based on experience with the company. At the end of her first year, the bonus equals $1500. The size of the bonus compounds at an annual rate of 6%. Mary decides to place two-thirds of her annual bonus in a fund that pays 10% compounded annually. How much will be in the fund immediately after the thirtieth deposit? (3.4.3)

25. Using a compound interest rate of $i\%$ per period, determine the present worth equivalent for a series of cash flows having the following form:

$$A_t = [A + G(t - 1)](1 + j)^{t-1} \quad \text{for } t = 1, \ldots, n$$

Note:

$$\sum_{t=1}^{n} t = n(n + 1)/2$$

$$\sum_{t=1}^{n} x^t = \frac{x(1 - x^n)}{1 - x} \quad \text{for } |x| \neq 1$$

$$\sum_{t=1}^{n} tx^t = \frac{x - (1 - x)(n + 1)x^{n+1} - x^{n+2}}{(1 - x)^2} \quad \text{for } |x| \neq 1$$

(3.4.3)

26. Yavuz wishes to make a single deposit P at $t = 0$ into a fund paying 12% compounded quarterly such that $1000 payments are received at $t = 1, 2, 3$, and 4 (periods are 3-month intervals), and a single payment of $7500 is received at $t = 12$. What single deposit is required? (3.5)

27. Dr. Shieh deposits $3000 in a money market fund. The fund pays interest at a rate of 12% compounded annually. Just 3 years after making the single deposit, he withdraws half the accumulated money in his account. Then, 5 years after the initial deposit, he withdraws all of the accumulated money remaining in the account. How much does he withdraw 5 years after his initial deposit? (3.5)

28. Lynn borrows $5000 at 18%/year compounded monthly. She wishes to repay the loan with 12 end-of-month payments. She wishes to make her first payment 3 months after receiving the $5000. She also wishes that, after the first payment, the size of each payment be 10% greater than the previous payment. What is the size of her sixth payment? (3.5)

29. Solve Problem 28 for the case in which the size of each payment is $50 greater than the previous payment. (3.5)

30. David borrows $25,000 at 12% compounded quarterly. He wishes to repay the money with 10 equal semiannual installments. What must be the size of the payment if the first payment is made 1 year after obtaining the $25,000? (3.5)

31. Kristin deposits $150/month into an account paying 8% compounded quarterly. Twelve monthly deposits are made. Determine how much will be accumulated in the account 1 year after the last deposit. (3.5)

32. Barbara makes four consecutive annual deposits of $2000 in a savings account that pays interest at a rate of 10% compounded semiannually. How much money will be in the account 2 years after the last deposit? (3.5)

33. Lynn makes monthly deposits of $500 in a savings account that pays interest at a rate equivalent to 8% compounded quarterly. How much money should be in the account immediately after the sixtieth deposit? If no interest is earned on money deposited during a quarter and the first deposit coincides with the beginning of a quarter, what will be the account balance immediately after the sixtieth deposit? (3.5)

34. Stephen makes semiannual deposits of $1500 into an account that pays interest equivalent to 12% compounded monthly. Determine the account balance immediately *before* the tenth deposit. (3.5)

35. Mary Lib purchases a house for $160,000; a down payment of $20,000 is made at the time of purchase; the balance is financed at 12% compounded monthly, with monthly payments made over a 10-year period. (3.5)
   (a) What is the size of the monthly payments?
   (b) If the loan period had been 20 years, what would have been the size of the monthly payments?

36. What is the effective interest rate of 12% compounded monthly? (3.5)

37. Marvin borrows $2500 from the Shady Dealings Loan Company. He is told the interest rate is merely 2%/month and his payment is computed as follows:

Loan period	= 30 months
Interest = 30(0.02)($2500)	= $1500
Credit investigation and insurance	= $ 50
Total amount owed = $2500 + $1500 + $50	= $4050
Size of payment = $4050/30	= $135/month

What effective interest rate is paid for the transaction? (3.5)

38. Lynne borrows $15,000 at 1.5%/month. She desires to repay the money using equal monthly payments over 36 months. Lynne makes four such payments and decides to pay off the remaining debt with one lump sum payment at the time for the fifth payment. What should be the size of the payment if interest is truly compounded at a rate of 1.5%/month? (3.5)

39. Approximately how long will it take a deposit to triple in value if money is worth 9% compounded monthly? (3.5)

40. Ken borrows $5000 and pays the loan off, with interest, after 2 years. He pays back $6000. What is the effective interest rate for this transaction? (3.5)

41. An automobile agency advertises lease arrangements available for a new model car. The lease period is 48 months. A total of $949.11 will be due at the beginning of the lease, including a refundable $225 security deposit, a $500 down payment, and taxes and fees. The size of the monthly payments is advertised to be $224.10. The 4-year lease allows 60,000 miles to be driven; for each additional mile driven, there is a penalty of $0.06 per mile. To purchase the car will cost $9876.50 initially; after four years use, its resale value will be approximately $3000. Assuming no more than 60,000 miles are driven during the 4-year period and the operating and maintenance costs will be the same under leasing and purchasing, what is the effective interest rate for the lease? (3.5)

42. Ludo deposits $1000 now, $2000 2 years from now, and $5000 5 years from now into a fund paying 12% compounded semiannually. (3.5)
    (a) What sum of money will be accumulated at the end of the sixth year?
    (b) What equal deposits of size A, made every 6 months (with the first deposit at $t = 0$ and the last deposit at the end of the sixth year), are equivalent to the three deposits?

43. Semiannual deposits of $500 are made into a fund paying 8% compounded continuously. What is the accumulated value in the fund after 10 such deposits? (3.6.1)

44. Four equal, quarterly deposits of $1000 each are made at $t = 0$, 1, 2, and 3 (time periods are 3-month intervals) into a fund that pays 8% compounded *continuously*. Then, at $t = 7$ and $t = 10$, withdrawals of size A are made so that the fund is depleted at $t = 10$. What is the size of the withdrawals? (3.6.1)

45. Jaime receives an annual bonus from his employer. He wishes to deposit the bonus in a fund that pays interest at a rate of 6% compounded continuously. His first bonus is $1000. The size of his bonus is expected to increase at a rate of 4%/year. How much money will be in the fund immediately *before* the tenth deposit? (3.6)

46. Operating and maintenance costs for year k are given as $C_k = (e^{0.10})C_{k-1}$, $k = 2$, 3, . . . , 15, with $C_1 = \$1000$. Determine the equivalent uniform annual operating and maintenance cost based on continuous compounding with a nominal interest rate of (a) 10% and (b) 20%. (3.6.1)

47. A person borrows $10,000 and wishes to pay it back with 10 equal annual payments. What will the size of the payments be if the interest charge is 10% compounded (a) annually, (b) semiannually, and (c) continuously? (3.6.1)

48. For the cash flow series given below, determine (a) the present worth equivalent, (b) the future worth equivalent, and (c) the uniform annual series equivalent using a 10% continuous compound interest rate. (3.6.1)

EOY	Cash Flow	EOY	Cash Flow
0	−$10,000	6	$5,000e^{0.10}$
1	1,000	7	$5,000e^{0.20}$
2	2,000	8	$5,000e^{0.30}$
3	3,000	9	$5,000e^{0.40}$
4	4,000	10	$5,000e^{0.50}$
5	5,000		

49. Labor costs occur continuously during the year. At the end of each year a new labor contract becomes effective for the following year. Let $\bar{A}_j$ denote the cumulative labor cost occurring uniformly during year j, where $\bar{A}_j = 1.05\,\bar{A}_{j-1}$. If money is worth 15% compounded continuously and $\bar{A}_1 = \$25,000$, determine the present worth equivalent for 5 years of labor costs. (3.6.2)

50. Operating and maintenance costs for a production machine occur continuously during the year. If the total annual operating and maintenance cost is $24,000 and money is worth 25% compounded continuously, what single sum of money at the present is equivalent to 5 years of operating and maintenance costs? (3.6.2)

51. Consumption of fuel oil occurs continuously in a manufacturing plant. The total monthly cost in January is $25,000. The demand for fuel oil increases at a monthly continuous compound rate of 1.5%. Determine the equivalent end-of-year (future worth) cost for one year's consumption using a monthly nominal continuous compound rate of 2.0%. (3.6.2)

52. For the continuous cash flows shown below, determine (a) the present worth equivalent and (b) the future worth equivalent using a continuous compound interest rate of 20%. (3.6.2)

Year	Continuous Cash Flow, $\bar{A}$
1	$10,000
2	12,000
3	14,000
4	16,000
5	18,000

53. Solve Problem 52 for the following cash flow series (3.6.2):

Year	Continuous Cash Flow, $\bar{A}$
1	$10,000
2	$10,000e^{0.20}$
3	$10,000e^{0.40}$
4	$10,000e^{0.60}$
5	$10,000e^{0.80}$

54. A firm buys a new computer that costs $100,000. It may either pay cash now or pay $20,000 down and $30,000/year for 3 years. If the firm can earn 15% on investments, which would you suggest? (3.7)

55. Kareem invests $10,000 in a venture that returns him $500(0.80)^{t-1} at the end of year t for $t = 1, \ldots , 20$. With an interest rate of 10%, what is the equivalent uniform annual profit (cost) for the venture? (3.7)

56. Hugh is considering two investment alternatives. Alternative A requires an initial investment of $10,000; it will yield incomes of $3000, $3500, $4000, and $4500 over its 4-year life. Alternative B requires an initial investment of $12,000; it is anticipated that the revenue received will increase at a compound rate of 10%/year. Based on an interest rate of 10% compounded annually, what must be the revenue the first year for B in order for A and B to be equivalent? (3.7)

57. It is desired to determine the size of the uniform series over the time period [2,5] that is equivalent to the cash flow profile shown below using an interest rate of 10%. (3.7)

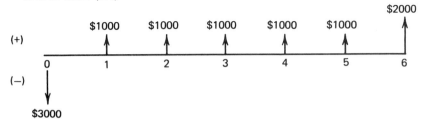

58. Given the following cash flows, what single sum at $t = 4$ is equivalent to the given data. Assume $i = 15\%$. (3.7)

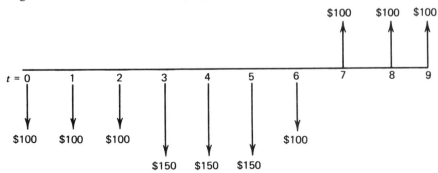

59. A machine is purchased at $t = 0$ for $40,000 (including installation costs). Net annual revenues resulting from operating the machine are $15,000. The machine is sold at the end of 8 years, $t = 8$, for $5000. The cash flow series for the machine is equivalent to what single sum, X, at (a) $t = 6$ and (b) $t = 8$ if money is worth 10% compounded continuously? (3.7)

60. Assume the following two investment plans.

   (a) Purchase for $4000 (a negative cash flow) and receive (1) $400 at the end of each 6 months for 4 years $(t = 1, 2, \ldots , 8)$, and (2) a single payment of $2000 at the end of the fourth year.

   (b) Purchase for $3850 and receive (1) $900 at the *beginning* of each year for the 4-year period, and (2) a single payment of $X at the end of the fourth year.

If money is worth 8% compounded semiannually, what is the value of X so that the two investments are equivalent? (3.7)

61. Gerardo invests $5000 and receives $600 each year for 10 years, at which time he sells out for $1000. With a 10% interest rate, what equal annual cost (profit) is equivalent to the venture? (3.7)

62. Consider the following cash flow series.

EOY	CF
0	−$10,000
1	6,500
2	6,000
3	5,500
4	5,000
5	4,500
6	4,000
7	3,500
8	3,000

At 10% annual compound interest, what uniform annual cash flow is equivalent to the above cash flow series? (3.7)

63. Susan borrows $1000 from a bank at $t = 0$ at 10% simple interest for 2 years. She pays the total interest due for the 2-year period at $t = 0$ and thus receives $800 at $t = 0$. If she pays back $1000 at $t = 2$, Susan is, in effect, paying an interest rate of $X\%$ compounded annually. Solve for X. (3.7)

64. Assume payments of $2000, $3000, and $5000 are received at $t = 3$, 4, and 5, respectively. What five equal payments occurring at $t = 1$, 2, 3, 4, and 5, respectively, are equivalent if $i = 10\%$ compounded per period? (3.7)

65. What single deposit of size $X into a fund paying 8% compounded annually is required at $t = 0$ in order to make withdrawals of $500 each at $t = 4, 5, 6,$ and 7 and a single withdrawal of $1000 at $t = 10$? (3.7)

66. If the withdrawals in Problem 65 are immediately placed into another fund paying 12% compounded annually, what amount will be accumulated in this fund at $t = 10\%$ (3.7)

67. Ms. Torro-Ramos borrows $6000 and repays the loan with four quarterly payments of $600 during the first year and four quarterly payments of $1500 during the second year after receiving the $6000 loan. Determine the effective interest rate of the loan transaction. (3.7)

68. Quarterly deposits of $500 are made at $t = 1, 2, 3, 4, 5, 6,$ and 7. Then withdrawals of size A are made at $t = 12, 13, 14,$ and 15, and the fund is depleted with the last withdrawal. If the fund pays 8% compounded quarterly, what is the value of A? (3.7)

69. Given the cash flow diagram shown below and a nominal interest rate of 10% compounded continuously, solve for the value of an equivalent amount at (a)

$t = 5$, (b) $t = 10$, and (c) $t = 15$. (*Note:* Upward arrows represent cash inflows and downward arrows represent cash outflows.)

70. What single sum of money at $t = 4$ is equivalent to the cash flow profile shown below? Use a 12% interest rate in your analysis. (3.7)

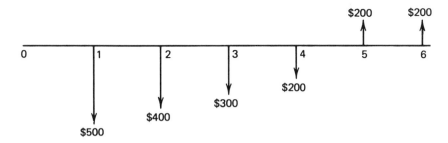

71. Determine the value of X so that the following cash flow series are equivalent at 8% interest. (3.7)

EOY	CF(A)	CF(B)
0	−$8,000	−$15,000
1	6,000	4,000
2	5,000	3,000 + X
3	4,000	2,000 + 2X
4	5,000	3,000 + 3X
5	6,000	4,000 + 4X
6	5,000	3,000 + 5X

72. Given the cash flow profiles shown below, determine the value of X so that the two cash flow profiles are equivalent at a 10% interest rate (3.7)

EOY	CF(A)	CF(B)
1	−$12,000	−$10,000
2	1,000	7,000
3	3,000	6,000 + 0.5X
4	5,000	5,000 + 1.0X
5	7,000	4,000 + 1.5X
6	9,000	3,000 + 2.0X

73. Given the two cash flow profiles shown below, for what value of X are the two series equivalent using an interest rate of 10% (3.7)

EOY	CF(A)	CF(B)
0	-$200,000	-$140,000
1	24,000	16,000
2	32,000	16,000
3	40,000	16,000 + X
4	48,000	16,000 + 2X
5	56,000	16,000 + 3X
6	64,000	16,000 + 4X
7	72,000	16,000 + 5X
8	80,000	16,000 + 6X

74. Given the cash flow profiles shown below, determine the value of X such that the two cash flow profiles are equivalent at 20% compounded annually. (3.7)

EOY	CF(A)	CF(B)
1	-$12,000	$ -X
2	1,000	7,000
3	4,000	9,000
4	6,000	10,000
5	7,000	10,000
6	5,000	7,000

75. Given the two cash flow profiles shown below, for what value of X are the two series equivalent using an interest rate of 12%? (3.7)

EOY	CF(A)	CF(B)
0	-$100,000	-$70,000
1	12,000	8,000
2	16,000	8,000
3	20,000	8,000
4	24,000	8,000 + X
5	28,000	8,000 + 2X
6	32,000	8,000 + 3X
7	36,000	8,000 + 4X
8	40,000	8,000 + 5X

76. Given the three cash flow profiles shown below, determine the values of X and Y so that all three cash flow profiles are equivalent at an annual interest rate of 15%. (3.7)

EOY	CF(A)	CF(B)	CF(C)
0	−$1,000	−$2,500	$ Y
1	X	3,000	Y
2	1.5X	2,500	Y
3	2.0X	2,000	2Y
4	2.5X	1,500	2Y
5	3.0X	1,000	2Y

77. Given the cash flow profiles shown below, determine the value of X such that the two cash flow profiles are equivalent at 20% *compounded continuously*. (3.7)

EOY	CF(A)	CF(B)
1988	−$75,000.00	$ −X
1989	10,000.00	5,000
1990	12,500.00	10,000
1991	15,625.00	15,000
1992	19,531.25	20,000
1993	24,414.06	25,000
1994	30,517.58	30,000

78. Given the two cash flow profiles shown below, for what value of X are the two series equivalent using an interest rate of 10%? (3.7)

EOY	CF(A)	CF(B)
0	−$60,000	−$80,000
1	15,000	15,000
2	$15,000(1 + x)$	20,000
3	$15,000(1 + x)^2$	25,000
4	$15,000(1 + x)^3$	30,000
5	$15,000(1 + x)^4$	35,000

79. The Ajax Chemical Company sells a particular chemical to the A. C. Works. A 5-year contract has been negotiated with the following terms. A constant volume, V_t, measured in cubic feet will be supplied uniformly and continuously during year t, with $V_t = 1.10V_{t-1}$. The price per cubic foot supplied during year t will be p_t, where $p_t = 1.08p_{t-1}$. The Ajax Chemical Company wishes to determine the equivalent *present value* of the contract using a continuous compounding rate of 10%/year. Express the equivalent present value in terms of p_1 and V_1. (3.7)

80. If a machine costs $10,000 and lasts for 10 years, at which time it is sold for $2000, what equal annual cost over its life is equivalent to these two cash flows with a 20% annual interest rate? (3.7)

81. Given the following profile of continuous flows, determine the equivalent series of annual discrete cash flows using a continuous compounding rate of 15%/year. (3.7)

EOY	$\bar{A}$
1	$10,000
2	12,000
3	14,000
4	16,000
5	18,000

82. Alison makes five annual deposits of $1000 in a money market account that pays interest at a rate of 8% compounded annually. One year after making the last deposit, the interest rate changes to 10% compounded annually. Five years after the last deposit, the accumulated money is withdrawn from the account. How much is withdrawn? (3.8)

83. Jerry deposits $10,000 in a money market account paying 8% compounded annually for the first 2 years and 6% compounded annually for the next 2 years. Four annual withdrawals are made from the savings account. The size of the withdrawal increases by $1000/year, with the first withdrawal occurring 1 year after the deposit. Determine the size of the last withdrawal, which depletes the balance of the account. (3.8)

84. Assume that six equal deposits of $500 are made at $t = 1, 2, 3, 4, 5,$ and 6 (3-month periods) into a fund paying 8% *compounded quarterly*. This interest applies until $t = 16$. The accumulated sum is withdrawn at this time and immediately deposited into a fund paying 10% *compounded continuously*. Beginning with $t = 17$, determine the amount of three equal amounts, A, that may be withdrawn such that the fund is depleted at $t = 19$. (3.8)

85. Sam borrows $10,000 and repays the loan with four equal annual payments. The interest rate for the first 2 years of the loan is 8% compounded continuously, and for the third and fourth years of the loan it is 10% compounded continuously. Determine the size of the annual payment. (3.8)

86. Linda deposits $5000 in a savings account. One year after the initial deposit, $1000 is withdrawn. Two years after the first withdrawal, $4000 is deposited in the account. Three years after the second deposit, $2000 is withdrawn from the account. Four years after the second withdrawal, all funds are withdrawn from the account. During the period of time the savings account was in use, the bank paid 8% compounded continuously for the first 2 years, 8% compounded annually for the next 4 years, and 8% compounded quarterly for the remainder of the time the account was in use. Determine the amount of the final withdrawal. (3.8)

87. John borrows $5000 and repays the loan with three equal annual payments. The interest rate for the first year of the loan is 8% compounded annually, for the second year of the loan, it is 10% compounded annually, and for the third year of the loan it is 12% compounded annually. Determine the size of the equal annual payment. (3.8)

88. Ken deposits $1000 in a fund that pays interest at a continuous compound rate of r_t for the tth year after the initial deposit, where

$$r_t = \begin{matrix} 0.06 + 0.005t, & t = 1, 2, 3, 4, 5, 6 \\ 0.09 & t = 7, 8, 9, 10, \ldots \end{matrix}$$

Thus a continuous compound rate of 6½% is earned during the first year, 7% is earned during the second year, 7½% is earned during the third year, and so on. A maximum of 9% compounded continuously is earned during the sixth and each successive year. If Ken withdraws $500 2 years after the initial deposit, how much money will be in the fund 8 years after the initial deposit? (3.8)

89. Robert deposits $2500 in an account that pays interest at a rate of 7% compounded annually. Two years after the deposit, the savings account begins paying interest at a rate of 8% compounded continuously. Five years after the deposit, the savings account begins paying interest at a rate of 9% compounded semiannually.

   a. How much money should be in the savings account 10 years after the initial deposit?

   b. What annual compound interest rate is equivalent to the interest pattern of the savings account over the 10-year period? (3.8)

90. Deborah deposits $1000 in a fund each year for a 10-year period. The fund initially pays 8% compounded annually. Immediately after she makes the sixth deposit, the fund begins paying 10% compounded annually. Deborah removes the money from the fund 3 years after her last deposit. How much should she be able to withdraw at that time? (3.8)

91. Based on discrete cash flows and continuous compounding, compute the present worth and annual worth for the following situation: (3.8)

EOY	CF	Nominal Rate During Year
0	−$10,000	0
1	1,000	0.08
2	3,000	0.09
3	5,000	0.10
4	7,000	0.11
5	9,000	0.12

92. Solve problem 91 for the case of discrete compounding. (3.8)

93. Maria deposits $2000 in a savings account that pays 7% compounded annually; 2 years after the deposit, the interest rate increases to 8% compounded annually. A second deposit of $2000 is made immediately after the interest rate changes to 8%. How much will be in the fund 5 years after the second deposit? (3.8)

94. Based on continuous cash flows and continuous compounding, compute the present worth for the following situation: (3.8)

Year	$\bar{A}$	Nominal Interest Rate During Year
1	$2,000e^{0.05}$	0.25
2	$2,000e^{0.10}$	0.20
3	$2,000e^{0.15}$	0.15
4	$2,000e^{0.20}$	0.10
5	$2,000e^{0.25}$	0.05

95. Labor costs over a 4-year period have been forecast in then-current dollars as follows: $10,000, $12,000, $15,000, and $17,500. The general inflation rate for the 4 years is forecast to be 9%. Determine the constant worth dollar amounts of labor over each of the 4 years. (3.9)

96. Yearly labor costs of a highway maintenance group are currently $420,000/ year. If labor rates escalate at a 13% rate and general inflation increases at 9%, determine for each of the next 4 years the labor cost in then-current and constant worth dollars. (3.9)

97. Big Screen, Inc. televisions are selling today for $2000. If the yearly growth rate is −3% and the general inflation rate is 10%, what will a Big Screen television set cost 5 years from now in then-current and constant worth dollars. (3.9)

98. If you desire a real return of 4% on your money, excluding inflation, and inflation is running at 8%, what combined discount rate should you be seeking? (3.9)

99. Mellin Transformers Co. uses a required return of 15% in all alternative evaluations. Inflation is running at 10%. What real discount rate, exclusive of inflation, are they implicitly using? (3.9)

100. The following material costs are anticipated over a 5-year period: $9000, $11,000, $14,000, $18,000, and $23,000. It is estimated that a 9% inflation rate will apply over the time period in question. The material costs given above are expressed in then-current dollars. The time value of money, excluding inflation, is estimated to be 4%. Determine the present worth equivalent for material cost using the following. (3.9)

   a. Then-current costs

   b. Constant worth costs

101. Zero Defects Electronics has current labor and materials costs of $200,000 and $300,000 per year, respectively. Over the next 3 years, labor costs will grow at a rate of 13%; material costs will grow at a rate of −6% (deescalation). Inflation will average 9%, and a real return of 5% is desired. Determine the following. (3.9)

   a. Then-current labor and material costs for each year and their discounted present worth

   b. Constant-worth labor and material costs for each year and their discounted present worth

102. A landfill has a first cost of $270,000. Annual operating and maintenance costs for the first year will be $40,000. These costs will increase at 11%/year. Income for dumping rights at the landfill will be held fixed at $120,000/year. The landfill will be in operation for 10 years. Inflation will average 8%, and a real return of 3.6% is desired. (3.9)

   a. Determine the present worth of this project using then-current dollars.

   b. Determine the present worth of this project using constant worth dollars.

103. The inflation rates for 4 years are forecast to be 9%, 7%, 10%, and 10%. The interest rate exclusive of inflation is anticipated to be 6%, 5%, 4%, and 5% over this same period. If labor is projected to be $1000, $1500, $2000, and $1000 in then-current dollars, during those years, determine the present worth equivalent for labor cost. (3.9)

104. Labor costs total $1.5 million currently. Labor costs are expected to have an overall growth rate of 10%/year. The yearly inflation rate is expected to be 7%. Determine the labor cost in 5 years, measured in today's dollars. (3.9)

105. The unit price of microcomputers, measured in constant dollars, is expected to decrease at an annual rate of 10%. However, the number of microcomputers purchased by the company is expected to increase over a 5-year period at an annual rate of 30%. The unit price of a microcomputer is currently $2000. This year the company will purchase 100 microcomputers at the current price. Using a real interest rate of 8% and an inflation rate of 6%, determine the following:

   a. the yearly expenditures for microcomputers over the 5-year period, measured in constant worth dollars.

   b. the yearly expenditures for microcomputers over the 5-year period, measured in then-current dollars.

   c. the single sum equivalent at the end of the current year of the expenditures on microcomputers over the 5-year period, measured in current dollars.

   d. the future worth equivalent of the expenditures on microcomputers at the end of the 5-year period, measured in then-current dollars. (3.9)

106. Inflation occurs continuously, rather than at discrete points in time. Solve Problem 104 based on an overall continuous nominal growth rate of 10%/year and a continuous nominal inflation rate of 7%/year. (3.9)

107. $90,000 is invested in a program to reduce the material requirements in a production process. As a result of the investment, the annual material requirement is reduced by 10,000 pounds. The present unit cost of the material is $2 per pound. The price of a pound of the material is expected to increase at an annual rate of 8%, due to inflation. Determine the combined interest rate that equates the present worth of the savings to the present worth of the investment over a 5-year period; based on the result obtained, determine the real interest rate that equates the two present worths. (3.9)

108. Due to productivity improvements and learning effects over the life of an electronic assembly, the labor content per assembly decreases at a rate of 0.50% per unit assembled. During the first year, 100,000 assembled units are required; the annual requirement increases at an annual rate of 20%. If the first unit required two hours to assemble and the cost of assembly labor totaled $25, what will be the total labor cost for those units assembled during the fifth year? Assume labor costs, on an hourly basis, remain constant. (3.9)

109. In Problem 108, suppose the annual inflation rate is 8% and the hourly cost for labor increases at an annual rate of 10%. Determine the total labor cost for those units assembled during the fifth year, (a) measured in constant worth dollars and (b) measured in then-current dollars. (3.9)

110. At $t = 0$, Martin borrows $5000 at 12%/period. Six equal payments are used to repay the loan at $t = 1, \ldots, 6$. Determine the amount of interest included in the fourth payment. (3.10)

111. Chris and Debbie borrow $10,000 at 12% compounded monthly. The loan is to be paid off with 48 equal monthly payments. One month after making the thirtieth payment, they decide to pay off the unpaid balance on the note. How much should be repaid? (3.10)

112. Tom and Dale purchase a boat for $150,000; the down payment is $15,000; the balance is financed over a 5-year period. Equal monthly payments are made. Determine the amount of interest paid the first month if the nominal monthly interest rate is ¾%. (3.10)

113. Dale borrows $8000 at 10% compounded annually. Four equal annual payments are used to repay the loan, with the first payment occurring 2 years after receiving the $8000. Determine the amount of principal included in each payment. (3.10)

114. Dr. Schultz is considering purchasing a bond having a face value of $2500 and a bond rate of 12% payable semiannually. The bond has a remaining life of 12 years. How much should she pay for the bond in order to earn a return on investment of 14% compounded semiannually? Assume the bond will be redeemed for face value. (3.11)

115. A $5000, 10-year, 12% semiannual bond is purchased at $t = 0$ by Dr. Rich for par value. After receiving the twelfth dividend, Dr. Rich sells the bond to Dr. Ruth at a price sufficient to yield a return on the original purchase price equivalent to 10% compounded semiannually. What was Dr. Rich's selling price? If Dr. Ruth keeps the bond until maturity and redeems it for $5000, what approximate annual effective yield rate will Dr. Ruth receive? (3.11)

116. Dr. Ramirez wishes to purchase a bond having a face value of $10,000 and a bond rate of 12% payable annually. The bond has a remaining life of 8 years. In order to earn an 18% rate of return on the investment, what amount should be paid for the bond? (3.11)

117. Elizabeth wishes to sell a bond that has a face vale of $1000. The bond has a bond rate of 9% with bond premiums paid annually. Four years ago, she paid $900 for the bond. At least a 12% return on investment is desired. What must be the minimum selling price for the bond in order to earn the desired return on investment? (3.11)

118. Dr. Schultz purchases a $10,000 face value bond for $10,500. It matures in 15 years after paying $1200 at the end of each year. What yield rate was earned on the investment? (3.11)

119. Kristin buys a $2000 bond for $1800. The bond has a bond rate of 12% with bond premiums paid annually. If the bond is kept for 8 years and sold for par value, determine the equivalent annual yield rate (rate of return) for Kristin's bond investment. (3.11)

120. Dr. Bozer is considering purchasing a bond having a face value of $2500 and a bond rate of 8% payable semiannually. The bond has a remaining life of 8 years. How much should Dr. Bozer pay for the bond in order to yield a return of 10% compounded semiannually? Assume the bond will be redeemed for face value. (3.11)

CHAPTER

4

Measuring the Worth
of Investments

4.1 INTRODUCTION

Chapter One recommended that engineers and systems analysts solve problems by formulating and analyzing the problem, generating a number of feasible solutions (alternatives) to the problem, comparing the investment alternatives, selecting the preferred solution, and implementing the solution. The process of evaluating the investment alternatives and selecting the preferred solution is the subject of the remainder of this book.

In this chapter we develop a number of methods and criteria for measuring the worth of investments. This material forms the basis for techniques used in Chapter Five to compare alternative investments and select those preferred from the economic point of view. The methods and criteria of this chapter must be understood and properly applied if valid economic analyses are to be assured.

Before proceeding with a treatment of the various methods of measuring the economic worth of investment alternatives, it is important to note that all but one of the measures of worth of an investment we treat involve the use of a compound interest rate. The interest rate used is variously referred to as a *minimum attractive rate of return (MARR), hurdle rate, required rate of return, return on investment,* and *discount rate.* (Recall, in Chapter Three we referred to the interest rate as the *yield rate* for a bond.) Where compound interest calculations are used, the measures of investment worth are referred to as *discounted cash flow (DCF) measures;* the methods used to

147

compute the values of investment worth are called *discounted cash flow methods.*

Various methods exist for determining the value of the interest rate to use. However, as explained subsequently in Chapter Five, we will consider the interest rate to represent a minimum attractive rate of return (*MARR*) on investment capital. If an investment proves to be profitable after considering the *MARR,* then it will be a viable candidate for investment. Furthermore, if money is not invested in the alternative in question, then we assume it will be invested elsewhere and earn a return at least equal to the *MARR.*

4.2 METHODS OF MEASURING INVESTMENT WORTH

As noted in Chapter Three in discussing equivalence, the economic worth of an investment can be measured in a number of ways. Present worth (PW) and annual worth (AW) methods are two commonly used approaches. Among the several methods of measuring investment worth are

1. Present worth method (PW).
2. Annual worth method (AW).
3. Future worth method (FW).
4. Internal rate of return method (IRR).
5. External rate of return method (ERR).
6. Savings/investment ratio method (SIR).
7. Payback period method (PBP).
8. Capitalized worth method (CW).

Each of the above measures of merit or measures of effectiveness has been used numerous times in evaluating real-world investments. They may be described briefly as follows:

1. Present worth method converts all cash flows to a single sum equivalent at time zero using $i = MARR$.
2. Annual worth method converts all cash flows to an equivalent uniform annual series of cash flows over the planning horizon using $i = MARR$.
3. Future worth method converts all cash flows to a single sum equivalent at the end of the planning horizon using $i = MARR$.
4. Internal rate of return method determines the interest rate that yields a future worth (or present worth or annual worth) of zero.
5. External rate of return method determines the interest rate that yields a future worth of zero, explicitly assuming reinvestment of recovered funds at the *MARR*.
6. Savings/investment ratio method determines the ratio of the present worth of savings to the present worth of the investment.

7. Payback period method determines how long at a zero interest rate it will take to recover the initial investment.

8. Capitalized worth method determines the single sum at time zero that is equivalent at $i = MARR$ to a cash flow pattern that continues indefinitely.

With the exception of the payback period and capitalized worth methods, all of the measures listed are equivalent methods of measuring investment worth. Hence, applying each of the first six measures of merit to the same investment alternative will yield the same recommendation.

Since the present worth, annual worth, future worth, internal rate of return, external rate of return, and savings/investment ratio methods are equivalent, why do more than one method exist? The primary reason for having different, but equivalent, measures of effectiveness for economic alternatives appears to be the differences in preferences among managers. Some individuals (and firms) prefer to express the net economic worth of an investment alternative as a single sum amount; hence, either the present worth method or the future worth method is used. Other individuals prefer to see the net economic worth spread out uniformly over the planning horizon, so the annual worth method is used by them. Yet another group of individuals wishes to express the net economic worth as a rate or percentage; consequently, one of the rate-of-return methods would be preferred. Finally, some individuals prefer to see the net economic worth expressed as a percentage of the investment required; the savings/investment ratio is one method of providing such information.

Since many organizations have established procedures for performing economic analyses, it seems worthwhile to consider in this chapter the more popular measures of merit that are used. Among those listed, it appears that the present worth, rate-of-return, and payback period methods are currently the most popular. However, the U.S. Postal Service and a number of governmental agencies have adopted some version of the savings/investment (or benefit/cost) ratio method for purposes of comparing investment alternatives; hence, it is popular in some sectors.

4.2.1 Present Worth Method

The present worth of investment alternative j can be represented as

$$PW_j(i) = \sum_{t=0}^{n} A_{jt}(1 + i)^{-t} \qquad (4.1)$$

with $PW_j(i)$ = present worth of Alternative j using $MARR$ of $i\%$
  n = planning horizon
  A_{jt} = net cash flow for Alternative j at the end of period t
  $i = MARR$

The *present worth method* is the most popular measure of merit available. When it is used in Chapter Five to compare alternatives, the one having the greatest present worth is the alternative recommended.

Example 4.1

A pressure vessel was purchased for $16,000, kept for 5 years, and sold for $3000. Annual operating and maintenance costs were $4000. Using a 12% minimum attractive rate of return, what was the present worth for the investment?

$$PW_1(12\%) = -\$16,000 - \$4000(P|A\ 12,5) + \$3000(P|F\ 12,5)$$
$$= -\$16,000 - \$4000(3.6048) + \$3000(.5674)$$
$$= -\$28,717$$

We can see that a single expenditure of $28,717 at time zero is equivalent to the cash flow profile for the pressure vessel.

Example 4.1 deals with an investment where no returns other than the salvage value are shown, resulting in a negative present worth. Example 4.2 considers an investment having a series of positive net cash flows.

Example 4.2

Improved tooling for numerical control machinery will cost $10,000, last 6 years, and have no salvage value at that time. Due to this investment, net income will increase by $2525 during each of the first 3 years and by $3840 during each of the remaining 3 years. Using a 15% *MARR*, what is the present worth for the investment?

$$PW_1(15\%) = -\$10,000 + \$2525(P|A\ 15,3) + \$3840(P|A\ 15,3)(P|F\ 15,3)$$
$$= -\$10,000 + \$2525(2.2832) + \$3840(2.2832)(.6575)$$
$$= \$1529.70$$

A single sum of $1529.70 at time zero is equivalent to the cash flow profile for this investment.

4.2.2 Annual Worth Method

The annual worth of investment alternative j can be computed as

$$AW_j(i) = \left[\sum_{t=0}^{n} A_{jt}(P|F\ i,t)\right](A|P\ i,n) \qquad (4.2)$$

or

$$AW_j(i) = PW_j(i)(A|P\ i,n) \tag{4.3}$$

where $AW_j(i)$ denotes the annual worth of alternative j using $i = MARR$. The *annual worth method* and the *present worth method* are related by a constant multiplier, $(A|P\ i,n)$, and are therefore equivalent measures of investment worth. The investment alternative having the greatest annual worth will be preferred when the annual worth method is used to compare alternatives in Chapter Five.

Example 4.3

Let us determine the annual worth for the pressure vessel of Example 4.1 in two different ways. Recall that $P = \$16,000$, $n = 5$, $F = \$3000$, annual operating and maintenance were \$4000, and $i = 12\%$.

$$\begin{aligned}AW_1(12\%) &= -\$16,000(A|P\ 12,5) - \$4000 + \$3000(A|F\ 12,5) \\ &= -\$16,000(.2774) - \$4000 + \$3000(.1574) \\ &= -7966.20/\text{year}\end{aligned}$$

Or, using Equation 4.3, and the answer to Example 4.1,

$$\begin{aligned}AW_1(12\%) &= (-\$28,717)(A|P\ 12,5) \\ &= (-\$28,717)(.2774) \\ &= -\$7966.10/\text{year}\end{aligned}$$

The minor difference in answers is due to round-off error. An expenditure of \$7966.20 at the end of each of 5 years is equivalent to the cash flow profile for the pressure vessel.

Example 4.4

We want to determine the annual worth for the improved numerical control tooling of Example 4.2. Recall that the tooling lasts 6 years and costs \$10,000, but net incomes increase by \$2525 during each of the first 3 years and by \$3840 thereafter. *MARR* is 15%.

$$\begin{aligned}AW_1(15\%) &= -\$10,000(A|P\ 15,6) + \$2525 + \$1315(F|A\ 15,3)(A|F\ 15,6) \\ &= -\$10,000(.2642) + \$2525 + \$1315(3.4725)(.1142) \\ &= \$404.48/\text{year}\end{aligned}$$

or

$$\begin{aligned}AW_1(15\%) &= PW_1(15\%)(A|P\ 15,6) \\ &= \$1529.70(.2642) \\ &= \$404.15/\text{year}\end{aligned}$$

Note that the difference between \$404.48/year and \$404.15/year is due to round-off errors.

___

4.2.3 Future Worth Method

The future worth of investment alternative j can be determined using the relationship

$$FW_j(i) = \sum_{t=0}^{n} A_{jt}(1 + i)^{n-t} \tag{4.4}$$

or

$$FW_j(i) = PW_j(i)(F|P\ i,n) \tag{4.5}$$

or

$$FW_j(i) = AW_j(i)(F|A\ i,n) \tag{4.6}$$

where $FW_j(i)$ is defined as the future worth of investment j using a $MARR$ of $i\%$. The *future worth method* is equivalent to the *present worth method* and the *annual worth method*, since the ratio of $FW_j(i)$ and $PW_j(i)$ equals a constant, $(F|P\ i,n)$, and the ratio of $FW_j(i)$ and $AW_j(i)$ equals a constant, $(F|A\ i,n)$. The alternative having the greatest future worth will be preferred when the future worth method is used in Chapter Five to compare alternatives.

___

Example 4.5

For the previous two examples, the future worth is given by

$$FW_1(15\%) = -\$10,000(F|P\ 15,6) + \$2525(F|A\ 15,6) + \$1315(F|A\ 15,3)$$
$$= -\$10,000(2.3131) + \$2525(8.7537) + \$1315(3.4725)$$
$$= \$3538.43$$

or

$$FW_1(15\%) = PW_1(15\%)(F|P\ 15,6)$$
$$= \$1529.70(2.3131)$$
$$= \$3538.35$$

or

$$FW_1(15\%) = AW_1(15\%)(F|A\ 15,6)$$
$$= \$404.48(8.7537)$$
$$= \$3540.70$$

Again, differences in answers are due to round-off errors.

4.2.4 Internal Rate of Return Method

The rate of return for an alternative can be defined as the interest rate that equates the future worth to zero. Letting i_j^* denote the rate of return for alternative j, Equation 4.4 becomes

$$0 = \sum_{t=0}^{n} A_{jt}(1 + i_j^*)^{n-t} \qquad (4.7)$$

This method of defining the rate of return is referred to in the economic analysis literature as the discounted *cash flow rate of return, internal rate of return*, and the *true rate of return*. We prefer the term *internal rate of return* (*IRR*).

Note that the present worth and the annual worth can be obtained by multiplying both sides of Equation 4.4 by the appropriate interest factor [i.e., $(P|F\, i_j^*, n)$ and $(A|F\, i_j^*, n)$]. Hence, the internal rate of return can also be defined to be the interest rate that yields either a present worth or an annual worth of zero. Depending on the form of a particular cash flow profile, it might be more convenient to use a present worth or an annual worth formulation to determine the internal rate of return for an alternative.

It is important to understand the definition of rate of return inherent in the use of the *IRR* method. In particular, the internal rate of return on an investment is defined as *the rate of interest earned on the unrecovered balance of an investment*. This concept was demonstrated in Chapter Three in discussing the amount of a loan payment that was principal. It is illustrated again in Table 4.1, where $10,000 is invested to obtain the receipts shown over a 6-year period. A_t denotes the cash flow at the *end* of period t, B_t represents the unrecovered balance at the *beginning* of period t, E_t is the unrecovered balance at the *end* of period t, and I_t is defined as the interest on the unrecovered balance *during* period t. The following relationships exist.

TABLE 4.1

Data Illustrating the Meaning of the Internal Rate of Return

t	A_t	B_t	I_t[a]	E_t
0	−10,000			−10,000
1	2,525	−10,000	−2,000	− 9,475
2	2,525	− 9,475	−1,895	− 8,845
3	2,525	− 8,845	−1,769	− 8,089
4	3,840	− 8,089	−1,618	− 5,867
5	3,840	− 5,867	−1,173	− 3,200
6	3,840	− 3,200	− 640	0

[a] Based on $i = 0.20$.

$$E_0 = A_0$$

$$B_t = E_{t-1} \qquad t = 1, \ldots, n$$

$$I_t = B_t i \qquad t = 1, \ldots, n$$

$$E_t = A_t + B_t + I_t \qquad t = 1, \ldots, n$$

If i is the internal rate of return, then E_n will equal zero. As indicated in Table 4.1, if i is 20%, E_n is approximately zero. Consequently, i^* is approximately 20%. The equivalence of E_n being zero and the future worth being zero is easily understood by recognizing that E_n is actually the future worth of the cash flow profile. To see why this is true, notice that

$$E_n = A_n + B_n + I_n$$

Employing the definition of I_n,

$$E_n = A_n + B_n(1 + i)$$

By the relationship between B_n and E_{n-1}, it is seen that

$$E_n = A_n + E_{n-1}(1 + i)$$

Since a similar relationship exists between E_{n-1} and E_{n-2}, we note that

$$E_n = A_n + A_{n-1}(1 + i) + E_{n-2}(1 + i)^2$$

Generalizing, the recursive relationship between E_t and E_{t-1} gives

$$E_n = A_n + A_{n-1}(1 + i) + A_{n-2}(1 + i)^2 + \cdots + A_0(1 + i)^n$$

Hence, we see that

$$E_n = FW(i\%)$$

as anticipated.

The above example illustrates that the time value of money operations involved in the *IRR* method are equivalent to assuming that all monies received are reinvested and earn interest at a rate equal to the internal rate of return. This can also be seen mathematically by letting

$$A_{jt} = \text{net cash flow for Investment } j \text{ in period } t$$

$$R_{jt} = \begin{cases} A_{jt}, & \text{if } A_{jt} \geq 0 \\ 0, & \text{otherwise} \end{cases}$$

$$C_{jt} = \begin{cases} -A_{jt}, & \text{if } A_{jt} < 0 \\ 0, & \text{otherwise} \end{cases}$$

r_t = reinvestment rate for positive cash flows occurring in period t

i' = the rate of return for negative cash flows

Then the following relationship can be defined.

$$\sum_{t=0}^{n} R_{jt}(1 + r_t)^{n-t} = \sum_{t=0}^{n} C_{jt}(1 + i')^{n-t} \qquad (4.8)$$

Note that the future worth of reinvested monies received must equal the future worth of investments.

If r_t equals i', Equation 4.8 becomes

$$0 = \sum_{t=0}^{n} (R_{jt} - C_{jt})(1 + i')^{n-t} \qquad (4.9)$$

Letting A_{jt} equal $R_{jt} - C_{jt}$ defines the *IRR* method given by Equation 4.7. Hence, we see that the rate of return obtained using the *IRR* method can be interpreted as the reinvestment rate for all recovered funds.

It should be noted that it is not essential that recovered funds be reinvested at the solving rate of return in order for the initial investment to have yielded the internal rate of return. For example, suppose an investment of $100,000 yields $10,000 annually for a 4-year period, and at the end of the fifth period the investment is terminated, and $110,000 is recovered. Clearly, the investment yielded an internal rate of return of 10%, regardless of what was done with the annual recovery of $10,000. The recovered amounts could have been invested in, say, the stock market and resulted in positive or negative results. However, the original investment of $100,000 did, in fact, yield a 10% return. [17]

Since the determination of the rate of return involves solving Equation 4.7 for i_j^*, it is seen that (for a given investment j) it is necessary to determine the values of x that satisfy the n degree polynomial $0 = A_0 x^n + A_1 x^{n-1} + \cdots + A_{n-1}x + A_n$ where $x = (1 + i_j^*)$. In general, there can exist n distinct roots (values of x) for an n-degree polynomial; however, most cash flow profiles encountered in practice will have a unique root (internal rate of return).

Descartes' rule of signs indicates the n-degree polynomial will have a single positive real root if there is a single sign change in the sequence of cash flows, $A_0, A_1, \ldots, A_{n-1}, A_n$. If there are two sign changes, then there will be either two or no positive real roots; if there are three sign changes, there will be either three or one positive real roots; if there are four sign changes, there will be either four, two, or no positive real roots; if there are five sign changes, there will be either five, three, or one positive real roots; and so forth. Since the typical cash flow pattern begins with a negative cash flow followed by a positive cash flow, a unique root normally exists. (Note, since $x = (1 + i_j^*)$, a positive real root can occur even when i_j^* is negative, e.g., x will be positive when $i_j^* > -1$.)

Example 4.6

As an illustration of a cash flow profile having multiple roots, consider the data given in Table 4.2. The future worth of the cash flow series will be zero using a 20, 40, or 50% interest rate.

$$FW_1(20\%) = -\$1000(1.2)^3 + \$4100(1.2)^2 - \$5580(1.2) + \$2520 = 0$$

$$FW_1(40\%) = -\$1000(1.4)^3 + \$4100(1.4)^2 - \$5580(1.4) + \$2520 = 0$$

$$FW_1(50\%) = -\$1000(1.5)^3 + \$4100(1.5)^2 - \$5580(1.5) + \$2520 = 0$$

A plot of the future worth for this example is given in Figure 4.1. The future worth polynomial is a third-degree polynomial, and there are three changes of sign in the ordered sequence of cash flows $(-, +, -, +)$. In this case there are three unique positive real roots, corresponding to $i = .20$, $i = .40$, and $i = .50$. This is seen by factoring the future worth polynomial, which can be written as

$$FW_1(i) = \$1000(1.2 - x)(1.4 - x)(1.5 - x)$$

where $x = (1 + i)$.

Multiple rates of return as seen in Example 4.6 are difficult to interpret properly without other information such as the future worth, present worth, or annual worth. For example, the pattern of cash flows shown in Table 4.2 is preferred to investing at the *MARR* if $0\% \leq MARR < 20\%$ or $40\% < MARR < 50\%$. This is easily seen by looking at Figure 4.1, but is not immediately obvious from the three roots of the future worth equation.

The difficulty of interpreting the polynomial roots in complex cash flow profiles is a severe drawback to the use of the *IRR* approach. For additional discussion of.the subject of multiple roots in internal rate of return calculations, as well as their interpretation, see Bernhard [4, 5, 6] and Bussey [7].

4.2.5 External Rate of Return Method

The possibility of multiple roots occurring in the internal rate of return calculation has led to the development of an alternative rate of return

TABLE 4.2

Cash Flow Profile

EOY	CF
0	−$1,000
1	4,100
2	− 5,580
3	2,520

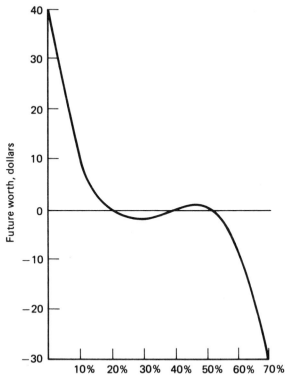

FIGURE 4.1. Plot of future worth for Example 4.6.

method, called the *external rate of return method*. The external rate of return (*ERR*) method consists of the determination of the value of i' that satisfies Equation 4.8, which is repeated here for convenience.

$$\sum_{t=0}^{n} R_{jt}(1 + r_t)^{n-t} = \sum_{t=0}^{n} C_{jt}(1 + i')^{n-t} \qquad (4.10)$$

In this equation, i' is now known as the external rate of return. As defined previously, R_{jt} and C_{jt} are positive and negative net cash flows for investment j during period t, respectively, and r_t is the reinvestment rate for positive cash flows occurring in period t. Normally, r_t equals the minimum attractive rate of return, since the *MARR* reflects the opportunity cost for money available for investment. That is, we assume that positive net cash flows can always be reinvested at the *MARR*.

The external rate of return method is one of several rate of return approaches that consider the explicit reinvestment of positive cash flows, positive net cash flows, accumulated cash flows, and so forth. The *ERR* method offers an advantage over the *IRR* approach, since a unique value of i' satisfies Equation 4.10. That is, there is only one external rate of return for a given cash flow profile. If i' exceeds the *MARR*, the investment is preferred over investing at the *MARR*.

As an exercise, you might want to show that the value of the external rate of return will always be between the values of the internal rate of return and the *MARR*. Because of this property, if you use either the *ERR* or the *IRR* method, your economic choice will be the same.

Example 4.7

Let us reconsider the cash flows of Example 4.6 in Table 4.2. The value of the *ERR* may be determined by solving the equation

$$4100(1 + r_t)^2 + 2520 = 1000(1 + i')^3 + 5580(1 + i')$$

for i', where $r_t = MARR$. This results in the following series of values for a wide range of the *MARR*.

MARR Value (r_t)	External Rate of Return (i')
.10	.101302
.15	.150458
.20	.200000
.25	.249817
.30	.299812
.35	.349898
.40	.400000
.45	.450053
.50	.500000
.55	.549795
.70	.697893

There is not much difference between the value of r_t and its corresponding value i'. This is simply a function of the particular cash flow profile and is not important (note that the future worths in Figure 4.1 did not deviate far from zero either). What *is* important is that the *ERR* exceeds the *MARR* in the range $0 \le MARR < 20\%$ and $40\% < MARR < 50\%$. These are precisely the same ranges for which the future worths were positive in Figure 4.1.

This example illustrates the consistency of the *ERR* method with the future worth, present worth, and annual worth methods when evaluating a single investment, even when that investment has a complex cash flow pattern.

Example 4.8

For the data provided in Table 4.1, suppose money received from the initial investment is reinvested and earns 15% interest. At the end of the sixth year the reinvested funds total

$$\$2525(F|A\ 15,6) + \$1315(F|A\ 15,3)$$
$$= \$2525(8.7537) + \$1315(3.4725)$$
$$= \$26,669.43$$

Consequently, the *external rate of return* is defined as the interest rate such that the future worth of the $10,000 investment equals $26,669.43. Thus,

$$\$10,000(1 + i')^6 = \$26,669.43$$

Taking the logarithm and solving for i yields a value of 17.76% as the external rate of return.

Example 4.9

Now, rework Example 4.8 assuming recovered funds are reinvested at 20%. The future worth of the reinvested funds will be

$$\$2525(F|A\ 20,6) + \$1315(F|A\ 20,3)$$
$$= \$2525(9.9299) + \$1315(3.6400)$$
$$= \$29,859.60$$

Setting the future worth of the $10,000 investment equal to $29,859.60 yields

$$\$10,000(1 + i')^6 = \$29,859.60$$

Solving for i' gives a value of 20% as the external rate of return, as anticipated.

4.2.6 Savings/Investment Ratio Method

The *savings/investment ratio method* can be defined in many ways; two typical approaches will be described. The *savings/investment ratio* to be used for analyzing and comparing alternatives in Chapter Five is given as

$$SIR_j(i) = \frac{\sum_{t=0}^{n} R_{jt}(1 + i)^{-t}}{\sum_{t=0}^{n} C_{jt}(1 + i)^{-t}} \tag{4.11}$$

where $SIR_j(i)$ is the savings/investment ratio for investment alternative j based on a *MARR* of i%. We again define A_{jt}, R_{jt}, and C_{jt} as follows:

$$A_{jt} = \text{net cash flow for investment } j \text{ in period } t$$

$$R_{jt} = \begin{cases} A_{jt}, & \text{if } A_{jt} \geq 0 \\ 0, & \text{otherwise} \end{cases}$$

$$C_{jt} = \begin{cases} -A_{jt}, & \text{if } A_{jt} < 0 \\ 0, & \text{otherwise} \end{cases}$$

Note that this *SIR* formulation is the present worth of net positive cash flows divided by the present worth of net negative cash flows. Therefore, for a project to be preferred over investing at the *MARR*, the ratio must be greater than one.

Another formulation of the savings/investment ratio is defined as follows:

$$SIR_j(i) = \frac{\sum\limits_{t=0}^{n} A_{jt}(1 + i)^{-t}}{\sum\limits_{t=0}^{n} C_{jt}(1 + i)^{-t}} \tag{4.12}$$

where A_{jt} and C_{jt} are as defined previously. Note that this formulation is simply the present worth of all cash flows divided by the present worth of all negative cash flows. A project having a ratio greater than zero is preferable to investing at the *MARR*.

Example 4.10

Consider the cash flows given in Table 4.1 and let the *MARR* be 15%. The present worth of the net positive cash flows is

$$\sum_{t=0}^{n} R_{1t}(1 + i)^{-t} = \$2525(P|A\ 15,3) + \$3840(P|A\ 15,3)(P|F\ 15,3)$$

$$= \$2525(2.2832) + \$3840(2.2832)(.6575)$$

$$= \$11,529.70$$

The present worth of the net negative cash flows is

$$\sum_{t=0}^{n} C_{1t}(1 + i)^{-t} = -\$10,000$$

The present worth of all net cash flows is

$$\sum_{t=0}^{n} A_{1t}(1 + i)^{-t} = -\$10,000 + \$11,529.70 = \$1529.70$$

The savings/investment ratio using Equation 4.11 is given by

$$SIR_1(15\%) = \frac{\$11,529.70}{\$10,000} = 1.15297$$

Using the savings/investment ratio given by Equation 4.12 results in

$$SIR_1(15\%) = \frac{\$1529.70}{\$10,000} = .15297$$

Note that these ratios are greater than one and zero, respectively, both of them indicating that an investment resulting in the cash flow profile of Table 4.1 is preferable to investing at the *MARR*.

An alternative label for the savings/investment ratio is the *benefit-cost ratio*. The savings are interpreted as benefits derived from a venture, and the investment corresponds to the cost of providing the benefits. The benefit-cost ratio is used extensively in government and public sector economic analyses. Like the savings/investment ratio, the benefit-cost ratio has many possible formulations. We explore benefit-cost analysis more fully in Chapter Seven.

4.2.7 Payback Period Method

Each of the preceding methods for measuring project worth is equivalent or consistent. That is, each will provide the same decision regarding the desirability of investing in a project versus investing at the *MARR*. The *payback period method,* unfortunately, is not an equivalent method. Yet, it is used frequently in economic analyses; therefore, we should know about its good points as well as its pitfalls.

The payback period method involves the determination of the length of time required to recover the initial investment based on a zero interest rate. Letting C_{j0} denote the initial cost of investment alternative j, and R_{jt} denote the net revenue received from investment j during period t, if we assume no other negative net cash flows occur, then the smallest value of m_j such that

$$\sum_{t=1}^{m_j} R_{jt} \geq C_{j0} \tag{4.13}$$

defines the payback period for investment j. The investment alternative having the smallest payback period is the preferred alternative using the payback period method.

Example 4.11

Based on the data given in Table 4.1,

$$\sum_{t=1}^{3} R_{1t} = \$2525 + \$2525 + \$2525 = \$7575 < C_{10} = \$10,000$$

and

$$\sum_{t=1}^{4} R_{1t} = \$2525 + \$2525 + \$2525 + \$3840 = \$11,415 > C_{10} = \$10,000$$

Consequently, m equals 4, indicating that 4 years are required to pay back the original investment.

A number of variations of the payback or payout method have been used by different organizations. For example, if negative net cash flows occur after year zero, the payback period might be defined as the smallest value of m_j in which the sum of the net cash flows $\sum_{t=0}^{m_j} A_{jt}$ first exceeds zero. Also, the payback period may be noninteger, assuming cash flows are uniformly distributed throughout the year. However, the basic deficiencies of the payback period method are present in all variations with which we are familiar; the *timing* of cash flows and the *duration* of the project are ignored.

To illustrate the deficiencies of the payback period, consider the cash flow profiles depicted in Figure 4.2a. Using the payback period, profile B would be preferred, even though we would most likely choose profile A. Likewise, in Figure 4.2b, profiles C and D are equally preferred using the payback period, even though profile D is clearly better.

Despite its obvious deficiencies, the payback period method continues to be one of the most popular methods of judging the desirability of investing in a project. The reasons for its popularity include the following:

1. Does not require interest rate calculations.
2. Does not require a decision concerning the discount rate (*MARR*) to use.
3. Is explained and understood easily.
4. Reflects a manager's attitudes when investment capital is limited.
5. Hedges against uncertainty of future cash flows.
6. Provides a rough measure of the liquidity of an investment.

The payback period method is recommended as a secondary method of measuring investment worth. In particular, it is suggested that the payback period method be used in addition to a method based on the time value of money.

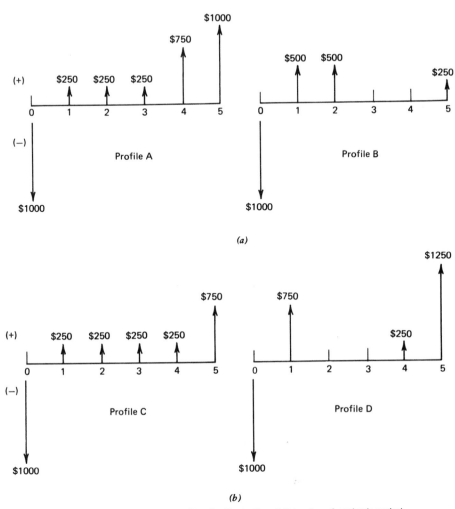

FIGURE 4.2. Four different cash flow profiles for illustrating deficiencies of payback period.

4.2.8 Perpetuities and Capitalized Worth Method

A specialized type of cash flow series is a *perpetuity*, a uniform series of cash flows which continues indefinitely. This is a special case, since an infinite series of cash flows would rarely be encountered in the business world; rather, a finite series of cash flows is the general rule. However, for such very long-term investment projects as bridges, highways, forest harvesting, or the establishment of endowment funds where the estimated life is 50 years or more, an infinite cash flow series may be appropriate.

If a present value P is deposited into a fund at interest rate i per period so that a payment of size A may be withdrawn each and every period forever, then the following relation holds between P, A, and i.

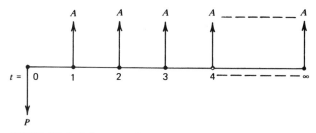

FIGURE 4.3. An infinite cash flow series.

$$Pi = A \qquad (4.14)$$

Thus, as depicted in Figure 4.3, P is a present value that will pay out equal payments of size A indefinitely if the interest rate per period is i. The present value P is termed the *capitalized worth* of A, the size of each of the perpetual payments.

Example 4.12

What deposit at $t = 0$ into a fund paying $9\frac{1}{2}\%$ annually is required in order to pay out $5000 each year forever? The solution is straightforward from

$$P = \frac{A}{i} = \$5000/.095 = \$52,631.58$$

By means of the subsequent example, let us now broaden the meaning of capitalized worth to be that present value that would pay for the first cost of some project and provide for its perpetual maintenance at an interest rate i.

Example 4.13

Project ABC consists of the following requirements:

1. A $50,000 first cost at $t = 0$.
2. A $5000 expense every year.
3. A $25,000 expense every third year forever, with the first expense occurring at $t = 3$.

What is the capitalized cost of Project ABC if $i = 15\%$ annually?

It will be instructive to determine the capitalized cost of each requirement separately and then sum the results. First, the capitalized cost of "$25,000

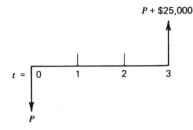

FIGURE 4.4. Capitalized cost of $25,000 every third year—first view.

every third year forever'' may be determined from any of three points of view. One view is that a value P is required at the beginning of a 3-year period such that, with interest compounded at 15% annually, a sum of P + $25,000 will accrue at the end of the 3-year period. Thus, $25,000 would be withdrawn; thereby leaving the value P to repeat the cycle indefinitely each 3-year period. This logic is illustrated in Figure 4.4.

Thus,

$$P(F|P\ 15,3) = P + \$25,000$$

or

$$1.5209P = P + \$25,000$$

and the capitalized worth (CW) is

$$CW = P = \$47,993.86$$

which is the capitalized cost of the requirement.

A second view is to reason that, for each 3-year period, three equal deposits of size A are required that will amount to $25,000 at the end of the third year. Then, if these payments of size A occur every year, $25,000 will be available every third year forever. The present value P that yields the required payment A is the solution, as shown in Figure 4.5.

The required payment A is given by

$$A = F(A|F\ 15,3) = \$25,000(0.2880)$$
$$= \$7200$$

Then,

$$CW = \frac{A}{i} = \frac{\$7200}{0.15} = \$48,000.00$$

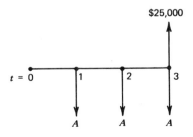

FIGURE 4.5. Capitalized cost of $25,000 every third year—second view.

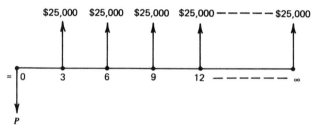

FIGURE 4.6. Capitalized cost of $25,000 every third year—third view.

which is approximately the same answer as from the first approach. The difference of $6.14 is due to rounding in the interest factor calculations.

A third point of view for the $25,000 requirement is to consider the infinite series depicted in Figure 4.6, such that $P = A/i$ would yield the desired result. However, the value of i required is the effective interest rate for a 3-year period. Thus,

$$i = [(1 + 0.15)^3 - 1] = 0.5209$$

and

$$CW = \frac{\$25,000}{0.5209} = \$47,996.16$$

The capitalized cost of Project ABC is computed as follows:

1. Capitalized cost of $50,000 first cost   = $ 50,000
2. Capitalized cost of $5000 every year = $5000/0.15 = $ 33,333
3. Capitalized cost of $25,000 every third year   = $ 48,000
   TOTAL CAPITALIZED COST $131,333

The capitalized worth of $131,333 would provide approximately $A = Pi = \$131,333(0.15) = \$19,700$ every year forever.

The capitalized worth may be looked upon as a present worth of some cash flow pattern that repeats indefinitely.

Example 4.14

Recall that the annual worth of the cash flows in Example 4.4 was calculated to be $404.48/year (these were the same cash flows later shown in Table 4.1). If this cash flow pattern were to repeat every 6 years forever, the capitalized worth would be

$$CW = \frac{A}{i} = \frac{\$404.48}{.15} = \$2699.20$$

FIGURE 4.7. Cash flow diagram of investment in an asset.

Note that this is not drastically higher than the present worth of $1529.70 calculated in Example 4.2 for the first 6 years. This is because the distant cash flows are substantially discounted at $i = 15\%$ in the capitalized worth calculation.

4.3 CAPITAL RECOVERY FORMULA

In engineering economic analyses, it is common to refer to the *capital recovery cost* of an asset. Figure 4.7 illustrates an investment of $P in an asset having a life of n years and disposed of for a salvage value of $F. The capital recovery cost is a uniform annual amount defined as

$$CR = P(A|P\ i,n) - F(A|F\ i,n) \qquad (4.15)$$

The salvage value at year n is an income, and therefore a negative cost.

Example 4.15

A stand-alone minicomputer is purchased for $P = \$82,000$, has a service life of $n = 7$ years, and a salvage value of $F = \$5000$. At an interest rate of 15%, the capital recovery cost is

$$
\begin{aligned}
CR &= \$82,000(A|P\ 15,7) - \$5000(A|F\ 15,7) \\
&= \$82,000(.2404) - \$5000(.0904) \\
&= \$19,260.80/\text{year}
\end{aligned}
$$

From Chapter Three and the interest tables we know that

$$(A|P\ i,n) = (A|F\ i,n) + i \qquad (4.16)$$

and similarly

$$(A|F\ i,n) = (A|P\ i,n) - i \qquad (4.17)$$

Substituting Equations 4.16 and 4.17 into Equation 4.15 leads to two more capital recovery cost formulas. They are, respectively,

$$CR = (P\text{-}F)(A|F\ i,n) + Pi \qquad (4.18)$$

and

$$CR = (P\text{-}F)(A|P\ i,n) + Fi \qquad (4.19)$$

Among the three methods of computing capital recovery cost, Equation 4.19 appears to be the most popular. We tend to use Equation 4.15 since it is a direct application of the interest factors to the actual cash flows. The relationship between capital recovery cost and depreciation or cost recovery is explained in Chapter Six.

Example 4.16

We can apply Equations 4.18 and 4.19 to the previous example problem having $P = \$82,000$, $n = 7$, $F = \$5000$, and $i = 15\%$. Using Equation 4.18, we have

$$CR = (\$82,000 - \$5000)(A|F\ 15,7) + \$82,000(.15)$$
$$= \$77,000(.0904) + \$82,000(.15)$$
$$= \$19,260.80/year$$

Using Equation 4.19 we have

$$CR = (\$82,000 - \$5000)(A|P\ 15,7) + \$5000(.15)$$
$$= \$77,000(.2404) + \$5000(.15)$$
$$= \$19,260.80/year$$

4.4 SUMMARY

In this chapter we developed the bases for estimating the worth of an investment. We considered selection of the minimum attractive rate of return, numerous measures of investment worth, bond problems, capital recovery, and the effects of inflation. Now, we are prepared to use these concepts in analyzing and selecting from among several alternative investments. This will be done in Chapter Five.

BIBLIOGRAPHY

1. Adler, M., "The True Rate of Return and the Reinvestment Rate," *The Engineering Economist, 15* (3), 1970, pp. 185–87.

2. Beaves, R. G., "Net Present Value and Rate of Return: Implicit and Explicit Reinvestment Assumptions," *The Engineering Economist, 33* (4), 1988, pp. 275–302.

3. Beenhakker, H. L., "Discounting Indices Proposed for Capital Investment Evaluation: A Further Examination," *The Engineering Economist, 18* (3), 1973, pp. 149–68.

4. Bernhard, R. H., "On the Inconsistency of the Soper and Sturm-Kaplan Conditions for Uniqueness of the Internal Rate of Return," *The Journal of Industrial Engineering, 18* (8), 1967, pp. 498–500.

5. Bernhard, R. H., "A Comprehensive Comparison and Critique of Discounting Indices Proposed for Capital Investment Evaluation," *The Engineering Economist, 16* (3), 1971, pp. 157–186.

6. Bernhard, R. H., "'Modified' Rates of Return for Investment Project Evaluation—A Comparison and Critique," *The Engineering Economist, 24* (3), 1979, pp. 161–68.

7. Bussey, L. E., *The Economic Analysis of Industrial Projects,* Prentice-Hall, 1978.

8. Canada, J. R., "Rate of Return: A Comparison Between the Discounted Cash Flow Model and a Model Which Assumes an Explicit Reinvestment Rate for the Uniform Income Flow Case," *The Engineering Economist, 9* (3), 1964, pp. 1–15.

9. de Faro, C., "On the Internal Rate of Return Criterion," *The Engineering Economist, 19* (3), 1974, pp. 165–94.

10. Freidenfelds, J., and Kennedy, M., "Price Inflation and Long-Term Present-Worth Studies," *The Engineering Economist, 24* (3), 1979, pp. 143–60.

11. Grant, E. L., "Reinvestment of Cash Flow Controversy—A Review of: FINANCIAL ANALYSIS IN CAPITAL BUDGETING by Pearson Hunt," *The Engineering Economist, 11* (3), 1966, pp. 23–29.

12. Jeynes, P. H., "Minimum Acceptable Return," *The Engineering Economist, 9* (4), 1964, pp. 9–25.

13. Jeynes, P. H., "Comment on Ralph O. Swalm's Letter," *The Engineering Economist, 10* (3), 1965, pp. 38–42.

14. Jeynes, P. H., "The Significance of Reinvestment Rate," *The Engineering Economist, 11* (1), 1966, pp. 1–9.

15. Kirshenbaum, P. S., "A Resolution of the Multiple Rate-of-Return Paradox," *The Engineering Economist, 10* (1), 1965, pp. 11–16.

16. Lin, S. A., "The Modified Internal Rate of Return and Investment Criterion," *The Engineering Economist, 21* (4), 1976, pp. 237–48.

17. Lohmann, J. R., "The IRR, NPV and the Fallacy of the Reinvestment Rate Assumptions," *The Engineering Economist, 18* (4), 1988, pp. 303–330.

18. Oakford, R. V., Bhimjee, S. A., and Jucker, J. V., "The Internal Rate of Return, the Pseudo Internal Rate of Return, and NPV and Their Use in Financial Decision Making," *The Engineering Economist, 22* (3), 1977, pp. 187–202.

19. Sharp, G. P., and Guzman-Garza, A., "Borrowing Interest Rate as a Function of the Debt-Equity Ratio in Capital Budgeting Models," *The Engineering Economist, 26* (4), 1981, pp. 293–315.

20. Sullivan, W. G., and Bontadelli, J. A., "The Industrial Engineer and Inflation," *Industrial Engineering, 12* (3), March 1980, pp. 24–33.

21. Swalm, R. O., "On Calculating the Rate of Return on an Investment," *The Journal of Industrial Engineering, 9* (2), 1958, pp. 99–103.

22. Swalm, R. O., "Comment on 'Minimum Acceptable Return' by Paul H. Jeynes," *The Engineering Economist, 10* (3), 1965, pp. 35–38.

23. Teichroew, D., Robicheck, A. A., and Montalbano, M., "Mathematical Analysis of Rates of Return Under Certainty," *Management Science, 11* (3), 1965, pp. 395–403.

24. Waters, R. C., and Bullock, R. L., "Inflation and Replacement Decisions," *The Engineering Economist, 21* (4), 1976, pp. 249–58.
25. Weaver, J. B., "False and Multiple Solutions by the Discounted Cash Flow Method for Determining Interest Rate of Return," *The Engineering Economist, 3* (4), 1958, pp. 1–31.
26. White, J. A., Case, K. E., and Agee, M. H., "Rate of Return: An Explicit Reinvestment Rate Approach," *Proceedings of the 1976 AIIE Conference,* American Institute of Industrial Engineers, Norcross, Ga., 1976.

PROBLEMS

1. Brock Associates invested $40,000 in a business venture with the following cash flow results:

EOY	CF	EOY	CF	EOY	CF
0	−$40,000	3	$11,000	6	$11,000
1	5,000	4	14,000	7	8,000
2	8,000	5	14,000	8	5,000

If *MARR* is 10%, determine the following.

(a) Present worth. (4.2.1)

(b) Annual worth. (4.2.2)

(c) Future worth. (4.2.3)

(d) Internal rate of return. (4.2.4)

(e) External rate of return. (4.2.5)

(f) Savings/investment ratio. (4.2.6)

(g) Payback period. (4.2.7)

2. Shrewd Endeavors, Inc. invested $70,000 in a business venture with the following cash flow results:

EOY	CF	EOY	CF	EOY	CF
0	−$70,000	7	$14,000	14	$7,000
1	20,000	8	13,000	15	6,000
2	19,000	9	12,000	16	5,000
3	18,000	10	11,000	17	4,000
4	17,000	11	10,000	18	3,000
5	16,000	12	9,000	19	2,000
6	15,000	13	8,000	20	1,000

Assuming *MARR* to be 20%, determine the following.

(a) Present worth. (4.2.1)

(b) Annual worth. (4.2.2)

(c) Future worth. (4.2.3)

(d) Internal rate of return. (4.2.4)

(e) External rate of return. (4.2.5)

(f) Savings/investment ratio. (4.2.6)

(g) Payback period. (4.2.7)

3. An investment of $20,000 is to be made on an analog computer that will last for 6 years and have a zero salvage value at that time. Operating and maintenance costs are projected to be $5000 the first 3 years and $6000 the last 3 years. The minimum attractive rate of return is specified to be 8%. Determine for this investment the following.

(a) Present worth. (4.2.1)

(b) Annual worth. (4.2.2)

(c) Future worth. (4.2.3)

4. Tulsa Precision, Inc. borrows $100,000 to purchase a new numerically controlled milling machine and pays the loan back over a 4-year period with equal payments. Interest on the loan is 15% compounded annually. The machine is estimated to have annual operating and maintenance costs of $24,000/year and have a life of 10 years. Salvage value is estimated to be $25,000. The firm has a *MARR* of 12%. Determine for this investment the following.

(a) Present worth. (4.2.1)

(b) Annual worth. (4.2.2)

(c) Future worth. (4.2.3)

5. A utility vehicle is purchased for $13,000, kept for 4 years, and sold for $2000. Annual operating and maintenance costs were $2500. Using a 10% minimum attractive rate of return, determine the following.

(a) Present worth. (4.2.1)

(b) Annual worth. (4.2.2)

(c) Future worth. (4.2.3)

6. An investment of $20,000 for a new condenser is being considered. Estimated salvage value of the condenser is $5000 at the end of an estimated life of 8 years. Annual gross income each year for the 8 years is $8500. Annual operating expenses are $2300. Assume money is worth 25% compounded annually. Determine the following measures of investment worth; for each, state whether or not your results indicate the new condenser should be purchased.

(a) Present worth. (4.2.1)

(b) Annual worth. (4.2.2)

(c) Future worth. (4.2.3)

(d) Internal rate of return. (4.2.4)

(e) External rate of return. (4.2.5)

(f) Savings/investment ratio. (4.2.6)

7. Smith Investors places $10,000 in an investment fund. One year after making the investment, Smith receives $1500 and continues to receive $1500 annually until

10 such amounts are received. Smith receives nothing further until 15 years after the initial investment, at which time $15,000 is received. Over the 15-year period, what are each of the following if $MARR = 10\%$?

(a) Present worth. (4.2.1)

(b) Annual worth. (4.2.2)

(c) Future worth. (4.2.3)

(d) Internal rate of return. (4.2.4)

(e) External rate of return. (4.2.5)

(f) Savings/investment ratio. (4.2.6)

(g) Payback period. (4.2.7)

8. Many Chemicals, Unlimited purchases a computer-controlled filter for $100,000. Half of the purchase price is borrowed from a bank at 15% compounded annually. The loan is to be paid back with equal annual payments over a 3-year period. The filter is expected to last 10 years, at which time it will have a salvage value of $4000. Over the 10-year period the operating and maintenance costs are anticipated to equal $20,000/year; however, by making the investment, annual fines of $45,000 for pollution will be avoided. The firm expects to earn 12% on its investments. Determine each of the following measures of investment worth and state whether or not the filter purchase was economically sound.

(a) Present worth. (4.2.1)

(b) Annual worth. (4.2.2)

(c) Future worth. (4.2.3)

(d) Internal rate of return. (4.2.4)

(e) External rate of return. (4.2.5)

(f) Savings/investment ratio. (4.2.6)

9. Owners of an economy motel chain are considering building a new 200-unit motel. The present worth cost of building the motel is $8,000,000; the firm estimates furnishings for the motel will cost an additional $700,000 and will require replacement every 5 years. Annual operating and maintenance costs for the facility are estimated to be $800,000. The average rate for a unit is anticipated to be $40/day. A 15-year planning horizon is used by the firm in evaluating new ventures of this type; a terminal salvage value of 15% of the original building cost is anticipated; furnishings are estimated to have no salvage value at the end of each 5-year replacement interval. Assuming average daily occupancy percentages of 50%, 60%, 70%, 80% for years 1 through 4, respectively, and 90% for the fifth and each remaining year, $MARR$ of 12%, 365 operating days/year, and ignoring the cost of the land, should the motel be built? Base your decision upon the following values.

(a) Present worth. (4.2.1)

(b) Annual worth. (4.2.2)

(c) Future worth. (4.2.3)

(d) External rate of return. (4.2.5)

(e) Savings/investment ratio. (4.2.6)

10. Growth Fertilizer Company purchases a gravity settling tank by borrowing the $30,000 purchase price. The loan is to be repaid with four equal annual payments at an annual compound rate of 12%. It is anticipated that the tank will be used for 9 years and then be sold for $2000. Annual operating and maintenance expenses are estimated to be $9000/year. A savings of $15,000/year is realized over the present filtration system. The firm uses a *MARR* of 15% for its economic analyses. Determine the following.

   (a) Present worth. (4.2.1)

   (b) Annual worth. (4.2.2)

   (c) Future worth. (4.2.3)

   (d) Internal rate of return. (4.2.4)

   (e) External rate of return. (4.2.5)

   (f) Savings/investment ratio. (4.2.6)

   (g) Payback period. (4.2.7)

11. A firm purchases an analog computer for $15,000; the computer is used for 4 years and is then sold for $2000. Annual disbursement for operating and maintenance costs equals $3000/year over the 4-year period. The computer replaced an older manual system that cost $8000/year. On the basis of a 15% *MARR*, determine if the decision to buy the analog computer was economically sound. Use the following measures of investment worth.

   (a) Present worth. (4.2.1)

   (b) Annual worth. (4.2.2)

   (c) Future worth. (4.2.3)

   (d) Internal rate of return. (4.2.4)

   (e) External rate of return. (4.2.5)

   (f) Savings/investment ratio. (4.2.6)

12. If a fund pays 12% compounded annually, what deposit is required today such that $1000 can be withdrawn every year forever? (4.2.8)

13. If a fund pays 12% compounded annually, what deposit is required today such that $5000 can be withdrawn every 4 years forever? (4.2.8)

14. A scholarship fund pays $2000, $1000, $2000, and $1000 at the end of quarters 1, 2, 3, and 4 each year. If the scholarship fund earns 12% compounded quarterly, how much money must be in the fund for this series of payments to continue forever? (4.2.8)

15. A scholarship fund pays $2000, $1000, $2000, and $1000 at the *beginning* of quarters 1, 2, 3, and 4 each year. If the scholarship fund earns 8% compounded quarterly, how much money must be in the fund for this series of payments to continue forever? (4.2.8)

16. Maintenance on a reservoir is cyclic with the following *costs* occurring over a 5-year period: $5000, $2000, $7000, $0, and $2000. It is anticipated that the sequence will repeat itself ever 5 years forever. Determine the capitalized cost of the maintenance costs based on a time value of money of 12%. (4.2.8)

17. A flood control project has a construction cost at $t = 0$ of $2,000,000, an annual maintenance cost of $50,000, and a major repair at 5-year intervals projected to cost $250,000 with the first such repair occurring at $t = 5$. If interest is 10% annually, determine the amount of money needed at $t = 0$ to provide for construction and perpetual upkeep. (4.2.8)

18. A firm can invest in a venture which costs $200,000 and returns $70,000/year at the end of each of 4 years. This investment can be renewed perpetually every 4 years. If the firm's *MARR* is 20%, determine the capitalized worth of an infinite series of these investments. (4.2.8)

19. By investing $80,000 in special-purpose production equipment, the required number of units can be manufactured in 1 year for a one-time order and generate net income of $1 million, including the salvage value for the equipment. Alternatively, by using the present production methods, the net income will be $500,000 per year for 2 years. Hence, an $80,000 investment will generate $500,000 in extra income the first year, but will cause $500,000 in income to be lost the second year. Determine the internal rate(s) of return for the incremental cash flows. (4.2.4)

20. In Problem 19, determine the external rate of return when the MARR is (a) 15% and (b) 30%. (4.2.5)

21. An individual is faced with two investment possibilities. In the first case, an investment of $105,000 can be made in a limited partner agreement involving land speculation and development. One year following the investment it is expected $45,000 in income will be obtained; incomes for each of the next four years are estimated to be $45,000, $255,000, $0, and $60,000 as various parcels are developed and sold. As an alternative, an investment of $100,000 can be made; two years following the investment, an income of $200,000 is expected; two years later another $200,000 is expected. The cash flow profile for the difference in the two investments is given below.

End of Year	Difference in Cash Flows
0	−$ 5,000
1	45,000
2	− 155,000
3	255,000
4	− 200,000
5	60,000

Determine the internal rate(s) of return for the difference in cash flows. (Hint: Solve Problem 22 first.) (4.2.4)

22. In Problem 21, determine the external rate of return when the MARR is (a) 0%, (b) 100%, and (c) 200%. (4.2.5)

23. A robot is installed in the painting area of an assembly line at a cost of $50,000. After 8 years, the robot is donated to a high school, costing the company $2000 for removal and transportation. Operating and maintenance costs began in year 1

at \$3000/year and increased by \$450 each year thereafter. The company has a 10% MARR. Supply the values of U, V, W, X, Y, and Z in the following equation for computing the equivalent uniform annual cost of the robot. (4.3)

$$(U - V)(A|P\ 10,W) + (-2000)(X) + 3000 + Y(A|Z\ 10,8).$$

24. The Acme Manufacturing Company purchased an automatic transfer machine for \$25,000 installed. At the end of 8 years the company paid \$4000 to have the machine removed and junked. The operating and maintenance-cost history was as follows.

EOY	O&M Cost	EOY	O&M Cost
1	\$2000	6	3500
2	2000	7	3500
3	2000	8	3500
4	2000	9	5000
5	3500	10	5000

The Acme Manufacturing Company has a 15% MARR. Supply the values of U, V, W, X, and Y in the following equation for computing the equivalent uniform annual cost for the machine. (4.3)

$$(\$25,000 - U)(A|F\ 15\%,10) + V(0.15) + W + \$1500[(F|A\ 15\%,6)$$
$$+ (F|A\ 15\%,X)](Y,\ 15\%,10)$$

25. A distillation column is purchased for \$300,000. Operating and maintenance costs for the first year are \$30,000. Thereafter, operating and maintenance costs increase by 10%/year over the previous year's costs. At the end of 8 years the column is sold for \$50,000. During the life of the investment, revenue was produced that could be related directly to the investment in the column. The revenue the first year was \$75,000. Thereafter, revenue increased by \$10,000 over the previous year's revenue. Using a MARR of 10%, determine the equivalent annual worth for the investment. (4.3)

26. Determine the capital recovery cost of a bulldozer costing \$200,000, having a life of 7 years, and a salvage value of \$50,000. Use an interest rate of 15%. Solve using each of three different capital recovery formulas. (4.3)

27. Determine the capital recovery cost for chemical processing equipment costing \$100,000, having a life of 6 years, and a salvage value of $-\$25,000$. Use an interest rate of 15%. (4.3)

28. John A. Smooth purchases an automobile costing \$21,000. It will be kept for 3 years and then sold. How much must he receive for the car in order for the capital recovery cost to be \$5000/yr. if an interest rate of 12% is used? (4.3)

29. Determine the capital recovery cost of a ditching machine having a first cost of \$35,000, a salvage value of \$3000, and a life of 6 years. Use an interest rate of 20%. Solve using each of the three different capital recovery formulas. (4.3)

30. Crush Autosmashers can purchase a new electromagnet for moving cars at a cost of \$12,000. When scrapped out, the electromagnet will be worth \$500. If Crush's MARR is 15%, how many years must the unit last so that its capital recovery cost will be \$2500/yr. or less? (4.3)

CHAPTER
5

Comparison of Alternatives

5.1 INTRODUCTION

In this chapter, we apply time value of money concepts and measures of investment worth to the comparison of engineering investment alternatives. Although multiple objectives are often involved in performing a comparison of alternatives, for now we concentrate on the comparison of investment engineering alternatives on the basis of monetary considerations alone. We will assume that our set of investment alternatives is collectively exhaustive and mutually exclusive. By collectively exhaustive, we mean that we did not leave out any alternatives; by mutually exclusive, we mean it is an either–or situation; no more than one alternative can be chosen. The adage of "not being able to have one's cake and eat it too" illustrates the notion of mutual exclusivity.

A systematic approach that can be used in comparing the economic worths of engineering investment alternatives is summarized as follows:

1. Define the set of feasible, mutually exclusive alternatives to be compared.
2. Define the planning horizon to be used in the comparison.
3. Develop the cash flow profiles for each alternative.
4. Specify the MARR to be used.
5. Compare the alternatives using a specified measure of worth.
6. Perform supplementary analyses.
7. Select the preferred alternative.

The procedures for comparing engineering investment alternatives outlined in this chapter are intended to aid in making better measurements of the *quantitative aspects* of capital investment alternatives. Following development of the 7-step approach, more detail is provided on comparing alternatives having unequal lives. Two special classes of alternatives are examined further. These include alternatives having no positive cash flows and those involving the possible replacement of existing assets.

The comparison of engineering investment alternatives involves the consideration of many factors, not the least of which are technological and monetary aspects of the engineering design. Quality, safety, performance, delivery, risk, flexibility, appearance, marketing, environmental, employment, engineering and construction capability, and competitive pressures, for example, might dominate the final selection of the design. However, for now, the focus will be on comparing engineering alternatives on the basis of their economic impact on the organization. Approaches that can be used to assimilate multiple objectives are treated in Chapter Nine.

Regardless of the type of alternatives being considered, it cannot be too strongly emphasized that *no economic evaluation can replace the sound judgement of experienced managers concerned with both the quantitative and nonquantitative aspects of investment alternatives.*

5.2 DEFINING INVESTMENT ALTERNATIVES

An individual investment alternative selected from a set of alternatives can be made up of several *investment proposals*. Investment proposals are distinguished from investment alternatives by Thuesen et al. [32] by noting that investment alternatives are decision options; investment proposals are single projects or undertakings that are being considered as investment possibilities.

Example 5.1

As an illustration of the distinction between investment proposals and investment alternatives, consider a distribution center that receives pallet loads of product, stores the product, and ships pallet loads of product to various customer locations. A new distribution center is to be constructed, and the following proposals have been made:

1. Proposed methods of moving materials from receiving to storage and from storage to shipping include:
    a. Conventional lift trucks for operating in 12-foot aisles.
    b. Narrow-aisle lift trucks for operating in 5-foot aisles.
    c. Automated guided vehicle system.
    d. Towline conveyor system.
    e. Pallet conveyor system.

2. Proposed methods of placing materials in and removing materials from storage include:
   a. Conventional lift trucks for operating in 12-foot aisles.
   b. Narrow-aisle lift trucks for operating in 5-foot aisles.
   c. Narrow-aisle, operator driven, rail-guided storage/retrieval vehicle.
   d. Narrow-aisle, automated, rail-guided storage/retrieval vehicle.
3. Proposed methods of storing materials include:
   a. Stacking pallet loads of material (8 feet high, 12-foot aisles).
   b. Conventional pallet rack (20 feet high, 12-foot aisles).
   c. Flow rack (20 feet high, 12-foot aisles).
   d. Narrow-aisle, pallet rack (20 feet high, 5-foot aisles).
   e. Narrow-aisle, flow rack (20 feet high, 5-foot aisles).
   f. Medium height, pallet rack (35 feet high, 5-foot aisles).
   g. High rise, pallet rack (70 feet high, 5-foot aisles).

Given the set of proposals, alternative designs for the material handling system can be obtained by combining a proposed method of moving materials from receiving to storage, a proposed method of placing materials in storage, a proposed method of storage, a proposed method of removing materials from storage, and a proposed method of transporting materials from storage to shipping. Some of the combinations of proposals will be eliminated because of their incompatibility. For example, lift trucks requiring 12-foot aisles cannot be used to place materials in and remove materials from storage when 5-foot aisles are used. Other combinations might be eliminated because of budget limitations; a desire to minimize the variation in types of equipment due to maintainability, availability, reliability, flexibility, and operability considerations; ceiling height limitations; physical characteristics of the product (crushable product might require the use of storage racks); and a host of other considerations. Characteristically, experience and judgment are used to trim the list of possible combinations to a manageable number.

Example 5.2

To illustrate the formation of investment alternatives from a set of investment proposals, consider a situation involving m investment proposals. Let x_j be defined to be 0 if proposal j is not included in an alternative and let x_j be defined to be 1 if proposal j is included in an alternative. Using the binary variable x_j we can form 2^m alternatives. Thus, if there are three investment proposals, we can form eight investment alternatives, as depicted in Table 5.1.

Among the alternatives formed, some might not be feasible, depending on the restrictions or constraints placed on the problem. To illustrate, there

TABLE 5.1

Developing Investment Alternatives from Investment Proposals

Alternative	Proposals			Explanation
	x_A	x_B	x_C	
1	0	0	0	Do nothing (proposals A, B, and C not included)
2	0	0	1	Accept proposal C only
3	0	1	0	Accept proposal B only
4	0	1	1	Accept proposals B and C only
5	1	0	0	Accept proposal A only
6	1	0	1	Accept proposals A and C only
7	1	1	0	Accept proposals A and B only
8	1	1	1	Accept all three proposals

might be a budget limitation that precludes the possibility of combining all three proposals; thus, Alternative 8 would be eliminated. Additionally, some of the proposals might be *mutually exclusive proposals*. For example, Proposals A and B might be alternative computer designs, and only one is to be selected; in this case Alternative 7 would be eliminated from consideration. Other proposals might be *contingent proposals* so that one proposal cannot be selected unless another proposal is also selected. As an illustration of a contingent proposal, Proposal C might involve the procurement of computer terminals, which require the selection of the computer design associated with Proposal B. In such a situation, Alternatives 2 and 6 would be infeasible. Thus, depending on the restrictions present, the number of feasible alternatives that result can be considerably less than 2^m.

In many organizations there is a rather formalized hierarchy for determining how the organization will invest its funds. Typically, the entry point in this hierarchy involves an individual analyst or engineer who is given an assignment to solve a problem; the problem may be one requiring the design of a new product, the improvement of an existing manufacturing process, or the development of an improved system for performing a service. The individual performs the steps involved in the problem-solving procedure and recommends the preferred solution to the problem. In arriving at the preferred solution, a number of alternative solutions are normally compared.

The preferred solution is usually forwarded to the next level of the hierarchy for approval. In fact, one would expect many preferred solutions to various problems to be forwarded to the second level of the hierarchy for approval. Each preferred solution becomes an investment *proposal,* the resulting set of mutually exclusive investment *alternatives* are formed, and the process of comparing economic investment alternatives is repeated. This sequence of operations is usually performed in various forms at each level of the hierarchy until, ultimately, the preferred solution by the individual analyst or engineer is accepted or rejected. In this textbook we concentrate on the process of comparing investment alternatives at the first level of the hierarchy; however, the need for such comparisons at many levels of the organization should be kept in mind.

One danger inherent in the hierarchical approval approach is the use of *size gates*. In particular, companies often delegate approval authority over investment alternatives based on the level of funding required. Depending on the size of the investment, a different *approval gate* might be required. For example, a plant manager might have the authority to make decisions on investments involving less than $50,000 in capital; for those investments requiring from $50,000 to $250,000 in capital, division-level approval might be required; and for those requiring in excess of $250,000 in capital, corporate-level approval is required.

Due to the use of size gates, piecemeal investment strategies are often used. Namely, major investments are subdivided into smaller pieces in order to obtain approval at lower levels of the hierarchy. Rather than seek approval to undertake the entire project in one year, a multiyear strategy is developed—simply because of the "gamesmanship" used to cope with the size gates! The adage of "eating the elephant one bite at a time" becomes the strategy for obtaining approval for major system changes.

Unfortunately, the benefits of an integrated system are postponed for years, due to the incremental approach. Further, it is often the case that one of the "bites" is not approved, simply because of the number of competing proposals that particular year, or because of the difficulty of quantifying the benefits of some critical element of the overall system. As a result, small incremental changes are made, rather than major system-wide improvements. (For further consideration of the impact of approval hierarchies on the generation of investment alternatives, see Kaplan [12].)

Another aspect of integrated systems merits consideration. Namely, in justifying computer integrated manufacturing (CIM) systems, it is usually difficult to quantify the *synergistic benefits* of systems integration. Aristotle is credited with having said that "the whole is greater than the sum of its parts." Whether or not Aristotle made such a claim is not the issue; instead, while it is easy to accept in principle that two plus two is greater than four in an integrated system, the application of incremental justification techniques often belies the claim. As a result, many firms have found that the best strategy is to "design the whole, sell or justify the whole, and then implement the pieces." In particular, the total system is justified and sold to management; where capital is limited, a phased implementation plan is used and individual components of the total system are implemented over time.

Obviously, the alternatives to be considered must be defined carefully so as not to prematurely eliminate viable candidates. At the same time, care must be taken to include only those alternatives that appear to satisfy the noneconomic considerations that will influence the ultimate selection.

Before concluding our consideration of the generation of engineering investment alternatives, a further word is required concerning the use of the "do nothing" alternative. As in Example 5.2, it is frequently the case that the do-nothing alternative will be included in the set of feasible investment alternatives to be compared. Such an alternative is intended to represent business as usual or maintaining the status quo; however, it is rarely the case that business conditions stand still. Doing nothing does not mean that nothing will be done; rather, it might mean that management has opted to pass up the opportunity to influence future events. It is often used as a baseline, against which other investment alternatives are compared.

In this chapter, as a matter of convenience, we often associate zero incremental costs to doing nothing. In practice, when the do-nothing alternative is feasible, extreme care must be taken not to underestimate the cost of doing nothing. For many firms, business as usual is their most expensive alternative; "standing pat" or maintaining the status quo for too long can prove to be a disasterous course for many businesses. For, *while the firm is doing nothing, its competition is generally doing something!*

Over a period of time, a series of decisions to not make capital investments or to do nothing can result in a firm's losing market share due to obsolete products and/or processes. Considered individually, each decision might seem to be the most economic at the time. However, collectively, the decisions might prove to be disasterous to the long-term well-being of the firm, due to their being based on too narrow a perspective. In the defense of the decisions to do nothing, they are made in the absence of perfect knowledge of the future.

Ironically, it is because management cannot accurately predict the economic outcome of change that the absence of change occurs and doing nothing is preferred. Yet, there is also uncertainty concerning the economic impact of doing nothing—of failing to change. One must guard against taking too narrow a view of the economic impact of choosing to follow the path of doing nothing. In considering the long-term impact of doing nothing, managers must decide if they want to be "change masters" or to be mastered by change.

In the case of manufacturing and distribution modernizations, a common reason for engineers' failing to view the do-nothing alternative from a broad perspective is the propensity to restrict their thinking to the costs that arise *inside* the manufacturing plant or distribution facility. It is seldom the case that an engineer looks beyond the walls of the facility and assesses the impact of a do-nothing decision on, say, the marketplace. However, when the engineer enlists the assistance of persons from the marketing organization in analyzing the impact of doing nothing, it is frequently the case that automated manufacturing and distribution systems are justified and installed successfully.

5.3 DEFINING THE PLANNING HORIZON

In comparing engineering investment alternatives, it is important to compare them over a common period of time. In Chapter Four we define that period of time to be the *planning horizon*. In the case of investments in, say, equipment to perform a required service, the period of time over which the service is required might be used as the planning horizon. Likewise, in one-shot investment alternatives, the period of time over which receipts continue to occur might define the planning horizon.

In a sense, the planning horizon defines the width of the "window" through which the economic performance of each alternative will be viewed. In making the selection, consider the person who looks only a foot ahead when traveling along an unfamiliar path; that person might encounter dead ends and might be required to backtrack many times. Likewise, the person who looks a mile ahead and fails to see the short term obstacles is very likely to stumble and fall along the journey. Adopting too short a planning horizon might result in the elimination of those alternatives that require radical change, but will produce dramatic payoffs in the long run; however, adopting too long a planning horizon might result in the firm going out of business before realizing the promised benefits.

It is important to distinguish between the length of the planning horizon, the working life of equipment, and the depreciable life of equipment. The working life is the actual period of time the equipment is capable of being used (e.g., 20 years); the depreciable life of the equipment is the allowable period of time for depreciating the asset. The planning horizon might have no relationship to the other two periods of time; it is simply the time frame to be used in comparing the alternatives and should realistically represent the period of time over which reasonably accurate cash flow estimates can be provided. Also, to make an objective evaluation, *the same window must be used in viewing each alternative.*

In some cases the planning horizon is easily determined; in other cases the duration of one or more subjects is sufficiently uncertain to cause concern over the time period to use. Some commonly used methods for determining the planning horizon to use in economy studies include:

1. Least common multiple of lives for the set of feasible alternatives, denoted $\hat{T}$.
2. Shortest life among alternatives, denoted T_s.
3. Longest life among alternatives, denoted T_l.
4. Some other period of time.

The *least common multiple of lives* is the most common method of selecting the planning horizon. In this method, each alternative's cash flow profile is assumed to repeat in the future until a time is reached when all alternatives under consideration conclude at the same time. Using such a procedure when three alternatives are being considered and the individual lives are 3 years, 6 years, and 5 years yields a planning horizon of $\hat{T} = 30$ years. If the lives had been 3 years, 6 years, and 6 years, $\hat{T} = 6$ years. Clearly, strict reliance on $\hat{T}$ as the planning horizon is not advisable. Unfortunately, selection of this method is usually made implicitly, not explicitly, when one simply calculates and compares the annual worths of unequal-lived alternatives. This practice is covered in more detail in Section 5.10.

In using the *shortest life among alternatives* to define the planning horizon, estimates are required for the values of the unused portions of the lives of the remaining alternatives. Thus, for the situation considered above, with $T_s = 3$ years, the salvage or residual values at the end of 3 years' use must be assessed for the other two alternatives.

If the *longest life among alternatives*, T_l, is used in determining the planning horizon, some difficult decisions must be made concerning the period of time between T_s and T_l. If the alternative selected is to provide a necessary service, that service must continue throughout the planning horizon, regardless of the alternative selected. Consequently, when the shortest life alternative reaches the end of its project life, it must be replaced with some other asset capable of performing the required service. However, since technological developments will probably take place during the period of time T_s, new and improved candidates will be available for selection at time T_s. Thus, the specification of the cash flows for the shortest life alternative during the period of time from T_s to T_l is a difficult undertaking. As a result, T_l is seldom used as the planning horizon.

Many organizations have adopted a *standard planning horizon* for all economic alternatives in order to ensure some consistency in economic analyses. Letting T denote the planning horizon specified by the organization, different approaches are recommended, depending on whether $T < T_s$, $T_s < T < T_l$, or $T > T_l$. If $T < T_s$, each alternative's cash flow profile must be truncated with salvage or residual values estimated at time T. If $T_s < T < T_l$, alternatives having a life less than T require that plans and cash flows be specified up through time T; those having lives longer than T must be truncated. If $T > T_l$, all alternatives require that plans and cash flows be specified through the lesser of $\hat{T}$ and T. That is, when the planning horizon is greater than or equal to the least common multiple of lives, it is recommended that the economic analysis be based on a period of time equal to the least common multiple of lives. The reason is that at time $\hat{T}$ a new economic analysis can be performed based on the alternatives available at that time. After $\hat{T}$ years new alternatives might be available; furthermore, we can more accurately estimate the values of cash flows occurring after $\hat{T}$ if we wait until nearer time $\hat{T}$ to make the estimates.

Example 5.3

To illustrate the difficulties associated with the selection of the planning horizon, consider the three mutually exclusive alternatives whose cash flow profiles are given in Table 5.2. Alternatives 1, 2, and 3 have anticipated lives of 3, 6, and 5 years, respectively. Each has revenues and costs which are used to determine the net cash flows. Alternative 1 involves used equipment

TABLE 5.2

Cash Flow Profiles for Three Mutually Exclusive Investment
Alternatives Having Unequal Lives

End of Year, t	Revenues, R_t	Costs, C_t	Net Cash Flows, $R_t - C_t$	Salvage Value if Sold at Time t
Alternative 1				
0	—	—	—	$0
1–3	$27,500	$23,000	$4,500	0
Alternative 2				
0	—	$50,000	−$50,000	$50,000
1	$30,000	10,000	20,000	35,000
2	30,000	10,000	20,000	25,000
3	30,000	10,000	20,000	15,000
4	30,000	10,000	20,000	5,000
5	30,000	10,000	20,000	0
6	30,000	10,000	20,000	0
Alternative 3				
0	—	$75,000	−$75,000	$75,000
1	$27,500	7,500	20,000	55,000
2	32,500	7,500	25,000	40,000
3	37,500	7,500	30,000	25,000
4	42,500	7,500	35,000	10,000
5	47,500	7,500	40,000	0

having a useful life of 3 years. Much of this older equipment is readily
available at virtually no cost other than for operating and maintenance.
Alternatives 2 and 3 involve the purchase and operation of new equipment
which will last for up to 6 and 5 years, respectively; estimated salvage values
for the equipment are shown as a function of time.

Using a least common multiple of lives approach, a planning horizon of
$\hat{T} = 30$ years would be used. Using a 30-year planning horizon requires
answers to the following questions. What cash flows are anticipated for
years 4 to 30 if Alternative 1 is selected? What will be the cash flows for
Alternative 2 for years 7 to 30? What about Alternative 3 for years 6 to 30?

As shown in Table 5.3, the traditional approach is to assume that Alternative 1 will be repeated 10 times, Alternative 2 will be repeated 5 times,
Alternative 3 will be repeated 6 times, and that identical cash flows occur
during these repeating life cycles. Inflation effects, as well as technological
improvements and consumer demand changes tend to invalidate such assumptions.

If the shortest life approach is used, a planning horizon of 3 years would be
used. In such a case, the estimated salvage values of Alternatives 2 and 3
should be indicated at the end of year 3, as denoted in Table 5.3 by an
asterisk.

TABLE 5.3

Cash Flow Profiles for Various Planning Horizons

End of Year, t	Net Cash Flows for Alternatives		
	A_{1t}	A_{2t}	A_{3t}
$T = \hat{T} = 30$ years			
0	$4,500	−$50,000	−$75,000
1	4,500	20,000	20,000
2	4,500	20,000	25,000
3	4,500	20,000	30,000
4	4,500	20,000	35,000
5	4,500	20,000	−75,000 + 40,000
6	4,500	−50,000 + 20,000	20,000
7	4,500	20,000	25,000
8	4,500	20,000	30,000
.	.	.	.
.	.	.	.
.	.	.	.
29	4,500	20,000	35,000
30	4,500	20,000	40,000
$T = T_s = 3$ years			
0	—	−$50,000	−$75,000
1	$4,500	20,000	20,000
2	4,500	20,000	25,000
3	4,500	20,000 + 15,000[a]	30,000 + 25,000[a]
$T = T_l = 6$ years			
0	—	−$50,000	−$75,000
1	$4,500	20,000	20,000
2	4,500	20,000	25,000
3	4,500	20,000	30,000
4	4,500	20,000	35,000
5	4,500	20,000	40,000
6	4,500	20,000	4,500
$T = 2$ years $< T_s$			
0	—	−$50,000	−$75,000
1	$4,500	20,000	20,000
2	4,500	20,000 + 25,000[a]	25,000 + 40,000[a]
$T_s < T = 5$ years $< T_l$			
0	—	−$50,000	−$75,000
1	$4,500	20,000	20,000
2	4,500	20,000	25,000
3	4,500	20,000	30,000
4	4,500	20,000	35,000
5	4,500	20,000 + 0[a]	40,000

TABLE 5.3 Continued
Cash Flow Profiles for Various Planning Horizons

End of Year, t	Net Cash Flows for Alternatives		
	A_{1t}	A_{2t}	A_{3t}
$T = 8$ years $> T_l$			
0	—	$-\$50,000$	$-\$75,000$
1	\$4,500	20,000	20,000
2	4,500	20,000	25,000
3	4,500	20,000	30,000
4	4,500	20,000	35,000
5	4,500	20,000	$40,000 - 75,000$
6	4,500	$20,000 - 50,000$	20,000
7	4,500	20,000	25,000
8	4,500	$20,000 + 25,000^a$	$30,000 + 25,000^a$

[a] Denotes salvage value.

Using the longest life approach yields a 6-year planning horizon. In this instance a decision must be made concerning the cash flows of Alternative 1 in years 4, 5, and 6 and Alternative 3 in year 6. In the case of Alternative 1, we assume that another piece of used equipment will be employed, resulting in resumption of the \$4500 net cash flows during years 4, 5, and 6. For Alternative 3, we *could* assume that another new unit is purchased for \$75,000 at the end of year 5. This would provide service and net income of \$20,000 during year 6, and then have a salvage value of \$55,000. This 1-year purchase is obviously not economically sound. Therefore, we assume that a piece of readily available used equipment is employed for year 6, earning \$4500.

Suppose a standard planning horizon of 2, 5, or 8 years is used. The same questions that arise under planning horizons of $\hat{T}$, T_s, or T_l apply when one of these other horizons is used. As shown in Table 5.3, Alternatives 1, 2, and 3 are either truncated at, or repeated until, the end of the planning horizon at which time estimates of terminal salvage values are provided.

Although the use of a standard planning horizon has the benefit of a consistent approach in comparing investment alternatives, there are also some dangers that should be recognized. In some cases the major benefits associated with an alternative might occur in the later stages of its project life. If the planning horizon is less than the project life, such alternatives would seldom be accepted. Just such a practice caused one major textile firm to lose its strong position in the industry. A major modernization of the processing departments had been proposed, but its benefits would not be realized until the bugs had been worked out of the new system, all personnel were trained under the new system, and the marketing people had regained the lost customers. Unfortunately, the planning horizon specified by the firm was too short in duration and the modernization plan was not approved.

Example 5.4

As a second illustration of the planning horizon selection process, consider the two cash flow diagrams given in Figure 5.1. The two alternatives are *one-shot* investments. We are unable to predict what investment alternatives will be available in the future, but we do anticipate that recovered capital can be reinvested and earn a 15% return.

For this type of situation, a 6-year planning horizon is suggested, with zero cash flows occurring in years 5 and 6 with Alternative 1. At the end of 6 years the net future worths for the two alternatives will be

$$
\begin{aligned}
FW_1(15\%) &= \$4500(F|P\ 15,2) + \$3500(P|A\ 15,3)(F|P\ 15,6) \\
&\quad - \$4000(F|P\ 15,6) \\
&= \$4500(1.3225) + \$3500(2.2832)(2.3131) \\
&\quad - \$4000(2.3131) \\
&= \$15,183.29
\end{aligned}
$$

$$
\begin{aligned}
FW_2(15\%) &= \$1000(F|A\ 15,6) + \$1000(A|G\ 15,6)(F|A\ 15,6) \\
&\quad - \$5000(F|P\ 15,6) \\
&= \$1000(8.7537) + \$1000(2.0972)(8.7537) \\
&\quad - \$5000(2.3131) \\
&= \$15,546.46
\end{aligned}
$$

Thus, we would recommend Alternative 2.

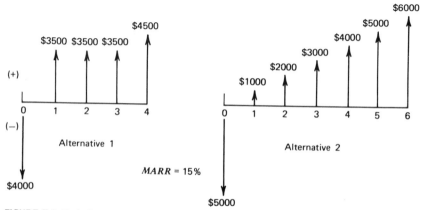

FIGURE 5.1. Cash flow diagrams for the example problem.

If we did not give careful thought to the situation involved and simply calculated the annual worth over 4- and 6-year horizons, respectively, values of $3000.15/year and $1776.20/year would result in favor of Alternative 1. This would be comparable to blindly assuming a least common multiple of lives planning horizon of 12 years, with identical cash flows in repeating life cycles. Hence, it is important to consider the particular situation involved and specify the planning horizon instead of employing a rule of thumb for establishing planning horizons that does not consider the nature of the investments.

It appears that the preferred approach would be to have a "flexible" standard planning horizon. All of the "routine" economic analyses would be based on the standard planning horizon of, say, 5 to 10 years; nonroutine economic analyses would be based on a planning horizon that was appropriate for the situation.

5.4 DEVELOPING CASH FLOW PROFILES

Once the set of investment alternatives has been specified and the planning horizon decision has been made, the cash flow profiles can be developed for the alternatives. As has been emphasized, the cash flow profiles should be developed by giving careful consideration to *future* conditions instead of relying completely on *past* cash flows. The cash flows for an investment alternative are obtained by aggregating the cash flows for all investment proposals included in the investment alternative.

Example 5.5

To illustrate the approach to be taken, suppose a planning horizon of 5 years is used and there are three investment proposals. Cash flow profiles for the proposals are given in Table 5.4. A budget limitation of $50,000 is available for investment among the proposals. Proposal B is contingent on Proposal

TABLE 5.4
Cash Flow Profiles for Three Investment Proposals

End of Year, t	Net Cash Flows for Proposals		
	CF_{At}	CF_{Bt}	CF_{Ct}
0	−$20,000	−$30,000	−$50,000
1	− 4,000	4,000	− 5,000
2	2,000	6,000	10,000
3	8,000	8,000	25,000
4	14,000	10,000	40,000
5	25,000	20,000	10,000

TABLE 5.5
Developing Investment Alternatives

Feasible Alternatives	Proposals X_A	X_B	X_C	Investment Required
0	0	0	0	$ 0
1	0	0	1	50,000
~~0~~	~~1~~	~~0~~		30,000
~~0~~	~~1~~	~~1~~		80,000
2	1	0	0	20,000
~~1~~	~~0~~	~~1~~		70,000
3	1	1	0	50,000
~~1~~	~~1~~	~~1~~		100,000

A, and Proposals A and C are mutually exclusive. Based on the restrictions associated with the combinations of proposals, only four investment alternatives are to be considered. These are developed in Table 5.5. Alternative 0 is the "do-nothing" alternative; Alternative 1 involves Proposal C alone; Alternative 2 involves Proposal A alone; and Alternative 3 involves a combination of Proposals A and B. The cash flow profiles for the four alternatives are given in Table 5.6.

In developing the cash flow profiles for each alternative, best estimates are needed of the annual costs and income anticipated over the planning horizon, including the salvage values of all assets at the end of the planning horizon. Again, care must be taken to assess the impact of each alternative on the economic performance of the firm, including the economic impact of the do-nothing alternative, if it is a feasible candidate.

In most, if not all, economic evaluations, it is not necessary to develop detailed forecasts of all items of cost, revenue, and investment associated

TABLE 5.6
Cash Flow Profiles for Four Investment Alternatives

End of year, t	Net Cash Flows for Alternatives A_{0t}	A_{1t}	A_{2t}	A_{3t}
0	0	−$50,000	−$20,000	−$50,000
1	0	− 5,000	− 4,000	0
2	0	10,000	2,000	8,000
3	0	25,000	8,000	16,000
4	0	40,000	14,000	24,000
5	0	10,000	25,000	45,000

with an alternative. Costs and revenues that will be the same regardless of the alternative selected can be omitted. If a cost reduction alternative will not affect sales revenues, no forecast of such revenues need be developed. Attention is focused on the items of cost and revenue that will be affected by the alternative selected. However, as noted previously, before concluding that, say, revenues will not be affected by the decision, a careful assessment should be made—including consideration of the competition.

Too frequently, incremental decisions appear to be made in a vacuum. Considerations of the competition and the market environment are seldom incorporated in engineering economic analyses, but they should be! Often, it is blindly assumed that the only cash flows associated with the replacement of existing equipment with state-of-the-art equipment are the costs associated with each alternative. The fact that installing the newest equipment might influence customers to select your product rather than the competition's product is often overlooked.

In determining the costs associated with each alternative, one should not blindly accept cost reduction estimates from a cost accounting system. As noted in Chapter Two, the fundamental and historical purpose of accounting is to maintain a consistent *historical* record of the financial results of the operations of the organization. Accounting figures are based on definitions derived consistent with this objective. Accounting methods are not designed to determine the economic worth of *future* alternative courses of action. Furthermore, certain accounting practices can create obstacles to capital investment.

For example, cost accounting systems typically include a proration of overhead on the basis of direct labor. As a result, claimed cost reductions through the elimination of direct labor should be examined carefully to ensure that credit is not being given for reductions in overhead, when, in fact, overhead might be increased by the reduction in direct labor.

Another problem that frequently arises from the allocation of overhead on the basis of direct labor is illustrated by investments in automation. If the use of automation reduces direct labor costs and/or increased indirect labor, the overhead percentage for a plant will increase. Hence, if corporate management places limits on the overhead percentage for a given plant, then the plant might be prevented from investing in automation.

Again, in using a cash flow approach, the actual costs and savings to be realized should be included in the cash flows for each period. Also, the cash flow estimates should reflect the impact of the alternative on the entire system, rather than just the impact on an individual department.

Finally, accurate estimates of cash flows should be used. Too frequently, management suspects that engineers use whatever numbers are needed to satisfy the approval criteria established. As an example, in a conversation with the treasurer of a major U.S. corporation, we learned that he did not believe he could rely on the cash flow estimates obtained from their engineers. Past experience indicated that few of the approved investments came close to delivering the economic performances promised. He noted that for the most important investments made by the corporation, top management depended more on their intuition than the cash flow estimates provided. In many companies, a credibility problem exists regarding the use of discounted cash flow methods. Some managers are arguing for the elimination of the use of DCF methods.

However, the situation that exists is akin to "shooting the messenger because you don't like the message!" The problem is not with the DCF methods, but with their use. DCF methods are GIGO (garbage in, garbage out) methods; the result obtained is no better than the inputs provided.

A primary area of abuse in using DCF methods is in estimating the cash flows associated with an investment. One major firm audited a dozen automation investments and found that 75% of them failed to yield the promised return on investment. In each case, the difficulties of implementing the new equipment had been underestimated. The engineers had shown a single investment occurring in year zero followed by uniform annual savings. Due to late delivery and start-up, the investment occurred over a 2-year period. Furthermore, it took some time for the operators to become familiar with the equipment and to produce quality products at the required quantities. As a result, the promised savings were slower to materialize than had been promised. Due to the poor economic performance of the automation investments, management of the firm became reluctant to approve similar investments.

Depending on the organization, specially designed forms are often provided for aiding the analyst in developing cash flow profiles and conducting the analysis. Sample forms used by one major industrial organization are provided in Figures 5.2 to 5.4. This particular company compares investment alternatives using the internal rate of return method; they refer to it as the discounted cash flow rate of return method.

A particularly difficult class of benefits to estimate are those that are categorized as *intangible benefits*. Among the so-called intangible benefits are reduced inventory, increased quality, reduced space, increased flexibility, reduced lead time, increased throughout, reduced scrap, improved tracking of material, and increased safety. We believe that every effort should be made to quantify the benefits of such intangibles, recognizing that precise measurements might be difficult, if not impossible. In those cases where explicit consideration of intangible benefits is not allowed by the firm, then (as noted in Section 5.7) they certainly should be included in the supplementary analyses to be performed.

5.5 SPECIFYING THE *MARR*

The comparison of alternatives on the basis of the time value of money requires an explicit determination of the time value to be used. As noted in Chapter Four, the value of money used in discounted cash flow methods is variously referred to as a minimum attractive rate of return, a hurdle rate, a required rate of return, a return on investment, and a discount rate. Regardless of its label, a specified interest rate is needed to convert the mix of cash

NET CASH FLOW SCHEDULE

ALTERNATE **A** OR **B**
CIRCLE ONE

TITLE (PROJECT) _____
TITLE (THIS ALTERNATE) _____

PROJECT NO. _____   DATE _____

LINE	EXPENSES *	0	1	2	3	4	5	6	7	8	TOTAL
1	NET BOOK VALUE (BEGINNING OF YEAR)	X									
2	DEPRECIATION (___ YEARS LIFE)										
3	RENTAL / LEASE COST										
4	IN-PLANT LABOR WITH FRINGE BENEFITS @ ___ %										
5	PURCHASED LABOR										
6	REWORK / SCRAP										
7	TOOLING / ANCILLARY EQUIPMENT										
8	MAINTENANCE & REPAIR										
9	POWER & FUEL / UTILITIES EXPENSE										
10	MATERIALS & SUPPLIES										
11	PRODUCT SHIPPING COSTS										
12	PERSONAL PROPERTY TAX										
13	REAL ESTATE TAX										
14	IMPLEMENTING COST										
15											
16											
17	TOTAL CASH EXPENSES (LINES 2 THRU 16)										
18	AFTER TAX ANNUAL COST ** (50% × LINE 17)										
19	INVESTMENT/EXISTING ASSET VALUE										
20	INVESTMENT TAX CREDIT/REBATE										
21	SALVAGE										
22											
23	NET CASH FLOW (ALGEBRAIC SUM OF LINES 18 THRU 22 AND ADD BACK DEPRECIATION)										

(Period column header spans columns 0–8)

FIGURE 5.2.

DISCOUNTED CASH FLOW SUMMARY

PERIOD	"A" NCF	"B" NCF	"A"-"B" △ NCF 0% INTEREST	CUMULATIVE CASH FLOW BACK AMOUNT	%	15% INTEREST FACTOR	PW	25% INTEREST FACTOR	PW	40% INTEREST FACTOR	PW	60% INTEREST FACTOR	PW
-1						1.150		1.250		1.400		1.600	
0						1.000	✕	1.000	✕	1.000	✕	1.000	✕
TOTAL "X" (DISBURSEMENTS)													
1						.870		.800		.714		.625	
2						.756		.640		.510		.391	
3						.658		.512		.364		.244	
4						.572		.409		.260		.153	
5						.497		.328		.186		.095	
6						.432		.262		.133		.060	
7						.376		.210		.095		.037	
8						.327		.168		.048		.023	
9						.284		.134		.048		.015	
10						.247		.107		.035		.009	
11						.215		.086		.025		.006	
12						.187		.069		.018		.004	

TOTAL "Y" (RECEIPTS)

RATIO "X"/"Y"

PROPOSED INVESTMENT _____

DCF / ROR _____ %

PAYOUT @ 0% _____ YRS.

DIVISION _____

PROJECT _____

PROJECT NO. _____ J.O. NO. _____

PREPARED BY _____ DATE _____

APPROVED BY _____ DATE _____

PAYOUT CHART

CUMULATIVE PERCENT RETURN OF INVESTMENT

YEARS TO PAY OUT

INTERPOLATION CHART

DCF / ROR

RATIO X/Y

FIGURE 5.3.

194

DCF / ROR DATA SHEET

PROJECT TITLE _____ _____ PROJECT NO._____ DATE _____
PROJECT LIFE IS ____ YEARS, DETERMINED BY: _____
_____
_____

OTHER ALTERNATIVES CONSIDERED AND REASONS FOR REJECTION: _____
_____
_____
_____
_____

EXPLANATIONS AND CALCULATIONS	LINE REFERENCE	ALTERNATE A	LINE REFERENCE	ALTERNATE B
		TITLE _____ METHOD DESCRIPTION _____ _____ _____ _____		TITLE _____ METHOD DESCRIPTION _____ _____ _____
DEPRECIATION DESCRIBE EXISTING / PROPOSED EQUIPMENT AND EXPLAIN DEPRECIATION	2		2	
EXPENSES (LINES 3–16) NOTE ZERO YEAR EXPENSES SUCH AS IMPLEMENTATION AND TRAINING, MAINTENANCE/MAJOR OVERHAUL, OR ANY UNUSUAL EXPENSES. NOTE IF ONLY DELTA COSTS OF ONE ALTERNATIVE ARE BEING SHOWN.				
INVESTMENT SHOW DERIVATION OF EXISTING ASSET VALUE AND PROPOSED CAPITAL INVESTMENT. IDENTIFY ANY GOVT. EQUIP. STATE IF OLD EQUIP TO BE KEPT.	19		19	
INVESTMENT TAX CREDIT OR REBATE SHOW PERCENT USED.	20		20	
SALVAGE CALCULATE RESIDUAL AFTER TAX VALUE OF ASSETS	21		21	
OTHER CONSIDERATIONS INCLUDES INTANGIBLES, RISK AND LIKELIHOOD, OTHER ANALYSES AND RECOMMENDATION.				

FIGURE 5.4.

flows occurring throughout the planning horizon to a meaningful economic measure that can be used rationally to compare engineering alternatives.

A number of common factors influence the selection of the interest rate to be used. Although some firms treat the *MARR* as a parameter whose value depends on the type of investment in question, others use a fixed value for the *MARR*. Regardless of the basis used for determining the value of the *MARR*, its value should be greater than the *cost of securing capital;* likewise, it should represent the *opportunity cost* associated with investing in the candidate alternative as opposed to other available alternatives. Furthermore, care should be taken to ensure that arbitrarily specified values for the *MARR* do not counteract management's desire to maintain state-of-the-art competency.

5.5.1 The Cost of Capital

Except where other intangible benefits are involved, the discount rate should be greater than the cost of securing additional capital. Indeed it should be greater than the cost of capital by an amount that will cover the unprofitable investments that a firm must make for nonmonetary reasons. Examples of the latter would include investments in antipollution equipment, safety devices, and recreational facilities for employees. The discount rate that is specified establishes the firm's minimum attractive rate of return (*MARR*) in order for the investment to be justified.

It costs money to obtain money for investment. The cost of capital was first discussed in Chapter Two. There, it was noted that funds for financing projects may consist of *debt capital* and *equity* (or ownership) *capital*. Most often, a major project is financed by a mix of both. In trying to establish the time value of money, it is important that we understand why money has a cost, even when it is our own money that we use.

Debt capital such as a loan results in a specific obligation to pay back both interest and principal. A short-term loan may be obtained from a bank or an insurance company. The interest paid for the use of the money is a cost of capital. Fortunately, not all of the interest is lost. Interest is deductible from income for tax purposes.[1] For example, even though we pay I in interest during a year, that I may be subtracted from taxable income, resulting in a tax savings of $I \times T$ where T is our tax rate (34% for large companies). That is, the effective interest payment is $I(1 - T)$. Therefore, for a simple loan, the effective after-tax interest rate (k_l) is the product of the effective interest rate and $(1 - T)$. It is given by

$$k_l = \left[\left(1 + \frac{r}{m} \right)^m - 1 \right] (1 - T) \qquad (5.1)$$

where r is the loan's nominal annual interest rate and m is the number of payment periods per year.

---

[1] Tax effects will be only briefly considered here. Income taxes are treated in considerable detail in Chapter Six.

Example 5.6

A company has taken out a short-term loan with the bank at a rate of 18%/year with quarterly payments. Assuming the tax rate is 0.34, what is the cost of capital for this debt funding?

Answer:

$$k_l = \left[\left(1 + \frac{r}{m}\right)^m - 1\right](1 - T)$$

$$= \left[\left(1 + \frac{.18}{4}\right)^4 - 1\right](1 - 0.34) = .127$$

or 12.7%.

Debt capital can also be raised through the sale of bonds. As noted in Chapter Three, common debenture bonds have a stated face value, annual rate of interest, payment interval, and maturation period (e.g., $1000, 12%, semiannual, and 15 years). When a bond is sold, not necessarily for its face value, the firm receives money for immediate investment. In return, it gives a promise to pay interest on the face value m times per year at an annual rate r. In addition, the firm promises to repay the face value of the bond at the end of a specified period of time. Much like a loan, bond interest paid by a firm is deductible from income for tax purposes. If the amount paid for the bond equals the face value, and if any costs of preparing and selling the bond are neglected, the effective after-tax interest rate (k_b) is

$$k_b = \left[\left(1 + \frac{r}{m}\right)^m - 1\right](1 - T) \qquad (5.2)$$

Equity capital is sometimes called *ownership capital*. One source of equity capital is the sale of stock. Money raised through a stock issue is available to the company for investment. Surprisingly, there is no guarantee of a return to an investor in stock. Of course, the investor expects to receive periodic dividends and also hopes that the trading price of the stock will increase, resulting in a gain when the stock is sold to someone else in the future. Company management usually decides upon the amount of dividends to declare for stockholders after the earnings of the firm are known for the year. Management desires to seek a suitable balance between the amount of dividends paid to keep stockholders happy and the amount of earnings retained in the company for investment in the future. Perhaps the most simplistic estimate of the cost of equity capital involving stock (k_s) is

$$k_s = \frac{D}{P_s} \qquad (5.3)$$

where D is the dividend per share of stock and P_s is the present trading price or current value of the stock. Since dividends are paid out of after-tax earnings, k_s is already an after-tax rate. This estimate assumes that dividends will remain constant into the future.

Example 5.7

A company issues dividends of $6/share on stock selling at $50 per share on the financial market. What is the cost of capital for this equity funding?

Answer:

$$k_s = \frac{D}{P_s}$$

$$= \frac{\$6}{\$50} = .12$$

or 12.0%.

Another source of equity capital is the earnings retained within the company and not declared as dividends. Even though this money appears free, there is a cost of capital. Remember, stockholders are owners of the company and therefore the retained earnings are theirs. The cost of using that money for investment carries an opportunity cost, and that is the cost of dividends forgone by the stockholders from their own earnings. For this reason, one estimate of the cost of capital for retained earnings (k_r) is

$$k_r = \frac{D}{P_s} = k_s \tag{5.4}$$

The overall weighted average cost of capital (k_a) is simply the individual source costs of capital weighted by the fraction of total funding from that source. That is,

$$k_a = k_l p_l + k_b p_b + k_s p_s + k_r p_r \tag{5.5}$$

where p_l, p_b, p_s, and p_r are the fractions of funding from loans, bonds, stocks, and retained earnings, respectively.

Example 5.8

The capital structure of our firm is made up as follows:

	Funding	Fraction
Loans	$ 1,000,000	1/13
Bonds	1,800,000	1.8/13
Stock	8,000,000	8/13
Retained earnings	2,200,000	2.2/13
	$13,000,000	13/13

The costs of capital for these various sources are 10.4, 7.2, 12.0, and 12.0%, respectively. What is the overall weighted average cost of capital for the company?
Answer:

$$k_a = .104(1/13) + .072(1.8/13) + .120(8/13) + .120(2.2/13)$$
$$= .1121$$

or 11.21%.

The foregoing analysis implicitly assumed either zero inflation or an inflation-based cost of capital. Kaplan [12] noted that studies of returns to investors in equity and fixed-income markets during the period from 1926 to 1984 indicated that the average total return from a diversified portfolio of common stock was 11.7%/year; fixed-income securities averaged nominal before-tax returns of less than 5%/year. However, both figures included the effect of inflation over the same period. Factoring out the impact of inflation, the returns were approximately 8.5% and 1.5%, respectively. Kaplan concluded that "a mixture of debt and equity financing produces a total real cost of capital of less than 8%." He contended that one of the primary reasons American industry has tended to underinvest in capital improvements, such as automation and computer integrated manufacturing, is the use of an overly large discount rate in discounting the future cash flows anticipated from such investments.

It is important for the cost of capital to reflect the true cost of capital to the firm during the future investment period. For this reason, we prefer to determine the cost of capital *absent* inflation effects.

We briefly introduced tax and inflation considerations in order to describe more realistically the cost of capital. Further discussion of taxes will be deferred until Chapter Six where much more complete coverage is given. Likewise, in Chapter Six we build on the discussion of inflation given in Chapter Three.

5.5.2 The Minimum Attractive Rate of Return

Some firms establish a standard discount rate or minimum attractive rate of return to be used in all economy studies; others maintain a flexible posture. As one firm described it,

The XYZ Company return on investment (ratio of earnings to gross investment) has been, on the average over the past 5 years, approximately equal to a rate of return of 15% per year. Accordingly, 15% per year is being established as a tentative *minimum requirement for investment alternatives* whose results are primarily measurable in quantitative dollar terms. The principle being applied is that such alternatives should be expected to maintain or to improve overall return-on-investment performance for the XYZ Company. The minimum requirement is based on overall XYZ financial results (as opposed to results for various parts of the company) in order to avoid situations in which investment alternatives of a given level of attractiveness are unknowingly undertaken in one part of the company and rejected in another.

The 15% minimum attractive rate-of-return standard is intended as a guide, rather than as a hard-and-fast decision rule. Furthermore, it is intended to apply to alternatives having risks of the kind usually associated with investments which are primarily in plant and facilities. For alternatives involving expenditures with substantially lower risks, such as those solely for inventories or those in buy-or-lease alternatives, a lower minimum requirement may apply.

Other approaches used to establish the *MARR* include the following:

1. Add a fixed percentage to the firm's cost of capital.
2. Average rate of return over the past 5 years is used as this year's *MARR*.
3. Use different *MARR* for different planning horizons.
4. Use different *MARR* for different magnitudes of initial investment.
5. Use different *MARR* for new ventures than for cost-improvement projects.
6. Use as a management tool to stimulate or discourage capital investments, depending on the overall economic condition of the firm.
7. Use the average stockholder's return on investment for all companies in the same industry group.

There are a number of different approaches used by companies in establishing the discount or hurdle rate to be used in performing economic analyses. The issue is not a simple one.

Kaplan [12] observed that the use of discounted cash flow methods ''most often goes wrong when companies set arbitrarily high hurdle rates for evaluating new investment projects.'' He noted that ''the discounting function serves only to make cash flows received in the future equivalent to cash flows received now. For this narrow purpose—the only purpose, really, of discounting future cash flows—companies should use a discount rate based on the project's opportunity cost of capital (that is, the return available in the capital markets for investments of the same risk)'' [12, pp. 87–88].

Although some advocate increasing the value of the *MARR* as the risks associated with the project increase, an alternative approach to dealing with risk seems preferable. Specifically, as noted in Section 5.7 and as covered in Chapter Eight, risk analyses can be performed explicitly. In such cases, probabilities are assigned to the various possible outcomes of the economic investment, and, using a ''normal'' *MARR*, the expected present worth or the probability of the rate of return being greater than the *MARR* is determined. The *explicit* consideration of risk is preferred to the *implicit* approach of arbitrarily increasing the value of the *MARR*, based on a qualitative assessment of risk.

Regardless of how one chooses to deal with the risks associated with investing capital in anticipation of future returns, the proper determination of the discount rate is not a simple process. Even though it has been the subject of controversy for many years, in some ways we are only slightly closer to agreement today than we were 30 years ago. In many cases, a particular alternative will be preferred over a range of possible discount rates; in other cases, the alternative preferred will be quite sensitive to the discount rate used. Hence, depending upon the situation under study, it might not be necessary to specify a particular value for the discount rate—a range of possible values might suffice. We examine this situation in more depth in Chapter Eight.

Subsequently, it will be convenient to refer to the interest rate, hurdle rate, or discount rate used as the *MARR* and to interpret its value using the opportunity cost concept. The argument will be made that money should not be invested in an alternative if it cannot earn a return at least as great as the *MARR*, since it is reasoned that other opportunities for investment exist that will yield returns equal to the *MARR*.

In the case of the public sector, a different interpretation is required in determining the discount rate to use. Since this chapter emphasizes economic analysis in the private sector, Chapter Seven discusses establishing the discount rate for the public sector.

5.6 COMPARING THE INVESTMENT ALTERNATIVES

We are now ready to begin comparing investment alternatives on the basis of economic considerations. This chapter has dealt with structuring investment alternatives, establishing a planning horizon, specifying the *MARR,* and defining the cash flow profiles. The time value of money and one or more methods of measuring the worth of investments, both developed in Chapter Four, may now be applied to select the preferred alternative. Recall that the measures of investment worth include (1) present worth, (2) annual worth, (3) future worth, (4) internal rate of return, (5) external rate of return, (6) savings/investment ratio, (7) payback period, and (8) capitalized worth.

Up to this point, there has been little mention of income taxes and depreciation. Certainly, these affect the cash flow patterns of our alternatives. In order to simplify the comparison of investment alternatives, however, we will assume that either the cash flows are after-tax cash flows or a before-tax study is desired. In Chapter Six we address the subject of income taxes and their effects on the preference among investment alternatives.

To illustrate the use of the various methods of comparing investment alternatives, consider the four alternatives having the cash flow profiles given in Table 5.7 for a planning horizon of 5 years. Note that to do nothing is presented here as an explicit alternative. A minimum attractive rate of return of 15% will be used in subsequent analyses.

TABLE 5.7

Cash Flow Profiles for Four Investment Alternatives

End of Year, t	Net Cash Flows for Alternatives			
	A_{0t}	A_{1t}	A_{2t}	A_{3t}
0	$0	$ 0	−$50,000	−$75,000
1	0	4,500	20,000	20,000
2	0	4,500	20,000	25,000
3	0	4,500	20,000	30,000
4	0	4,500	20,000	35,000
5	0	4,500	20,000	40,000

5.6.1 Present Worth Method

The *present worth method* can be used to compare investment alternatives by simply computing the present worth of each investment alternative and recommending the one having the greatest present worth. Alternatively, you can compute the present worth of each *increment of investment* and select or reject "increments" depending upon whether or not they have positive or negative present worths. A flowchart of the incremental approach is given in Figure 5.5.

Example 5.9

Ranking Approach

For the data given in Table 5.7, the present worth of Alternative 0, the do-nothing alternative, is zero. For Alternative 1, the present worth is

$$PW_1(15\%) = \$4500(P|A\ 15,5)$$
$$= \$4500(3.3522)$$
$$= \$15,085$$

For Alternative 2, the present worth is

$$PW_2(15\%) = -\$50,000 + \$20,000(P|A\ 15,5)$$
$$= -\$50,000 + \$20,000(3.3522)$$
$$= \$17,044$$

For Alternative 3, the present worth is

$$PW_3(15\%) = -\$75,000 + \$20,000(P|A\ 15,5) + \$5000(P|G\ 15,5)$$
$$= -\$75,000 + \$20,000(3.3522) + \$5000(5.7751)$$
$$= \$20,920$$

Since Alternative 3 has the greatest present worth, it would be recommended.

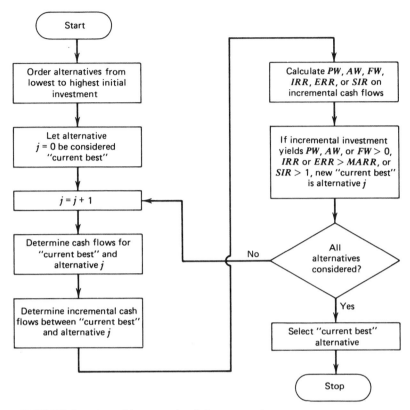

FIGURE 5.5. Flowchart of incremental cash flow approach.

Incremental Approach

Using an incremental approach, since no other alternative requires a smaller initial investment than Alternative 0, it will be used as the base for incremental comparison. Based on a rank ordering of alternatives by the size of the initial investment required, the first incremental comparison will be between Alternatives 0 and 1. If the present worth of the incremental difference is positive, then Alternative 1 will be preferred to Alternative 0, and it will become the new base for incremental comparison; if the present worth of the incremental difference is negative, then Alternative 0 will remain the base, and Alternative 1 will be discarded from further consideration.

$$PW_{1-0}(15\%) = \$4500(P|A\ 15,5)$$
$$= \$4500(3.3522)$$
$$= \$15,085$$

Since $PW_{1-0} > 0$, Alternative 1 is preferred to Alternative 0. Next, we look at the increment of investment required to go from Alternative 1 to Alternative 2. The present worth of the incremental investment will be

$$PW_{2-1}(15\%) = -\$50,000 + \$15,500(P|A\ 15,5)$$
$$= \$1959$$

Since $PW_{2-1} > 0$, Alternative 2 is preferred to Alternative 1. Next, we look at the incremental investment needed to go to Alternative 3 from Alternative 2. The incremental present worth will be

$$PW_{3-2}(15\%) = -\$25,000 + \$5000(P|G\ 15,5)$$
$$= \$3876$$

Since $PW_{3-2} > 0$, Alternative 3 is preferred to Alternative 2. Since no additional alternatives are available, we conclude that Alternative 3 is the most economic choice.

Notice that the incremental approach always compares the "next available alternative," based on rank ordering by initial investment, against the best alternative available to date—what we called the *base alternative*. If, in the example, PW_{2-1} had been less than zero, then Alternative 3 would have been compared to Alternative 1; if the incremental present worth equals zero, then we are indifferent between the two alternatives.

5.6.2   Annual Worth Method

The *annual worth method* is used in the same way as the present worth method, except annual worths are computed rather than present worths. It would be used when it is preferable to think in terms of annual worths or dollars per year.

Example 5.10

Ranking Approach

For the data given in Table 5.7, the annual worth of Alternative 0, the do-nothing alternative, is zero. For Alternative 1, the annual worth is $4500. For Alternative 2, the annual worth is

$$AW_2(15\%) = -\$50,000(A|P\ 15,5) + \$20,000$$
$$= -\$50,000(0.2983) + \$20,000$$
$$= \$5085/\text{year}$$

For Alternative 3, the annual worth is

$$AW_3(15\%) = -\$75,000(A|P\ 15,5) + \$20,000 + \$5000(A|G\ 15,5)$$
$$= -\$75,000(0.2983) + \$20,000 + \$5000(1.7228)$$
$$= \$6242/\text{year}$$

Alternative 3 is recommended since it has the greatest annual worth.

Incremental Approach

$$AW_{1-0}(15\%) = \$4500 > 0 \qquad \text{(Prefer Alternative 1 to 0)}$$

$$AW_{2-1}(15\%) = -\$50,000(A|P\ 15,5) + \$15,500$$
$$= \$585/\text{year} > 0 \qquad \text{(Prefer Alternative 2 to 1)}$$

$$AW_{3-2}(15\%) = -\$25,000(A|P\ 15,5) + \$5000(A|G\ 15,5)$$
$$= \$1157/\text{year} > 0 \qquad \text{(Prefer Alternative 3 to 2)}$$

Since no additional alternatives are available, we conclude that Alternative 3 is the most economic choice.

5.6.3 Future Worth Method

The *future worth method* is used in the same way as the present worth and annual worth methods, except future worths are computed instead of present worths and annual worths. Although it is easy to apply, it is not used as often as the present worth and annual worth methods for economic analyses of investment alternatives.

Example 5.11

Ranking Approach

For the data given in Table 5.7, the future worth of Alternative 0, the do-nothing alternative, is zero. For Alternative 1, the future worth is $4500.

$$FW_1(15\%) = \$4500(F|A\ 15,5)$$
$$= \$30,341$$

For Alternative 2, the future worth is

$$FW_2(15\%) = -\$50,000(F|P\ 15,5) + \$20,000(F|A\ 15,5)$$
$$= \$34,278$$

For Alternative 3, the future worth is

$$FW_3(15\%) = -\$75,000(F|P\ 15,5) + \$20,000(F|A\ 15,5)$$
$$+ \$5000(P|G\ 15,5)(F|P\ 15,5)$$
$$= \$42,073$$

We prefer Alternative 3 since it has the greatest future worth.

Incremental Approach

$$FW_{1-0}(15\%) = \$30,341 > 0 \qquad \text{(Prefer Alternative 1 to 0)}$$

$$FW_{2-1}(15\%) = -\$50,000(F|P\ 15,5) + \$15,500(F|A\ 15,5)$$
$$= \$3937 > 0 \qquad \text{(Prefer Alternative 2 to 1)}$$

$$FW_{3-2}(15\%) = -\$25,000(F|P\ 15,5) + \$5000(P|G\ 15,5)(F|P\ 15,5)$$
$$= \$7795 > 0 \quad \text{(Prefer Alternative 3 to 2)}$$

Alternative 3 is the last surviving alternative; therefore, it is preferred.

5.6.4 Internal Rate-of-Return Method

The *internal rate-of-return method* is one of the most popular, yet misused, methods of comparing alternatives. Although we recommend using an incremental cash flow approach in applying the *IRR* method, an alternative approach is possible; it is referred to as the *aggregate cash flow approach*.

The philosophy underlying the aggregate cash flow approach is to maximize the aggregate (or total) return obtained by investing in one of the investment alternatives *and the "reserve account."* The reserve account consists of the uninvested funds, which are assumed to be earning a return at a rate equal to the *MARR*.

Example 5.12

To illustrate the difference in the aggregate cash flow approach and the incremental cash flow approach, suppose two investment alternatives, A and B, are being considered: A requires an investment of $10,000, and B requires an investment of $50,000. Suppose A will return $2000/year forever; hence, an *IRR* of 20% is obtained. Suppose B will return $8000/year forever; in this case, an *IRR* of 16% is obtained. Now, suppose the *MARR* is 12%.

The aggregate cash flow approach proceeds in the following way. For B to be a feasible investment alternative, $50,000 must be available for investment. Investing in A means that $10,000 will return $2000/year and the reserve account will return 12% of the difference in the amount invested and the amount available for investment (i.e., (0.12)($40,000) or $4800/year). By investing in A, the aggregate return will be $6800/year from an aggregate investment of $50,000. The aggregate rate of return will be 13.6%. By investing the full amount in B, the aggregate rate of return will be 16%. In this case, B is preferred because it has the greatest aggregate rate of return.

Using an incremental cash flow approach, we begin with the alternative requiring the smallest initial investment, A, and compute its *IRR*; if it is greater than the *MARR,* then A is an acceptable base for incremental comparison. In this case, $IRR_A = 20\% > MARR$; so, A is an acceptable base. An incremental investment of $40,000 is required to step up to B from A; the incremental returns resulting from the incremental investment will be $6000/year; thus, the incremental *IRR* will be 15%, which is greater than the *MARR* of 12%. Hence, B is preferred to A. (From the results obtained, it should be obvious that A would have been preferred if the *MARR* had been 15% or greater.)

Now, let us apply the incremental cash flow approach in selecting the preferred alternative from among those described in Table 5.7.

Example 5.13

Between Alternatives 1 and 0, there is no incremental investment, but there are incremental returns of $4500 in each of years 1 to 5. The incremental *IRR* is therefore $i_{1-0} = \infty\%$ (Prefer 1 over 0).
Between Alternatives 2 and 1,

$$FW_{2-1}(i) = \$0 = -\$50,000(F|P\ i,5) + \$15,500(F|A\ i,5)$$

$@i = 15\%$ $\$0 \neq -\$50,000(2.0114) + \$15,500(6.7424) = \$3937.20$

$@i = 20\%$ $\$0 \neq -\$50,000(2.4883) + \$15,500(7.4416) = -\$9070.20$

Interpolating, $i_{2-1} = 16.51\% > MARR$ of 15% (Prefer 2 over 1).
Between Alternatives 3 and 2.

$$FW_{3-2}(i) = \$0 = -\$25,000(F|P\ i,5) + \$5,000(P|G\ i,5)(F|P\ i,5)$$

$@i = 15\%$ $\$0 \neq -\$25,000(2.0114) + \$5,000(5.7751)(2.0114) = \$7795.18$

$@i = 20\%$ $\$0 \neq -\$25,000(2.4883) + \$5,000(4.9061)(2.4883) = -\$1168.26$

Interpolating, $i_{3-2} = 19.35\% > 15\%\ MARR$ (Prefer 3 over 2).
We again prefer Alternative 3 since it prevailed in the pairwise comparison of alternatives in terms of *IRR* on incremental cash flows. Note that Alternative 3 did not have the highest incremental *IRR*! That is not important; what is important is that in the final pairwise comparison its rate exceeded the *MARR* of 15%.

As mentioned earlier, decision makers often use the *IRR* method incorrectly. They compute the *IRR* for individual alternatives and choose the one having the greatest rate. In so doing, they ignore the fact that any remaining funds will earn only the *MARR*. In the case of the above example, an infinite return is earned on Alternative 1 since no investment is required; the individual rate of return for Alternative 2 is approximately 28.73%; and the individual rate of return for Alternative 3 is approximately 24.90%. Thus, by ranking alternatives on the basis of their individual rates of return, Alternative 1 would be preferred to Alternative 2, which would be preferred to Alternative 3. On the other hand, by examining the incremental returns and investing so long as each increment is justified, we found (correctly) that Alternative 3 was the most profitable alternative.

The incremental cash flow approach is recommended when using the *IRR* method. It is not only correct, but much easier to use than the aggregate cash flow approach.

5.6.5 External Rate-of-Return Method

The *external rate-of-return method* is rarely used, but it has advantages over the *IRR* method.

Example 5.14

Let us now compare the alternatives of Example 5.9, Table 5.7 using the external rate-of-return method and incremental cash flows approaches. The positive net cash flows will be assumed to be reinvested at the $MARR = 15\%$ and Equation (4.15) will be employed.

Between Alternatives 1 and 0, there is no incremental investment, but there are incremental returns of $4500 in each of years 1 to 5. The incremental *ERR* is therefore $i'_{1-0} = \infty\%$ (Prefer 1 over 0).

Between Alternatives 2 and 1,

$$\$15,500(F|A\ 15,5) = \$50,000(\ + i')^5$$

$$\$15,500(6.7424) = \$50,000(1 + i')^5$$

Solving, $i'_{2-1} = 15.89\% > MARR$ (Prefer 2 over 1).

Between Alternatives 3 and 2,

$$\$5000(P|G\ 15,5)(F|P\ 15,5) = \$25,000(1 + i')^5$$

$$\$5000(5.7751)(2.0114) = \$25,000(1 + i')^5$$

Solving, $i' = 18.36\% > MARR$ (Prefer 3 over 2).

Again, Alternative 3 is preferred on the basis of its winning the series of pairwise comparisons using the *ERR* method and the incremental cash flow approach. Note that Alternative 3 did not have the highest *ERR* on its incremental investment; its *ERR* was, however, higher than the *MARR* in the final pairwise comparison. The incremental cash flow approach is recommended when using the *ERR*, since it is easily applied.

5.6.6 Savings/Investment Ratio Method

The *savings/investment ratio method* is most often used in public sector projects; it is also known as the benefit/cost ratio.

Example 5.15

We will now analyze the four alternatives of Example 5.9, Table 5.7 using the *SIR* formulation given in Equation (4.11). Again, the $MARR = 15\%$.

Between Alternatives 1 and 0, there is no incremental investment, but there are incremental returns of $4500 in each of three years 1 to 5. The incremental *SIR* is therefore $SIR_{1-0}(15\%) = \infty$ (Prefer 1 to 0).

Between Alternatives 2 and 1,

$$SIR_{2-1}(15\%) = \frac{\$15,500(P|A\ 15,5)}{\$50,000}$$

$$= \frac{\$15,500(3.3522)}{\$50,000}$$

$$= 1.04 > 1.00 \qquad \text{(Prefer 2 to 1)}$$

Between Alternatives 3 and 2,

$$SIR_{3-2}(15\%) = \frac{\$5,000(P|G\ 15,5)}{\$25,000}$$

$$= \frac{\$5,000(5.7751)}{\$25,000}$$

$$= 1.16 > 1.00 \qquad \text{(Prefer 3 to 1)}$$

Alternative 3 is preferred since it won the final pairwise comparison by having an *SIR* greater than 1.00 on the incremental investment.

5.6.7 Payback Period Method

The *payback period method* is not consistent with the previous six methods presented, and as such may yield entirely different selections or rankings of alternatives. Neither the total nor the incremental cash flow approach as presented is applicable. The PBP method is recommended for use only as an auxilliary or secondary criterion in alternative selection.

Example 5.16

Let us now use the payback period method to analyze the alternatives presented in Example 5.9, Table 5.7.

For Alternative 0, it is not meaningful to talk about payback. Since Alternative 1 requires no investment, its payback is instantaneous.

For Alternative 2.

$$\sum_{i=1}^{2} R_{2t} = \$20,000 + \$20,000 = \$40,000 < C_{20} = \$50,000$$

$$\sum_{i=1}^{3} R_{2t} = \$20,000 + \$20,000 + \$20,000 = \$60,000 > C_{20} = \$50,000$$

The payback period is 2½ (or 3) years.

For Alternative 3,

$$\sum_{i=1}^{2} R_{3t} = \$20,000 + \$25,000 + \$30,000 = \$75,000 = C_{30}$$

The payback period is exactly 3 years.

Note that the ranking based upon PBP for our Alternatives is 1, 2, 3 which is exactly in reverse order of preference dictated by the six consistent methods.

5.6.8 Capitalized Worth Method

The *capitalized worth method* is applicable only if there is reason to believe that a series of cash flows will repeat indefinitely into the future. As such, it is not a method comparable to those used previously in this section. Neither the total nor the incremental cash flow approach as presented is applicable.

Example 5.17

Let us assume that the cash flows as presented in Example 5.9, Table 5.7 actually do repeat indefinitely into the future. We will continue using the *MARR* of 15%. Also, we will make use of the annual worths calculated in Example 5.10.

For Alternative 0,

$$CW_0(15\%) = \$0$$

For Alternative 1,

$$CW_1(15\%) = \frac{AW_1(15\%)}{.15}$$

$$= \frac{\$4500}{.15}$$

$$= \$30,000$$

For Alternative 2,

$$CW_2(15\%) = \frac{AW_2(15\%)}{.15}$$

$$= \frac{\$5085}{.15}$$

$$= \$33,900$$

For Alternative 3,

$$CW_3(15\%) = \frac{AW_3(15\%)}{.15}$$

$$= \frac{\$6241.50}{.15}$$

$$= \$41,610$$

We prefer Alternative 3 on the basis that it has the highest capitalized worth. Had we been calculating the capitalized cost where costs are positive, we would have preferred the smallest value.

5.7 PERFORMING SUPPLEMENTARY ANALYSES

The sixth step in performing economic comparisons of engineering investment alternatives is to perform supplementary analyses. Among the supplementary analyses that might be performed are break-even analyses, sensitivity analyses, and risk analyses. Explicit consideration of break-even, sensitivity, and risk analyses is reserved for Chapter Seven. However, it is useful to consider each analysis technique here, albeit briefly.

A break-even analysis is performed when the value of one or more parameters is not known with certainty, but a judgment can be made as to whether its value is greater than or less than some break-even value. Many of the equivalence calculations performed in Chapter Three were examples of break-even analyses.

A sensitivity analysis involves an explicit consideration of the economic impact of changes in the values of one or more parameters. The objective is to determine how sensitive the selection decision is to the values of the parameters used in the analysis. The parameters subject to change might include the planning horizon, the *MARR*, and any of the cash flows. The process of providing estimates of cash flows and decisions concerning the planning horizon and *MARR* is not a precise, errorless process. Sensitivity analyses are performed in an attempt to cope with the inexact aspects of economic justifications.

Risk analyses are performed by explicitly representing the range of possible values of parameters in the form of probability distributions. By treating the parameters as random variables, analytical or simulation models can be developed of the economic performance of the alternatives. Subsequent comparisons between alternatives can then be made on the basis of the probability distributions for the measure of merit being used (e.g. probability distribution for present worth).

Too often, management will not accept estimates of cost savings or items not easily quantified and measured. As a result, savings and benefits due to inventory reductions, energy savings, quality improvements, space reductions, safety improvements, cycle-time reductions, throughput improvement, lead-time reductions, and increased flexibility are often ignored. Every effort should be made to capture the monetary impact of these benefits. Even though they cannot be measured precisely, their value is certainly greater than zero. As Kaplan observed, "Although intangible benefits are difficult to quantify, there is no reason to value them at zero in a capital expenditure analysis. Zero is, after all, no less arbitrary than any other number. Conservative accountants who assign zero values to many intangible benefits prefer being precisely wrong to being vaguely right. Managers need not follow their example" [12, p. 92]. In the end a rough estimate of intangible benefits is better than no estimate at all! Supplementary analyses provide one mechanism for assessing the impact of the selection decision on the values of difficult-to-measure benefits.

As noted previously, many firms have well-defined procedures for performing economic justifications. Hence, a supplementary analysis might not be required. Even so, we strongly recommend that they be performed—if for no one else's benefit than yours. Supplementary analyses are performed in order to answer a number of "what if?" questions that will either be asked by management or be of concern to management; in a sense, they provide an "insurance policy"; they are performed to provide a higher level of confidence in the results of the economic justifications.

In summary, we have found it to be particularly useful to do more than *just what was required by the client or the company* in performing economic justifications. The benefits of performing supplementary analyses included increasing our own confidence level regarding our recommendations, as well as increasing management's confidence in our analysis.

5.8 SELECTING THE PREFERRED ALTERNATIVE

The seventh, and final, step in performing a comparison of engineering investment alternatives involves the selection of the preferred alternative. The final selection decision might be quite different from the one recommended as "the economic choice." Multiple criteria typically exist, rather than the single criterion of maximizing the economic worth. (However, it should be noted that any long-term decision process that consistently selects alternatives that do not earn a rate of return at least equal to the cost of capital will result in financial ruin for the firm. Hence, even when multiple criteria exist, the alternative ultimately selected should not be uneconomic.)

The presence of multiple criteria, coupled with the risks and uncertainties associated with estimating future outcomes, result in the selection process being quite complicated. To make the process easier, the engineer is encouraged to address as many as possible of management's concerns in comparing the investment alternatives. To the extent that management's concerns have been addressed, the selection decision will agree with the engineer's recommendation.

Our discussion in this chapter has concentrated solely on the economic factor; we have been concerned with determining the most economical alternative. The final decision may be based on a host of criteria instead of on the single criterion of economics. In Chapter Nine we examine the decision process in the face of multiple objectives, and we extend the analyses of Chapter Eight to include decision making in the face of risk and uncertainty.

As pointed out in Chapter One, the selection or rejection of the recommended solution is heavily dependent on the sales ability of the individual presenting the recommendation to management. Since the corporate decision makers are normally presented with many more investment alternatives than can be funded, it is important to communicate effectively in order to compete favorably for the company's limited capital. Toward this end, Klausner [13] provides four specific suggestions for the engineer or systems analyst.

1. He should recognize that the decision-makers' perspective is broad and develop his proposal accordingly. The project's capital requirements should be related to previous and estimated future capital requirements on similar investments. The project's Discounted Cash Flow return or Net Present Value should be compared to other projects with which the decision-makers are familiar. The proposal should be shown to fit in with long-range corporate plans and support short-range objectives. Comparison should be made with similar investments by competition and competitive advantage (if any) shown. Ancillary marketing, public relations and/or political benefits that the company will derive from the investment should be pointed out. The effect that the investment will have on other functional activities of the company should be noted and overall benefits stressed. In other words, the investment proposal should be related to the well-being of the total enterprise.

2. He should recognize that this investment proposal is only one of several that the decision-makers are reviewing and that not all proposals will be accepted. The engineer should know how the decision-makers classify investments and what competition there is for available capital resources, he should know the relative strengths and weaknesses of his investment proposal vis-à-vis competing investment proposals and deal with each in the presentation. If the proposed investment is relatively risk-free, that point should be made strongly to possibly offset less desirable aspects of the proposal (e.g., low return on investment).

3. He should know the decision-makers and tailor the investment proposal accordingly. The engineer should, for example, know and use the measure of merit they prefer, and support it with any other measures of merit that may be necessary or helpful. For example, if the decision-makers favor Discounted Cash Flow return on investment and the proposal has one that barely meets minimum return standards but does have a large Net Present Value, the engineer must make sure that the potential contribution to profit is clearly pointed out. If it is known that the decision-makers still calculate the payout period of each investment proposal, the investment's payout period should be stated—either to support the proposal or to point out its irrelevance in hopes of minimizing its impact on the ultimate investment decision. The engineer should always use technical and economical terms that the decision-makers will understand and relate to.

The engineer should become familiar with the values that the decision-makers attach to different aspects of investment proposals—particularly economic uncertainty. Low risk should be emphasized in proposals to decision-makers who tend to be risk-avoiders; potential economic gain should be highlighted for the risk-taking decision-maker. This, of course, does not suggest that the engineer should be less than honest in his proposal, but only makes the point that the communication of the investment proposal should be developed with the decision-maker in mind and emphasis varied accordingly.

4. He should not oversell the technical engineering aspects of the investment proposal. It should be remembered that decision-makers are primarily interested in the economic aspects of the proposal. The engineer must resist the temptation to overstate the complicated and sophisticated technology that might underlie a proposal. The less decision-makers understand about the technical/engineering aspects of a proposal, the more uneasy they become. This apprehension becomes part of the uncertainty which the decision-makers subjectively assign to the proposal and the result could be its rejection in favor of a proposal with which they are more familiar, or at least, more comfortable. The engineer should keep in mind the background and interests of the decision-makers and use technical and commercial terms with which they are familiar.

5.9 ANALYZING ALTERNATIVES WITH NO POSITIVE CASH FLOWS

The previous analysis of an investment decision involved four mutually exclusive alternatives, including the do-nothing alternative, which included positive-valued cash flows. However, there exist solutions in which no positive-valued cash flows are present.

Example 5.18

Consider a case in which two new cost reduction alternatives have been proposed. The present method, which we refer to as the *do-nothing* alternative, is also a feasible alternative. Thus, three mutually exclusive alternatives are considered. Cash flow profiles for the alternatives are given in Table 5.8.

Alternative 0 is the do-nothing alternative in which the present expenditure of $12,000/year is continued. Alternative 1 involves an initial investment of $10,000 in order to reduce the annual expenditures by $3000 over the 5-year period. Alternative 2 requires an initial investment of $15,000 in order to obtain decreasing annual expenditures. The minimum attractive rate of return is specified to be 10%.

TABLE 5.8

Cash Flow Profiles for Three Investment
Alternatives with No Positive Cash Flows

t	A_{0t}	A_{1t}	A_{2t}
0	0	−$10,000	−$15,000
1	−$12,000	− 9,000	− 9,000
2	− 12,000	− 9,000	− 8,000
3	− 12,000	− 9,000	− 7,000
4	− 12,000	− 9,000	− 6,000
5	− 12,000	− 9,000	− 5,000

The PW, AW, or FW methods using the ranking approach, or any measure of investment worth using the incremental cash flow approach, may be applied to this problem. Let us analyze these alternatives in three different ways.

Annual Worth Method—Total Cash Flow Approach

$$AW_0(10\%) = -\$12,000/\text{year}$$

$$AW_1(10\%) = -\$10,000(A|P\ 10,5) - \$9000$$
$$= -\$10,000(.2638) - \$9000$$
$$= -\$11,638/\text{year}$$

$$AW_2(10\%) = -\$15,000(A|P\ 10,5) - \$9000 + \$1000(A|G\ 10,5)$$
$$= -\$15,000(.2638) - \$9000 + \$1000(1.8101)$$
$$= -\$11,146.90/\text{year}$$

Alternative 2 is preferred since it has the highest annual worth (or lowest annual cost).

Annual Worth Method—Incremental Cash Flow Approach

$$AW_{1-0}(10\%) = -\$10,000(A|P\ 10,5) + \$3000$$
$$= -\$10,000(.2638) + \$3000$$
$$= \$362/\text{year} \qquad (\text{Prefer 1 over 0})$$

$$AW_{2-1}(10\%) = -\$5,000(A|P\ 10,5) + \$1000(A|G\ 10,5)$$
$$= \$5,000(.2638) + \$1000(1.8101)$$
$$= \$491.10/\text{year} \qquad (\text{Prefer 2 over 1})$$

Again, Alternative 2 is preferred.

Internal Rate of Return Method—Incremental Cash Flow Approach

$$FW_{1-0}(i) = \$0 = -\$10,000(F|P\ i,5) + \$3,000(F|A\ i,5)$$

$$@i = 15\% \quad \$0 \ne -\$10,000(2.0114) + \$3,000(6.7424) = \$113.20$$

$$@i = 20\% \quad \$0 \ne -\$10,000(2.4883) + \$3,000(7.4416) = -\$2558.20$$

Interpolating, $i_{1-0} = 15.21\% > MARR$ (Prefer 1 over 0).

$$FW_{2-1}(i) = \$0 = -\$5000(F|P\ i,5) + \$1000(P|G\ i,5)(F|P\ i,5)$$

$$@i = 15\% \quad \$0 \ne -\$5000(2.0114) + \$1000(5.7751)(2.0114) = \$1559.04$$

$$@i = 20\% \quad \$0 \ne -\$5000(2.4883) + \$1000(4.9061)(2.4883) = -\$233.65$$

Interpolating, $i_{2-1} = 19.35\% > MARR$ (Prefer 2 over 1).

Again, Alternative 2 is preferred.

The *IRR* method deserves additional explanation. Since the sum of the cash flows (ignoring the time value of money) is negative for each alternative, positive-valued rates of return among the three alternatives do not exist. However, even though Alternative 1 does not have a rate of return by itself, in comparison with Alternative 0, it is clear that the investment of the $10,000 yields annual savings of $3000 in annual expenditures. Thus, on an incremental basis, the $10,000 incremental investment produces positive-valued cash flows of $3000/year for each of the 5 years.

The return on the incremental investment, using the internal rate-of-return method, is found to be approximately 15.21%. Since we would do better to invest the $10,000 in Alternative 1 and earn 15.21% than to choose Alternative 0 and only earn the minimum attractive rate of return on the remaining money available for investment, Alternative 1 is preferred to Alternative 0.

Of course, we have $15,000 to invest (otherwise Alternative 2 would be unfeasible). Thus, at this point we are willing to invest $10,000 in Alternative 1 and earn 15.21% and invest the remaining $5000 in some other opportunity and earn the minimum attractive rate of return of 10%. The question now considered is, "Should we use the additional $5000 and pool it with the $10,000 to invest in Alternative 2?" The rate of return on the incremental investment required to obtain Alternative 2 is calculated to be approximately 19.35%.

At this point, the question is "Should we invest $10,000 in Alternative 1 and invest $5000 elsewhere at the *MARR*, or should we invest $15,000 in Alternative 2?" Investing in Alternative 2 yields a return of 15.21% on the $10,000 increment and 19.35% on the $5000 increment. Consequently, Alternative 2 would be preferred to the investment of $10,000 in Alternative 1 and investing the remaining $5000 to earn only 10%.

5.10 MORE ON DEALING WITH UNEQUAL LIVES

Recall that in Section 5.3 we recommended that a planning horizon be specified, cash flows over the planning horizon be given explicitly, and the evaluation be performed over the common planning horizon time frame. While this method is becoming better known and accepted, many still compare alternatives having unequal lives on the basis of the individual life cycles. This approach either implicitly assumes a least common multiple of lives approach, or it implicitly establishes a salvage value which may or may not be reasonable. Enough background material has now been presented to permit us to clarify these points and demonstrate their drawbacks.

Example 5.19

Consider two mutually exclusive alternatives having cash flow profiles as depicted in Figure 5.6. Using a minimum attractive rate of return of 15%, the following annual worths are obtained:

$$AW_1(15\%) = -\$5000(A|P\ 15,5) - \$2000 + \$1000(A|F\ 15,5)$$
$$= -\$5000(.2983) - \$2000 + \$1000(.1483)$$
$$= -\$3343.20$$

$$AW_2(15\%) = -\$7000(A|P\ 15,6) - \$1500 + \$1500(A|F\ 15,6)$$
$$= -\$7000(.2642) - \$1500 + \$1500(.1142)$$
$$= -\$3178.10$$

Implicit in the comparison of these alternatives using annual worths based on individual life cycles is the assumption that a 30-year planning horizon is being used. As shown in Table 5.9, if the present worths are computed for a 30-year period, the following values will be obtained:

$$PW_1(15\%) = -\$21,952.10$$

$$PW_2(15\%) = -\$20,868.30$$

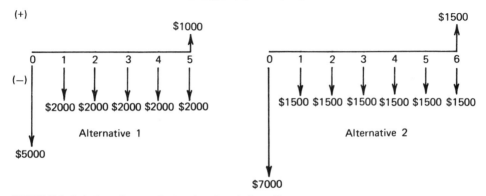

FIGURE 5.6. Cash flow diagrams for two investment alternatives.

TABLE 5.9

Cash Flows for Two Investment Alternatives,
with a 30-year Planning Horizon

EOY	CF(1)	CF(2)
0	−$ 5,000	−$ 7,000
1–4	− 2,000	− 1,500
5	− 6,000	− 1,500
6	− 2,000	− 7,000
7–9	− 2,000	− 1,500
10	− 6,000	− 1,500
11	− 2,000	− 1,500
12	− 2,000	− 7,000
13–14	− 2,000	− 1,500
15	− 6,000	− 1,500
16–17	− 2,000	− 1,500
18	− 2,000	− 7,000
19	− 2,000	− 1,500
20	− 6,000	− 1,500
21–23	− 2,000	− 1,500
24	− 2,000	− 7,000
25	− 6,000	− 1,500
26–29	− 2,000	− 1,500
30	− 1,000	0
Present worth	−$21,952.10	−$20,868.30

Converting the present worths to annual worths yields

$$AW_1(15\%) = -\$21,952.10(A|P\ 15,30)$$
$$= -\$21,952.10(.1523)$$
$$= -\$3343.30$$

$$AW_2(15\%) = -\$20,868.30(A|P\ 15,30)$$
$$= -\$20,868.30(.1523)$$
$$= -\$3178.24$$

which are the values obtained (except for round-off error) using individual life cycles.

An alternative assumption that could be made is that a 5-year planning horizon is being used and the salvage value for Alternative 2 is such that an annual worth of −$3178.10 will still be obtained. Hence, S_5, the salvage value at the end of the fifth year, for Alternative 2 must be

$$AW_2(15\%) = -\$7000(A|P\ 15,5) - \$1500 + S_5(A|F\ 15,5)$$

$$-\$3178.10 = -\$7000(.2983) - \$1500 + S_5(.1483)$$

$$S_5 = \$2764.67$$

This salvage value may or may not have any relationship to reality.

Example 5.20

Consider the alternatives depicted in Figure 5.7. They will be used in a more glaring example to illustrate the shortcoming of an assumption that the salvage value for unused portions of an asset's life will be such that the annual worth will be unchanged. The annual worths for individual life cycles are found to be:

$$AW_1(15\%) = -\$5000(A|P\ 15,5) - \$3000 + \$1000(A|F\ 15,5)$$
$$= -\$5000(.2983) - \$3000 + \$1000(.1483)$$
$$= -\$4343.20$$

$$AW_2(15\%) = -\$6000(A|P\ 15,6) - \$1000 - \$1000(A|G\ 15,6)$$
$$\qquad + \$1000(A|F\ 15,6)$$
$$= -\$6000(.2642) - \$1000 - \$1000(2.0972) + \$1000(.1142)$$
$$= -\$4568.20$$

If a 5-year planning horizon is used, to obtain an annual worth of $-\$4568.20$ for Alternative 2 requires that the salvage value at the end of the fifth year, F_5, be such that

$$-\$4568.20 = -\$6000(A|P\ 15,5) - \$1000 - \$1000(A|G\ 15,5) + S_5(A|F\ 15,5)$$

$$-\$4568.20 = -\$6000(.2983) - \$1000 - \$1000(1.7228) + S_5(.1483)$$

$$S_5 = -\$374.92$$

Thus, instead of having a $1000 salvage value at the end of the sixth year, a salvage value of $-\$374.92$ must exist at the end of the fifth year to yield the annual worth obtained by treating individual life cycles.

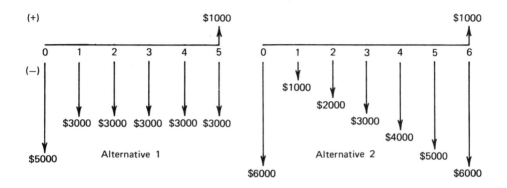

$$MARR = 15\%$$

FIGURE 5.7. Cash flow diagrams for two investment alternatives.

We do not recommend that alternatives having unequal lives be blindly compared on the basis of the annual worths for individual life cycles. Such an approach assumes *implicitly* that either a least common multiple of lives planning horizon is appropriate or the salvage values for unused portions of an asset's life are such that the annual worth is unchanged. We prefer to make *explicit* all assumptions concerning the planning horizon and the salvage values.

5.11 REPLACEMENT ANALYSIS

Replacement analysis is one of the most important and most common types of alternative comparisons encountered in practice. In a replacement analysis, one of the feasible alternatives involves maintaining the status quo; the remaining alternatives provide various replacement options that are available. In some organizations, replacement analyses are performed routinely in an effort to ensure that the best equipment and facilities are in use, compared to their possible successors.

The reasons for considering replacement are numerous. First, the current asset (defender) may have a number of deficiencies including high set-up cost, excessive maintenance, declining production efficiency, heavy energy consumption, and physical impairment. For example, when you are confronted with a car that is expensive to operate and maintain, or one soon to need a major overhaul, you begin to consider replacing the car.

Second, potential replacement assets (challengers) may take advantage of new technology and be easily set up, maintained at low cost, high in output, energy efficient, and possess increased capabilities, perhaps at a vastly reduced cost. For example, some new-generation, computer-controlled manufacturing equipment has rendered many old machines economically obsolete. Also, we can relate to the phenomenal accomplishments with which calculators and personal computers have resulted in increased capabilities, vastly lower prices, and economic obsolescence for equipment only a few years old.

Finally, the environment affects replacement decisions. Consumer demand preferences change, and present equipment may be unable to adapt to the new designs dictated by consumers. Demand levels may also cause equipment capacity to be relatively high or low, resulting in inefficiency or inability to perform. Rental firms specializing in equipment, automobiles, furniture, and so forth, may affect ownership decisions by offering lease options. Also, outside contractors may be able to perform some tasks cheaper and better than in-house facilities. All of the above considerations and more may affect the decision to keep or replace an asset.

Replacement analyses are basically just another type of alternative comparison. That is, they follow the same systematic seven-step approach outlined and developed in this chapter. All of the six consistent measures of investment worth are also applicable. Unfortunately, replacement decisions are often confounded by our nearness to the existing asset. That is, we may have an "emotional" attachment to present equipment, particularly if we are the ones who recommended its installation previously. This attachment often results in attempts to recover *sunk costs,* identified in Chapter Two as past costs that are *unrecoverable.* Sunk costs should not be included in a replacement analysis (or any other alternative evaluation); to include sunk costs penalizes or burdens the potential replacement asset and unfairly favors the current asset.

Two approaches are commonly used in replacement analyses. One is the *cash flow approach* in which actual cash flows associated with keeping, purchasing, or leasing an asset are used directly. The other is the *outsider viewpoint approach* in which the cash flow profiles faced by an objective outsider are used. Which approach is used in a replacement analysis is strictly a matter of preference. Both are mathematically equivalent and yield consistent decisions.

5.11.1 Cash Flow Approach

The *cash flow approach* might properly be called the *insider viewpoint approach*. Often, there is no additional capital cost if the present asset is kept. If a replacement is purchased, there is often a trade-in allowance given for the present asset. In order to develop the cash flow profiles, the decision maker should ask "How much money will be spent and received if I adopt this alternative?" As noted above, past costs should be viewed from the proper perspective; unrecoverable past costs are *sunk costs* and are not to be included in economy studies that deal with the future, except as those sunk costs may affect income taxes if a present asset is disposed.

The planning horizon to be used is again at the discretion of the decision maker. As in any alternative evaluation, the current asset and its proposed replacements must be evaluated over a common planning horizon, with cash flow profiles extending throughout but not beyond that horizon for each alternative. Since the remaining life of the present asset is usually shorter than that of a new asset, the *shortest life among alternatives* is often chosen.

Example 5.21

To illustrate a replacement analysis, consider a situation involving a chemical plant that owns a filter press that was purchased 3 years ago for $30,000. Actual operating and maintenance (O & M) expenses (excluding labor) for the press have been $4000, $5000, and $6000 each of the past 3 years, as depicted in Table 5.10. It is anticipated that the filter press can be used for 5 more years and salvaged for $2000 at that time. The undepreciated worth of the press as indicated on our accounting books is $12,600; however, technological developments of the past 3 years have resulted in excellent new competitive filter presses. As a result, the current market value for the used

TABLE 5.10

Data for a Replacement Alternative

	Alternative 1		Alternative 2	
End of Year, t	Operating and Maintenance Costs	End of Year, t	Operating and Maintenance Costs	Salvage Value, S_t
-3	—	0	—	$36,000
-2	$-$ 4,000	1	—	30,000
-1	$-$ 5,000	2	$-$1,000	24,600
0	$-$ 6,000	3	$-$ 2,000	19,800
1	$-$ 7,000	4	$-$ 3,000	15,600
2	$-$ 8,000	5	$-$ 4,000	12,000
3	$-$ 9,000	6	$-$ 5,000	9,000
4	$-$ 10,000	7	$-$ 6,000	6,600
5	$-$ 11,000	8	$-$ 7,000	4,800
		9	$-$ 8,000	3,600
		10	$-$ 9,000	3,000

filter press is only $9000. If the old filter press is retained, annual operating and maintenance costs are anticipated to be as shown in Table 5.10.

A new filter press is available and can be purchased for $36,000. It has an anticipated life of 10 years and is expected to have annual operating and maintenance costs as also given in Table 5.10. Based upon historical data concerning salvage values of filter presses, estimated salvage values (S_t) for the new press are also given in the table. Alternative 1 is defined to be "keep the old press," while Alternative 2 is defined to be "replace the old press with the new press."

Since the "old" press can only be used for 5 more years, a planning horizon of 5 years is specified; the salvage value for the "new" press is estimated to be $12,000 in 5 years. Cash flows for each alternative are given in Table 5.11. Note that the $9000 market value of the existing press is

TABLE 5.11

Cash Flows for a Replacement Alternative—Cash Flow Approach

End of Year, t	Alternative 1 Net Cash Flows, A_{1t}	Alternative 2 Net Cash Flows, A_{2t}	Difference in Cash Flows, $A_{2t} - A_{1t}$
0	$0	$-$36,000 + 9,000	$-$27,000
1	$-$ 7,000	0	7,000
2	$-$ 8,000	$-$1,000	7,000
3	$-$ 9,000	$-$2,000	7,000
4	$-$ 10,000	$-$3,000	7,000
5	$-$11,000 + 2,000	$-$4,000 + 12,000	17,000

applied as a positive cash flow for the challenger since the old unit will be sold if the new unit is purchased.

Computing the annual worths for each alternative using a minimum attractive rate of return of 15% yields the following results:

$$AW_1(15\%) = -\$7000 - \$1000(A|G\ 15,5) + \$2000(A|F\ 15,5)$$
$$= -\$7000 - \$1000(1.7228) + \$2000(.1483)$$
$$= -\$8426.20/\text{year}$$

$$AW_2(15\%) = -\$27,000(A|P\ 15,5) - \$1000(A|G\ 15,5) + \$12,000(A|F\ 15,5)$$
$$= -\$27,000(.2983) - \$1000(1.7228) + \$12,000(.1483)$$
$$= -\$7997.30/\text{year}$$

On the basis of the annual worth difference of \$428.90/year, it is recommended that the new filter press be purchased and the old press be sold.

In listing the cash flows for each alternative in Table 5.11, we ignored the operating and maintenance costs that will occur after the fifth year if the new filter press is purchased. The argument against including these operating and maintenance costs is based on the specification of a planning horizon of 5 years. The 5-year planning horizon was based on the maximum useful life for the old filter press. If the old press is retained, then it will have to be replaced in 5 years (if not before, because of the possible development of attractive replacement alternatives in the future). Consequently, in 5 years we might have available an alternative that will yield even greater operating and maintenance savings than the filter press currently being considered. Once the planning horizon has been established, it is not fair to include any cash flows that might occur later for one alternative without including similar estimates for the other alternative(s).

The above interpretation defines the planning horizon as a window through which only the cash flows that occur during the planning horizon can be seen. This window should *include a terminal (salvage) value for any alternative having a life longer than the planning horizon,* even though the alternative might not be physically replaced at that time. The end of the planning horizon defines a point in time at which another replacement study is planned. At that time, the future savings and costs will be compared against other available replacement candidates.

It should also be noted that neither the \$30,000 first cost nor the \$12,600 book value of the existing press appears as a cash flow. The \$30,000 is a past cost which is now irrelevant (except as it affects taxes as seen in Chapter Six), and the \$12,600 was never a cash flow. The difference between the value on our accounting books (\$12,600) and the market value (\$9000) is \$3600. This \$3600 is a sunk cost that analysts are often tempted to add to the first cost of the challenger in an attempt to recover it. To do so is absolutely incorrect, and biases the decision maker against the proposed replacement item. For example, if we burdened Alternative 2 by adding a sunk cost of

$3600 to its first cost, its annual worth would be decreased by $-\$3600(A|P$ $15,5) = -\$3600(.2983) = -\$1073.88/\text{year}$ to a total of $-\$7997.30/\text{year} - \$1073.88/\text{year} = -\$9071.18/\text{year}$. On this basis, we would mistakenly select Alternative 1 and continue to operate an economically inferior filter press.

Example 5.22

If, in Example 5.21, a 10-year planning horizon is desired, we recommend that consideration be given to the replacement of the "old" filter press in 5 years. Based on a forecast of the growth of filter press technology, suppose we anticipate that at the end of 5 years a filter press will be available at a cost of $31,000. Net operating and maintenance costs, and a terminal salvage value are anticipated to be as depicted in Table 5.12. •

Calculating the annual worth of these alternatives results in:

$$
\begin{aligned}
AW_1(15\%) &= [-\$7000(P|A\ 15,5) - \$1000(P|G\ 15,5) - \$29,000(P|F\ 15,5) \\
&\quad -\$1000(P|G\ 15,5)(P|F\ 15,5) \\
&\quad + \$15,000(P|F\ 15,10)](A|P\ 15,10) \\
&= (-\$7000(3.3522) - \$1000(5.7751) - \$29,000(.4972) \\
&\quad -\$1000(5.7751)(.4972) + \$15,000(.2472)](.1993) \\
&= -\$8534.56/\text{year}
\end{aligned}
$$

$$
\begin{aligned}
AW_2(15\%) &= -\$27,000(A|P\ 15,10) - \$1000(A|G\ 15,10) \\
&\quad + \$3000(A|F\ 15,10) \\
&= -\$27,000(.1993) - \$1000(3.3832) + \$3000(.0493) \\
&= -\$8616.40/\text{year}
\end{aligned}
$$

TABLE 5.12

Cash Flows for a Replacement Alternative Using a 10-Year Planning Horizon-Cash Flow Approach

End of Year, t	Alternative 1 Net Cash Flows, A_{1t}	Alternative 2 Net Cash Flows, A_{2t}
0	$ 0	$-\$36,000 + \$9,000$
1	$-\$ 7,000$	0
2	$-$ 8,000	$-1,000$
3	$-$ 9,000	$-2,000$
4	$-$ 10,000	$-3,000$
5	$-11,000 + 2,000 - 31,000$	$-4,000$
6	0	$-5,000$
7	$-$ 1,000	$-6,000$
8	$-$ 2,000	$-7,000$
9	$-$ 3,000	$-8,000$
10	$-4,000 + 15,000$	$-9,000 + 3,000$

These computations establish that the difference in annual worths for the two alternatives is $81.84/year, with Alternative 1 being more economic. Thus, on the basis of technological forecasts of filter press alternatives in 5 years, it would appear advantageous to postpone the replacement. However, the degree of uncertainty in our forecast of the cash flows for a projected future replacement candidate would cause us to question the merits of postponing the replacement because of a difference of $81.84/year. These kinds of considerations are explored more fully in Chapters Eight and Nine.

Often, there is a discrepancy between trade-in allowances if several potential replacement assets are being considered. Further, these trade-in values have little relationship to the true market value that could be realized if the existing asset were sold separately. In this case, when using the cash flow approach, the appropriate question to ask is still, "How much money will be spent and received if I adopt this alternative?"

Example 5.23

In Example 5.21, Alternative 1 was to keep our existing filter press having a market value of $9000. Alternative 2 included selling the current press and buying a new one for $36,000. Suppose, however, that the dealer has offered to allow a $10,000 trade-in value for the old press. Also two additional alternatives have been identified. Alternative 3 is a new asset costing $40,000, having a salvage value after 5 years of $13,000, and having annual operating and maintenance costs as given in Table 5.13. A trade-in value of $12,000 is offered for the current press in Alternative 3. Alternative 4 is to lease a press for $7500/year payable at the beginning of each year during the 5-year horizon. If the lease is taken, the existing press will be sold on the open market. The complete cash flow profile for each alternative is given in Table 5.13.

TABLE 5.13

Cash Flows for Several Replacement Alternatives—Cash Flow Approach

End of Year, t	Alternative 1 Net Cash Flows, A_{1t}	Alternative 2 Net Cash Flows, A_{2t}	Alternative 3 Net Cash Flows, A_{3t}	Alternative 4 Net Cash Flows, A_{4t}
0	0	$-\$36,000 + \$10,000$	$-\$40,000 + 12,000$	$-\$7,500 + 9,000$
1	$-\$ 7,000$	0	$- 500$	$-7,500$
2	$- 8,000$	$-1,000$	$-1,000$	$- 7,500 - 800$
3	$- 9,000$	$-2,000$	$-1,500$	$- 7,500 - 1,600$
4	$- 10,000$	$-3,000$	$-2,000$	$- 7,500 - 2,400$
5	$-11,000 + 2,000$	$-4,000 + 12,000$	$-2,500 + 13,000$	$-3,200$

Calculating the annual worths of these alternatives at a *MARR* of 15% results in

$$AW_1(15\%) = -\$7000 - \$1000(A|G\ 15,5) + \$2000(A|F\ 15,5)$$
$$= -\$7000 - \$1000(1.7228) + \$2000(.1483)$$
$$= -\$8426.20/\text{year}$$

$$AW_2(15\%) = -\$26{,}000(A|P\ 15,5) - \$1000(A|G\ 15,5) + \$12{,}000(A|F\ 15,5)$$
$$= -\$26{,}000(.2983) - \$1000(1.7228) + \$12{,}000(.1483)$$
$$= -\$7699.00/\text{year}$$

$$AW_3(15\%) = -\$28{,}000(A|P\ 15,5) - \$500 - \$500(A|G\ 15,5)$$
$$+ \$13{,}000(A|F\ 15,5) = -\$28{,}000(.2983) - \$500$$
$$- \$500(1.7228) + \$13{,}000(.1483) = -\$7785.90/\text{year}$$

$$AW_4(15\%) = -\$7500(F|P\ 15,1) + \$9000(A|P\ 15,5) - \$800(A|G\ 15,5)$$
$$= -\$7500(1.15) + \$9000(.2983) - \$800(1.7228)$$
$$= -\$7318.54/\text{year}$$

Based upon these calculations, the lease alternative (Alternative 4) appears to be economically most favorable.

5.11.2 Outsider Viewpoint Approach

The *outsider viewpoint approach* is preferred by many because it forces the decision maker to view both the existing asset and its challengers from an objective point of view as would an "outsider." That is, the outsider is assumed to have no existing asset. The outsider is then free to choose either a used asset (the defender) available for the price of its market value, or any of the potential replacement assets (the challengers).

In essence, the outsider viewpoint approach considers the salvage value of the existing asset to be its investment cost if it is retained in service. Such an approach is consistent with the opportunity cost concept described in Chapter Two. Since the retention of the defender is equivalent to a decision to forego the receipt of its salvage value, then an opportunity cost is assigned to the defender.

Example 5.24

Let us again consider Example 5.20, this time using the outsider viewpoint. Recall that the present filter press has a market value of $9000, a life of 5 years, and a salvage value of $2000 at that time. The challenger has a cost of $36,000, a life of 10 years, and estimated salvage values at any point in time as given in Table 5.14. A 5-year planning horizon is to be used and the *MARR* is 15%. Cash flows from the outsider's point of view are as given in Table 5.14.

TABLE 5.14

Cash Flows for a Replacement Alternative—Outsider's
Viewpoint Approach

End of Year, t	Alternative 1 Net Cash Flows, A_{1t}	Alternative 2 Net Cash Flows, A_{2t}	Difference in Cash Flows, $A_{2t} - A_{1t}$
0	−$ 9,000	−$36,000	$27,000
1	− 7,000	0	7,000
2	− 8,000	− 1,000	7,000
3	− 9,000	− 2,000	7,000
4	− 10,000	− 3,000	7,000
5	−11,000 + 2,000	−4,000 + 12,000	17,000

We can see by comparing Tables 5.11 and 5.14 that the differences in cash flows between alternatives are the same; therefore, the cash flow approach and the outsider viewpoint approach are equivalent. As further verification, we can calculate the annual worth of each alternative.

$$AW_1(15\%) = -\$9000(A|P\ 15,5) - \$7000 - \$1000(A|G\ 15,5)$$
$$+ \$2000(A|F\ 15,5)$$
$$= -\$9000(.2983) - \$7000 - \$1000(1.7228) + \$2000(.1483)$$
$$= -\$11,110.90/\text{year}$$

$$AW_2(15\%) = -\$36,000(A|P\ 15,5) - \$1000(A|G\ 15,5) + \$12,000(A|F\ 15,5)$$
$$= -\$36,000(.2983) - \$1000(1.7228) + \$12,000(.1483)$$
$$= -\$10,682.00/\text{year}$$

Note that the difference in annual worths is $428.90/year in favor of the new filter press. This is the identical conclusion reached in Example 5.21.

In using the outsider viewpoint approach, we must be careful to use "rational" first costs for each alternative. Example 5.23 showed that the trade-in allowances may be different between alternatives, and that there may be little correspondence between the trade-in value and an asset's true market value. Usually, a high trade-in value indicates the seller is using an inflated selling price. In reality, this inflated selling price would most likely be decreased if a straight purchase with no trade-in were made. Therefore, for analysis purposes, when using the outsider's viewpoint approach, *we will decrease the inflated selling price of the asset by the difference between the trade-in value and a lower market value.* Presumably, no adjustment is necessary if the trade-in allowance is less than the market value since we would simply choose to dispose of the defender separately at the market value price.

Example 5.25

Now, let us consider Example 5.23 from the outsider viewpoint. Recall that Alternatives 2 and 3 cost $36,000 and $40,000, respectively. Also, recall that the market value for the current asset is $9000, and the trade-in allowances under Alternatives 2 and 3 are $10,000 and $12,000, respectively. Since these trade-in allowances are inflated by $1000 and $3000, respectively, the corresponding costs of Alternatives 2 and 3 will be reduced to $35,000 and $37,000. A planning horizon of 5 years and a *MARR* of 15% are still in effect. Table 5.15 presents the cash flow profiles from the outsider's point of view.

Calculating the annual worths of these alternatives results in

$$
\begin{aligned}
AW_1(15\%) &= -\$9000(A|P\ 15,5) - \$7000 - \$1000(A|G\ 15,5) \\
&\quad + \$2000(A|F\ 15,5) \\
&= -\$9000(.2983) - \$7000 - \$1000(1.7228) + \$2000(.1483) \\
&= -\$11,110.90/\text{year}
\end{aligned}
$$

$$
\begin{aligned}
AW_2(15\%) &= -\$35,000(A|P\ 15,5) - \$1000(A|G\ 15,5) + \$12,000(A|F\ 15,5) \\
&= -\$35,000(.2983) - \$1000(1.7228) + \$12,000(.1483) \\
&= -\$10,383.70/\text{year}
\end{aligned}
$$

$$
\begin{aligned}
AW_3(15\%) &= -\$37,000(A|P\ 15,5) - \$500 - \$500(A|G\ 15,5) \\
&\quad + \$13,000(A|F\ 15,5) \\
&= -\$37,000(.2983) - \$500 - \$500(1.7228) + \$13,000(.1483) \\
&= -\$10,470.60/\text{year}
\end{aligned}
$$

$$
\begin{aligned}
AW_4(15\%) &= -\$7500(F|P\ 15,1) - \$800(A|G\ 15,5) \\
&= -\$7500(1.15) - \$800(1.7228) \\
&= -\$10,003.24/\text{year}
\end{aligned}
$$

Again, the lease alternative is preferred based upon annual worth calculations. Note that the above annual worth differences between alternatives using the outsider viewpoint approach are *exactly* the same as the annual worth differences between alternatives in Example 5.23, in which the cash

TABLE 5.15

Cash Flows for Several Replacement Alternatives—Outsider's Viewpoint Approach

End of Year, t	Alternative 1 Net Flows, Cash A_{1t}	Alternative 2 Net Flows, Cash A_{2t}	Alternative 3 Net Flows, Cash A_{3t}	Alternative 4 Net Flows, Cash A_{4t}
0	−$ 9,000	−$35,000	−$37,000	−$7,500
1	− 7,000	0	− 500	− 7,500
2	− 8,000	− 1,000	− 1,000	− 7,500 − 800
3	− 9,000	− 2,000	− 1,500	− 7,500 − 1,600
4	− 10,000	− 3,000	− 2,000	− 7,500 − 2,400
5	−11,000 + 2,000	−4,000 + 12,000	−2,500 + 13,000	−3,200

flow approach was used. This further substantiates the equivalence of the two approaches.

5.11.3 Optimum Replacement Interval

Although logical arguments can be given for treating a replacement decision as just another economic investment alternative, many firms fail to subject their existing equipment to careful scrutiny on a periodic basis to determine if replacement is required. Despite the fact that replacement studies can yield significant reductions in costs, it still remains that many firms postpone replacing assets beyond the "optimum" time for replacement, perhaps because a decision to replace an asset involves a change, and resistance to change is inherent in most individuals. For example, an engineer who 2 years ago successfully argued that compressor Z should be replaced by compressor Y may now find that compressor X is more economical than compressor Y. If X is now championed it may be viewed by management as an admission that the wrong compressor was selected as a replacement for Z.

Some reasons for delaying the replacement of assets beyond the economic replacement time are:

1. The firm is making a profit with its present equipment.
2. The present equipment is operational and is producing an acceptable quality product.
3. There is risk or uncertainty associated with predicting the expenses of a new machine, whereas one is relatively certain about the expenses of the current machine.
4. A decision to replace equipment is a stronger commitment, for a period of time into the future, than keeping the existing equipment.
5. Management tends to be conservative in decisions regarding the replacement of costly equipment.
6. There may be a limitation on funds available for purchasing new equipment, but no limitation on funds for maintaining existing equipment.
7. There may be considerable uncertainty concerning the future demands for the services of the equipment in question.
8. Sunk costs psychologically affect decisions to replace equipment.
9. An anticipation that technological improvements in the future might render obsolete equipment available currently; a wait-and-see attitude prevails.
10. Reluctance to be a pioneer in adopting new technology; instead of replacing now, wait for the competition to act.

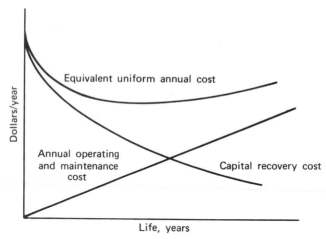

FIGURE 5.8. Portrayal of components of equivalent annual costs.

As a unit of equipment ages, operating and maintenance costs increase. At the same time, the capital recovery cost decreases with prolonged use of the equipment. The combination of decreasing capital recovery costs and increasing annual operating and maintenance costs results in the equivalent uniform annual cost taking on a form similar to that depicted in Figure 5.8.

By forecasting the operating and maintenance costs for each year of service, as well as the anticipated salvage values for various replacement ages, one can determine the "optimum replacement interval" for equipment.

Example 5.26

Suppose a small compressor can be purchased for $1000; the salvage value for the compressor is assumed to be negligible, regardless of the replacement interval. Annual operating and maintenance costs are expected to increase by $75/year, with the first year's cost anticipated to be $150. Using a minimum attractive rate of return of 20%, the following equivalent uniform annual costs are obtained:

$$EUAC(n = 1) = \$1000(A|P\ 20,1) + \$150 + \$75(A|G\ 20,1)$$
$$= \$1350.00/\text{year}$$

$$EUAC(n = 2) = \$1000(A|P\ 20,2) + \$150 + \$75(A|G\ 20,2)$$
$$= \$838.59/\text{year}$$

$$EUAC(n = 3) = \$1000(A|P\ 20,3) + \$150 + \$75(A|G\ 20,3)$$
$$= \$690.63/\text{year}$$

$$EUAC(n = 4) = \$1000(A|P\ 20,4) + \$150 + \$75(A|G\ 20,4)$$
$$= \$631.87/\text{year}$$

$$EUAC(n = 5) = \$1000(A|P\ 20,5) + \$150 + \$75(A|G\ 20,5)$$
$$= \$607.44/\text{year}$$

$$EUAC(n = 6) = \$1000(A|P\ 20,6) + \$150 + \$75(A|G\ 20,6)$$
$$= \$599.11/\text{year}$$

$$EUAC(n = 7) = \$1000(A|P\ 20,7) + \$150 + \$75(A|G\ 20,7)$$
$$= \$599.17/\text{year}$$

$$EUAC(n = 8) = \$1000(A|P\ 20,8) + \$150 + \$75(A|G\ 20,8)$$
$$= \$603.77/\text{year}$$

$$EUAC(n = 9) = \$1000(A|P\ 20,9) + \$150 + \$75(A|G\ 20,9)$$
$$= \$610.83/\text{year}$$

$$EUAC(n = 10) = \$1000(A|P\ 20,10) + \$150 + \$75(A|G\ 20,10)$$
$$= \$619.04/\text{year}$$

$$EUAC(n = 11) = \$1000(A|P\ 20,11) + \$150 + \$75(A|G\ 20,11)$$
$$= \$627.80/\text{year}$$

As n increases beyond 6 years, the equivalent uniform annual cost increases. Hence, for this example, a replacement interval of 6 years is indicated.

Certain assumptions are inherent in optimum replacement interval calculations. First, we implicitly assume that the planning horizon is an integer multiple of the replacement interval selected. (Recall our discussion of the use of the annual worth based on one life cycle and a planning horizon of the least common multiple of lives.) Second, we have assumed that each time the compressor is replaced it will be replaced with a compressor having an *identical cash flow profile*. If neither assumption is valid, then the above approach is not v 'id.

Example 5.27

Suppose in the previous case an 11-year planning horizon is appropriate. For simplicity, we will continue to assume that replacements will have identical cash flow profiles. If the original compressor is kept for 11 years, the present worth equivalent will be

$$PW(n = 11) = -\$627.80(P|A\ 20,11)$$
$$= -\$2716.55$$

If the compressor is to be replaced at some intermediate point during the planning horizon, say after k years, and the replacement is to be kept until the end of the planning horizon, the following present worth calculations result for $k = 6$ or 7:

$$PW(k = 6) = -\$599.11(P|A\ 20,6) - \$607.44(P|A\ 20,5)(P|F\ 20,6)$$
$$= -\$2600.72$$

$$PW(k = 7) = -\$599.17(P|A\ 20,7) - \$631.87(P|A\ 20,4)(P|F\ 20,7)$$
$$= -\$2616.30$$

Thus, the compressor should be replaced after 6 years, and the replacement should be kept until the end of the planning horizon. Why is it unnecessary to consider the remaining values of k (i.e., $k = 5, 4, 3, 2,$ and 1)? In order to answer this, consider the case of $k = 5$ and compare the cash flow profiles over the 11-year planning horizon using $k = 5$ and $k = 6$. As shown in Figure 5.9, a consideration of the time value of money eliminates the possibility of $k = 5$ having a lower present worth than $k = 6$. Similar comparisons can be used to eliminate the cases of $k = 1, 2, 3,$ and 4. Likewise, $k = 8, 9,$ and 10 may be eliminated since costs have already begun to increase from $k = 6$ to $k = 7$.

For this particular example, failure to replace the equipment at precisely the optimum time will not result in a significantly lower annual *worth* or present *worth*. This is often the case with many replacement situations; the resulting measure of effectiveness is relatively insensitive to deviations from the optimum strategy. Hence, the firm might establish an operating policy of reviewing the *actual* operating and maintenance costs for the equipment at the anticipated "optimum" time of replacement and then, perform a replace-

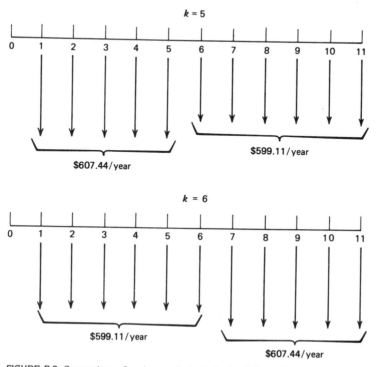

FIGURE 5.9. Comparison of replacement strategies involving replacement at the end of year 5 versus the end of year 6.

ment study, using the seven-step procedure for comparing investment alternatives.

The subject of replacement analysis is popular among engineering economists. In fact, sufficient literature is available on the subject to devote an entire book to it. Such books do exist (e.g., [31]). For further treatment of replacement analysis, you may with to consult the *Engineering Economist* publication.

5.12 SUMMARY

The process of comparing investment alternatives was described in this chapter. A cash flow approach was recommended, as was a planning horizon approach. Methods for comparing alternatives included the present worth, annual worth, future worth, internal rate of return, external rate of return, savings/investment ratio, payback period, and capitalized worth. The total cash flow and incremental cash flow approaches were considered in detail. Replacement problems were treated as alternative comparison problems. The discussion of the comparison of investment alternatives did not consider either the effects of income taxes, the public sector requirements for measuring benefits in economic units, or the effects of risk and uncertainty; such considerations are reserved for the following chapters.

BIBLIOGRAPHY

1. American Telephone and Telegraph Company, *Engineering Economy,* Third Edition, McGraw-Hill, 1977.

2. Baldwin, R. H., "How to Assess Investment Proposals," *Harvard Business Review, 37* (3), 1959, pp. 98–99.

3. Bernhard, R. H., "Discount Methods for Expenditure Evaluation—A Clarification of Their Assumptions," *The Journal of Industrial Engineering, 13* (1), 1962, pp. 19–27.

4. Bussey, L. E., *The Economic Analysis of Industrial Projects,* Prentice-Hall, 1978.

5. Canada, J. R., and White, J. A., *Capital Investment Decision Analysis for Management and Engineering,* Prentice-Hall, 1980.

6. DeGarmo, P. E., Canada, J. R., and Sullivan, W. G., *Engineering Economy,* 6 ed. Macmillan, 1979.

7. *Economic Analysis Handbook,* Publication P-442, Department of the Navy, Naval Facilities Engineering Command, Washington, D.C., 1971.

8. Estes, C. B., and Jalali-Yazdi, A., "Now Replace Equipment on Actual Cost Basis," *Industrial Engineering, 10* (11), November 1978, pp. 34–37.

9. Fleischer, G. A., "Two Major Issues Associated with the Rate of Return

Method for Capital Allocation: The 'Ranking Error' and 'Preliminary Selection,'" *The Journal of Industrial Engineering, 17* (4), 1966, pp. 202–8.

10. Grant, E. L., Ireson, W. G., and Leavenworth, R. S., *Principles of Engineering Economy, 7* ed. Wiley, 1982.

11. Heebink, D. V., "Rate of Return, Reinvestment and the Evaluation of Capital Expenditures," *The Journal of Industrial Engineering, 13* (1), 1962, pp. 48–49.

12. Kaplan, R. S., "Must CIM be Justified by Faith Alone?," *Harvard Business Review, 64* (2) March–April 1986, pp. 87–95.

13. Klausner, R. F., "Communicating Investment Proposals to Corporate Decision-Makers," *The Engineering Economist, 17* (1), 1971, pp. 45–55.

14. Kulonda, D. J., "Replacement Analysis with Unequal Lives—The Study Period Method," *The Engineering Economist, 23* (3), 1978, pp. 171–80.

15. Leautaud, J. L. L., "On the Fundamentals of Economic Evaluation," *The Engineering Economist, 19* (2), 1974, pp. 105–26.

16. Lutz, R. P., "How to Justify the Purchase of Material Handling Equipment to Management," *Proceedings of the 1976 MHI Material Handling Seminar,* The Material Handling Institute, Pittsburgh, Pa., 1976.

17. Mao, J. C. T., "The Internal Rate of Return as a Ranking Criterion," *The Engineering Economist, 11* (4), 1966, pp. 1–13.

18. Mao, J. C. T., "An Analysis of Criteria for Investment and Financing Decisions Under Certainty: A Comment," *Management Science, 13* (3), 1966, pp. 289–91.

19. Mao, J. C. T., *Quantitative Analysis of Financial Decisions,* Macmillan, 1969.

20. Morris, W. T., *Engineering Economic Analysis,* Reston Publishing Company, 1976.

21. Newnan, D. G., *Engineering Economic Analysis,* Engineering Press, 1976.

22. Oakford, R. V., *Capital Budgeting: A Quantitative Evaluation of Investment.* Ronald Press, 1970.

23. Renshaw, E., "A Note on the Arithmetic of Capital Budgeting Decisions," *The Journal of Business, 30* (3), 1957, pp. 193–204.

24. Riggs, J. L., *Engineering Economics,* Second Edition, McGraw-Hill, 1982.

25. Shore, B., "Replacement Decisions Under Capital Budgeting Constraints," *The Engineering Economist, 20* (4), 1975, pp. 243–56.

26. Solomon, E., "The Arithmetic of Capital Budgeting Decisions," *The Journal of Business, 29* (2), 1956, pp. 124–29.

27. Stevens, G. T., Jr., *Economic and Financial Analysis of Capital Investments,* Wiley, 1979.

28. Szonyi, A. J., Fenton, R. G., White, J. A., Agee, M. H., and Case, K. E., *Principles of Engineering Economic Analysis,* Canadian Edition, Wiley Canada Limited, 1982.

29. Tarquin, A. J., and Blank, L. T., *Engineering Economy, 2* ed. McGraw-Hill, 1983.

30. Teichroew, D., Robicheck, A. A., and Montalbano, M., "An Analysis of Criteria for Investment and Financing Decisions Under Certainty," *Management Science, 12* (3), 1965, pp. 151–79.

31. Terborgh, G., *Business Investment Management,* Machinery and Allied Products Institute, 1958.
32. Thuesen, H. G., Fabrycky, W. J., and Thuesen, G. J., *Engineering Economy,* 5 ed. Prentice-Hall, 1977.

PROBLEMS

1. The ETG Corporation Capital Budget Committee has identified three major proposals, A, B, and C, to enhance customer service. Proposals A and C are mutually exclusive; Proposal C is contingent on Proposal B. The initial investments required for Proposals A, B, and C are $70,000, $80,000, and $35,000, respectively. A total of $120,000 is available for investment. Clearly specify all feasible alternatives to be considered. (5.2)

2. Marlow Electronics has identified four investment proposals, A, B, C, and D. Proposals A and B are mutually exclusive and Proposals A and D are mutually exclusive. Proposal C is contingent on either Proposal A or B. To do nothing is not a viable alternative. Clearly specify all feasible alternatives to be considered. (5.2)

3. Consider the net cash flows (NCF) and salvage values (SV) for each of Alternatives 1, 2, and 3 having lives of 3, 8, and 4 years, respectively.

	Alternative 1		Alternative 2		Alternative 3	
EOY	NCF_1	SV_1	NCF_2	SV_2	NCF_3	SV_3
0	-$60,000	$60,000	-$100,000	$100,000	-$140,000	$140,000
1	35,000	30,000	30,000	60,000	65,000	80,000
2	35,000	10,000	30,000	40,000	65,000	40,000
3	35,000	0	30,000	25,000	65,000	20,000
4			30,000	15,000	65,000	10,000
5			30,000	10,000		
6			30,000	5,000		
7			30,000	5,000		
8			30,000	5,000		

Assume each alternative can be renewed indefinitely with the same NCF and SV profiles.

(a) If a least common multiple of lives approach is to be used, specify the planning horizon and the complete set of cash flows for each alternative.
(b) Repeat part (a) using the shortest life among alternatives.
(c) Repeat part (a) using the longest life among alternatives.
(d) Repeat part (a) using a planning horizon of 2 years.
(e) Repeat part (a) using a planning horizon of 6 years.
(f) Repeat part (a) using a (tentative) planning horizon of 30 years. (5.3)

4. Greasy Petroleum Company has available four proposals, A, B, C, and D. Proposal B is contingent on the acceptance of either Proposal A or Proposal D. Also, Proposal D is contingent on Proposal A, while Proposal A is contingent on either Proposal B or Proposal C. The initial investments for these proposals are $420,000, $60,000, $45,000, and $300,000, respectively. If Greasy has a budget limitation of $600,000, clearly identify all feasible alternatives to be considered. (5.2)

5. Alternatives 1, 2, and 3 have lives of 3, 4, and 6 years, respectively. Their net cash flow (NCF) and salvage value (SV) profiles are as follows:

EOY	Alternative 1		Alternative 2		Alternative 3	
	NCF_1	SV_1	NCF_2	SV_2	NCF_3	SV_3
0	−$20,000	—	−$40,000	$40,000	−$70,000	$70,000
1	8,000	—	20,000	30,000	30,000	50,000
2	8,000	—	20,000	20,000	30,000	30,000
3	28,000	—	20,000	10,000	30,000	20,000
4			20,000	0	30,000	10,000
5					30,000	5,000
6					30,000	2,000

The NCF profile for Alternative 1 is due to a $20,000/year lease payable at the beginning of each year, plus an end-of-year net revenue of $28,000. This lease arrangement may be renewed indefinitely; however, premature cancellation of the lease costs $10,000. All other cash flows and salvage values are expected to repeat indefinitely as shown.

(a) If a least common multiple of lives approach is to be used, specify the planning horizon and the complete set of cash flows for each alternative.

(b) Repeat part (a) using the shortest life among alternatives.

(c) Repeat part (a) using the longest life among alternatives.

(d) Repeat part (a) using a planning horizon of 2 years.

(e) Repeat part (a) using a planning horizon of 5 years.

(f) Repeat part (a) using a planning horizon of 9 years.

(g) Repeat part (a) using a (tentative) planning horizon of 18 years. (5.3)

6. Two alternatives have the following net cash flow (NCF) and salvage value (SV) profiles *for the first cycle of each.*

EOY	Alternative 1		Alternative 2	
	NCF_1	SV_1	NCF_2	SV_2
0	−$50,000	$50,000	−$80,000	$80,000
1	25,000	25,000	35,000	50,000
2	30,000	10,000	45,000	20,000
3	35,000	0	50,000	10,000
4			55,000	0
5			60,000	0

Alternatives 1 and 2 can both repeat; however, the cash flow profiles and salvage values will change. Renewal of Alternative 1 will cost 40% more for the initial investment, increasing to $70,000. Alternative 1 salvage values will decrease in the same proportion to the initial investment as shown above (e.g., $70,000, $35,000, $14,000, $0). Further renewals of Alternative 1 will cause initial investments and salvage values to increase 40% over the previous cycle's values, and salvage values will again decrease in the same proportions as illustrated in the table. Annual net revenues are projected to continue increasing at $5000/year (e.g., $40,000/year, $45,000/year, etc.) indefinitely. Renewal of Alternative 2 will cost 60% more for the initial investment with each renewal, and salvage values will decrease in proportion to those shown. Also, annual net revenues will continue increasing at $5000/year.

(a) Specify the complete set of cash flows for each alternative if a planning horizon of 2 years is to be used.

(b) Repeat part (a) using a planning horizon of 3 years.

(c) Repeat part (a) using a planning horizon of 4 years.

(d) Repeat part (a) using a planning horizon of 5 years.

(e) Repeat part (a) using a planning horizon of 10 years. (5.3)

7. Three investment proposals have been selected for further evaluation. Proposals A and C are mutually exclusive; Proposal A is contingent on Proposal B; and Proposal B is contingent on Proposal C. A budget limitation of $200,000 exists. The cash flow profiles for the three investment proposals are given below for the 4-year planning horizon.

EOY	NCF(A)	NCF(B)	NCF(C)
0	−$60,000	−$40,000	−$80,000
1–4	20,000	10,000	15,000
4	—	15,000	20,000

(a) Specify the alternatives to be considered. (5.2)

(b) Clearly show the net cash flow profile associated with each alternative. (5.4)

8. Four investment proposals, A, B, C, and D, are available. Proposals A and B are mutually exclusive, and Proposals A and C are mutually exclusive. Proposal D is contingent on either Proposal A or B. Funds available for investment are limited to $200,000.

EOY	NCF(A)	NCF(B)	NCF(C)	NCF(D)
0	−$80,000	−$125,000	−$100,000	−$80,000
1–5	30,000	40,000	35,000	32,000

(a) Specify the alternatives to be considered. (5.2)

(b) Clearly show the net cash flow profile associated with each alternative. (5.4)

9. A firm has available four proposals, A, B, C, and D. Proposal A is contingent on acceptance of either Proposal C or Proposal D. In addition, Proposal C is contingent on Proposal D, while Proposal D is contingent on either Proposal A or Proposal B. The firm has a budget limitation of $500,000. The cash flows are as follows:

EOY	NCF(A)	NCF(B)	NCF(C)	NCF(D)
0	−$250,000	−$350,000	−$50,000	−$38,000
1	85,000	119,000	10,000	3,000
2	85,000	122,000	10,000	8,000
3	85,000	125,000	10,000	13,000
4	85,000	128,000	10,000	18,000
5	150,000	170,000	10,000	23,000

(a) Specify the alternatives to be considered. (5.2)

(b) Clearly show the net cash flow profile associated with each alternative. (5.4)

10. Metal Salvage, Inc. has available three investment proposals A, B, and C, having the cash flow profiles shown below. Proposals B and C are mutually exclusive and Proposal C is contingent on Proposal A being chosen.

	NCF(A)	NCF(B)	NCF(C)
Initial investment	$400,000	$600,000	$300,000
Life	8 years	12 years	6 years
Annual receipts	$320,000	$380,000	$400,000
Annual disbursements	$230,000	$240,000	$300,000
Salvage value	$100,000	$200,000	$100,000

The firm is willing to use a planning horizon of 24 years and accepts the assumption of identical cash flow profiles for successive life cycles of a proposal.

(a) Determine the set of feasible alternatives. (5.2)

(b) Determine the net cash flow profile for each alternative. (5.4)

11. Four investment proposals, W, X, Y, and Z, are being considered by the Ajax Corporation. Proposals X and Z are mutually exclusive. Proposal Y is contingent on either X or Z. Proposals W and Y are mutually exclusive. A budget limitation of $400,000 exists. Either Proposal X or Proposal Z must be included in the alternative selected.

EOY	NCF(W)	NCF(X)	NCF(Y)	NCF(Z)
0	−$200,000	−$250,000	−$180,000	−$200,000
1–8	60,000	100,000	50,000	40,000
8	80,000	20,000	240,000	200,000

(a) Determine the set of feasible alternatives. (5.2)

(b) Determine the net cash flow for each alternative. (5.4)

12. Charles Machine Works is faced with three investment proposals, A, B, and C, having the cash flow profiles shown below over the planning horizon of 5 years. Proposals B and C are mutually exclusive, and Proposal C is contingent on Proposal A being selected. A budgetary limitation of $125,000 exists on the amount that can be invested initially.

EOY	NCF(A)	NCF(B)	NCF(C)
0	−$50,000	−$25,000	−$10,000
1	10,000	10,000	3,000
2	15,000	10,000	3,000
3	20,000	10,000	3,000
4	25,000	10,000	3,000
5	30,000	10,000	7,500

Specify the cash flow profiles for all feasible investment alternatives. (5.4)

13. Rex Electric Company has borrowed money from First National Bank on a short-term loan at 18% compounded monthly. The company's tax rate is 34%. What is the after-tax cost of capital for this money? (5.5.1)

14. Ajax Machinery borrows money at 16% compounded semiannually. Assume a tax rate of 34%.

   (a) Determine the effective before-tax cost of capital in percent.

   (b) Determine the effective after-tax cost of capital in percent. (5.5.1)

15. Precision Components, Inc. issues bonds promising semiannual interest payments at an annual nominal interest rate of 11%. The bonds are all sold for face value, and the costs of preparing and selling the bonds are relatively negligible. What is the after-tax cost of this debt funding to Precision Components if their tax rate is 34% (5.5.1)

16. Jericho Steel calculates that they must pay interest of $30 each quarter to holders of bonds having $1000 face value. Using a 34% tax rate, clearly and explicitly state any necessary assumptions and

   (a) Calculate the before tax interest rate of this debt funding.

   (b) Calculate the after-tax interest rate of this debt funding. (5.5.1)

17. Witherspoon, Inc. is a publicly held firm having a present trading price of $35 per share of common stock. The Board of Directors has just voted a dividend of $3 per share to be paid to all shareholders. What is the cost of this equity capital in percent? Is this a before-tax cost or an after-tax cost of capital? (5.5.1)

18. Dunbar Chemicals plans to invest its retained earnings in capital improvements and expansion ventures. The company's stock trades for $76 per share on the financial market, and dividends recently paid were $8 per share. Is there a cost to

Dunbar of using its own retained earnings? If so, what is the cost of capital in percent? (5.5.1)

19. The capital makeup of JWS Corporation is as follows:

	Amount	Cost of Capital, Percent
Loans	$ 800,000	9.6
Bonds	$1,400,000	8.2
Stock	$2,100,000	11.9
Retained earnings	$ 700,000	11.9

What is the overall average cost of capital for JWS? (5.5.1)

20. Systems Analysis, Inc. has the following capital structure:

	Amount	Other Information
Loans	$2,600,000	14% before-tax rate, semiannual compounding
Bonds	$4,100,000	12% bond interest rate, quarterly payment
Stock	$5,000,000	$6.50 dividends on selling price of $65 per share
Retained earnings	$3,300,000	

Systems' tax rate is 34%. Determine the firm's overall average after-tax cost of capital. (5.5.1)

21. Two investment alternatives are to be evaluated and the economically better alternative specified. The cash flow profiles for the alternatives are shown below.

EOY	NCF(1)	NCF(2)
0	−$10,000	−$15,000
1	1,000	1,500
2	1,500	2,000
3	—	500
4	2,500	3,000
5	3,500	4,000
6	—	500
7	4,500	5,000
8	6,000	11,500

(a) Assuming $15,000 is available to invest, determine the total cash flow associated with each alternative over each year of the planning horizon. Any excess capital is invested at $MARR = 15\%$, earning equal annual returns over the planning horizon of 8 years. (5.6.1)

(b) Determine the incremental cash flow profile between Alternatives 2 and 1 for each year of the planning horizon. (5.6.2)

22. Consider the following investment decision.

	Machine 1	Machine 2
First cost	$15,000	$20,000
Estimated life	5 years	10 years
Estimated annual revenues	$12,000	$14,000
Estimated annual operative costs	$ 6,000	$ 8,000
Estimated salvage value at end of 5 years	$ 1,500	$ 5,000

(a) Assuming $20,000 is available to invest, determine the total cash flow associated with each machine over each year of the planning horizon. Any excess capital is invested at *MARR* = 20% earning equal annual returns over the planning horizon of 5 years. (5.6.1)

(b) Determine the incremental cash flow profile between Machine 2 and Machine 1 for each year of the planning horizon of 5 years. (5.6.2)

23. Two mutually exclusive proposals, each with a life of 5 years, are under consideration. *MARR* is 15%. Each proposal has the following cash flow profile:

EOY	NCF(A)	NCF(B)
0	−$30,000	−$42,000
1	9,300	12,625
2	9,300	12,625
3	9,300	12,625
4	9,300	12,625
5	9,300	12,625

(a) Clearly specify the alternatives available and their net cash flow profiles. (5.4)

(b) Determine which alternative the decision maker should select. Use a *ranking approach* and the present worth method. (5.6.3)

(c) Determine which alternative the decision maker should select. Use an *incremental cash flow approach* and the present worth method. (5.6.3)

(d) Repeat part (b) using the annual worth method. (5.6.4)

(e) Repeat part (c) using the annual worth method. (5.6.4)

(f) Repeat part (b) using the future worth method. (5.6.5)

(g) Repeat part (c) using the future worth method. (5.6.5)

(h) Repeat part (c) using the internal rate of return method. (5.6.6)

(i) Repeat part (c) using the external rate of return method. (5.6.7)

(j) Repeat part (c) using the savings/investment ratio method. (5.6.8)

24. The Belt Company is considering three investment proposals, A, B, and C. Proposals A and B are mutually exclusive, and Proposal C is contingent on

Proposal B. The cash flow data for the investments over a 10-year planning horizon are given below. The Belt Company has a budget limit of $1 million for investments of the type being considered currently. $MARR = 25\%$.

	NCF(A)	NCF(B)	NCF(C)
Initial investment	$600,000	$800,000	$470,000
Life	10 years	10 years	10 years
Salvage values	$ 70,000	$130,000	$ 65,000
Annual receipts	$400,000	$600,000	$260,000
Annual disbursements	$130,000	$270,000	$ 70,000

(a) Clearly specify the alternatives available and their net cash flow profiles. (5.4)

(b) Determine which alternative Belt's decision maker should select. Use a *ranking approach* and the present worth method. (5.6.3)

(c) Determine which alternative Belt's decision maker should select. Use an *incremental cash flow approach* and the present worth method. (5.6.3)

(d) Repeat part (b) using the annual worth method. (5.6.4)

(e) Repeat part (c) using the annual worth method. (5.6.4)

(f) Repeat part (b) using the future worth method. (5.6.5)

(g) Repeat part (c) using the future worth method. (5.6.5)

(h) Repeat part (c) using the internal rate of return method. (5.6.6)

(i) Repeat part (c) using the external rate of return method. (5.6.7)

(j) Repeat part (c) using the savings/investment ratio method. (5.6.8)

25. A firm has available two mutually exclusive investment proposals, A and B. Their net cash flows are as shown in the table over a 10-year planning horizon. *MARR* is 15%.

EOY	NCF(A)	NCF(B)
0	−$40,000	−$30,000
1	8,000	9,000
2	8,000	8,500
3	8,000	8,000
4	8,000	7,500
5	8,000	7,000
6	8,000	6,500
7	8,000	6,000
8	8,000	5,500
9	8,000	5,000
10	8,000	4,500

(a) Clearly specify the alternatives available and their net cash flow profiles. (5.4)

(b) Determine which alternative the decision maker should select. Use a *ranking approach* and the present worth method. (5.6.3)

(c) Determine which alternative the decision maker should select. Use an *incremental cash flow approach* and the present worth method. (5.6.3)

(d) Repeat part (b) using the annual worth method. (5.6.4)

(e) Repeat part (c) using the annual worth method. (5.6.4)

(f) Repeat part (b) using the future worth method. (5.6.5)

(g) Repeat part (c) using the future worth method. (5.6.5)

(h) Repeat part (c) using the internal rate of return method. (5.6.6)

(i) Repeat part (c) using the external rate of return method. (5.6.7)

(j) Repeat part (c) using the savings/investment ratio method. (5.6.8)

26. A firm is faced with four investment proposals, A, B, C, and D, having the cash flow profiles shown below. Proposals A and C are mutually exclusive, and Proposal D is contingent on Proposal B being chosen. Currently, $750,000 is available for investment, and the firm has stipulated a *MARR* of 20%.

	NCF(A)	NCF(B)	NCF(C)	NCF(D)
Initial investment	$400,000	$400,000	$600,000	$300,000
Planning horizon	10 years	10 years	10 years	10 years
Annual receipts	$205,000	$215,000	$260,000	$230,000
Annual disbursements	$110,000	$125,000	$120,000	$150,000
Salvage value	$ 50,000	$ 50,000	$100,000	$ 50,000

(a) Clearly specify the alternatives available and their net cash flow profiles. (5.4)

(b) Determine which alternative the decision maker should select. Use a *ranking approach* and the present worth method. (5.6.3)

(c) Determine which alternative the decision maker should select. Use an *incremental cash flow approach* and the present worth method. (5.6.3)

(d) Repeat part (b) using the annual worth method. (5.6.4)

(e) Repeat part (c) using the annual worth method. (5.6.4)

(f) Repeat part (b) using the future worth method. (5.6.5)

(g) Repeat part (c) using the future worth method. (5.6.5)

(h) Repeat part (c) using the internal rate of return method. (5.6.6)

(i) Repeat part (c) using the external rate of return method. (5.6.7)

(j) Repeat part (c) using the savings/investment ratio method. (5.6.8)

27. Rich N. Smug is considering two mutually exclusive investment alternatives, each requiring an investment of $5000. Alternative 1 returns $1000 after 1 year and $6000 after 2 years. This alternative is not renewable. Alternative 2 returns $1000/year for the first 3 years and $5000 after 4 years. Mr. Smug has established a *MARR* of 10%. He is committed to make one of the two investments.

(a) Clearly specify the alternatives available and their net cash flow profiles. (5.4)

(b) Determine which alternative Mr. Smug should select. Use a *ranking approach* and the present worth method. (5.6.3)

(c) Determine which alternative Mr. Smug should select. Use an *incremental cash flow approach* and the present worth method. (5.6.3)

(d) Repeat part (b) using the annual worth method. (5.6.4)

(e) Repeat part (c) using the annual worth method. (5.6.4)

(f) Repeat part (b) using the future worth method. (5.6.5)

(g) Repeat part (c) using the future worth method. (5.6.5)

(h) Repeat part (c) using the internal rate of return method. (5.6.6)

(i) Repeat part (c) using the external rate of return method. (5.6.7)

(j) Repeat part (c) using the savings/investment ratio method. (5.6.8)

28. Insulation is being considered for installation on a steam pipe. Either 1-inch or 2-inch insulation is being figured. The annual heat loss from the pipe at present is estimated to be $2.50 per foot of pipe. The 1-inch insulation costs $1/foot and will reduce the heat loss by 65%; the 2-inch insulation costs $1.90/foot and will reduce the heat loss by 79%. The insulation will last 5 years with no salvage value. *MARR* is 25%.

(a) Clearly specify the alternatives available and their net cash flow profiles. (5.4)

(b) Determine which kind of insulation the decision maker should select. Use a *ranking approach* and the present worth method. (5.6.3)

(c) Determine which kind of insulation the decision maker should select. Use an *incremental cash flow approach* and the present worth method. (5.6.3)

(d) Repeat part (b) using the annual worth method. (5.6.4)

(e) Repeat part (c) using the annual worth method. (5.6.4)

(f) Repeat part (b) using the future worth method. (5.6.5)

(g) Repeat part (c) using the future worth method. (5.6.5)

(h) Repeat part (c) using the internal rate of return method. (5.6.6)

(i) Repeat part (c) using the external rate of return method. (5.6.7)

(j) Repeat part (c) using the savings/investment ratio method. (5.6.8)

29. Three investment alternatives have the cash flow profiles shown below. A *MARR* of 20% is used in the selection of a preferred alternative.

EOY	NCF(1)	NCF(2)	NCF(3)
0	−$40,000	−$25,000	$0
1	16,500	5,000	0
2	16,500	10,000	0
3	16,500	15,000	0
4	16,500	20,000	0
5	16,500	25,000	0

(a) Determine which alternative the decision maker should select. Use a *ranking approach* and the present worth method. (5.6.3)

(b) Determine which alternative the decision maker should select. Use an *incremental cash flow approach* and the present worth method. (5.6.3)

(c) Repeat part (b) using the annual worth method. (5.6.4)

(d) Repeat part (c) using the annual worth method. (5.6.4)

(e) Repeat part (b) using the future worth method. (5.6.5)

(f) Repeat part (c) using the future worth method. (5.6.5)

(g) Repeat part (c) using the internal rate of return method. (5.6.6)

(h) Repeat part (c) using the external rate of return method. (5.6.7)

(i) Repeat part (c) using the savings/investment ratio method. (5.6.8)

30. In Problem 27, Rich N. Smug has two alternatives to consider. What are the payback periods for each alternative? On the basis of payback period, which alternative does Mr. Smug prefer? (5.6.9)

31. Using the payback period, evaluate Alternatives 1 and 2 in Problem 29. (5.6.9)

32. Consider the net cash flows of the following three alternatives. None is renewable. *MARR* is 20%.

EOY	NCF(1)	NCF(2)	NCF(3)
0	−$3,000	−$4,000	−$6,000
1	1,000	4,000	3,000
2	1,500	2,000	3,000
3	0	0	3,000
4	4,000	0	0
5	6,000	0	0

(a) Compare these alternatives on the basis of the present worth method. (5.6.3)

(b) Compare these alternatives on the basis of the payback period method. (5.6.9)

33. Two alternatives for a recreational facility are being considered. Their cash flow profiles are as follows for lives of 5 and 3 years, respectively.

EOY	NCF(1)	NCF(2)
0	−$11,000	−$5,000
1	5,000	2,000
2	4,000	3,000
3	3,000	4,000
4	2,000	
5	1,000	

Each alternative is expected to repeat forever with roughly the same net cash flow profile. Using the capitalized worth method and a *MARR* of 10%, select the preferred alternative. (5.6.10)

34. Three bridge designs are under consideration for a Payne County road. Wood, steel, and concrete underlie the three designs. The wood design costs $25,000 installed, with $2000/year maintenance and $15,000 rework at the end of every 10 years. Steel costs $35,000 installed, with $1500/year maintenance. Concrete costs $40,000 installed, but costs only $1000/year in maintenance. If *MARR* is 8%, which alternative should be selected on the basis of capitalized cost, assuming profiles remain constant indefinitely? (5.6.10)

35. A consulting engineer is designing an irrigation system and is considering two pumps to meet a pumping demand for 15,000 gallons/minute at 15 feet total dynamic head. The specific gravity of the fertilizer mixture being pumped is 1.25. Pump A operates at 78% efficiency and costs $11,500; Pump B operates at 86% efficiency and costs $16,250. Energy costs $0.060/kilowatt-hour. Continuous pumping for 200 days/year is required (i.e., 24 hours/day). *MARR* is 15%, the planning horizon is 5 years, and equal salvage values are estimated for the two pumps. (*Note:* Dynamic head × gallons/minute × specific gravity × 0.746 ÷ 3960 = kilowatts.) (5.9)

(a) Clearly specify the net cash flow profiles for each mutually exclusive alternative.

(b) Determine which pump should be selected. Use a *ranking approach* and the present worth method.

(c) Repeat part (b) using the annual worth method.

(d) Repeat part (b) using the future worth method.

(e) Determine which pump should be selected. Use an *incremental cash flow approach* and the internal rate of return method.

(f) Repeat part (e) using the external rate of return method.

(g) Repeat part (e) using the savings/investment ratio method.

36. The motor on a numerically controlled machine tool must be replaced. Two different 20-horsepower (output) electric motors are being considered. Motor U sells for $1360 and has an efficiency rating of 90%; Motor V sells for $300 and has a rating of 84%. The cost of energy is $0.055/kilowatt-hour. An 8-year planning horizon is used, and zero salvage values are assumed for both motors. Annual usage of the motor averages approximately 4800 hours. A *MARR* of 20% is to be used. Assume the motor selected will be loaded to capacity. (*Note:* 0.746 kilowatt = 1 horsepower.) (5.9)

(a) Clearly specify the net cash flow profiles for each mutually exclusive alternative.

(b) Determine which motor should be selected. Use a *ranking approach* and the present worth method.

(c) Repeat part (b) using the annual worth method.

(d) Repeat part (b) using the future worth method.

(e) Determine which motor should be selected. Use an *incremental cash flow approach* and the internal rate of return method.

(f) Repeat part (e) using the external rate of return method.

(g) Repeat part (e) using the savings/investment ratio method.

37. A chemical plant is considering the installation of a storage tank for water. The tank is estimated to have an initial cost of $213,000, annual costs for maintenance are estimated to be $3200/year. As an alternative, a holding pond can be provided some distance away at an initial cost of $90,000 for the pond, plus $45,000 for pumps and piping; annual operating and maintenance costs for the pumps and holding pond are estimated to be $8500. The planning horizon is 20 years with neither alternative having any salvage value. *MARR* is 15%. (5.9)

(a) Clearly specify the net cash flow profiles for each mutually exclusive alternative.

(b) Determine which alternative should be selected. Use a *ranking approach* and the present worth method.

(c) Repeat part (b) using the annual worth method.

(d) Repeat part (b) using the future worth method.

(e) Determine which alternative should be selected. Use an *incremental cash flow approach* and the internal rate of return method.

(f) Repeat part (e) using the external rate of return method.

(g) Repeat part (e) using the savings/investment ratio method.

38. The state civil defense organization wishes to establish a communications network to cover the state. It is desired to maintain a specified minimum signal strength at all points in the state. Two alternatives have been selected for detailed consideration. Design I involves the installation of five transmitting stations of low power. The investment at each installation is estimated to be $40,000 in structure and $25,000 in equipment. Design II involves the installation of two transmitting stations of much higher power. Investment in structure will be approximately $130,000/installation; equipment at each installation will cost $180,000. Annual operating and maintenance costs per installation are anticipated to be $25,000 for Design I and $55,000 for Design II. Structures are anticipated to last for 20 years; equipment life is estimated to be 10 years, with replacements assumed to be identical in costs and performance. The planning horizon is 20 years, salvage values are nil at $t = 20$, and *MARR* is 12%. (5.9)

(a) Clearly specify the net cash flow profiles for each mutually exclusive alternative.

(b) Determine which design should be selected. Use a *ranking approach* and the present worth method.

(c) Repeat part (b) using the annual worth method.

(d) Repeat part (b) using the future worth method.

(e) Determine which design should be selected. Use an *incremental cash flow approach* and the internal rate of return method.

(f) Repeat part (e) using the external rate of return method.

(g) Repeat part (e) using the savings/investment ratio method.

39. Two compressors are being considered by the Ajax Company. Compressor A can be purchased for $9700; annual operating and maintenance costs are estimated to be $3500. Alternatively, Compressor B can be purchased for $7500; annual operating and maintenance costs are estimated to be $4200. An 8-year planning horizon is to be used; salvage values are estimated to be 15% of the original purchase price; a *MARR* of 12% is to be used. (5.9)

(a) Clearly specify the net cash flow profiles for each mutually exclusive alternative.

(b) Determine which compressor should be selected. Use a *ranking approach* and the present worth method.

(c) Repeat part (b) using the annual worth method.

(d) Repeat part (b) using the future worth method.

(e) Determine which compressor should be selected. Use an *incremental cash flow approach* and the internal rate of return method.

(f) Repeat part (e) using the external rate of return method.

(g) Repeat part (e) using the savings/investment ratio method.

40. The Ajax Manufacturing Company wishes to choose one of the following machines:

	Machine 1	Machine 2
First cost	$10,000	$12,000
Planning horizon	6 years	6 years
Salvage value	$ 1,000	$ 1,000
Operating and maintenance cost for year k	$800 + 80k$	$200 + 100k$

MARR is 12%. (5.9)

(a) Clearly specify the net cash flow profiles for each mutually exclusive alternative.

(b) Determine which machine should be selected. Use a *ranking approach* and the present worth method.

(c) Repeat part (b) using the annual worth method.

(d) Repeat part (b) using the future worth method.

(e) Determine which machine should be selected. Use an *incremental cash flow approach* and the internal rate of return method.

(f) Repeat part (e) using the external rate of return method.

(g) Repeat part (e) using the savings/investment ratio method.

41. The manager of the distribution center for the southeastern territory of a major pharmaceutical manufacturer is contemplating installing an improved material handling system linking receiving and storage, as well as storage and shipping. Two designs are being considered. The first consists of a conveyor system that is tied into an automated storage/retrieval system. Such a system is estimated to cost $1,125,000 initially, have annual operating and maintenance costs of $72,000, and a salvage value of $75,000 at the end of the 10-year planning horizon.

The second design consists of manually operated narrow-aisle, high-stacking lift trucks. To provide service comparable to that provided by the alternative design, an initial investment of $490,000 is required. Annual operating and maintenance costs of $160,000 are anticipated. An estimated salvage value of $30,000 is expected at the end of the planning horizon. *MARR* is 12%. (5.9)

(a) Clearly specify the net cash flow profiles for each mutually exclusive alternative.

(b) Determine which alternative should be selected. Use a *ranking approach* and the present worth method.

(c) Repeat part (b) using the annual worth method.

(d) Repeat part (b) using the future worth method.

(e) Determine which alternative should be selected. Use an *incremental cash flow approach* and the internal rate of return method.

(f) Repeat part (e) using the external rate of return method.

(g) Repeat part (e) using the savings/investment ratio method.

42. A manufacturing plant in Minnesota has been contracting snow removal at a cost of $400/day. The past 4 years have yielded unusually heavy snowfalls, resulting in the cost of snow removal being of concern to the plant manager. The plant engineer has found that a snow-removal machine can be purchased for $53,000; it is estimated to have a useful life of 10 years, and a zero salvage value at that time. Annual costs for operating and maintaining the equipment are estimated to be $15,000. *MARR* is 20% and the estimated demand for the equipment is equal to 80 days/year. (5.9)

(a) Clearly specify the net cash flow profiles for each mutually exclusive alternative.

(b) Determine which alternative should be selected. Use a *ranking approach* and the present worth method.

(c) Repeat part (b) using the annual worth method.

(d) Repeat part (b) using the future worth method.

(e) Determine which alternative should be selected. Use an *incremental cash flow approach* and the internal rate of return method.

(f) Repeat part (e) using the external rate of return method.

(g) Repeat part (e) using the savings/investment ratio method.

43. A firm is considering two compressors. One must be chosen; the relevant data follow:

	Compressor 1	Compressor 2
Initial investment	$15,000	$30,000
Life	6 years	7 years
Salvage value	$ 1,000	$ 7,500
Annual disbursements	$ 8,000	$ 4,500

MARR is 20% and the cash flow profiles are believed to be repeatable over a planning horizon of 42 years. (5.9)

(a) Clearly specify the net cash flow profiles for each mutually exclusive alternative.

(b) Determine which compressor should be selected. Use a *ranking approach* and the present worth method.

(c) Repeat part (b) using the annual worth method.

(d) Repeat part (b) using the future worth method.

(e) Determine which compressor should be selected. Use an *incremental cash flow approach* and the internal rate of return method.

(f) Repeat part (e) using the external rate of return method.

(g) Repeat part (e) using the savings/investment ratio method.

44. Two submergible pumps are under consideration. Their cost data follow:

	Pump 1	Pump 2
First cost	$9,000	$14,000
Salvage value	0	$ 1,000
Annual operating cost	$1,600	$ 1,100
Life	5 years	10 years

Either pump is needed over a 10-year planning horizon. The company uses a *MARR* of 15%. (5.9)

(a) Clearly specify the net cash flow profiles for each mutually exclusive alternative.

(b) Determine which pump should be selected. Use a *ranking approach* and the present worth method.

(c) Repeat part (b) using the annual worth method.

(d) Repeat part (b) using the future worth method.

(e) Determine which pump should be selected. Use an *incremental cash flow approach* and the internal rate of return method.

(f) Repeat part (e) using the external rate of return method.

(g) Repeat part (e) using the savings/investment ratio method.

45. Two numerically controlled drill presses are being considered by the production department of a major corporation; one must be selected. Both machines meet the quality and safety standards of the firm. Comparative data are as follows. *MARR* is 15%. (5.9)

	Drill Press X	Drill Press Y
Initial investment	$26,000	$39,000
Estimated life	6 years	6 years
Salvage value	$ 4,000	$ 7,000
Annual operating cost	$12,000	$ 7,500
Annual maintenance cost	$ 3,000	$ 5,000

(a) Clearly specify the net cash flow profiles for each mutually exclusive alternative.

(b) Determine which alternative should be selected. Use a *ranking approach* and the present worth method.

(c) Repeat part (b) using the annual worth method.

(d) Repeat part (b) using the future worth method.

(e) Determine which alternative should be selected. Use an *incremental cash flow approach* and the internal rate of return method.

(f) Repeat part (e) using the external rate of return method.

(g) Repeat part (f) using the savings/investment ratio method).

46. Machine 1 initially costs $160,000. Its resale value at the end of the kth year of service equals $160,000 - 12,000k$ for $k = 1, 2, . . . , 10$. Operating and maintenance expenses equal $35,000/year. Machine 2 initially costs $76,000. Its resale value at the end of the kth year, S_k, equals $76,000(0.80)^k$ for $k = 1, 2, . . . , 5$. Operating and maintenance expenses equal $40,000/year. Using the annual worth method and a *MARR* of 15%, determine which machine is preferred assuming machine cash flows are repeatable. (5.9)

(a) Use a planning horizon of 5 years.

(b) Use a planning horizon of 20 years.

47. The Ajax Manufacturing Company wishes to choose one of the following machines.

	Machine 1	Machine 2	Machine 3
First cost	$10,000	$15,000	$18,000
Planning horizon	5 years	5 years	5 years
Salvage value	—	$ 2,000	$ 3,000
Operating and maintenance cost for year k, $k = 1, . . . , n$	$600(1.10)^{k-1}$	$400(1.08)^{k-1}$	$200 + 100k$

MARR is 15% and the planning horizon is 5 years. Based upon the annual worth method, determine the preferred machine. (5.9)

48. A firm will either lease an office copier at an end-of-year cost of $10,000 for a 6-year period, or they will purchase the copier at an initial cost of $66,117. If purchased, the copier will have a zero salvage value at the end of its 6-year life. No other costs are to be considered. *MARR* is 12%. Should the firm lease or buy? Justify your answer on the basis of the following. (5.9)

(a) The present worth method.

(b) The internal rate of return method.

49. A company can construct a new warehouse for $1 million, or it can lease an equivalent building for $100,000/year for 25 years with the option of purchasing the building for $1 million at the end of the 25-year period. Lease payments are due at the *beginning* of each year. The company can earn 20%/year before taxes on its invested capital. Indicate the preferred alternative based upon the following. (5.9)

(a) The annual worth method.

(b) The savings/investment ratio method.

50. In Problem 49, for what beginning-of-year lease payment will the company be indifferent between leasing and buying? (5.9)

51. E-M Systems Company must decide if they should purchase a small computer or lease the computer. The computer costs $48,000 initially and will last 5 years, having a $5000 salvage value at that time. If the computer is purchased, all maintenance costs must be paid by E-M. Maintenance costs are $2,500/year over the life of the equipment. The E-M Company uses an interest rate of 15% in evaluating investment alternatives. For what beginning-of-year annual leasing charge is the firm indifferent between purchasing and leasing over the 5-year period? (5.9)

(a) Use the annual worth method.

(b) Use the external rate of return method.

52. A firm is considering either leasing or buying a microcomputer system. If purchased, the initial cost will be $250,000; annual operating and maintenance costs will be $80,000/year. Based on a 6-year planning horizon, it is anticipated the computer will have a salvage value of $30,000 at that time. If the computer is leased, annual operating and maintenance costs in excess of the annual lease payment will be $60,000/year. Based on an interest rate of 20%, what annual beginning-of-year lease payment will make the firm be indifferent between leasing and buying? (5.9)

(a) Use the present worth method.

(b) Use the internal rate of return method.

53. Thermo-D, Inc. is considering investing $14,000 in a heat exchanger. The heat exchanger will last 6 years, at which time it will be sold for $2000. Maintenance costs for the exchanger are estimated to increase by $300/year over its life. The maintenance cost for the first year is estimated to be $2000, As an alternative, the company may lease the equipment for $X/year, including maintenance. For what value of X should the company lease the heat exchanger? The company expects to earn 12% on its investments. Assume beginning-of-year lease payments. (5.9)

(a) Use the future worth method.

(b) Use the external rate of return method.

54. Cowboy Creek in Stillwater has a history of flooding during heavy rainfalls. Two improvement alternatives are under consideration by the city engineer.

1. Leave the existing 36-inch corrugated steel culvert in place and install another of the same size alongside at an installed cost of $7000.

2. Remove the culvert in place and replace it with a single 50-inch culvert at an installed cost of $11,000.

If the first alternative is adopted, the existing 36-inch culvert will need to be replaced in 10 years. It is assumed that a replacement at that time will cost $12,000. A newly installed culvert will last for 20 years; replacement prior to the end of 20 years will result in a zero salvage value. Use a planning horizon of 20 years and a 10% *MARR* to determine the preferred alternative. (5.11)

55. Pear Computers is considering replacing its material handling system and either purchasing or leasing a new system. The old system has an annual operating and maintenance cost of $32,000, a remaining life of 8 years, and an estimated salvage value of $5000 at that time.

A new system can be purchased for $280,000; it will be worth $60,000 in 8 years; and it will have annual operating and maintenance costs of $18,000/year. If the new system is purchased, the old system can be traded in for $20,000.

Leasing a new system will cost $26,000/year, payable at the beginning of the year, plus operating costs of $9000/year, payable at the end of the year. If the new system is leased, the old system will be sold for $10,000.

MARR is 15%. Compare the annual worths of keeping the old system, buying a new system, and leasing a new system based upon a planning horizon of 8 years.

(a) Use the cash flow approach. (5.11.1)

(b) Use the outsider viewpoint approach. (5.11.2)

56. Fluid Dynamics Company owns a pump that it is contemplating replacing. The old pump has annual operating and maintenance costs of $8000/year, it can be kept for 4 years more and will have a zero salvage value at that time.

The old pump can be traded in on a new pump. The trade-in value is $4000. The new pump will cost $18,000 and have a value of $9000 in 4 years and will have annual operating and maintenance costs of $4500/year.

Using a *MARR* of 20%, evaluate the investment alternative based upon the present worth method and a planning horizon of 4 years.

(a) Use the cash flow approach. (5.11.1)

(b) Use the outsider viewpoint approach. (5.11.2)

57. A company owns a 5-year old turret lathe that has a book value of $20,000. The present market value for the lathe is $16,000. The expected decline in market value is $2000/year to a minimum market value of $4000. Maintenance plus operating costs for the lathe equal $4200/year.

A new turret lathe can be purchased for $45,000 and will have an expected life of 8 years. The market value for the turret lathe is expected to equal $45,000(0.70)^k at the end of year k. Annual maintenance and operating cost is expected to equal $1600. Based on a 15% before-tax *MARR*, should the old lathe be replaced now? Use an equivalent uniform annual cost comparison, a planning horizon of 7 years, and the outsider viewpoint approach. (5.11.2)

58. Esteez Construction Company has an overhead crane that has an estimated remaining life of 7 years. The crane can be sold for $14,000. If the crane is kept in service it must be overhauled immediately at a cost of $6000. Operating and maintenance costs will be $5000/year after the crane is overhauled. After overhauling it, the crane will have a zero salvage value at the end of the 7-year period. A new crane will cost $36,000, will last for 7 years, and will have a $8000 salvage value at that time. Operating and maintenance costs are $2500 for the new crane. Esteez uses an interest rate of 15% in evaluating investment alternatives. Should the company buy the new crane based upon an annual cost analysis?

(a) Use the cash flow approach. (5.11.1)

(b) Use the outsider viewpoint approach. (5.11.2)

59. A firm is considering replacing a compressor that was purchased 4 years ago for $70,000. Currently, the compressor has a book value of $40,000, based on straight-line depreciation. If the compressor is retained for 4 years more, it is estimated to have a salvage value of $10,000. If the old compressor is retained for 8 years more, it is estimated that its salvage value will be negligible at that time. Operating and maintenance costs for the compressor have been increasing at a rate of $750/year, with the cost during the past year being $11,000.

A new compressor can be purchased for $60,000. It is estimated to have uniform annual operating and maintenance costs of $6000/year. The salvage value for the compressor is estimated to be $40,000 after 4 years and $15,000 after 8 years. If a new compressor is purchased, the old compressor will be traded in for $25,000.

Using a before-tax analysis, a *MARR* of 20%, and an annual worth comparison, determine the preferred alternative using a planning horizon of (a) 4 years, and (b) 8 years. Use the cash flow approach. (5.11.1)

60. A small foundry is considering the replacement of a No. 1 Whiting cupola furnace that is capable of melting gray iron only with a reverberatory-type furnace that can melt gray iron and nonferrous metals. Both furnaces have approximately the same melting rates for gray iron in pounds per hour. The foundry company plans to use the reverberatory furnace, if purchased, primarily for melting gray iron, and the total quantity melted is estimated to be about the same with either furnace. Annual raw material costs would therefore be about the same for each furnace. Available information and cost estimates for each furnace is given below.

Cupola furnace. Purchased used and installed 8 years ago for a cost of $20,000. The present market value is determined to be $8000. Estimated remaining life is somewhat uncertain but, with repairs, the furnace should remain functional for 7 years more. If kept 7 years more, the salvage value is estimated as $2000 and average annual expenses expected are:

Fuel	$35,000
Labor (including maintenance)	$40,000
Payroll taxes	10% of direct labor costs
Taxes and insurance on furnace	1% of purchase price
Other	$16,000

Reverberatory furnace. This furnace costs $32,000. Expenses to remove the cupola and install the reverberatory furnace are about $2400. The new furnace has an estimated salvage value of $3000 after 7 years of use and annual expenses are estimated as:

Fuel	$29,000
Labor (operating)	$30,000
Payroll taxes	10% of direct labor costs

Taxes and insurance on furnace   1% of purchase price

Other $16,000

In addition, the furnace must be relined every 2 years at a cost of $4000/occurrence. If the foundry presently earns an average of 20% on invested capital before income taxes, should the cupola furnace be replaced by the reverberatory furnace?

(a) Use the cash flow approach. (5.11.1)

(b) Use the outsider viewpoint approach. (5.11.2)

61. Metallic Peripherals, Inc. has received a production contract for a new product. The contract lasts for 5 years. To do the necessary machining operations, the firm can use one of its own lathes, which was purchased 3 years ago at a cost of $16,000. Today the lathe can be sold for $8000. In 5 years the lathe will have a zero salvage value. Annual operating and maintenance costs for the lathe are $4000/year. If the firm uses its own lathe it must also purchase an additional lathe at a cost of $12,000, its value in 5 years will be $3000. The new lathe will have annual operating and maintenance costs of $3500/year.

As an alternative, the presently owned lathe can be traded in for $10,000 and a new lathe of larger capacity purchased for a cost of $24,000; its value in 5 years is estimated to be $8000, and its annual operating and maintenance costs will be $6000/year.

An additional alternative is to sell the presently owned lathe and subcontract the work to another firm. Company X has agreed to do the work for the 5-year period at an annual cost of $12,000/end-of-year.

Using a 15% interest rate, determine the least-cost alternative for performing the required production operations.

(a) Use the cash flow approach. (5.11.1)

(b) Use the outsider viewpoint approach (5.11.2)

62. A machine was purchased 5 years ago at $12,000. At that time, its estimated life was 10 years with an estimated end-of-life salvage value of $1200. The average annual operating and maintenance costs have been $14,000 and are expected to continue at this rate for the next 5 years., However, average annual revenues have been and are expected to be $20,000. Now, the firm can trade in the old machine for a new machine for $5000. The new machine has a list price of $15,000, an estimated life of 10 years, annual operating plus maintenance costs of $7500, annual revenues of $13,000, and salvage values at the end of the jth year according to

$$S_j = \$15,000 - \$1500j, \quad \text{for } j = 0, 1, 2, 3, 4, 5, 6, 7, 8, 9, 10$$

Determine whether to replace or not by the annual worth method using a *MARR* equal to 15% compounded annually. Use a 5-year planning horizon and the outsider viewpoint approach. (5.11.2)

63. A building supplies distributor purchased a gasoline-powered fork-lift truck 4 years ago for $8000. At that time, the estimated useful life was 8 years with a salvage value of $800 at the end of this time. The truck can now be sold for $2500. For this truck, average annual operating expenses for year j have been

$$C_j = \$2000 + \$400(j - 1)$$

Now the distributor is considering the purchase of a smaller battery-powered truck for $6500. The estimated life is 10 years, with the salvage value decreasing by $600 each year. Average annual operating expenses are expected to be $1200. If a *MARR* of 10% is assumed and a 4-year planning horizon is adopted, should the replacement be made now? (5.11.2)

64. Kwik-Kleen Car Wash has been experiencing difficulties in keeping its equipment operational. The owner is faced with the alternative of overhauling the present equipment or replacing it with new equipment. The cost of overhauling the present equipment is $8500. The present equipment has annual operating and maintenance costs of $7500. If it is overhauled, the present equipment will last for 5 years more and be scrapped at zero value. If it is not overhauled, it has a trade-in value of $3200 toward the new equipment.

New equipment can be purchased for $28,000. At the end of 5 years the new equipment will have a resale value of $12,000. Annual operating and maintenance costs for the new equipment will be $3000.

Using a *MARR* of 12%, what is your recommendation to the owner of the car wash? Base your recommendation on a present worth comparison and the outsider viewpoint approach. (5.11.2)

65. A firm is contemplating replacing a computer they purchased 3 years ago for $400,000. It will have a salvage value of $20,000 in 4 years. Operating and maintenance costs have been $75,000/year. Currently the computer has a trade-in value of $100,000 toward a new computer that costs $300,000 and has a life of 4 years, with a salvage value of $50,000 at that time. The new computer will have annual operating and maintenance costs of $80,000.

If the current computer is retained, another small computer will have to be purchased in order to provide the required computing capacity. The smaller computer will cost $150,000, has a salvage value of $20,000 in 4 years, and has annual operating and maintenance costs of $30,000.

Using an annual worth comparison before taxes, with a *MARR* of 15%, determine the preferred course of action. (5.11.2)

66. National Chemicals has an automatic chemical mixture that it has been using for the past 4 years. The mixer originally cost $18,000. Today the mixer can be sold for $10,000. The mixer can be used for 10 years more and will have a $2500 salvage value at that time. The annual operating and maintenance costs for the mixer equal $6000/year.

Because of an increase in business, a new mixer must be purchased. If the old mixer is retained, a new mixer will be purchased at a cost of $25,000 and have a $4000 salvage value in 10 years. This new mixer will have annual operating and maintenance costs equal to $5000/year.

The old mixer can be sold and a new mixer of larger capacity purchased for $32,000. This mixer will have a $6000 salvage value in 10 years and will have annual operating and maintenance costs equal to $8000/year.

Based on a *MARR* of 15%, what do you recommend? (5.11.2)

67. The Ajax Specialty Items Corporation has received a 5-year contract to produce a new product. To do the necessary machining operations, the company is considering two alternatives.

Alternative A involves continued use of the currently owned lathe. The lathe was purchased five years ago for $20,000. Today the lathe is worth $8000 on the used machinery market. If this lathe is to be used, special attachments must be purchased at a cost of $3500. At the end of the 5-year contract, the lathe (with attachments) can be sold for $2000. Operating and maintenance costs will be $7000/year if the old lathe is used.

Alternative B is to sell the currently owned lathe and buy a new lathe at a cost of $25,000. At the end of the 5-year contract, the new lathe will have a salvage value of $13,000. Operating and maintenance costs will be $4000/year for the new lathe.

Using an annual worth analysis, should the firm use the currently owned lathe or buy a new lathe? Base your analysis on a minimum attractive rate of return of 20% (5.11.2)

68. The Telephone Company of America purchased a numerically controlled production machine 5 years ago for $300,000. The machine currently has a trade-in value of $70,000. If the machine is continued in use, another machine, X, must be purchased to supplement the old machine. Machine X costs $200,000, has annual operating and maintenance costs of $40,000, and will have a salvage value of $30,000 in 10 years. If the old machine is retained, it will have annual operating and maintenance costs of $55,000 and will have a salvage value of $15,000 in 10 years.

As an alternative to retaining the old machine, it can be replaced with Machine Y. Machine Y costs $400,000, has anticipated annual operating and maintenance costs of $70,000, and has a salvage value of $140,000 in 10 years.

Using a *MARR* of 20% and a present worth comparison, determine the preferred economic alternative. (5.11.2)

69. A highway construction firm purchased an item of earth-moving equipment 3 years ago for $125,000. The salvage value at the end of 8 years was estimated to be 35% of first cost. The firm earns an average annual gross revenue of $105,000 with the equipment and the average annual operating costs have been and are expected to be $65,000.

The firm now has the opportunity to sell the equipment for $70,000 and subcontract the work normally done by the equipment over the next 5 years. If the subcontracting is done, the average annual gross revenue will remain $105,000 but the subcontractor charges $85,000/end-of-year for these services.

If a 15% rate of return before taxes is desired, determine by the annual worth method whether or not the firm should subcontract. (5.11.2)

70. Bumps Umlimited, a highway contractor, must decide whether to overhaul a tractor and scraper or replace it. The old equipment was purchased 5 years ago for $130,000; it had a projected life of 12 years, with a $15,000 salvage value at that time. If traded in on a new tractor and scraper, it can be sold for $60,000. Overhauling the old equipment will cost $20,000. If overhauled, operating and maintenance costs will be $18,000/year, and the overhauled equipment will have a newly projected salvage value of $0 in 7 years.

A new tractor and scraper can be purchased for $150,000; it will have annual operating and maintenance costs of $10,000, and will have a salvage value of $40,000 in 7 years.

Using an annual worth comparison with a *MARR* of 15%, should the equipment be overhauled or replaced? (5.11.2)

71. A new automatic insertion machine has just been introduced on the market. The

Electronics Products Company (EPC) has learned that it will do the work of four existing machines. Currently EPC has 8 automatic insertion machines. Four operators and a supervisor are responsible for servicing the machines. The current annual labor cost for an operator is $24,000 and a supervisor is $32,000. Labor costs are expected to increase at an annual rate of 10%. The current machines require maintenance totaling $80,000/year, with the cost expected to increase at an annual rate of 12%.

If the 8 current machines are replaced with two new machines, only one operator will be required; the supervisor will no longer be required, since the operator would be assigned to another supervisor. Also, replacement of the current machines will release 3000 square feet of floor space needed for surface mount machines on order. Floor space costs $2.50 per square foot per year.

Currently, the average investment in work in process inventory totals $1 million. With the new machines, the average inventory investment will be reduced to $600,000; however, due to inflation, it is anticipated that the investment in work in process inventory will increase at an annual rate of 6%. It is anticipated that the maintenance costs for the new machines will total $20,000/year during the first year and increase at an annual rate of 10%.

The old machines can be sold for a total of $50,000. The two new machines can be installed for a total cost of $500,000. Using a 5-year planning horizon, the cumulative salvage value for the two machines will be approximately $50,000 in 5 years.

EPC believes the discount rate used with then-current estimates of cash flows should be 15%, including inflation. If constant worth cash flows are used, then a *MARR* of 8% should be used.

Determine the economic viability of purchasing the two new automatic insertion machines, first, by ignoring the inventory savings and, second, by including the savings realized by reducing work in process inventory. (5.11.2)

72. A firm is presently using a machine that has a market value of $11,000 to do a specialized production job. The requirement for this operation is expected to last only 6 years more, after which it will no longer be done. The predicted costs and salvage values for the present machine are:

Year	1	2	3	4	5	6
Operating cost	$1,500	$1,800	$2,100	$2,400	$2,700	$3,000
Salvage value	8,000	6,000	5,000	4,000	3,000	2,000

A new machine has been developed that can be purchased for $17,000 and has the following predicted cost performance.

Year	1	2	3	4	5	6
Operating cost	$ 1,000	$ 1,100	$ 1,200	$1,300	$1,400	$1,500
Salvage value	13,000	11,000	10,000	9,000	8,000	7,000

If interest is at 0%, when should the new machine be purchased? (5.11.3)

73. A particular unit of production equipment has been used by a firm for a period of time sufficient to establish very accurate estimates of its operating and mainte-nance costs. Replacements can be expected to have identical cash flow profiles in successive life cycles if constant worth dollar estimates are used. The appro-priate discount rate is 30%. Operating and maintenance costs for a unit of equip-ment in its tth year of service, denoted by C_t, are as follows:

t	C_t	t	C_t
1	$ 6,000	6	$15,000
2	7,500	7	17,250
3	9,150	8	19,650
4	10,950	9	22,200
5	12,900	10	24,900

Each unit of equipment costs $45,000 initially. Because of its special design, the unit of equipment cannot be disposed of at a positive salvage value following its purchase; hence, a zero salvage value exists, regardless of the replacement interval used.

(a) Determine the optimum replacement interval assuming an infinite planning horizon. (Maximum feasible interval = 10 years.)

(b) Determine the optimum replacement interval assuming a finite planning hori-zon of 15 years, with $C_{t+1} = C_t + \$1500 + \$150(t - 1)$ for $t = 10, 11, \ldots$.

(c) Solve parts (a) and (b) using a discount rate of 0%.

(d) Based on the results obtained, what can you conclude concerning the effect the discount rate has on the optimum replacement interval? (5.11.3)

74. Given an infinite planning horizon, identical cash flows profiles for successive life cycles, and the following functional relationships for C_t, the operating and maintenance cost for the tth year of service for the unit of equipment in current use, and F_n, the salvage value at the end of n years of service:

$$C_t = \$4000(1.10)^t \qquad t = 1, 2, \ldots, 12$$

$$F_n = \$44,000(0.50)^n \qquad n = 0, 1, 2, \ldots, 12$$

Determine the optimum replacement interval assuming a *MARR* of (a) 0%, (b) 15%. (Maximum life = 12 years.) (5.11.3)

75. Solve Problem 74 given the following functional relationships. (5.11.3)

$$C_t = \$4000(1.40)^t \qquad t = 1, 2, \ldots, 12$$

$$F_n = \$44,000(0.90)^n \qquad n = 0, 1, 2, \ldots, 12$$

CHAPTER

6

Depreciation and Income Tax Considerations

6.1 INTRODUCTION

Depreciation and taxes are particularly important in engineering economic analyses. Although depreciation allowances are not cash flows, their magnitudes and timing do affect taxes. Tax dollars *are* cash flows, and therefore it is necessary to include them in a thorough economic analysis, just as costs of wages, equipment, materials, and energy are included. Proper knowledge and application of the tax laws can make the economic difference between accepting and rejecting a project, as well as between profit and loss on the corporate bottom line.

The Tax Reform Act of 1986 virtually rewrote the book on depreciation and income taxes. This follows the major revisions of only a few years ago in the Economic Recovery Tax Act of 1981, which itself vastly changed the pre-1981 depreciation and tax laws. Because of these changes, many companies today have assets that are simultaneously being depreciated under very different systems. This chapter presents the Modified Accelerated Cost Recovery System (MACRS—post-1986), the Accelerated Cost Recovery System (ACRS—1981 through 1986), some pre-1981 depreciation methods, and other still-allowable depreciation strategies. It also illustrates corporate income tax law, including the effects of depreciation method, recovery period, interest on borrowed money, and the expense deduction on after-tax cash flows. Inflation considerations and depletion round out the chapter. Capital gains and losses and the investment tax credit are covered, although they are not *currently* a factor in after-tax analyses.

261

First-time readers may want to become familiar only with topics that apply to new investments. This can be done efficiently by skipping those sections dealing with the ACRS (6.5), pre-1981 depreciation methods (6.6), comparisons of depreciation methods (6.7), and capital gains and losses (6.14). An asterisk (*) has been placed by each such section and related homework problems. Occasional reference to material in these sections should not cause any readability problems. Although the investment tax credit is not currently a factor in after-tax analyses, it is sufficiently important and likely to be reinstated that it should be studied.

No attempt is made in this chapter to discuss the minute details of tax law. Rather, the objective is to communicate the major aspects of corporate tax treatment so that you can perform economic analyses on an after-tax basis and recognize when to seek assistance. You are definitely advised to seek assistance from corporate tax personnel when performing formal after-tax economic analyses.

6.2 THE MEANING OF DEPRECIATION

Most property decreases in value with use and time. That is, it *depreciates*. In determining taxable income, the law permits deduction of a reasonable allowance for wear and tear, natural decay or decline, exhaustion, or obsolescence of property used in a trade or business or of property held for the production of income [1]. The amount of the annual allowance depends upon several factors, including (1) cost basis or investment in the property, (2) date placed in service, (3) property class or estimated useful life, (4) salvage value for certain methods, and (5) method of depreciation used [2].

The federal income tax is based on net income that results by deducting certain items such as expenses from gross income. For tax purposes, investment in a depreciable asset is treated as a prepaid expense, and the depreciation allocates that expense over time. It is important to note that depreciation is not a cash flow, but is merely treatable as an expense for income tax purposes. A larger deduction in a year decreases net taxable income and hence income taxes, making more money available for reinvestment or dividends. Because of the time value of money, it is generally desirable to take larger allowable deductions in the early years and smaller deductions in the later years of an asset's life.

6.3 SOME DEFINITIONS USED IN DEPRECIATION ACCOUNTING

Depreciable property meets these requirements:

1. It must be used in business or held for the production of income.
2. It must have a life that can be determined, and that life must be longer than one year.
3. It must be something that wears out, decays, gets used up, becomes obsolete, or loses value from natural causes.

In general, if property does not meet all three of these conditions, it is not depreciable [3].

Depreciable property may be tangible or intangible. *Tangible property* can be seen or touched. *Intangible property*, such as a copyright or franchise, is not tangible. Depreciable property may also be real or personal. *Personal property* is property, such as machinery or equipment, that is not real estate. Real property is land and generally anything that is erected on, growing on, or attached to land. Land itself, however, is never depreciable [3]. Most engineering economic analyses involve tangible personal property; some include tangible real property.

Depreciation is determined by the Modified Accelerated Cost Recovery System (MACRS) for most tangible depreciable property placed in service *after* 1986, while the Accelerated Cost Recovery System (ACRS) applies to similar property from 1981 through 1986. The MACRS and ACRS both exclude some tangible properties, including property depreciated on a basis other than years, such as the *units of production method;* they also exclude certain public utility property, films, video tapes, and so forth.

The cost basis of the property is recovered over the recovery period of the asset. The cost basis is essentially the taxpayer's investment. In most cases, this is the cost of the property plus the cost of additions to that property, including installation cost. It is also known as the *unadjusted basis.* The cost basis less capital recovered, such as depreciation allowances, is known as the *unrecovered investment, adjusted basis,* or *book value.* The *recovery period* is simply the time over which the cost basis can be recovered, and is keyed to 3, 5, 7, 10, 15, 20, 27.5, or 31.5 years, depending upon the type of property involved [1]. It is usually shorter than the inherent physical life of the asset.

Depreciation on assets placed in service before 1981, and assets specifically excluded from the MACRS and ACRS, requires an explicit estimate of the salvage value and the useful life. *Salvage value* is an estimate of the market value at the end of the asset's useful life. *Useful life* is an estimate of time during which the asset may actually be used by a taxpayer. These estimates are used explicitly under the straight-line, sum of the years' digits, or declining balance depreciation methods. The useful life is not used in either the MACRS or ACRS methods. Technically speaking, salvage value is used in the MACRS calculations, but always assumes a value of zero.

Nearly all engineering economic analyses involving new depreciable tangible property make use of the MACRS. Imbedded within the MACRS system is the use of the 200% and 150% declining balance, and straight-line methods. The MACRS straight-line method, referred to as *alternative MACRS,* is required for some properties and is optional for others. Replacement analyses may well involve ACRS and pre-1981 depreciation methods, depending on the date an asset was placed in service. Regular straight-line depreciation still applies in the case of intangible property. Other depreciation methods may be needed to determine cash flows on assets when depreciation is based upon something other than years. Obviously, the topic is very complex.

6.4 MODIFIED ACCELERATED COST RECOVERY SYSTEM (MACRS)

The MACRS applies to most depreciable property placed in service after 1986 (it was possible to *elect* it for most depreciable property placed in service after July 31, 1986). The specific depreciation method and recovery period depend upon the property class into which the asset falls.

6.4.1 MACRS Property Classes

Depreciable tangible property may be assigned to one of the following eight classes [1]:

1. 3-Year Property. Included, with some important exceptions, is qualifying property with a class life[1] of *4 years or less.* This involves tractor units for over-the-road use; special handling devices for the manufacture of food and beverages; special tools for the manufacture of rubber, finished plastic, glass, and fabricated metal products; and special tools for the manufacture of motor vehicles. Specifically *excluded* are automobiles and light, general-purpose trucks, which had been 3-year property under the ACRS, but are now 5-year property.

2. 5-Year Property. Included is property with a class life of *more than 4 years* but *less than 10 years,* plus automobiles; light, general-purpose trucks; computer-based telephone central office switching equipment; research and experimentation property; certain thermal, solar, and wind-energy property; and certain biomass property. Some properties included in the class life interval are computers and peripheral equipment; office machinery (typewriters, calculators, copiers, and so on); heavy, general-purpose trucks; motor transport assets; offshore, oil, and gas well-drilling assets; construction assets; timber-cutting and sawing assets; and many assets used for the manufacture of knitted goods, carpets, medical and dental supplies, chemicals, and electronic components.

3. 7-Year Property. Included is property with a class life of *10 years or more* but *less than 16 years,* property without any class life and not specifically included in the 27.5- and 31.5-year categories to follow, and several specifically named items. Some assets in this interval include office furniture, fixtures and equipment; oil and gas exploration assets; theme and amusement park assets; and most assets used for manufacturing such things as food products, spun yarn, wood products and furniture, pulp and paper, rubber products, finished plastic products, leather products, glass products, foundry products, fabricated metal products, motor vehicles, aerospace products, athletic goods, and jewelry.

---

[1] "Class life" refers to the class life guideline under the old Class Life Asset Depreciation Range (CLADR) system. The updated 1983 version of this guideline is reproduced in [1].

4. 10-Year Property.   This class includes property with a class life of *16 years or more* but *less than 20 years.* Some assets included are vessels, tugs, and similar water transportation equipment; assets used in petroleum refining; and assets used in the manufacture of grain, sugar, and vegetable oil products.

5. 15-Year Property.   This class includes property with a class life of *20 years or more* but *less than 25 years,* plus sewage treatment plants, and telephone distribution plant and comparable equipment used for the two-way exchange of voice and data communications. Some assets included in the interval are land improvements such as sidewalks, roads, drainage facilities, sewers, bridges, fencing, and landscaping; assets used in the manufacture of cement; some pipeline transportation assets; some liquified natural gas equipment; and certain utility property.

6. 20-Year Property.   Property with a class life of *25 years or more,* other than real property with a class life of 27.5 years or more, plus municipal sewers, is included. Such assets are farm buildings; some railroad electric generating equipment; certain transmission lines, pole lines, buried cable, and repeaters; and much other utility property.

7. 27.5-Year Residential Rental Property.   Included is a rental building or structure for which 80% or more of the gross rental income for the tax year is rental income from dwelling units. A dwelling unit is a house or an apartment used to provide living accommodations in a building or structure, but *not* a unit in a hotel, motel, inn, or other establishment in which more than one-half of the units are used on a transient basis.

8. 31.5-Year Nonresidential Real Property.   Depreciable property that has a class life of *27.5 years or more* and is not 27.5-year residential rental property generally falls in this class.

It is obvious from the abbreviated property class descriptions above that determining the class to which a given property belongs is a complex task. Professional tax guidance and a complete set of CLADR Class Life Tables are recommended.

6.4.2   Calculating Regular MACRS Deductions for 3-, 5-, 7-, 10-, 15-, and 20-Year Property

For property in the 3-, 5-, 7-, and 10-year classes, the depreciation method used is *200% declining balance switching to straight-line at the optimum time with a half-year convention (200%DBSLH).* The 15- and 20-year classes depreciate similarly using 150% *(150%DBSLH).* These classes may use *alternate MACRS* which is *straight-line depreciation with a half-year convention (alternate SLH).* The final two classes, 27.5-year residential rental and 31.5-year nonresidential real property, are required to use *straight-line de-*

preciation with a mid-month convention (SLM). There is also an alternate MACRS method for these classes, which is again *straight-line depreciation with a mid-month convention spread over 40 years (alternate SLM).* In all cases, the unrecovered investment will be reduced to the salvage value of zero, as required by MACRS, when the asset is depreciated over its full term.

To figure the 200% declining balance part of 200%DBSLH depreciation, first determine the *rate* of depreciation, $1/n$, where n is the recovery period. For a $100,000 7-year property, n is 7 and the rate is $1/7$ or 14.29%. This rate is then multiplied by 2 for 200%DBSLH, resulting in $p = .2857$ or 28.57%, the *allowable percentage* for the 7-year property class. The basis of the property at time zero, $P = \$100,000$, is then multiplied by half this figure, or 14.29%, due to the half-year convention. The depreciation for year 1, D_1, is then $0.5pP = 0.5(.2857)(\$100,000) = \$14,290$. The adjusted cost basis or book value at the end of year 1 is then $B_1 = P - D_1 = \$100,000 - \$14,290 = \$85,710$. Depreciation for year 2 is $D_2 = p(P - D_1) = pB_1 = 0.2857(\$85,710) = \$24,487$. Likewise, $D_3 = p(P - D_1 - D_2) = pB_2$, and so on. Since the half-year convention treats year 1 as only one-half year, it is necessary to carry depreciation over into year 8 in order to realize the full recovery period. In general, n + 1 years are involved in full depreciation of a property in any of the first 6 classes. The above discussion applies fully to 150%DBSLH, except the rate is multiplied by 1.5 instead of 2.

Since each year's depreciation involves multiplying B_t times p, the book value will never reach zero under the 200% declining balance method alone. However, the MACRS requires switching to straight-line depreciation at the optimum year t' which will ensure depreciation to a book value of zero. The optimum year is when the straight-line depreciation first exceeds the 200% declining balance allowance. Straight-line depreciation for year $t = 2$ through n is simply the remaining book value divided by the number of years left in the recovery period. That is, $D_t = B_{t-1}/(n - t + 1.5)$. Note that $n - t + 1.5$ is the recovery period remaining, considering the half-year convention, counting year t. Once straight-line depreciation is begun, D_t remains the same each year through year n, followed by $.5D_n$ in year $n + 1$.

Example 6.1

Consider the $100,000 7-year property discussed above. Using the approach described, calculate the 200%DBSLH depreciation for each year of the recovery period. It is already established that $p = .2857$. The 200% declining balance and straight-line components are shown separately in Table 6.1. The year 6 200%DB depreciation is calculated as $.2857(\$22,313) = \$6,375$. Straight-line depreciation for year 6 is calculated as $\$22,313/(7 - 6 + 1.5) = \$8,925$.

TABLE 6.1

200% Declining Balance Switching to Straight-Line
Depreciation with a Half-Year Convention

End of Year, t	200% Declining Balance Depreciation, D_t	Straight-Line Depreciation on Remainder of Recovery Period, D_t	Book Value, B_t
0	—	—	$100,000
1	$14,290[a,b]	7,143	85,710
2	24,487[b]	13,186	61,223
3	17,491[b]	11,131	43,732
4	12,494[b]	9,718	31,238
5	8,925[b]	8,925	22,313
6	6,375	8,925[b]	13,388
7		8,925[b]	4,463
8		4,463[b]	0

[a] Half-year convention used in year 1.
[b] Indicates depreciation allowance actually used. Switching to straight-line occurs in year 6.

The MACRS expressions for 200% declining balance depreciation may be summarized as follows:

$$D_1 = 0.5pP = d_1P$$

$$D_2 = pB_1 = p(P - D_1) = p(P - d_1P) = p(1 - d_1)P = d_2P$$

$$D_3 = pB_2 = p(P - D_1 - D_2) = p(P - d_1P - d_2P)$$
$$= p(1 - d_1 - d_2)P = d_3P$$

and, in general,

$$D_t = p\left(1 - \sum_{j=1}^{t-1} d_j\right) P = d_tP; \qquad t = 1, 2, 3, \ldots \qquad (6.1)$$

Note that D_t is the actual amount of depreciation, and d_t is the depreciation rate or fraction. For straight-line depreciation at the optimal time t',

$$D_{t'} = B_{t'-1}/(n - t' + 1.5) = \left(1 - \sum_{j=1}^{t'-1} d_j\right) P/(n - t' + 1.5) = d_{t'}P$$

Since D_t will be the same from t' through n,

$$D_t = d_{t'}P = d_tP; \qquad t = t', t' + 1, t' + 2, \ldots, n \qquad (6.2)$$

and

$$D_t = 0.5d_{t'}P = d_{n+1}P; \qquad t = n + 1 \qquad (6.3)$$

Note that each year's depreciation can be expressed as a factor d_t times the original cost basis. This observation holds true for both 200% and 150%DBSLH depreciation.

The regular MACRS deduction may be easily calculated by applying the appropriate percentage ($d_t \times 100\%$), for a specific property class and year, to the cost basis P. These percentages are found in Table 6.2. The allowable deduction in year t, D_t, is given by

$$D_t = d_t P \qquad (6.4)$$

Note that the percentages in Table 6.2 add up to 100% for each property class. This means the entire cost basis of a property can be depreciated over its recovery period, even if its useful life is much longer. The unrecovered investment B_t at the end of year t is given by

$$B_t = P - \sum_{j=1}^{t} D_j = P \left(1 - \sum_{j=1}^{t} d_j \right) \qquad (6.5)$$

This represents the amount of the cost basis yet to be recovered.

TABLE 6.2

MACRS Percentages ($d_t \times 100\%$) for 3, 5, 7, and 10-year Property are 200%DBSLH, and 15- and 20-year Property are 150%DBSLH

EOY	3-Year Property	5-Year Property	7-Year Property	10-Year Property	15-Year Property	20-Year Property
0						
1	33.33%	20.00%	14.29%	10.00%	5.00%	3.750%
2	44.45	32.00	24.49	18.00	9.50	7.219
3	14.81	19.20	17.49	14.40	8.55	6.677
4	7.41	11.52	12.49	11.52	7.70	6.177
5		11.52	8.93	9.22	6.93	5.713
6		5.76	8.92	7.37	6.23	5.285
7			8.93	6.55	5.90	4.888
8			4.46	6.55	5.90	4.522
9				6.56	5.91	4.462
10				6.55	5.90	4.461
11				3.28	5.91	4.462
12					5.90	4.461
13					5.91	4.462
14					5.90	4.461
15					5.91	4.462
16					2.95	4.461
17						4.462
18						4.461
19						4.462
20						4.461
21						2.231

Source: This table obtained on 9/28/87 from the IRS (prior to public release).

TABLE 6.3

MACRS Deduction and Unrecovered
Investment on Computer in
Example 6.2

End of Year, t	MACRS Deduction, D_t	Unrecovered Investment, B_t
0	—	$82,000.00
1	$16,400.00	65,600.00
2	26,240.00	39,360.00
3	15,744.00	23,616.00
4	9,446.40	14,169.60
5	9,446.40	4,723.20
6	4,723.20	0.00
7	0.00	0.00

Example 6.2

A company is purchasing a stand-alone computer with color graphics display, printer, plotter, and interfacing equipment. First cost including installation will be $82,000 with an estimated salvage value of $5000 at the end of a projected useful life of 7 years. This is 5-year property, and the percentages in Table 6.2 have been used to figure depreciation. The allowable MACRS deduction and resulting unrecovered investment for each year are given in Table 6.3.

Note in Example 6.2 that neither the estimated salvage value nor the projected useful life are involved at this point. We will have to account for the actual salvage value received, but this will be illustrated later. Further, the entire cost basis was recovered over 6 years, rather than over the 7-year useful life.

In Example 6.2 it is assumed that the asset is kept throughout the recovery period. If it is sold or otherwise disposed of, the depreciation allowance in the year of disposition may be affected. For 3-, 5-, 7-, 10-, 15-, and 20-year property, a half-year convention applies in the year of disposition.

Example 6.3

In Example 6.2, the MACRS deduction for our $82,000 computer in year 4 is $9446.40. What would be the year 4 deduction if our computer were sold sometime late in year 4? Immediately after year 4?

Answers: $4723.20 and $9446.40, respectively.

6.4.3 Calculating Regular MACRS Deductions for 27.5- and 31.5-Year Property

To figure the straight-line depreciation with a mid-month convention, for use with 27.5-year residential rental or 31.5-year nonresidential real property, first determine the *rate* of depreciation, $1/n$, where n is the recovery period. For a $1 million nonresidential real property, n is 31.5, and the rate is $1/31.5 = 0.031746$ or 3.1746%. The rates 3.174 and 3.175 will be the MACRS percentages ($d_t \times 100\%$) applied to the cost basis to determine depreciation in all years except the first and the last of the recovery period. If the property is placed in service during the fourth month of the tax year, the required mid-month convention allows a $8.5/12 = 0.7083$ share of a whole year's depreciation during the first year. Therefore, the allowable percentage in the first year is $0.7083(3.1746\%) = 2.249\%$. Equations (6.4) and (6.5) hold for SLM depreciation. The allowable percentages for 27.5- and 31.5-year properties are given in Tables 6.4 and 6.5, respectively.

Example 6.4

A firm having a tax year of January 1 to December 31 (a calendar year taxpayer) will place a $220,000 27.5-year residential rental property in service on May 1. The allowable MACRS deductions for each year are given in Table 6.6.

In Example 6.4, recovery takes place over a period of 28 years. Note that the entire cost basis is recovered. This would have been true even if an estimated salvage value had been declared.

In Example 6.4 it is assumed that the asset is kept throughout the recovery period. If it is sold or otherwise disposed of, the depreciation allowance in the year of disposition may be affected. If so, for 27.5- and 31.5-year property, a mid-month convention applies [4].

Example 6.5

In Example 6.4, the MACRS deduction in year 8 is $7999.20. What would be the year 8 deduction if the rental property were sold on October 1?
   Answer: $(9.5/12) \times \$7999.20 = \$6332.70$.

6.4.4 Alternate MACRS Election

A taxpayer may elect to claim alternate MACRS deductions instead of the regular allowances presented in the earlier sections. The alternate MACRS

TABLE 6.4

MACRS Percentages ($d_t \times 100\%$) for 27.5-year Residential Rental Property

Year		Month in Tax Year Property Placed in Service										
	1	2	3	4	5	6	7	8	9	10	11	12
1	3.485%	3.182%	2.879%	2.576%	2.273%	1.970%	1.667%	1.364%	1.061%	0.758%	0.455%	0.152%
2–9	3.636	3.636	3.636	3.636	3.636	3.636	3.636	3.636	3.636	3.636	3.636	3.636
Even 10–26	3.637	3.637	3.637	3.637	3.637	3.637	3.636	3.636	3.636	3.636	3.636	3.636
Odd 11–27	3.636	3.636	3.636	3.636	3.636	3.636	3.637	3.637	3.637	3.637	3.637	3.637
28	1.970	2.273	2.576	2.879	3.182	3.485	3.636	3.636	3.636	3.636	3.636	3.636
29							0.152	0.455	0.758	1.061	1.364	1.667

Source: This table obtained on 9/28/87 from the IRS (prior to public release).

TABLE 6.5

MACRS Percentages ($d_t \times 100\%$) for 31.5-year Nonresidential Real Property

Year												
1												
2–7												
Even 8–30												
Odd 9–31												
32												
33												

Source: This table obtained on 9/28/87 from the IRS (prior to public release).

					Month in Tax Year Property Placed in Service						
1	2	3	4	5	6	7	8	9	10	11	12
3.042%	2.778%	2.513%	2.249%	1.984%	1.720%	1.455%	1.190%	0.926%	0.661%	0.379%	0.132%
3.175	3.175	3.175	3.175	3.175	3.175	3.175	3.175	3.175	3.175	3.175	3.175
3.175	3.174	3.175	3.174	3.175	3.174	3.175	3.174	3.175	3.174	3.175	3.174
3.174	3.175	3.174	3.175	3.174	3.175	3.174	3.175	3.174	3.175	3.174	3.175
1.720	1.984	2.249	2.513	2.778	3.042	3.175	3.174	3.175	3.174	3.175	3.174
						0.132	0.397	0.661	0.926	1.190	1.455

TABLE 6.6

MACRS Deduction and Unrecovered
Investment on 27.5-year Residential
Rental Property (Example 6.4)

End of Year, t	MACRS Deduction D_t	Unrecovered Investment B_t
0		220,000.00
1	5,000.60	214,999.40
2–9	7,999.20	.
Even 10–26	8,001.40	.
Odd 11–27	7,999.20	.
.	.	
.	.	
.	.	
28	7,000.40	0.00

method is simply straight-line depreciation with either a half-year (alternate SLH) or a mid-month (alternate SLM) convention as appropriate. For most property, the straight-line method is applied over the class life[2] of the property with a half-year convention. For residential rental and nonresidential real property, a 40-year period is used with a mid-month convention. Some other properties do not follow the guidelines and are assigned an explicit period. For example, automobiles, light, general-purpose trucks, and any computer or peripheral equipment to which the straight-line method is applied must use a period of 5 years with a half-year convention.

The alternate MACRS method is *required* for use on some property, including property (1) used predominately outside the U.S., (2) leased to a tax-exempt entity or foreign person, (3) financed by tax-exempt bonds, or (4) imported from a country that maintains trade restrictions. Only in rare circumstances would a taxpayer elect to use alternate MACRS depreciation if not otherwise required to do so.

In general, SLH depreciation over an n-year period may be expressed as having rate

$$d_t = 1/n \tag{6.6}$$

and depreciation allowances of

$$D_t = 0.5d_tP; \qquad t = 1, n + 1 \tag{6.7}$$

and

$$D_t = d_tP; \qquad t = 2, 3, \ldots, n \tag{6.8}$$

---

[2] The *class life* was discussed in the presentation of MACRS property classes. It is generally longer than the recovery period.

If the property is disposed of during the period, standard MACRS rules apply regarding the half-year and mid-month conventions.

Example 6.6

A press forming machine is purchased for the manufacture of glass windows for $250,000. Its CLADR class life is 14 years, even though it is 7-year property. The firm elects to use alternate MACRS in order to defer depreciation deductions into more profitable years. The firm's allowable recovery during each year is given in Table 6.7. Note that $d_t = \frac{1}{14} = 0.071429$.

Example 6.7

Reconsider Example 6.4 with the 27.5-year residential rental property purchased for $220,000 on May 1 by a calendar year taxpayer. If we elect alternate MACRS and then sell the property on October 1 of year 8, the schedule of allowable deductions is as presented in Table 6.8. Recall that a 40-year period must be used and, therefore, $d_t = \frac{1}{40} = .025$.

*6.5 ACCELERATED COST RECOVERY SYSTEM

Capital costs for most depreciable tangible property placed in service after 1980 and before 1987 must be recovered using the ACRS. Knowledge of this method may be of use in estimating the cash flows of existing assets being considered for replacement. In many respects, the ACRS is similar to the MACRS, but there are some important differences.

TABLE 6.7

Alternate MACRS Deductions on Press Forming Machine, Example 6.6 ($n = 14$)

End of Year, t	Alternate SLH Deduction, D_t
0	—
1	$ 8,928.57
2–14	17,857.14
15	8,928.57

TABLE 6.8
Alternate MACRS Deductions on Residential
Rental Property, Example 6.7 ($n = 40$)

End of Year, t	Alternate SLM Deduction, D_t
0	—
1	$(7.5/12)(1/40)(\$220,000) = \$3,437.50$
2–7	$5,500.00$
8	$(9.5/12)(1/40)(\$220,000) = \$4,354.17$

*6.5.1 Recovery Property Classes

The ACRS refers to *recovery property,* which, for practical purposes, may be thought of as the same tangible depreciable property to which the MACRS applies. It also refers to *Section 1245 class property,* which may be thought of as depreciable tangible personal property such as equipment, machinery, tooling, office furniture, or vehicles used in trade or business. Further, the ACRS refers to *Section 1250 class property,* which is primarily depreciable tangible real property such as buildings, fixtures, fences, and parking lots.

Each item of recovery property may be assigned to one of the following five classes:

1. *3-Year Property*—Included is Section 1245 class property having a class life of 4 years or less. Also included is Section 1245 class property used in connection with research and experimentation. Examples of items having class lives of 4 years or less are automobiles, taxis, light, general-purpose trucks, over-the-road tractor units, and certain special tools and devices for manufacturing.

2. *5-Year Property*—Included is Section 1245 class property that is not 3- or 10-year property, or 15-year public utility property. This recovery property class contains the bulk of equipment, machinery, tooling, and office furniture. Also included are single-purpose agricultural structures, storage facilities used for petroleum distribution, and public utility property having a class life of 5 to 18 years, inclusive.

3. *10-Year Property*—Included is public utility property, other than Section 1250 class property or 3-year property, having a class life of more than 18 but not more than 25 years. Also included is Section 1250 class property having a class life of 12.5 years or less (theme and amusement parks). Residential manufactured homes, including mobile homes, belong to this class.

4. *15-Year Public Utility Property*—Included is public utility property, other than Section 1250 class property or 3-year property, having a class life of more than 25 years.

5. *15-, 18-, or 19-Year Real Property*—Included is Section 1250 class property with a class life of more than 12.5 years. This covers all real property, such as buildings, fixtures, fences, and parking lots not specifically designated in an earlier class. The 15-, 18-, or 19-year recovery period depends on the specific date from 1981 to 1986 the property was placed in service.

These five recovery property classes dictate the percentage of the cost basis to be recovered each year under the ACRS.

*6.5.2 Calculating ACRS Deductions

The ACRS deductions and unrecovered investment are calculated in the same way as given for the MACRS in Equations (6.4) and (6.5). The ACRS percentages ($d_t \times 100$) are found in Table 6.9 for 3-, 5-, 10-, and 15-year public utility property. Also in Table 6.9 is one column of the 15-year real property tables, assuming an asset placed in service in the first month of the tax year. Complete tables for 15-, 18-, and 19-year real property are readily available.

Example 6.8

Recall our company of Example 6.2. Suppose we had purchased the stand-alone computer for $82,000 in a year to which the ACRS applies. Let the

TABLE 6.9

ACRS Percentages ($d_t \times 100\%$) for 3-, 5-, and 10-Year Property, 15-Year Public Utility Property, and 15-Year Real Property Placed in Service During the First Month of the Tax Year

Year, t	3-Year Property	5-Year Property	10-Year Property	15-Year Public Utility Property	15-Year Real Property (Month 1)
1	25	15	8	5	12
2	38	22	14	10	10
3	37	21	12	9	9
4		21	10	8	8
5		21	10	7	7
6			10	7	6
7			9	6	6
8			9	6	6
9			9	6	6
10			9	6	5
11				6	5
12				6	5
13				6	5
14				6	5
15				6	5

TABLE 6.10
ACRS Deduction and Unrecovered Investment on
Computer, Example 6.8

End of Year, t	ACRS Deduction, D_t	Unrecovered Investment, B_t
0	— ·	$82,000.00
1	$12,300.00	69,700.00
2	18,040.00	51,660.00
3	17,220.00	34,440.00
4	17,220.00	17,220.00
5	17,220.00	0.00
6	0.00	0.00
7	0.00	0.00

$5000 salvage value and 7-year useful life estimates continue to hold. The allowable ACRS deduction and unrecovered investment for each year are given in Table 6.10. Note that this is 5-year recovery property.

Example 6.9

Reconsider the calendar year taxpaying firm of Example 6.3. Suppose the $220,000 real rental property had been placed in service on January 12 of a year to which 15-year ACRS applies. The allowable ACRS deductions and the unrecovered investment for each year are given in Table 6.11.

In Example 6.8 and 6.9, it assumed that the assets are kept throughout the recovery period. If they are sold or otherwise disposed of, the recovery allowance in the year of disposition may be affected. For personal property in the 3-, 5-, 10-, and 15-year public utility property classes, no recovery deduction is allowed in the year of disposition. An ACRS deduction is permitted for 15-, 18-, and 19-year real property, reflecting the number of months the property was in service in the year of disposition (a nearest whole month convention is illustrated in Example 6.11).

Example 6.10

In Example 6.8, the ACRS deduction in year 4 is $17,220.00. What would be the year 4 deduction if our computer were sold sometime late in year 4? Immediately after the end of year 4?
Answers: $0.00 and $17,220.00, respectively.

TABLE 6.11

ACRS Deduction and Unrecovered Investment on
Real Estate, Example 6.9

End of Year, t	ACRS Deduction, D_t	Unrecovered Investment, B_t
0	—	$220,000.00
1	$17,600.00	202,400.00
2	24,200.00	178,200.00
3	22,000.00	156,200.00
4	17,600.00	138,600.00
5	15,400.00	123,200.00
6	15,400.00	107,800.00
7	13,200.00	94,600.00
8	13,200.00	81,400.00
9	11,000.00	70,400.00
10	11,000.00	59,400.00
11	11,000.00	48,400.00
12	11,000.00	37,400.00
13	11,000.00	26,400.00
14	11,000.00	15,400.00
15	11,000.00	4,400.00
16	4,400.00	0.00

Example 6.11

In Example 6.9, the ACRS deduction in year 8 is $13,200.00. What would be
the year 8 deduction if the real estate were sold on September 28? Answer:
$9/12 \times \$13,200.00 = \$9,900.00$.

TABLE 6.12

Optional ACRS Straight-Line Recovery
Periods

Recovery Property Class	Optional Recovery Periods in Years, n
3-Year	3, 5, or 12
5-Year	5, 12, or 25
10-Year	10, 25, or 35
15-Year public utility	15, 35, or 45
15-Year real	15, 35, or 45
18-Year real	18, 35, or 45
19-Year real	19, 35, or 45

*6.5.3 Straight-Line Election

A taxpayer may elect to claim ACRS straight-line deductions instead of the regular allowances presented in the earlier section. The optional straight-line recovery periods are given in Table 6.12. Note that these periods are *not* the class life periods as appear in the alternate MACRS method.

The half-year convention applies to 3-, 5-, and 10-year property, and 15-year public utility property. Therefore, the formulas in Equations 6.6, 6.7, and 6.8, are applicable. If the property is sold during the recovery period, standard ACRS rules apply, in that no deduction is allowed in the year of sale.

Example 6.12

A sheet metal stamping machine was purchased for $25,000 on December 1 of an ACRS year. The firm elected to use the ACRS straight-line option with a recovery period of 12 years in order to defer depreciation deductions. The firm's allowable recovery during each year is given in Table 6.13.

For 15-, 18-, and 19-year real property, the first year's deduction and the year of disposition deduction must reflect the number of months in use.

Example 6.13

Suppose the $220,000 real rental property in Example 6.9 was placed in service on May 5 and recovered using ACRS straight-line over 35 years. If the property is eventually sold on September 28 of year 8, the schedule of allowable deductions is as presented in Table 6.14.

TABLE 6.13
ACRS Straight-Line Recovery on Stamping Machine, Example 6.12 ($n = 12$)

End of Year, t	Straight-Line ACRS Deduction, D_t
0	—
1	$25,000/12 \times 1/2 = \$1,041.67$
2–12	$25,000/12 \quad\quad = \$2,083.33$
13	$25,000/12 \times 1/2 = \$1,041.67$

TABLE 6.14

ACRS Straight-Line Recovery on Real Estate, Example 6.13 ($n = 35$)

End of Year, t	ACRS Straight-Line Deduction, D_t
0	—
1	$\$220{,}000/35 \times 8/12 = \$4{,}190.48$
2–7	$\$220{,}000/35 \qquad = \$6{,}285.71$
8	$\$220{,}000/35 \times 9/12 = \$4{,}714.29$

*6.6 OTHER DEPRECIATION METHODS BASED ON YEARS

Post-1980 depreciation on assets first placed in service before 1981 is to be computed according to the method selected at that time. Prior to 1981, the three most popular depreciation strategies included the (1) straight-line, (2) sum of the years' digits, and (3) declining balance methods. All three methods utilize a useful life (n) instead of a recovery period. They also require an estimate of the salvage value (F) at the end of the useful life. Under each of these methods, total depreciation taken over the useful life of the property may not exceed the cost basis less the salvage value ($P - F$). Knowledge of these methods may be of use in estimating after-tax cash flows of existing assets being considered for replacement.

*6.6.1 Straight-Line Depreciation

The *straight-line method* provides for the uniform write-off of an asset. The depreciation allowed at the end of each year t is equal throughout the property's useful life and is given by

$$D_t = \frac{P - F}{n} \tag{6.9}$$

The unrecovered investment at the end of year t is given by

$$B_t = P - \left(\frac{P - F}{n}\right) t \tag{6.10}$$

The straight-line method was very popular for property placed in service prior to 1981, and those assets will continue to be depreciated as originally set up. It is still to be used for depreciating intangible property.

Example 6.14

Reconsider the stand-alone computer of Examples 6.2 and 6.8 in which the cost basis is $82,000 with an estimated salvage value of $5000 and a useful

TABLE 6.15

Straight-Line Depreciation and Unrecovered Investment, Example 6.14

End of Year, t	Depreciation, D_t	Unrecovered Investment, B_t
0	—	$82,000.00
1	$11,000.00	71,000.00
2	11,000.00	60,000.00
3	11,000.00	49,000.00
4	11,000.00	38,000.00
5	11,000.00	27,000.00
6	11,000.00	16,000.00
7	11,000.00	5,000.00

life of 7 years. Had it been placed in service prior to 1981, the straight-line depreciation and unrecovered investment for each year would be as given in Table 6.15.

*6.6.2 Sum of the Years' Digits Depreciation

The *sum of the years' digits method* is known for its accelerated write-off of assets. That is, it provides relatively high depreciation allowances in the early years and lower allowances throughout the rest of the property's useful life. The name *sum of the years' digits* comes from the fact that the sum

$$1 + 2 + \cdots + (n - 1) + n = \frac{n(n + 1)}{2}$$

is used directly in the calculation of depreciation. The depreciation allowance during any year t is expressed as

$$D_t = \frac{n - (t - 1)}{n(n + 1)/2} (P - F) \tag{6.11}$$

The unrecovered investment at the end of each year t is given by

$$B_t = P - \sum_{j=1}^{t} \frac{n - (j - 1)}{n(n + 1)/2} (P - F)$$

which reduces to

$$B_t = (P - F) \frac{(n - t)(n - t + 1)}{n(n + 1)} + F \tag{6.12}$$

Sum of the years' digits depreciation is allowed if it was elected for pre-1981 tangible property having a useful life of 3 years or more and that was new in use (special rules apply to depreciable realty).

TABLE 6.16

Sum of the Years' Digits Depreciation and Unrecovered Investment, Example 6.15

End of Year, t	Value of $\dfrac{n-(t-1)}{n(n+1)/2}$	Depreciation, D_t	Unrecovered Investment, B_t
0	—	—	$82,000.00
1	7/28	$19,250.00	62,750.00
2	6/28	16,500.00	46,250.00
3	5/28	13,750.00	32,500.00
4	4/28	11,000.00	21,500.00
5	3/28	8,250.00	13,250.00
6	2/28	5,500.00	7,750.00
7	1/28	2,750.00	5,000.00

Example 6.15

Again let us consider the computer of Examples 6.2 and 6.8 having $P = \$82,000$, $F = \$5000$, and $n = 7$. Had it been placed in service prior to 1981, the sum of the years' digits depreciation and unrecovered investment for each year would be as given in Table 6.16.

*6.6.3 Declining Balance Depreciation

The *declining balance method,* like sum of the years' digits depreciation, is known for its accelerated write-off of assets. In this method, the depreciation allowed at the end of each year t is a constant fraction (p) of the unrecovered investment at the end of the previous year. That is,

$$D_t = pB_{t-1} \tag{6.13}$$

The unrecovered investment at the end of each year t is given by

$$B_t = P(1 - p)^t \tag{6.14}$$

Substituting Equation 6.14 into Equation 6.13 allows us to calculate the year t depreciation directly as

$$D_t = pP(1 - p)^{t-1} \tag{6.15}$$

Note that in the declining balance method of depreciation, the estimated salvage value need not come into play in figuring the deduction; however, the unrecovered investment must never fall below the salvage value.

Twice the straight-line rate, or $2/n$, is the maximum constant fraction permissible under law. When $p = 2/n$, this method is known as *200% declining balance* depreciation. It can be used only if it was elected for pre-1981 tangible property having a useful life of 3 years or more and that was new in use (special rules apply to depreciable realty). For used tangible property, either *150%* or *125% declining balance* depreciation is the maximum permitted.

Declining balance depreciation may be used alone; however, switching from either the 200% or 150% declining balance method to straight-line depreciation is permitted. The optimum switch takes place whenever straight-line depreciation on the unrecovered portion of the asset exceeds the declining balance allowance. That is, we should switch to straight-line at the first year for which

$$\frac{B_{t-1} - F}{n - (t - 1)} > pB_{t-1} \tag{6.16}$$

The estimated salvage value is used in determining the straight-line depreciation component, even though it is neglected in the double declining balance method. Switching to straight-line depreciation will never be desirable if the estimated salvage value F exceeds the declining balance unrecovered investment for the last year B_n, causing depreciation to be truncated.

Example 6.16

Let us again work with the computer of Examples 6.2 and 6.8 in which $P = \$82,000$, $F = \$5000$, and $n = 7$. Had it been placed in service prior to 1981, the 200% declining balance, with a switch to straight-line, depreciation and unrecovered investment for each year would be as given in Table 6.17. The value of p is $2/7$.

TABLE 6.17
200% Declining Balance Switching to Straight-Line Depreciation, and Unrecovered Investment, Example 6.16

End of Year, t	200% Declining Balance Depreciation, D_t	Straight-Line Depreciation on Remaining Life, D_t	Unrecovered Investment, B_t
0	—	—	$82,000.00
1	$23,428.57[a]	$11,000.00	58,571.43
2	16,734.69[a]	8,928.57	41,836.73
3	11,953.35[a]	7,367.35	29,883.38
4	8,538.11[a]	6,220.85	21,345.27
5	6,098.65[a]	5,448.42	15,246.62
6	4,356.18	5,123.31[a]	10,123.31
7	3,111.56	5,123.31[a]	5,000.00

[a] Indicates depreciation allowance actually used. Switching to straight line occurs in year 6.

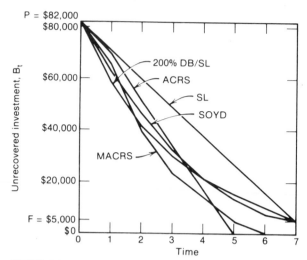

FIGURE 6.1. Value-time curves of five different depreciation methods.

*6.7 COMPARISON OF DEPRECIATION METHODS

MACRS, ACRS, straight-line (SL), sum of the years' digits (SOYD), and 200% declining balance switching to straight-line (DDB/SL) depreciation methods can be compared to provide insight into their relative performance.

*6.7.1 Value-Time Curves

Value-time curves are plotted in Figure 6.1 using the stand-alone computer data from Examples 6.2, 6.8, 6.14, 6.15, and 6.16. Note that the method of depreciation determines the path of unrecovered investments, which can vary widely.

*6.7.2 Depreciation and the *EUAC*

Calculation of the equivalent uniform annual cost (*EUAC*) is presented in previous chapters. The *EUAC* represents the equivalent uniform annual cost of capital recovered plus a return on the unrecovered investment. Using the stand-alone computer having $P = \$82,000$, $F = \$5000$, $n = 7$, and interest rate $i = 15\%$, the annual cost is[3]

$$EUAC = P(A|P\ 15,7) - F(A|F\ 15,7)$$
$$= \$82,000(.2404) - \$5000(.0904)$$
$$= \$19,257.75/\text{year}$$

---

[3] In Chapter Six, some calculations are performed using interest factors determined and retained within a computer for reasons of precision. Use of the interest tables in the Appendix will provide answers which differ insignificantly.

Note that *EUAC* is calculated here without respect to any depreciation method.

Table 6.18 presents the straight-line capital recovered plus return on the unrecovered investment from Example 6.14, covering the 7-year useful life of the computer.

Calculating the uniform annual cost of capital recovered plus return at $i = 15\%$ yields

$$
\begin{aligned}
A &= [23,300](P|F\ 15,1) + 21,650(P|F\ 15,2) + \cdots + 13,400(P|F\ 15,7)] \\
&\quad [A|P\ 15,7] \\
&= [23,300(.8696) + 21,650(.7561) + \cdots + 13,400(.3759)][.2404] \\
&= \$19,257.75/\text{year}
\end{aligned}
$$

This is the same as the *EUAC*! Had the *SOYD* or 200% *DB/SL* methods of Examples 6.15 and 6.16 been used, the result would have been the same, $A = \$19,257.75/\text{year}$.

Even using the MACRS and ACRS methods of Examples 6.2 and 6.8 will yield the same result if we are careful. Remember, the MACRS and ACRS allow us to recover the entire cost basis $P = \$82,000$, rather than just $P - F = \$77,000$. This $5000 over-recovery will eventually be lost in the form of "depreciation recapture" (discussed in Sec. 6.16) at the time of sale. If this $5000 is subtracted from the MACRS and ACRS Deduction (D_t) and capital recovered plus return $(D_t + iB_{t-1})$ in year 7, the uniform annual cost of capital recovered plus return will be $A = \$19,257.75/\text{year}$.

In summary, the equivalent uniform annual cost of an asset before taxes remains the same, regardless of the depreciation method used. It is most easily calculated using the formula for *EUAC*.

6.8 OTHER DEPRECIATION METHODS NOT BASED ON YEARS

Most depreciation methods are tied to the passage of time. Occasionally, however, recovery of cost over time would have little relation to an asset's

TABLE 6.18
Straight-Line Capital Recovery Plus Return

End of Year, t	SL Depreciation (Capital Recovered), D_t	Book Value (Unrecovered Investment), B_t	Return on Unrecovered Investment, iB_{t-1}	Capital Recovered Plus Return, $D_t + iB_{t-1}$
0	—	$82,000.00	—	—
1	$11,000.00	71,000.00	$12,300.00	$23,300.00
2	11,000.00	60,000.00	10,650.00	21,650.00
3	11,000.00	49,000.00	9,000.00	20,000.00
4	11,000.00	38,000.00	7,350.00	18,350.00
5	11,000.00	27,000.00	5,700.00	16,700.00
6	11,000.00	16,000.00	4,050.00	15,050.00
7	11,000.00	5,000.00	2,400.00	13,400.00

use or its production of income. That is why the IRS allows taxpayers to utilize other consistent methods of depreciation. Some recognized methods are briefly presented below. Assets that we elect to depreciate in any of these ways are specifically excluded from the definitions of property to which the MACRS and ACRS apply.

6.8.1 Units of Production Method

This procedure allows equal depreciation per each unit of output, regardless of the lapse of time involved. The allowance for year t is equal to the total depreciable amount $(P - F)$ times the ratio of units produced during the year (U_t) to the total units that are expected to be produced during the useful life of the asset (U). That is,

$$D_t = (P - F)\frac{U_t}{U} \tag{6.17}$$

This method has proven suitable for depreciating equipment used in exploiting natural resources such as mines, wells, etc.

6.8.2 Operating Day (Hour) Method

This is similar to the previous method, in that year t depreciation is based on the ratio of days (hours) used during the year (Q_t) to total days (hours) expected in a useful life (Q). Depreciation is expressed as

$$D_t = (P - F)\frac{Q_t}{Q} \tag{6.18}$$

6.8.3 Income Forecast Method

This method is applicable to depreciate the cost of rented property such as video tapes, sound recordings, and motion picture films. The ratio of year t rental income (R_t) to the total useful life income (R) is multiplied by the total lifetime depreciation, or

$$D_t = (P - F)\frac{R_t}{R} \tag{6.19}$$

6.9 TAX CONCEPTS

The taxes paid by a corporation represent a real cost of doing business and, consequently, affect the cash flow profile. For this reason, it is wise to perform economic analyses on an *after-tax* basis. After-tax analysis procedures are identical to the before-tax evaluation procedures studied already; however, the cash flows are adjusted for taxes paid or saved.

There are numerous kinds of taxes, including *ad valorem* (property), *sales, excise* (a tax or duty on the manufacture, sale, or consumption of various commodities), and *income taxes*. Income taxes are usually the only significant taxes to be considered in an economic analysis. Income taxes are assessed on gross income less certain allowable deductions, incurred both in the normal course of business as well as on gains resulting from the disposal of property.

Federal and state income tax regulations are not only detailed and intricate, but they are subject to change over time. During periods of recession and inflation, there is a tendency for the tax laws to be changed in order to improve the state of the economy. For this reason, only the general concepts and procedures for calculating after-tax cash flow profiles and performing after-tax analyses are emphasized here. Furthermore, only federal income tax will be considered because of the diversity of state laws. Practitioners involved in an actual analysis should seek the assistance of corporate legal counsel regarding any uncertainties about tax laws in effect at that time.

6.10 CORPORATE INCOME TAX—ORDINARY INCOME

Corporate income tax is not limited to business organizations. Engineers, lawyers, doctors, and other professionals may be treated as corporations if they have formally organized as professional corporations or associations. Associations, business trusts, joint stock companies, insurance firms, and certain limited partnerships are also taxed as corporations.

Ordinary federal income tax imposed on these corporations is presented in Table 6.19. The tax rates shown are those current as of 1987 for tax years beginning July 1, 1987 and beyond. Companies such as many small businesses with taxable incomes less than $75,000 are subject to significantly reduced tax rates. Note, however, that the 39% rate from $100,000 to $335,000 effectively eliminates the benefit of graduated rates for larger corporations. Corporations with taxable income in excess of $335,000 will, in effect, pay a flat tax of 34% on every dollar. In the remainder of this chapter, an applicable tax rate of 34% will be used unless otherwise specified. This

TABLE 6.19

Corporate Income Tax Rates for Tax Years Beginning July 1, 1987 and Beyond

Taxable Income, in Dollars	Tax Rate, Percent
$0 < \text{TI} \le 50,000$	15%
$50,000 < \text{TI} \le 75,000$	25
$75,000 < \text{TI} \le 100,000$	34
$100,000 < \text{TI} \le 335,000$	39
$335,000 < \text{TI}$	34

implicitly assumes we are dealing with a company already having over $335,000 in taxable income.

Example 6.17

Our small company is currently forecasting a taxable income of $50,000 for this year. We are considering another investment that will increase taxable income by $45,000. If we take on the project, what will be our increased income tax liability? What would it be if we were currently forecasting a taxable income of $350,000 for this year?

Answer 1: Without the project the tax is 0.15($50,000) = $7500. With the project the tax is 0.15($50,000) + 0.25 ($25,000) + 0.34($20,000) = $20,550. The tax increase is $20,550 − $7500 = $13,050 or 29% of the $45,000.

Answer 2: Without the project the tax is 0.15($50,000) + 0.25($25,000) + .34($25,000) + 0.39($235,000) + 0.34($15,000) = $119,000 or 34% of $350,000. Since every dollar of the additional $45,000 will be taxed at 34%, the increase will be 0.34($45,000) = $15,300 for a total tax of $134,300.

Taxable income must first be determined before any tax rate can be applied. Basically, *taxable income* is gross income less allowable deductions. *Gross income* is income in a general sense less any monies specifically exempt from tax liability. Corporate deductions are subtracted from gross income and commonly include items such as salaries, wages, repairs, rent, bad debts, taxes (other than income), charitable contributions, casualty losses, interest, and depreciation. Interest and depreciation are of particular interest, since we can control them to some extent through financing arrangements and accounting procedures.

Taxable income is represented pictorially in Figure 6.2, which shows that taxable income for any year is what is left after deductions, including interest on borrowed money and depreciation allowance, are subtracted from gross income. These components are not all cash flows, since the depreciation allowance is simply treated as an expense in determining taxable income.

FIGURE 6.2. Pictorial representation of taxable income.

FIGURE 6.3. Pictorial representation of after-tax cash flow.

6.11 AFTER-TAX CASH FLOW

We have now looked at the basic elements needed to calculate after-tax cash flows. These elements are summarized in Figure 6.3. This shows that the after-tax cash flow is the amount remaining after income taxes and deductions, including interest but excluding depreciation allowance, are subtracted from gross income.

In many of the tables to follow, we simplify our terminology by speaking of before-tax cash flows. The term *before-tax cash flow* is used when no borrowed money is involved, and it equals gross income less deductions, not including depreciation. *Before-tax and loan cash flow* is used when borrowed money is involved, and it equals gross income less deductions, not including either depreciation or principal or interest on the loan.

Example 6.18

Let us again consider the stand-alone computer of Example 6.2 for which $P = \$82,000$, $F = \$5000$, and $n = 7$. It will be monitoring and controlling manufacturing equipment, and we expect it to reduce operating expenses by $23,500 year. Therefore, before-tax cash flow will increase by $23,500 during each year of the computer's useful life. We plan to depreciate it using the MACRS, and desire to know the after-tax cash flows. We will then apply the present worth method as an effectiveness measure to see whether the alternative to purchase the computer yields at least a 15% after-tax rate of return. The cash flow calculations are given in Table 6.20.

Calculating the present worth of after-tax cash flows, we have

$$PW(15) = -\$82,000 + \$21,086(P|F\ 15,1) + \$24,431.60(P|F\ 15,2)$$
$$+ \cdots + \$15,510(P|F\ 15,7) + \$3300(P|F\ 15,7)$$
$$= \$3010.50$$

Year 7 is listed twice, once to record the normal cash flows and once to show the salvage value transaction. Note that the $5000 salvage value is considered a "depreciation recapture" and taxed as ordinary income. This

TABLE 6.20

After-Tax Cash Flow Profile Using the MACRS for Example 6.18

End of Year, A	Before-Tax Cash Flow, B	MACRS Deduction, C	Taxable Income B − C, D	Tax D × 0.34, E	After-Tax Cash Flow, B − E, F
0	−$82,000.00				−$82,000.00
1	23,500.00	$16,400.00	$ 7,100.00	$2,414.00	21,086.00
2	23,500.00	26,240.00	−2,740.00	−931.60	24,431.60
3	23,500.00	15,744.00	7,756.00	2,637.04	20,862.96
4	23,500.00	9,446.40	14,053.60	4,778.22	18,721.78
5	23,500.00	9,446.40	14,053.60	4,778.22	18,721.78
6	23,500.00	4,723.20	18,776.80	6,384.11	17,115.89
7	23,500.00	0.00	23,500.00	7,990.00	15,510.00
7	5,000.00 (salvage)	0.00	5,000.00 (depreciation recapture)	1,700.00	3,300.00

is because the entire cost basis (P = $82,000) had already been recovered through MACRS deductions. Depreciation recapture is covered in Section 6.15 later in this chapter.

Example 6.18 shows that the after-tax cash flow profile is nothing more than lumps of money flowing in or out at different points in time over a planning horizon. Although the present worth effectiveness measure is used in the example, a rate of return, future worth, or other equivalent analysis could have been applied. Since the present worth of the discounted cash flows in our example is positive ($3010.50), we are able to conclude that the alternative as described indeed yields more than a 15% after-tax rate of return.

In Example 6.18, we also see that an alternative's taxable income and tax can be negative for a particular year. It is assumed that such values are used to offset positive tax liabilities from other corporate activities. When there are insufficient positive taxable incomes to be offset, the taxable income losses can be carried back 3 years or forward 15 years in order to reduce positive taxable income. That is, if an operating loss occurs in 1991, we can reopen books beginning in 1988, then 1989, and 1990. If the loss is not fully absorbed against positive taxable incomes, the remainder may be used in 1992, 1993, and so on for as many as 15 years into the future. We will continue to assume that negative taxable incomes are used to offset positive values for the year in question.

6.12 EFFECT OF DEPRECIATION METHOD AND RECOVERY PERIOD

When we look at the taxable income and after-tax cash flow representations in Figures 6.2 and 6.3, it is natural to wonder about the tax effects of different depreciation methods, recovery periods, and interest on borrowed money. We will show that these factors have a substantial effect on taxes and thus on cash flow profiles.

We have seen how some depreciation methods provide for a higher depreciation allowance during the early years of an asset's life and a correspondingly lower allowance in later years. This places a lower tax burden on the asset during the early years followed by a higher yearly burden. *In most cases the total undiscounted ordinary income tax dollars paid will be the same,* regardless of the depreciation method used. This is always true when the tax rate remains constant and the total depreciation allowance for each method is the same or an adjustment for under- or over-recovery is made at ordinary income tax rates as was done in Example 6.18.

In order to illustrate the effects of depreciation method and recovery period, alternate MACRS and pre-1981 sum of the years' digits depreciation effects will be presented in detail. Abbreviated results for regular ACRS and pre-1981 straight-line depreciation are also given.

Example 6.19

Suppose instead of the computer we have been discussing, we are interested in a different asset used in the manufacture of medical and dental supplies. Let it, like our computer, have a cost basis of $82,000, and an estimated salvage value of $5000 at the end of a useful life of 7 years. Also, it will reduce operating expenses (and therefore increase before-tax cash flows) by $23,500/year. It is MACRS 5-year property; however, let us depreciate it using alternate MACRS. We are now required to use the asset's class life, and a search of the CLADR tables reveals that this is 9 years. The cash flow calculations are given in Table 6.21. Calculating the present worth of the after-tax cash flows at $i = 15\%$, we have $-\$1778.68$.

Table 6.21 is different from Table 6.20 in several respects. The manufacturing asset is assumed to be sold toward the end of year 7, before the recovery period is completed. According to MACRS rules, the half-year convention is applied in the year of sale. Also, only $4555.56(2) + \$9111.11(5) = \$54,666.67$ is recovered through alternate MACRS deductions, leaving $27,333.33 unrecovered. Another $5000 is recovered through salvage. The remaining $82,000 - \$54,666.67 - \$5000 = \$22,333.33$ is considered a Section 1231 loss and is assumed to be applied against other Section 1231 gains elsewhere in the company, resulting in a $22,333.33(0.34) = \$7593.33$ tax savings or negative tax. Section 1231 gains and losses are covered later in this chapter (see Sec. 6.15).

TABLE 6.21

After-Tax Cash Flow Profile Using MACRS Straight-Line Depreciation with 9-Year Recovery Period, Example 6.19

End of Year, A	Before-Tax Cash Flow, B	MACRS SLH Deduction, C	Taxable Income B − C, D	Tax D × 0.34, E	After-Tax Cash Flow B − E, F
0	−$82,000.00				−$82,000.00
1	23,500.00	$ 4,555.56	$18,944.44	$6,441.11	17,058.89
2	23,500.00	9,111.11	14,388.89	4,892.22	18,607.78
3	23,500.00	9,111.11	14,388.89	4,892.22	18,607.78
4	23,500.00	9,111.11	14,388.89	4,892.22	18,607.78
5	23,500.00	9,111.11	14,388.89	4,892.22	18,607.78
6	23,500.00	9,111.11	14,388.89	4,892.22	18,607.78
7	23,500.00	4,555.56	18,944.44	6,441.11	17,058.89
7	5,000.00 (salvage)	27,333.33 (unrecovered through MACRS deductions)	−22,333.33 (Section 1231 loss)	−7,593.33	12,593.33

A quick look at the −$1778.68 present worth of after-tax cash flows indicates that the purchase of this asset for manufacturing medical and dental supplies should be rejected. Yet, this asset *could* have been depreciated under regular MACRS just as was our computer in Example 6.18, which resulted in a $3010.50 present worth of after-tax cash flows; only the depreciation method and recovery period have changed! Comparing Tables 6.20 and 6.21, we can see that the standard MACRS deductions are higher in the early years (1 to 5), with correspondingly lower taxes and higher after-tax cash flows. The time value of money effect is seen in the present worth of the after-tax cash flows. Merely electing alternate MACRS over regular MACRS, thereby extending the recovery period and decreasing the allowable percentages d_t, made a difference of $3010.50 − (−$1778.68) = $4789.18 in the present worth of the after-tax cash flows at the MARR of 15%.

Example 6.20

Let us compare our post-1986 depreciation methods of Examples 6.18 and 6.19 with a popular pre-1981 method, sum of the years' digits depreciation. Again, $P = \$82,000$, $F = \$5000$, $n = 7$, and the before-tax cash flow is $23,500/year. The cash flow calculations are given in Table 6.22. Calculating the present worth of the after-tax cash flows at $i = 15\%$ results in $2107.85.

TABLE 6.22

After-Tax Cash Flow Profile Using Sum of the Years' Digits Depreciation, Example 6.20

End of Year, A	Before-Tax Cash Flow, B	SOYD Depreciation, C	Taxable Income B − C, D	Tax D × 0.34, E	After-Tax Cash Flow B − E, F
0	−$82,000.00				−$82,000.00
1	23,500.00	$19,250.00	$ 4,250.00	$1,445.00	22,055.00
2	23,500.00	16,500.00	7,000.00	2,380.00	21,120.00
3	23,500.00	13,750.00	9,750.00	3,315.00	20,185.00
4	23,500.00	11,000.00	12,500.00	4,250.00	19,250.00
5	23,500.00	8,250.00	15,250.00	5,185.00	18,315.00
6	23,500.00	5,500.00	18,000.00	6,120.00	17,380.00
7	23,500.00	2,750.00	20,750.00	7,055.00	16,445.00
7	5,000.00 (salvage)	5,000.00 (recovered through salvage)	0.00	0.00	5,000.00

In this example, the $5000 salvage value is not taxed. The total SOYD depreciation is only $P − F = \$77,000$ and the last $5000 is recovered through the salvage value. Since there is no over- or under-recovery, no extra tax treatment is necessary.

Had we been considering the computer (or the asset for manufacturing medical and dental supplies) for installation prior to 1981, an after-tax analysis would have resulted in a decision to invest. Similar calculations using regular ACRS and pre-1981 straight-line depreciation result in present worths of the after-tax cash flows of $2151.10 and −$32.23, respectively. Again, nothing about the computer is different; only the depreciation strategy has changed. By now, it should be coming clear why depreciation methods are of such concern to corporate taxpayers.

Examples 6.18 to 6.20 illustrate the important points that the method of depreciation and the recovery period affect cash flows and, therefore, the economic desirability of a project. It is interesting to note that the total capital recovery ($82,000), total taxes ($29,750), and total undiscounted after-tax cash flows ($57,750) are the same in each of Tables 6.20 to 6.22. Can we make any generalized statements *from these examples* about the preferability of depreciation methods and recovery periods with respect to effects on taxes and after-tax cash flows? No, because these examples are worked only for a particular set of conditions. It can, however, be shown mathematically that certain depreciation strategies are superior to others in that they provide a higher present worth of tax savings, assuming that the effective tax rate remains the same from year to year. For example, the regular MACRS percentages as given in Tables 6.2, 6.4, and 6.5 are always preferable to alternate MACRS. The same is true for regular ACRS versus ACRS straight-line. Further, if the ACRS straight-line method is used, the shortest recovery period is always superior to the longer alternatives. For pre-1981 property, both double declining balance switching to straight-line and sum of the years' digits are preferable to straight-line depreciation; however, double declining balance switching to straight-line is not always better than the sum of the years' method.

Quite often the MACRS percentages appear to be better than those of the ACRS and any of the pre-1981 depreciation methods. The examples and results mentioned in this section give support for such an argument. On the other hand, the recovery periods for many assets were lengthened significantly by the Tax Reform Act of 1986. Cars, light trucks, many machines, public utility property, and especially real estate have had their recovery periods increased. This has the effect of spreading out deductions, increasing the present worth of taxes paid, and reducing the present worth of after-tax cash flows.

6.13 EFFECT OF INTEREST ON BORROWED MONEY

Investment alternatives may be financed using equity (owner's) funds or debt (borrowed) funds. Until now we have implicitly assumed all financing to be through equity, although many companies use a mix of debt and equity for financing plant and equipment. Borrowed funds, including both principal and interest, must be repaid. The interest repaid each year affects taxable income and, consequently, taxes. Both the principal and interest payments affect after-tax cash flows.

There are four common (and many less common) ways in which money can be repaid. First is the periodic payment of interest over the stipulated repayment period with the entire principal being repaid at the end of that time. The second requires a periodic payment that uniformly repays the principal and also covers the periodic interest. In this method, the payments decrease as the interest on the unrepaid principal decreases. Third is the method requiring a uniform periodic payment for the sum of principal plus interest. In each payment the proportion of principal gradually increases as the proportion of interest decreases. The fourth method repays nothing, neither interest nor principal, until the end of a specified period.

Loan interest rates have varied widely in recent years. For example, during 1 calendar year, the prime interest rate charged by banks to their best corporate customers varied in the range from about 11% to 22%. Since neither the banks nor the customers want to get "stuck" in an unfavorable long-term agreement, loans with adjustable interest rates have become commonplace. For example, a typical small business loan might span 1 or 2 years, with interest adjusted to the current market rate and paid monthly, and with the entire principal due at the end of the loan period.

Example 6.21

Let us illustrate the four basic plans for repaying principal and interest on borrowed money. Assume that a business borrows $40,000 to be used in financing an alternative, and the interest rate on this loan is 18% compounded annually. The stipulated repayment period is 5 years.

A summary of all relevant components of our example is presented in Table 6.23. In method 1, the interest equals $40,000(.18) = $7200/year. Only the interest is paid, and only the principal of $40,000 is owed after each year's payment. Method 2 repays the principal in equal amounts of $40,000/5 = $8000 as well as the interest for year t, which is given by [$40,000 − $8000($t$ − 1)]0.18. Clearly, the interest payment, total payment, and total money owed after yearly payments are a decreasing gradient series. Method 3 requires equal annual total payments. This annual payment is equal to $40,000($A|P$ 18,5) = $12,791.11. The principal component of this annual payment for year t can be found quickly as $12,791.11($P|F$ 18,5 − t + 1), using the method given in Chapter Three. In method 4, the interest accrued during each year is added to the principal such that the total amount owed after t years is $40,000(1.18)^t. When payment is made at the end of year 5, everything over $40,000 is considered interest.

We have seen that the interest on borrowed money is deductible for tax purposes, whereas the principal repayment does not enter into taxable income. In addition, both the interest and principal portions of a payment are real and must be taken into account when calculating cash flows.

Example 6.22

Let us illustrate the effect of borrowed money by recalling the stand-alone computer of Example 6.2 having P = $82,000, F = $5000, a useful life of 7 years, and before-tax and loan cash flows of $23,500. Assume that $40,000 of the $82,000 paid for the computer is through debt funding. The loan is to be repaid in equal annual installments (method 3) at 18% over 5 years. The remaining $42,000 will be equity money. Applicable depreciation and loan activity (method 3) have previously been calculated in Examples 6.2 and 6.21, respectively. The resulting after-tax cash flow profile is detailed in Table 6.24.

Calculating the present worth of the after-tax cash flows at i = 15%, we have

TABLE 6.23

Illustration of Four Common Methods of Principal and Interest Repayment

End of Year	Interest Accrued During Year	Total Money Owed before Yearly Payment	Interest Payment	Principal Payment	Total Payment	Total Money Owed after Yearly Payment
Method 1						
0						$40,000.00
1	$ 7,200.00	$47,200.00	$ 7,200.00	$ 0	$ 7,200.00	40,000.00
2	7,200.00	47,200.00	7,200.00	0	7,200.00	40,000.00
3	7,200.00	47,200.00	7,200.00	0	7,200.00	40,000.00
4	7,200.00	47,200.00	7,200.00	0	7,200.00	40,000.00
5	7,200.00	47,200.00	7,200.00	40,000.00	47,200.00	0
Method 2						
0						$40,000.00
1	$ 7,200.00	$47,200.00	$ 7,200.00	$ 8,000.00	$15,200.00	32,000.00
2	5,760.00	37,760.00	5,760.00	8,000.00	13,760.00	24,000.00
3	4,320.00	28,320.00	4,320.00	8,000.00	12,320.00	16,000.00
4	2,880.00	18,880.00	2,880.00	8,000.00	10,880.00	8,000.00
5	1,440.00	9,440.00	1,440.00	8,000.00	9,440.00	0

Method 3

0		$47,200.00				$40,000.00
1	$ 7,200.00	40,602.49	$ 7,200.00	$ 5,591.11	$12,791.11	34,408.89
2	6,193.60	32,817.42	6,193.60	6,597.51	12,791.11	27,811.37
3	5,006.05	23,631.04	5,006.05	7,785.07	12,791.11	20,026.31
4	3,604.73	12,791.11	3,604.73	9,186.38	12,791.11	10,839.93
5	1,951.19		1,951.19	10,839.93	12,791.11	0

Method 4

0		$47,200.00				$40,000.00
1	$ 7,200.00	55,696.00	$ 0	$ 0	$ 0	47,200.00
2	8,496.00	65,721.28	0	0	0	55,696.00
3	10,025.28	77,551.11	0	0	0	65,721.28
4	11,829.83	91,510.31	0	0	0	77,551.11
5	13,959.20		51,510.31	40,000.00	91,510.31	0

TABLE 6.24

After-Tax Cash Flow Profile Using MACRS Depreciation Deductions and $40,000 Borrowed Money at 18%, Example 6.22

End of Year, A	Before-Tax and Loan Cash Flow, B	Loan Principal Payment, C	Loan Interest Payment, D	MACRS Deduction, E	Taxable Income B − D − E, F	Taxes F × 0.34, G	After-Tax Cash Flow B − C − D − G, H
0	−$82,000.00	−$40,000.00					−$42,000.00
1	23,500.00	5,591.11	$7,200.00	$16,400.00	−$100.00	−$34.00	10,742.89
2	23,500.00	6,597.51	6,193.60	26,240.00	−8,933.60	−3,037.42	13,746.31
3	23,500.00	7,785.07	5,006.05	15,744.00	2,749.95	934.98	9,773.90
4	23,500.00	9,186.38	3,604.73	9,446.40	10,448.87	3,552.62	7,156.27
5	23,500.00	10,839.93	1,951.19	9,446.40	12,102.41	4,114.82	6,594.06
6	23,500.00	0.00	0.00	4,723.20	18,776.80	6,384.11	17,115.89
7	23,500.00	0.00	0.00	0.00	23,500.00	7,990.00	15,510.00
7	5,000.00 (salvage)			0.00	5,000.00 (depreciation recapture)	1,700.00	3,300.00

$$PW(15) = -\$42,000 + \$10,742.89(P|F\ 15,1) + \$13,746.31(P|F\ 15,2)$$
$$+ \cdots + (\$15,510 + \$3300)(P|F\ 15,7) = \$6003.41$$

The value of our effectiveness measure jumped to \$6003.41 for this particular example. Note that the present worth calculation on cash flows was made using a discount rate equal to our 15% *MARR*. The 18% loan rate was used *only* in sizing the loan repayments. If we actually need to borrow the \$40,000 to implement the project, our estimates indicate that a handsome monetary return will be received on the \$42,000 equity investment. If, on the other hand, we really have at least the other \$40,000 available, borrowing allows us to invest that money as equity capital in another alternative that will earn a return at least equal to the *MARR*.

A relatively high rate of interest, 18%, was used for the loan in Example 6.22 for two reasons. First, it appears illogical for company management to borrow at 18% when they require only a 15% *MARR*. Remember though, the 15% *MARR* is an *after-tax* rate of return. Since loan interest is a deductible item, it is being paid with *before-tax* dollars. The after-tax interest rate or "cost of capital" for the loan is 18% × (1 − tax rate) = 18(1 − 0.34) = 11.88%. The second reason is to illustrate that borrowed money, even at relatively high rates, can often "leverage" equity capital such that a nice return is earned, as shown in Example 6.22.

We cannot conclude from the example that borrowing money is always favorable. The desirability of borrowed funds depends on the terms of the loan, including method of repayment, interest, and repayment period. Furthermore, collateral, which may be lost (after legal action) if the principal and interest cannot be paid on schedule, is frequently required. In summary, each alternative investment and financing strategy should be compared on its own merits.

We have used the same basic example to illustrate the implications of depreciation and financing strategies thus far throughout the chapter. The present worth of after-tax cash flows was seen to vary from −\$1778.68 using equity financing under alternate MACRS over 9 years, to \$6003.41 using a more sophisticated mix of debt and equity financing and the MACRS with a 5-year recovery period. Nothing about the basic incomes or costs relating to the asset were changed. By now the reader should have a good idea how to calculate and assess after-tax cash flows under differing depreciation schedules and financing arrangements.

*6.14 CAPITAL GAINS, LOSSES, AND TAX TREATMENT

The Tax Reform Act of 1986 repealed the favored tax treatment accorded capital gains in the past. Capital gains had been receiving preferential tax rates, being less severely taxed than ordinary income. It might, therefore, be argued that this section should be deleted from the text. Knowing how the blowing winds, the shape of the moon, and the re-election vote sway the minds in Congress, it is probably best to address the topic because it *is* important if the capital gains break is reinstated, an event likely to occur.

*6.14.1 Capital Gains and Losses

Examples of *capital assets* are stocks, stock rights, bonds, and stock received as a dividend. Real property and property used for the production of income, not used in a trade or business, are capital assets. Patents held for investment, inventions in the hands of the inventor, life estates, inherited jewelry, bank accounts, and cotton acreage allotments, for example, are also considered capital assets. Gold, silver, stamps, coins, gems, and so on, expand the list.

Specifically *excluded* are the things of most interest in a corporate engineering economic analysis—real property and depreciable property used in a trade or business! That is, machinery, plant, and equipment are not capital assets. Instead, they are classified as Section 1231 property (defined in the next section), which in the past has received the favored capital asset treatment. This explains our need to know about pre-1987 capital asset treatment.

It is interesting, however, to note that although corporate capital gain has lost its special tax status, the tax code's capital-gains statutory structure is retained *to make it easier to reinstate a capital gains rate differential if there is a future increase in the tax rates* [5]. There is considerable speculation that future tax rates will increase above 34%, but that the 34% rate will remain fixed for certain capital gains. This explains our potential need to know about capital asset treatment in the future.

Before applying capital treatment, the extent of the capital gain or loss must be determined. This requires a balancing of long- and short-term gains and losses. The excess of selling price over cost basis of a capital asset is a *capital gain*. If the selling price is less than the cost basis, then a *capital loss* has occurred. If the period during which the taxpayer holds the asset is 6 months or less, the gain or loss is referred to as a *short-term gain or loss*. If the asset is held longer than 6 months, a *long-term gain or loss* has occurred. A *net short-term gain* (*loss*) is the sum of the short-term gains minus the loss of short-term losses. Similarly, a *net long-term gain* (*loss*) is the sum of long-term gains minus long-term losses.

Example 6.23

A corporation has the following capital gains and losses during a tax year. The net short- and long-term gains and losses are summarized as follows:

Short-term capital gains	$19,300
Short-term capital losses	($27,600)
Net short-term capital loss	($ 8,300)

Long-term capital gains	$77,500
Long-term capital losses	($18,750)
Net long-term capital gain	$58,750

*6.14.2 Tax Treatment of Capital Assets

Short- and long-term capital gains may be consolidated for tax purposes. The result may be a net capital gain, or a net capital loss. The consolidated result dictates whether the tax treatment will be capital gains, ordinary income, or a loss to be carried back or over. The consolidation rules and tax treatments are summarized in Table 6.25.

Example 6.24

In Example 6.23 we determined that a corporation has a net short-term capital loss of $8300 and a net long-term capital gain of $58,750. Consolidating these in accordance with Table 6.25, we have a net capital gain of $58,750 − $8300 = $50,450 that is taxed as a capital gain.

TABLE 6.25

Results and Corporate Tax Treatment of Consolidated Long- and Short-Term Capital Gains and Losses

Net Short-Term Capital Gains and Losses	Net Long-Term Capital Gains and Losses		
	None	Long-Term Gain	Long-Term Loss
None	No capital treatment	Net capital gain taxed as capital gain	Net capital loss may be carried back or over
Short-term gain	Net capital gain taxed as ordinary income	Net capital gain. Long-term gain taxed as capital gain *and* short-term gain taxed as ordinary income	If a net capital loss, may be carried back or over. If a net capital gain, taxed as ordinary income
Short-term loss	Net capital loss may be carried back or over	If a net capital gain, taxed as capital gain. If a net capital loss, may be carried back or over	Net capital loss may be carried back or over

Note: Capital losses carried back and over are treated as short-term capital losses. The Tax Reform Act of 1986 caused "taxed as a capital gain" to be the same as "taxed as ordinary income."

As this chapter is written, capital gains are treated for tax purposes as ordinary income; total income is then taxed at up to 34% as illustrated in Section 6.10. Prior to 1987, however, the excess of net long-term capital gains over net short-term capital losses was excluded and taxed at a rate of only 28%, a lot lower than the pre-1987 ordinary income rates of up to 46%. Actually, the taxpayer had the choice of treating capital gains as either ordinary income (the *regular method*) or segregating the net capital gain to be taxed at 28% (the *alternate method*). This choice was useful when taxable income was sufficiently low that the regular tax rate (e.g., 15%) was less than the capital gains rate of 28%.

Example 6.25

Suppose a corporation has a $400,000 taxable ordinary income, plus $10,000 that is a net capital gain taxed as a capital gain. What would be the tax on the net capital gain under *current* law?
Answer: $10,000(0.34) = $3400.
What would be the tax on the net capital gain under *pre-1987* law?
Answer: Alternative 1—$10,000(0.46) = $4600; Alternative 2—$10,000(0.28) = $2800. Obviously, Alternative 2 would have been selected, saving $1800 in taxes.

Example 6.26

Suppose a corporation has a $40,000 taxable ordinary income, plus $10,000 that is a net capital gain taxed as a capital gain. What would be the tax on the net capital gain under *current* law?
Answer: $10,000(0.15) = $1500.

When capital losses are subtracted from capital gains and the result is negative, Table 6.25 shows that the net loss may be carried back or over. The carry-back period is 3 years and the carry-forward period is 5 years, during which time the capital loss is treated as a short-term capital loss and can be used to offset capital gains. Capital losses cannot be used to offset ordinary income.

6.15 DISPOSITION OF DEPRECIABLE PROPERTY

A disposition is the permanent withdrawal of property from use in trade or business or in the production of income. A withdrawal may be made by sale, exchange, retirement, abandonment, or destruction. There is generally a gain or loss to be recognized on the disposition of an asset by sale.

6.15.1 Section 1231 Gains and Losses and Depreciation Recapture

Much of the law regarding disposition refers to Section 1231 property. *Section 1231 property* includes property used in a trade or business or held for the production of rents or royalties and held for more than 6 months. Property used in trade or business includes real property and depreciable personal property, but not property held for sale to customers. Section 1231 property includes Sections 1245 and 1250 class property, referred to in section 6.5, plus some other assets. As such, *nearly any property we would be concerned with in an engineering economic analysis is Section 1231 property*.

When a property is disposed of, a gain occurs if the selling price exceeds the unrecovered investment, adjusted basis, or book value of the asset. If the reverse is true, the result is a loss. All gain on the disposition of tangible property, other than residential rental or nonresidential real, is treated as ordinary income so long as it does not exceed the total depreciation taken on the asset. This is often called *depreciation recapture*. Any gain in excess of depreciation taken is called a *Section 1231 gain*. A loss on such property is called a *Section 1231 loss*. This treatment applies to assets in the 3-, 5-, 7-, 10-, 15-, and 20-year property classes under the MACRS.

Gain on a disposition of residential rental or nonresidential real property is not recaptured as ordinary income. Rather, any gain or loss from disposition of such property is a Section 1231 gain or loss, respectively.

To help understand Section 1231 gains and losses and depreciation recapture, two figures are presented in which the following notation is used:

P = cost basis or original investment.
B_t = unrecovered investment at time t; this is the cost basis less allowable MACRS deductions.
F_t = salvage value actually received at time t.
t = any time, during or beyond the recovery period.

Depreciation recapture and Section 1231 gains and losses are summarized in Figures 6.4 and 6.5 in conjunction with Table 6.26.

6.15.2 Tax Treatment of Depreciable Properties

Gains and losses upon disposition of property can have a significant effect upon taxes paid. A special procedure is used to determine how dispositions will be treated for tax purposes in a given year. First, all Section 1231 gains and losses are grouped. Then, if the gains do not exceed the losses, each individual gain and loss is treated as if it were from the sale of a long-term capital asset. If the gains exceed the losses, an overall gain is recognized for that year. The net Section 1231 gain is then treated as ordinary income to the extent that it does not exceed nonrecaptured losses beginning after 1981. Nonrecaptured net Section 1231 losses are net Section 1231 losses deducted in tax years beginning after 1981 that have not yet been applied against any

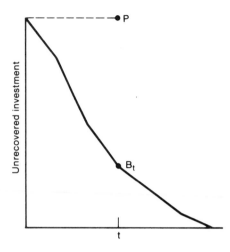

FIGURE 6.4. Unrecovered investment versus time for tangible property, other than residential rental or nonresidential real.

net Section 1231 gains for tax years beginning after 1984. Obviously, the tax law gets very complicated! Unfortunately, this is one of the clearer parts of the law.

Since it is impossible to know the performance of all Section 1231 assets in advance, a reasonable rule is necessary in order to treat dispositions of property in engineering economic analyses. Therefore, unless better information is available, Section 1231 losses are treated as ordinary losses for tax purposes, while Section 1231 gains are treated as capital gains. Of course, as noted in Section 6.14, at the time this book is written, capital gains receive no favored treatment; therefore, Section 1231 gains are now effectively treated and taxed as ordinary income. Should ordinary income tax rates rise (e.g., say to 40%), leaving capital gains to be taxed at a more favorable rate (e.g., say left at 34%), this favorable treatment would likely carry over to Section 1231 gains as well.

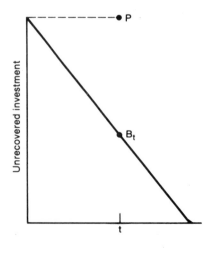

FIGURE 6.5. Unrecovered investment versus time for residental rental and non-residential real property.

TABLE 6.26

Rules for Gains and Losses on Depreciable Properties

Section 1231 depreciable property, other than residential rental or nonresidential real.

Rules (see Figure 6.4)

If	Then
$F_t > P$	$F_t - P$ is Section 1231 gain, and
	$P - B_t$ is depreciation recapture treated as ordinary income.
$B_t < F_t < P$	$F_t - B_t$ is depreciation recapture treated as ordinary income.
$F_t < B_t$	$B_t - F_t$ is Section 1231 loss.

Section 1231 residential rental or nonresidential real property.

Rules (see Figure 6.5)

If	Then
$B_t < F_t$	$F_t - B_t$ is Section 1231 gain.
$F_t < B_t$	$B_t - F_t$ is Section 1231 loss.

Example 6.27

In Examples 6.2 and 6.18, the stand-alone computer's cost basis $P = \$82,000$ was fully recovered under the MACRS by the end of year 6. How should the $5000 salvage value received at year 7 be treated for tax purposes.

The computer is tangible, depreciable, 5-year property. Referring to Tables 6.20 and 6.26 and Figure 6.4, $B_7 = \$0$ and $F_7 = \$5000$. Therefore, $F_7 - B_7 = \$5000$ is depreciation recapture and should be taxed as ordinary income in the amount of $\$5000(0.34) = \$1700$.

Example 6.28

Let us reconsider Example 6.19 in which we used alternate MACRS depreciation over a 9-year recovery period. Since the useful life of the asset was 7 years, only $54,666.67 of the cost basis $P = \$82,000$ had been recovered through depreciation. How should the disposal of the asset in year 7 be handled?

The asset is used in the manufacture of medical and dental supplies and is 5-year property. Referring to Tables 6.21 and 6.26 and Figure 6.4, $B_7 = \$82,000 - \$54,666.67 = \$27,333.33$ and $F_7 = \$5000$. Therefore, $B_7 - F_7 = \$27,333.33 - \$5000 = \$22,333.33$ is a Section 1231 loss. Since we have no other information regarding other Section 1231 gains and losses, this loss will be assumed to apply against ordinary income. As such, there should be a tax of $-\$22,333.33(0.34) = -\$7593.33$, or a negative tax to apply against other positive taxes in year 7. This agrees with the treatment in Example 6.19.

Example 6.29

Although unrealistic, suppose our computer of Example 6.18 having $P = \$82,000$, a useful life of 7 years, and MACRS recovery over 6 years is sold at $t = 7$ for $100,000. How should this salvage value be treated for tax purposes?

The computer is 5-year property. Referring to Table 6.26 and Figure 6.4, $B_7 = \$0$ and $F_7 = \$100,000$. Therefore, $F_7 - P = \$100,000 - \$82,000 = \$18,000$ which is a Section 1231 gain. Since we have no information on other Section 1231 gains and losses, this gain will be assumed to be taxed as a capital gain. The amount $P - B_7 = \$82,000$, represents depreciation recapture and is taxed as ordinary income. Thus, the total tax should be $\$18,000(0.34) + \$82,000(0.34) = \$34,000$. Note that the tax rate is 34% on each amount. In past years, when capital gains received favorable treatment, the $18,000 would have been taxed at a special low capital gains rate.

Example 6.30

Consider the residential rental property of Examples 6.4 and 6.5, placed in service on May 1, in which $P = \$220,000$. After 7 years and 5 months (this will be in the 8th tax year), on October 1, the property is sold for $150,000. How should this sale be taxed?

This is 27.5-year residential rental property. Referring to Table 6.6 and the calculations in Example 6.5, total depreciation to date is $\$5000.60 + \$7999.20(6) + \$6332.70 = \$59,328.50$. The unrecovered investment is $B_8 = \$220,000 - \$59,328.50 = \$160,671.50$. So, $F_8 - B_8 = \$150,000 - \$160,671.50 = -\$10,671.50$ which is a Section 1231 loss. Lacking information about other Section 1231 property, we will tax the $-\$10,671.50$ as an ordinary income loss, resulting in $-\$10,671.50(0.34) = -\$3628.31$ tax.

6.16 INVESTMENT TAX CREDIT

The *investment tax credit* (*ITC*) is designed to stimulate investment by providing reduced taxation in the year in which an asset is placed in service. The ITC has had a turbulent history dating back to 1962. It has been on-again, off-again during that time, with varying rules, percentages, and so forth. *At the time this book is written, the ITC has been repealed for property placed in service after 1985.* It was argued that with a lowering of the tax rate from 46% to 34%, a further tax break such as the ITC should be discontinued. Almost assuredly, the ITC will reappear when legislators believe it is in the best economic interest of the U.S., and/or their re-election efforts. For purposes of introducing readers to the concept of an ITC, the rules and percentages as they existed prior to the Tax Reform Act of 1986 will be presented.

The investment tax credit permitted taxpayers to claim a credit, not just a deduction, against income taxes for qualified investment in certain new or used depreciable property. The regular credit allowable was 10% of the eligible investment. The amount of eligible pre-1986 investment is summarized in Table 6.27.

The ITC was limited to $25,000 plus 85% of the tax liability over $25,000. The ITC was claimed for the taxable year in which the property was placed in service. Unused credits for a tax year could be carried back 3 tax years and then carried forward up to 15 tax years. If an asset was disposed of prior to completion of its recovery period, or its useful life if nonrecovery property, the ITC might need to be refigured. In this case, we should consult the tax law.

The ITC was available only for depreciable recovery property under the ACRS, or other depreciable property having a useful life of at least 3 years. It also had to be tangible personal property (other than an air-conditioning or space-heating unit), or other tangible property (except buildings and their structural components), used as an integral part of manufacturing, production, extraction, and so on. Further, it had to be placed in service during the year in a trade or business or for the production of income, and be used predominantly in the United States.

There was no limit on the amount of qualifying new property eligible for the ITC. The maximum amount of a taxpayer's investment in used property that was eligible for the ITC was $125,000.

TABLE 6.27

Amount of Pre-1986 Investment Eligible for ITC

ACRS Recovery Property Class	Proportion Eligible
3-year	60%
5-year	100%
10-year	100%
15-year public utility	100%
15-, 18-, 19-year real	None

Non-ACRS Recovery Property Life	Proportion Eligible
0, 1, 2	None
3, 4	1/3
5, 6	2/3
7 or more	100%

If the full ITC was taken, law required that half the amount of the ITC be used to reduce the cost basis of the asset, thereby lessening the allowable cost recovery. For example, if the 10% ITC was claimed, the cost basis of the asset was to be reduced immediately by 5%. For 3-year property, the ITC was only 6%, and the basis needed only to be reduced by 3%. As an alternative, if the taxpayer was willing to reduce the ITC from 10% to 8% or from 6% to 4%, the cost basis needed not be reduced at all.

Example 6.31

Suppose a firm invested in 1983 $1,500,000 in elaborate robotics equipment, $740,000 in an additional building and associated real property, and $40,000 in a fleet of automobiles. Tax liability for the year was $120,000. Applying the ITC, what was the reduction in tax liability?

The 3 properties were ACRS 5-year, 15-year real, and 3-year recovery property. The 15-year real property did not qualify for the ITC.

$$\text{Investment credit} = \$1,500,000(0.10) + \$40,000(0.06) = \$152,400$$

$$\begin{aligned}\text{Investment credit used} &= \$25,000 + (\$120,000 - \$25,000)(0.85) \\ &= \$105,750.\end{aligned}$$

Note that the tax credit of $152,400 - $105,750 = $46,650 could then be carried backward or forward to apply against taxes in other years.

Example 6.32

Let us suppose that the ITC is immediately reinstated exactly as described above, and we desire to take advantage of it with our stand-alone computer of Examples 6.2 and 6.18. Recall that $P = \$82,000$, $F = \$5000$, useful life is 7 years, before-tax cash flow is $23,500, and $i = 15\%$. What will be the present worth of after-tax cash flows?

The ITC will be $82,000(0.10) = $8200. The cost basis must be reduced by $4100, to $77,900. Table 6.28 summarizes the cash flow calculations. The present worth of the after-tax cash flows is $9178.85. Recall that the present worth in Example 6.18 without the ITC was only $3010.50.

It is obvious that the ITC can be a major economic factor for companies of any size. There will likely be extreme pressure on lawmakers to reinstate the ITC.

6.17 SECTION 179 EXPENSE DEDUCTION

The Section 179 expense deduction is designed to stimulate investment, primarily by smaller firms, by providing reduced taxation in the first year of

TABLE 6.28

After-Tax Cash Flows Using the Pre-1986 Investment Tax Credit

End of Year, A	Before-Tax Cash Flow, B	MACRS Deduction, C	Taxable Income B − C, D	Tax D × 0.34, E	Investment Tax Credit, G	After-Tax Cash Flow B − E + G, H
0	−$82,000.00				$8,200.00	−$82,000.00
1	23,500.00	$15,580.00	$7,920.00	$2,692.80	0.00	29,007.20
2	23,500.00	24,928.00	−1,428.00	−485.52	0.00	23,985.52
3	23,500.00	14,956.80	8,543.20	2,904.69	0.00	20,595.31
4	23,500.00	8,974.08	14,525.92	4,938.81	0.00	18,561.19
5	23,500.00	8,974.08	14,525.92	4,938.81	0.00	18,561.19
6	23,500.00	4,487.04	19,012.96	6,464.41	0.00	17,035.59
7	23,500.00	0.00	23,500.00	7,990.00	0.00	15,510.00
7	5,000.00 (salvage)		5,000.00 (depreciation recapture)	1,700.00		3,300.00

an asset's life. It was introduced in 1981 to begin with tax year 1982 and replaced the pre-1981 "additional first-year depreciation" that amounted to 20% of up to $10,000 of qualifying investment. It was strengthened by the Tax Reform Act of 1986.

Taxpayers may elect to treat the cost of certain property as an expense rather than as a capital expenditure. This property, called Section 179 property, must be recovery property acquired by purchase for use in a trade or business, and meet the eligibility criteria for the pre-1986 investment tax credit. The election to expense must take place in the year the property is placed in service. The expense deduction is limited to $10,000 for qualifying property placed in service after 1986. The deduction had previously been limited to $5000. The property's cost basis, P, must be reduced by the amount expensed.

There are some limits placed on the expense deduction. The $10,000 limit applies to each taxpayer, not to each asset. Also, the amount eligible to be expensed for any tax year is reduced, dollar for dollar, by the amount that the aggregate cost of qualifying property exceeds $200,000. That is, if the amount of eligible property placed in service during a tax year is $200,000, $204,000, or $210,000, the maximum amount that may be expensed is $10,000, $6000, or $0, respectively. Further, the amount eligible to be expensed cannot exceed the taxable income derived from the trade or business in which the property is used. Finally, conversion of the property to non-business use before the end of the property's recovery period results in recapture as ordinary income of the excess of the amount expensed over the MACRS deduction that would have been allowed.

Example 6.33

Let us replay Example 6.32, except that we will use the Section 179 expense deduction instead of the ITC. Recall that $P = \$82,000$, $F = \$5000$, useful life is 7 years, before-tax cash flow is $23,500, and $i = 15\%$. Assume that this is the company's only asset purchased and placed in service during the tax year. Also, assume that the company's taxable income is well above $10,000. What will be the present worth of after-tax cash flows?

The Section 179 expense deduction will be $10,000. The cost basis is therefore reduced to $72,000. Table 6.29 summarizes the cash flow calculations. The present worth of after-tax cash flows is $3620.46.

6.18 CARRY-BACK AND CARRY-FORWARD RULES

In several examples we saw that an alternative's taxable income and tax can be negative for a particular year. It was implicitly assumed that those values were used to offset positive tax liabilities from other corporate activities. When there are insufficient incomes or taxes to be offset, we can carry back

TABLE 6.29

After-Tax Cash Flows Using the Section 179 Expense Deduction

End of Year, A	Before-Tax Cash Flow, B	Section 179 Expense Deduction, C	MACRS Deduction, D	Taxable Income B − C − D, E	Tax E × 0.34, F	After-Tax Cash Flow B − F, G
0	−$82,000.00					−$82,000.00
1	23,500.00	$10,000.00	$14,400.00	−$900.00	−$306.00	23,806.00
2	23,500.00	0.00	23,040.00	460.00	156.40	23,343.60
3	23,500.00	0.00	13,824.00	9,676.00	3,289.84	20,210.16
4	23,500.00	0.00	8,294.40	15,205.60	5,169.90	18,330.10
5	23,500.00	0.00	8,294.40	15,205.60	5,169.90	18,330.10
6	23,500.00	0.00	4,147.20	19,352.80	6,579.95	16,920.05
7	23,500.00	0.00	0.00	23,500.00	7,990.00	15,510.00
7	5,000.00 (salvage)		0.00	5,000.00 (depreciation recapture)	1,700.00	3,300.00

or carry forward ordinary income losses and capital losses. We used to be able to carry the investment credit to past or future years' returns.

Ordinary income losses can be carried back 3 years, beginning with the earliest. If not entirely used to offset income in that year, it is carried to the second year back and so on. After applying losses to all 3 back years, remaining losses can be carried forward to as many as 15 years following the loss year. Capital losses may be carried back 3 years and forward 5 years. They are treated as short-term capital losses and used to offset capital gains. They cannot be used to offset ordinary income. The investment credit used to be carried back 3 years and forward 15 years to be applied directly against taxes on those returns.

6.19 AFTER-TAX ANALYSES UNDER INFLATION

We were introduced to inflation in Section 3.9. We learned that there are two procedures which may be used to perform economic analyses under inflation:

1. Express all cash flows in terms of their "then-current" amounts and use an interest rate (i) that combines the real discount rate (d) and the inflation rate (j): $i = d + j + dj$.

2. Express all cash flows in terms of their "constant worth" amounts and use the real discount rate (d) alone without an inflation rate component.

While these two procedures seem simple enough, it is easy to err in an after-tax economic analysis under inflation. A series of examples will illustrate the problems.

Example 6.34

We will again consider our computer example, but with some modification. The first cost of the computer installation is still $82,000 with an estimated salvage value of $5000 after a useful life of 7 years. Recall that a computer is 5-year property, and we will use regular MACRS depreciation. Again, MARR = 15%.

In past examples, the reduced operating expenses realized by purchasing the computer have been $23,500/year. Therefore, the before-tax cash flow column has shown this figure for each year of the 7-year useful life of the computer. Now, suppose we have reconsidered and changed the estimate of reduced operating expenses. We now believe they will be smaller than originally projected, having a value "today" of $15,500, but increasing at a 10% ($g = .10$) rate over the 7-year horizon. That is, the then-current before-tax cash flow will be $BTCF(t) = \$15,500(1.10)^t$, equivalent to Equation 3.60.

Up to now, we have given MARR = 15% little thought. Since we are considering then-current cash flows, it is apparent that MARR must be con-

TABLE 6.30

After-Tax Cash Flows Using Then-Current Procedure under Inflation, Example 6.34

End of Year, A	Before-Tax Cash Flow, B	MACRS Deduction, C	Taxable Income B − C, D	Tax D × 0.34, E	After-Tax Cash Flow B − E, F
0	−$82,000.00				−$82,000.00
1	17,050.00	$16,400.00	$650.00	$221.00	16,829.00
2	18,755.00	26,240.00	−7,485.00	−2,544.90	21,299.90
3	20,630.50	15,744.00	4,886.50	1,661.41	18,969.09
4	22,693.55	9,446.40	13,247.15	4,504.03	18,189.52
5	24,962.91	9,446.40	15,516.51	5,275.61	19,687.29
6	27,459.20	4,723.20	22,736.00	7,730.24	19,728.96
7	30,205.12	0.00	30,205.12	10,269.74	19,935.38
7	5,000.00	0.00	5,000.00	1,700.00	3,300.00
	(salvage)		(depreciation recapture)		

sidered a combined rate that includes both the real discount rate and the inflation rate. Suppose the company's real discount rate is 8.49% and the projected inflation rate is 6.00%. Then, $i = d + j + dj = 0.0849 + 0.0600 + 0.0849(0.0600) = .1500$ or MARR $= 15\%$. It is this 15% that will be used to discount all after-tax cash flows since they will be in then-current dollars. The calculations are shown in Table 6.30.

The present worth of the after-tax cash flows, properly discounted at 15%, is −$1335.43. This means the computer is not a good investment, owing to the revised BTCF estimates.

Note that we did nothing new or different in Table 6.30 to reflect an analysis under inflation. *This is because we have implicitly assumed all analyses in this chapter to be based on then-current dollars with a MARR that reflects both the real discount rate and the inflation rate.*

To provide confidence that the then-current analysis is correct, consider a constant-worth analysis. It should provide the same decision and, in fact, the same present worth of the after-tax cash flows.

Example 6.35

This example will show an INCORRECT constant-worth analysis to illustrate how easy it is to make a mistake and even a wrong decision using this

TABLE 6.31

Incorrect After-Tax Cash Flows While Attempting to Use Constant-Worth
Procedure under Inflation, Example 6.35

End of Year, A	Before-Tax Cash Flow, B	MACRS Depreciation, C	Taxable Income B − C, D	Tax D × 0.34, E	After-Tax Cash Flow B − E, F
0	−$82,000.00				−$82,000.00
1	16,084.91	$16,400.00	−$315.09	−$107.13	16,192.04
2	16,691.88	26,240.00	−9,548.12	−3,246.36	19,938.24
3	17,321.77	15,744.00	1,577.77	536.44	16,785.33
4	17,975.42	9,446.40	8,529.02	2,899.87	15,075.55
5	18,653.73	9,446.40	9,207.33	3,130.49	15,523.24
6	19,357.65	4,723.20	14,634.45	4,975.71	14,381.94
7	20,088.13	0.00	20,088.13	6,829.96	13,258.16
7	3,325.29	0.00	3,325.29	1,130.60	2,194.69
	(salvage)		(depreciation recapture)		

analysis procedure. Suppose everything is exactly as described in Example
6.34—$P = \$82,000$, $F = \$5000$, useful life = 7 years, 5-year property, depreciation using MACRS, MARR = 15%, $d = .0849$, $j = .0600$, and then-current
BTCFs of $15,500(1.10)^t$. To use a constant-worth analysis requires that the
BTCF be expressed in constant-worth dollars as in Equation 3.61. That is,
$BTCF(t) = \$15,500(1.10)^t(1.06)^{-t}$. Table 6.31 summarizes the cash flow cal-
culations.

The present worth of the after-tax cash flow, properly discounted at the
rate $d = .0849$, is $1774.26. But, this is different than the present worth
determined in Example 6.34! Even worse, based on these results, we would
decide that the computer is a good investment—we would be making an
incorrect decision! What happened? What went wrong?

In a constant-worth analysis, all cash flows *and charges* must be expressed in constant-worth dollars. In Table 6.31, the depreciation charges
were left unchanged. They were expressed in terms of the exact amounts to
be subtracted from BTCFs in future years. *That is, depreciation charges*
were left in then-current dollars.

Example 6.36

This example will correct the errors made in Example 6.35. All details remain the same—$P = \$82,000$, $F = \$5000$, useful life = 7 years, 5-year

TABLE 6.32

Correct After-Tax Cash Flows Using Constant-Worth Procedure under Inflation, Example 6.36

End of Year, A	Before-Tax Cash Flow, B	MACRS Depreciation, C	Taxable Income B − C, D	Tax D × 0.34, E	After-Tax Cash Flow B − E, F
0	−$82,000.00				−$82,000.00
1	16,084.91	$15,471.70	$613.21	$208.49	15,876.42
2	16,691.88	23,353.51	−6,661.62	−2,264.95	18,956.84
3	17,321.77	13,218.97	4,102.80	1,394.95	15,926.81
4	17,975.42	7,482.43	10,492.98	3,567.61	14,407.80
5	18,653.73	7,058.90	11,594.84	3,942.24	14,711.49
6	19,357.65	3,329.67	16,027.98	5,449.51	13,908.14
7	20,088.13	0.00	20,088.13	6,829.96	13,258.16
7	3,325.29	0.00	3,325.29	1,130.60	2,194.69
	(salvage)		(depreciation recapture)		

property, depreciation using MACRS, MARR $= 15\%$, $d = 0.0849$, $j = 0.0600$, and constant-worth BTCFs of $\$15,500(1.10)^t(1.06)^{-t}$. This time, however, depreciation charges are converted to a constant-worth amount $D_t(1.06)^{-t}$. The cash flow calculations are given in Table 6.32.

The present worth of the after-tax cash flows, properly discounted using $d = .0849$, is $-\$1335.43$.

The present worth of $-\$1335.43$ is exactly the same amount realized with the then-current analysis in Example 6.34. Of course, this is as it should be, since the two methods of analysis are equivalent. Note that every entry in Table 6.32 is simply the corresponding entry in Table 6.30 divided by 1.06^t.

6.20 DEPLETION

Depletion is a gradual reduction of minerals, gas and oil, timber, and natural deposits. In a sense, depletion is closely akin to depreciation. The difference is that while a depleting asset is losing value by actually being removed and sold, a depreciable asset is losing value through wear, tear, and obsolescence in the manufacture of goods to be sold. Money recovered through the depletion allowance is likely to be used in the exploration and development of depletable assets, just as depreciation reserves are reinvested for new equipment. As in most tax-related matters, laws relating to lessors, lessees, royalties, and sales are complex and will probably require expert assistance.

However, the importance of depletion and the alternative methods that can be used for figuring the allowance will be illustrated here.

Cost depletion is the basic method of computing the depletion deduction. First, for any given year, the number of units (tons, barrels, board-feet, etc.) remaining in the property must be estimated. Then, the adjusted cost basis (first cost less depletion deductions taken previously) is divided by the number of units remaining. This quotient, multiplied by the number of units sold during the year, determines the depletion allowance. It is expressed as

$$\text{Cost depletion deduction} = \frac{\text{Adjusted cost basis} \times \text{Number of units sold}}{\text{Number of units remaining in property}}$$

This calculated deduction may or may not be used, depending on whether or not the percentage depletion allowance provides a larger deduction.

Percentage depletion provides an allowance equal to a percentage of the gross income from the mineral property. The applicable percentage depends on the type of property depleted. Table 6.33 gives some recent depletion percentages. The percentage depletion deduction may not exceed 50% of the taxable income *before* allowance for depletion. The total of allowances under percentage depletion is in no way limited by the adjusted basis of the property. Thus, even though the basis of depleted property may be reduced to $0.00, percentage depletion is still allowed. Cost depletion, however, stops when the adjusted basis reaches $0.00. Whichever of the two depletion methods results in the higher allowance in a given year must be used [6].

TABLE 6.33
Depletion Percentage for Some Minerals and Similar Resources[a]

Type	Depletion Percentage
Oil and gas (percentage depletion is not allowed, except for certain gas and oil production)	
a. Oil and gas—independent producers and royalty owners, with some limitations	15
b. Natural gas—regulated natural gas and gas sold under a fixed contract, with some limitations	22
Sulphur and uranium; and, if from U.S. deposits, asbestos, bauxite, chromite, graphite, mica, quartz crystals, cadmium, cobalt, lead, mercury, nickel, tin, zinc, and certain other ores and minerals	22
Gold, silver, oil shale, copper, and iron ore if from U.S. deposits	15
Numerous other minerals such as diatomaceous earth, granite, limestone, marble, and so forth	14
Coal, lignite, perlite, and sodium chloride	10
Clay and shale used for sewer pipe and brick	7½
Clay (for roofing tile, flower pots, etc.), gravel, peat, pumice, sand, and stone	5

[a] As of 1987.

TABLE 6.34

Depletion Allowance, Tax, and After-Tax Cash Flows for Oil Well, Example, 6.37

A End of Year	B Barrels Sold	C Gross Income	D Operating Cost	E Taxable Income Before Depletion C − D	F 50% of Taxable Income Before Depletion E × 0.5	G Percentage Depletion at 15% C × 0.15	H Cost Depletion	I Adjusted Cost Basis	J Barrels of Oil Remaining	K Taxable Income E − max{[min(F, G)], H}	L Income Tax K × 0.34	M After-Tax Cash Flow E − L
0				−$140,000				$140,000	48,000			$140,000
1	15,000	300,000	$75,000	225,000	$112,500	$45,000[a]	$43,750	95,000	33,000	$180,000	$61,200	163,800
2	13,000	260,000	104,000	156,000	78,000	39,000[a]	37,424	56,000	20,000	117,000	39,780	116,220
3	10,000	200,000	110,000	90,000	45,000	30,000[a]	28,000	26,000	10,000	60,000	20,400	69,600
4	6,000	120,000	84,000	36,000	18,000	18,000[a]	15,600	8,000	4,000	18,000	6,120	29,880
5	3,000	60,000	51,000	9,000	4,500	9,000	6,000[a]	2,000	1,000	3,000	1,020	7,980
6	1,000	20,000	20,000	0	0	3,000	2,000[a]	0	0	−2,000	−680	680

[a] Indicates depletion allowance used.

317

Example 6.37

We own a small independent oil company and have purchased the rights, drilled, and developed a 48,000-barrel oil well on NW/4 35/19N/1W, Payne County, Oklahoma, for $140,000. Operating expenses, based on past experience, are equal to $5 + 3(t - 1)$/barrel, where t is the year in which the oil is removed and sold. This increase in cost per barrel over time indicates the increased difficulty of recovering the oil as the field is depleted. Our geologist expects the field to last 6 years, yielding 15,000, 13,000, 10,000, 6000, 3000, and 1000 barrels of oil over that time. Each barrel will be worth $20 to our company. The cash flow calculations are given in Table 6.34.

Using the cost method, ($140,000/48,000)(15,000) = $43,750 will be our depletion allowance during year 1. From Table 6.33, our depletion percentage is seen to be 15%. This results in a $300,000(0.15) = $45,000 deduction in year 1, so this figure will be used since it is larger than $43,750.

Note that in year 5, cost depletion of $6000 is chosen. This is because the $9000 percentage depletion is limited to $4500, or 50% of taxable income before depletion.

6.21 SUMMARY

This chapter has presented the most important elements of depreciation and tax law as they pertain to economic analyses. It is clear that depreciation (or depletion) method, recovery period, financing, various deductions, and inflation can have significant effects on the desirability of making an investment. Many of these factors are law-related, and changes are being made daily. Therefore, in cases of uncertainty regarding depreciation and tax treatment, it is wise to seek competent advice.

Capital gains and losses, as well as the investment tax credit, are not a factor at the time this chapter is written. They are covered, however, in the expectation that they will reappear in some recognizable form.

BIBLIOGRAPHY

1. Commerce Clearing House, *1987 Depreciation Guide*, June 18, 1987.

2. Department of the Treasury, Internal Revenue Service, *Depreciation*, Publication 534, Rev. Dec. 86.

3. Department of the Treasury, Internal Revenue Service, *Tax Guide for Small Business*, Publication 334, Rev. Dec. 86.

4. Staff, Joint Committee on Taxation, "General Explanation of the Tax Reform Act of 1986," *Federal Taxes*, Bulletin 20 Extra, Prentice-Hall, May 11, 1987.

5. "A Complete Guide to the Tax Reform Act of 1986," *Federal Taxes*, Bulletin 45 Extra, Prentice-Hall, October 18, 1986.

6. Commerce Clearing House, Inc., *Standard Federal Tax Reporter—1987*, Volume 5, 1986.

PROBLEMS

1. For each of the following assets, state whether the asset is tangible/intangible property, personal/real property, and depreciable/nondepreciable property. (6.3)

   (a) A gas chromatograph for research.

   (b) An apartment complex.

   (c) A fast-food franchise.

   (d) A desk.

   (e) An office building.

   (f) A van.

   (g) A photocopying machine.

   (h) A sidewalk.

   (i) A dozer used in construction.

   (j) Liquified natural gas equipment.

2. For each of the assets named in Problem 1, state both the MACRS property class, if applicable, and the specific regular post-1986 depreciation method to be used. (6.4.1)

3. A mold for manufacturing medical supplies (5-year property, 9-year class life) is purchased at the beginning of the fiscal year for $20,000. The estimated salvage value after 10 years is $2000. Calculate the depreciation deduction and the resulting unrecovered investment during each year of the asset's life.

   (a) Use regular MACRS allowances. (6.4.2)

   (b) Use alternate MACRS allowances. (6.4.4)

4. A small truck is purchased for $12,000. The truck is expected to be of use to the company for 6 years, after which it will be sold for $2500. Calculate the depreciation deduction and the resulting unrecovered investment during each year of the asset's life.

   (a) Use regular MACRS allowances. (6.4.2)

   (b) Use alternate MACRS allowances (class life = 4 years, but must be treated as 5 years as specially noted by law). (6.4.4)

5. A digitally-controlled plane for manufacturing furniture (7-year property, 10-year class life) is purchased on April 1 by a calendar-year taxpayer for $120,000. It is expected to last 12 years and have a salvage value of $10,000. Calculate the depreciation deduction during years 1, 4, and 8.

   (a) Use regular MACRS allowances. (6.4.2)

   (b) Use alternate MACRS allowances. (6.4.4)

6. Two terminals, a CPU, and peripherals are purchased in December by a calen-

dar-year taxpayer for $40,000. The computer will be used for 7 years and be worth $5000 at that time. Calculate the depreciation deduction during years 1, 4, and 7.

(a) Use regular MACRS allowances. (6.4.2)

(b) Use alternate MACRS allowances (class life = 6 years, but must be treated as 5 years as specially noted by law). (6.4.4)

7. Material-handling equipment used in the manufacture of grain products (10-year property, 17-year class life) is purchased and installed for $250,000. It is placed in service in the middle of the tax year. If it is removed just *before* the end of the tax year approximately 4.5 years from the date placed in service, determine the depreciation deduction during each of the tax years involved.

(a) Use regular MACRS allowances. (6.4.2)

(b) Use alternate MACRS allowances. (6.4.4)

8. Repeat Problem 7 if the material-handling equipment is removed just *after* the end of the tax year.

(a) Use regular MACRS allowances. (6.4.2)

(b) Use alternate MACRS allowances. (6.4.4)

9. Public utility property (20-year property, 30-year class life) is placed in service at a cost of $400,000. It is expected to last 30 years with a salvage value of $20,000. Determine the depreciation deduction and the resulting unrecovered investment during each of the first 3 tax years.

(a) Use regular MACRS allowances. (6.4.2)

(b) Use alternate MACRS allowances. (6.4.4)

10. A business building (31.5-year property) is placed in service by a calendar-year taxpayer on January 4 for $300,000. Calculate the depreciation deduction for years 1 and 10, assuming the building is kept longer than 10 years.

(a) Use regular MACRS allowances. (6.4.3)

(b) Use alternate MACRS allowances. (6.4.4)

11. A rental apartment complex (27.5-year property) is placed in service by a calendar-year taxpayer on January 2 for $180,000. If the apartments are kept for 5 years and 2 months (sold on March 6), determine the depreciation deduction during each of the 6 tax years involved.

(a) Use regular MACRS allowances. (6.4.3)

(b) Use alternate MACRS allowances. (6.4.4)

12. A building used for the overhaul of dewatering systems is placed in service on October 10 by a calendar-year taxpayer for $240,000. It is sold almost 4 years later on August 15. Determine the depreciation deduction during years 1, 3, and 5, and the unrecovered investment at the end of years 1, 3, and 5.

(a) Use regular MACRS allowances. (6.4.3)

(b) Use alternate MACRS allowances. (6.4.4)

13. A rental house is placed in service by a calendar-year taxpayer during July for $60,000. It is sold in July, 3 years later. Determine the depreciation deduction and resulting unrecovered investment for each applicable year.

    (a) Use regular MACRS allowances. (6.4.3)

    (b) Use alternate MACRS allowances. (6.4.4)

*14. For each of the assets named in Problem 1, state both the ACRS recovery property class, if applicable, and state whether it is Section 1245 class property, Section 1250 class property, or neither. (6.5.1)

*15. Work Problem 3 assuming 1981–1986 placement in service and 5-year property.

    a. Use regular ACRS allowances. (6.5.2)

    b. Use the ACRS straight-line election and the minimum permissible number of years in the recovery period. (6.5.3)

*16. Work problem 4 assuming 1981–1986 placement in service and 3-year property.

    a. Use regular ACRS allowances. (6.5.2)

    b. Use the ACRS straight-line election and the minimum permissible number of years in the recovery period. (6.5.3)

*17. Work problem 7 assuming 1981–1986 placement in service and 5-year property.

    a. Use regular ACRS allowances. (6.5.2)

    b. Use the ACRS straight-line election and the minimum permissible number of years in the recovery period. (6.5.3)

*18. Work problem 8 assuming 1981–1986 placement in service and 5-year property.

    a. Use regular ACRS allowances. (6.5.2)

    b. Use the ACRS straight-line election and the minimum permissible number of years in the recovery period. (6.5.3)

*19. Work problem 10 assuming 1981–1986 placement in service and 15-year real property.

    a. Use regular ACRS allowances. (6.5.2)

    b. Use the ACRS straight-line election and the minimum permissible number of years in the recovery period. (6.5.3)

*20. Work problem 11 assuming 1981–1986 placement in service and 15-year real property.

    a. Use regular ACRS allowances. (6.5.2)

    b. Use the ACRS straight-line election and the minimum permissible number of years in the recovery period. (6.5.3)

*21. Work problem 3 assuming pre-1981 placement in service.

    a. Use straight-line depreciation. (6.6.1)

    b. Use sum of the years' digits depreciation. (6.6.2)

    c. Use double declining balance switching to straight-line depreciation. (6.6.3)

*22. Work problem 4 assuming pre-1981 placement in service.

    a. Use straight-line depreciation. (6.6.1)

    b. Use sum of the years' digits depreciation. (6.6.2)

    c. Use double declining balance switching to straight-line depreciation. (6.6.3)

*23. Work problem 5 assuming pre-1981 placement in service.

    a. Use straight-line depreciation. (6.6.1)

    b. Use sum of the years' digits depreciation. (6.6.2)

    c. Use double declining balance switching to straight-line depreciation. (6.6.3)

*24. A special tool used in the manufacture of cars (3-year property) is placed in service for $6000 and has a useful life of 4 years, after which there is $0 salvage value. Interest is 15%. Determine for each year the capital recovered, the return on capital unrecovered, and the capital recovered plus return. Also, find the annual equivalent capital recovered plus return.

    (a) Use regular MACRS allowances. (6.7.2)

    (b) Use regular ACRS allowances. (6.7.2)

    (c) Use pre-1981 straight-line depreciation. (6.7.2)

*25. A copier for use in a print shop (5-year property) is placed in service for $40,000 and has a useful life of 6 years, after which there is a $4000 salvage value. Interest is 15%. Determine for each year the capital recovered, the return on capital unrecovered, and the capital recovered plus return. Also, find the annual equivalent capital recovered plus return.

    (a) Use regular MACRS allowances. (6.7.2)

    (b) Use regular ACRS allowances. (6.7.2)

    (c) Use pre-1981 sum of the years' digits depreciation. (6.7.2)

26. An automatic control mechanism is estimated to provide 7000 hours of service during its life. The mechanism costs $14,400 and has a salvage value of $400 after 7000 hours of use. Its use is projected over a 4-year period as follows:

Year	Hours of Use
1	2000
2	1400
3	3000
4	600

Calculate the depreciation charge for each year using a method similar to the "operating day" method, but based on operating hours. (6.8.2)

27. A utility trailer costs $3000 and is rented out by the hour, day, or week. It is expected to depreciate to $0 by the time it has been rented out for a total of

$15,000 in gross income. If its forecasted annual revenues are $5000, $5000, $3500, and $1500, what will be the appropriate annual depreciation charges? (6.8.2)

28. What is the federal income tax for each of the following corporate taxable incomes?

   (a) $15,000 (6.9)
   (b) $66,000 (6.9)
   (c) $88,000 (6.9)
   (d) $180,000 (6.9)
   (e) $400,000 (6.9)

29. A consulting engineer is seeking a new project that will increase taxable income significantly. Determine the tax increase and the percentage of the new taxable income that goes to pay the tax increase. The present and additional new taxable incomes are, respectively:

   (a) $40,000 and $36,000 (6.9)
   (b) $88,000 and $36,000 (6.9)
   (c) $350,000 and $36,000 (6.9)

30. A dust collector (5-year property, 6-year class life) with a first cost of $50,000 is placed in service. It will be used for 8 years, have an annual gross income less operating expenses of $14,000 and will have no salvage value. The effective tax rate is 34%. Determine the after-tax cash flow for years 0, 1, 5, and 7.

   (a) Use regular MACRS allowances. (6.11)
   (b) Use alternate MACRS allowances. (6.11)
   *(c) Use regular ACRS allowances. (6.11)
   *(d) Use pre-1981 straight-line depreciation. (6.11)

31. A special tool for manufacturing fabricated metal products (3-year property, 3-year class life) costs $20,000 and has a salvage value of $1000 after a useful life of 5 years. The tool generates a net savings of $7000/year. The effective tax rate is 34%. Find the after-tax cash flow for each year.

   (a) Use regular MACRS allowances. (6.11)
   (b) Use alternate MACRS allowances. (6.11)
   *(c) Use regular ACRS allowances. (6.11)
   *(d) Use sum of the years' digits depreciation. (6.11)

32. A firm purchases a centrifugal separator (5-year property) for $20,000; it is estimated to have a life of 6 years. Operating and maintenance costs increase by $1000/year, with the cost for the first year being $2000. A $3000 salvage value is anticipated. Assuming a 34% income tax rate, and a 15% after-tax MARR, determine the annual equivalent after-tax cost of the separator. (*Hint:* First calculate the after-tax cash flows, letting gross income equal $0)

   a. Use regular MACRS allowances. (6.11)

    *b. Use regular ACRS allowances. (6.11)

    *c. Use pre-1981 double declining balance switching to straight-line depreciation. (6.11)

33. An investment (3-year property, 4-year class life) has the following net cash flows before taxes:

EOY	0	1	2	3	4
BTCF	−$70,000	42,000	44,000	28,000	23,000 + 20,000 (salvage)

The effective tax rate is 34% and the after-tax MARR is 20%. The company is profitable in its other activities. Find the after-tax cash flows and the present worth of those cash flows.

    a. Use regular MACRS allowances. (6.11)

    *b. Use regular ACRS allowances. (6.11)

    *c. Use pre-1981 double declining balance switching to straight-line depreciation. (6.11)

34. A firm may either invest $20,000 in a numerically controlled lathe (7-year property) that will last for 8 years and have a zero salvage value at that time, or invest $X in a methods improvement study (this is not depreciable). Both investment alternatives yield an increase in income of $10,000/year for 8 years. With regular MACRS allowances for the investment, and with a 34% tax rate, for what value of X will the firm be indifferent between the two investment alternatives? Assume an after-tax MARR of 15%. (6.11)

35. Two mutually exclusive alternatives, A and B (both 5-year property), are available. Alternative A requires an original investment of $80,000, has a useful life of 6 years, annual operating costs of $2000, and a salvage value in year k given by $80,000(0.80)^k$. Alternative B requires an original investment of $120,000, has a life of 8 years, zero annual operating costs, and a salvage value in year k given by $120,000(0.85)^k$. The after-tax MARR is 15%; a 34% tax rate is applicable. Regular MACRS allowances are used. Perform an annual worth comparison and recommend the least-cost alternative.

    (a) Use a planning horizon of 6 years. (6.11)

    (b) Use a planning horizon of 8 years. (6.11)

    (c) Use a planning horizon of 24 years. (6.11)

36. A high-precision robotic soldering machine is purchased for $60,000. Installation cost is $12,000 due to the extreme care and provisions required. It is estimated that the asset can be sold for $8000 after a useful life of 12,000 hours of operation; however, before selling, it must be removed at a cost of $2000. Inspection and maintenance cost $33/operating hour. It requires ¼ hour to solder one unit with the robot, and 16,000 units are soldered per year. The after-tax MARR is 15%, and the effective tax rate is 34%. Depreciation is based on operating hours.

a. What is the equivalent annual after-tax cost of the soldering machine? (6.11)

b. What is the cost of soldering per unit? (6.11)

37. Equipment for the manufacture of dental supplies (5-year property, 9-year class life) is purchased for $36,000 with an estimated salvage value of $6000 at the end of 10 years. Annual revenues and operating costs, excluding depreciation, are $38,000 and $23,000, respectively. Assume that MARR = 20% and the firm's tax rate is 34%. (6.12)

a. Determine the total undiscounted amount of depreciation deductions for the 10-year period if (1) regular MACRS allowances are used, and (2) alternate MACRS allowances are used.

b. Determine the present worth of the depreciation deductions described in (a).

c. Determine the total undiscounted amount of taxes paid for the 10-year period if (1) regular MACRS allowances are used, and (2) alternate MACRS allowances are used.

d. Determine the present worth of the taxes described in (c).

e. What do your answers to (a) and (c) suggest about the relationship between depreciation method, the total amount of depreciation allowed, and the total amount of taxes paid?

f. What do your answers to (b) and (d) suggest about the effect of the depreciation method used and the present worths of depreciation and taxes?

38. The Esteez company is considering the purchase of an asset for manufacturing apparel (5-year property, 9-year class life) that will cost $80,000, have a life of 10 years, and no salvage value. It will generate a gross income of $86,000/year. The annual cost of operating and maintaining the asset will be $60,000/year. The tax rate is 34% and MARR is 15% after taxes.

a. Determine the total undiscounted amount of depreciation deductions for the 10-year period if (1) regular MACRS allowances are used, and (2) alternate MACRS allowances are used.

b. Determine the present worth of the depreciation deductions described in (a).

c. Determine the total undiscounted amount of taxes paid for the 10-year period if (1) regular MACRS allowances are used, and (2) alternate MACRS allowances are used.

d. Determine the present worth of the taxes described in (c).

e. What do your answers to (a) and (c) suggest about the relationship between depreciation method, the total amount of depreciation allowed, and the total amount of taxes paid?

f. What do your answers to (b) and (d) suggest about the effect of the depreciation method used and the present worths of depreciation and taxes?

39. The Nazerousivious company plans to invest in a water purification system (5-year property) requiring $500,000 capital. The firm has $300,000 available and must borrow the additional $200,000 at an interest rate of 15% per year. The system will last 6 years with a salvage value of $0. The before-tax and loan cash flow for each of years 1 to 6 is $150,000. Only the loan interest is paid each year, and the entire principal is repaid at the end of year 3. Regular MACRS depreciation is used; the applicable tax rate is 34%. Construct a table showing each of the following for each of the 6 years. (6.13)

(a) Before-tax and loan cash flow.

(b) Loan principal payment.

(c) Loan interest payment.

(d) MACRS deduction.

(e) Taxable income.

(f) Taxes.

(g) After-tax cash flow.

40. Rework Problem 39 assuming the loan is paid back over 3 years as follows.

(a) In equal annual amounts.

(b) The principal in equal annual amounts plus yearly interest.

41. Turner Precision, Inc. purchased a new digitally controlled machine (5-year property) used to precisely deposit materials on electronic chips. The machine cost $500,000 and has an expected life of 10 years, but Turner plans to keep it only 6 years with a $25,000 salvage value at that time. It will replace several manually operated machines, resulting in net productivity and quality increases of $150,000/year. It is financed at 16% for $210,000; the remaining $290,000 is equity money. Repayment occurs over 3 years with equal principal amounts of $70,000, plus interest on the unrecovered balance. Clearly and completely set up a table and determine the after-tax cash flow for each appropriate year. Be sure to include everything that helps determine the ATCF. The tax rate is 34%, and regular MACRS allowances are used. (6.13)

42. Complete the partial table below. (6.13)

End of Year	Before-Tax and Loan Cash Flow	Loan Principal Payment	Loan Interest Payment	MACRS Deduction	Taxable Income	Tax 0.34	After-Tax Cash Flow
0	(100,000.00)	(40,000.00)					(60,000.00)
1	20,000.00	0.00	4,800.00	14,290.00			
2	20,000.00	0.00	4,800.00	24,490.00			
3		0.00		17,490.00		2,621.40	
4		0.00				1,941.40	
5	40,000.00	0.00	4,800.00			8,931.80	
6	20,000.00	0.00	4,800.00				13,064.80
7		0.00	4,800.00		18,270.00		
8		0.00	4,800.00		30,740.00		
9	45,000.00	0.00	4,800.00				
10	30,000.00	40,000.00	4,800.00				

43. Mizer Corporation purchases a heat exchanger (5-year property) by borrowing the entire $15,000 purchase price. The loan is to be repaid with 3 equal annual payments at an annual compound rate of 12%. It is anticipated that the exchanger will be used for 6 years and then be sold for $1000. Annual operating and maintenance expenses are estimated to be $7500/year. Assume regular MACRS depreciation, a 34% tax rate, and an after-tax MARR of 15%. Compute the after-tax equivalent uniform annual cost for the heat exchanger. (6.13)

44. Branman World Industries borrows $20,000, paying back $5000/year on the principal plus 15% on the unpaid balance at the end of each year. The $20,000 is used to purchase a special metering device (3-year property) that has a life of 5 years and a $2000 salvage value at that time. Branman uses regular MACRS depreciation, requires a 15% after-tax MARR, and has a 34% tax rate. Determine the after-tax equivalent uniform annual cost of the device. (6.13)

*45. A company has the following capital gains and losses and ordinary taxable incomes:

Short-term capital gains	$ 40,000
Short-term capital losses	19,000
Long-term capital gains	118,000
Long-term capital losses	23,000
Ordinary taxable income	95,000

Suppose favorable treatment of capital gains is brought back, and the capital gains rate is 28%. Determine the federal taxes to be paid by the company.

(a) Use the regular method. (6.14.2)

(b) Use the alternate method. (6.14.2)

*46. Our company has short-term capital gains of $21,000, short-term capital losses of $15,000, long-term capital gains of $60,000, and long-term capital losses of $12,000. Ordinary taxable income is $90,000. Suppose favorable treatment of capital gains is brought back, and the capital gains rate is 28%. How much tax must be paid?

(a) Use the regular method. (6.14.2)

(b) Use the alternate method. (6.14.2)

*47. Suppose the company in Problem 46 had short-term capital losses of $40,000 (the other information stays the same). How much tax must be paid now? (6.14.2)

*48. Determine the amount of tax (or other appropriate handling) for each of the following situations. Assume favorable treatment of capital gains is brought back, the regular tax rate is 40%, and the capital gains rate is 28%. (6.14.2)

(a) Net short-term capital gains	$30,000
Net long-term capital gains	$40,000
(b) Net short-term capital gains	$20,000
Net long-term capital losses	$12,000
(c) Net short-term capital losses	$26,000
Net long-term capital losses	$14,000
(d) Net short-term capital losses	$15,000
Net long-term capital gains	$26,000

(e) Net short-term capital gains	$18,000
Net long-term capital losses	$27,000
(f) Net short-term capital losses	$21,000
Net long-term capital gains	$10,000

49. Consider an asset that is 5-year property with a class life of 9 years. The purchase price is $100,000, and the asset is sold during the sixth year for $15,000. How should the salvage value received be treated for tax purposes if the following methods are used and the tax rate is 34%?

(a) Use regular MACRS allowances. (6.15.2)

(b) Use alternate MACRS allowances. (6.15.2)

50. Reconsider the asset in Problem 49. Assume it is sold for $15,000 during the fourth year.

(a) Use regular MACRS allowances. (6.15.2)

(b) Use alternate MACRS allowances. (6.15.2)

51. A 7-year property with a 12-year class life is purchased for $200,000 and depreciated using regular MACRS allowances. Assuming we do not know the status of other Section 1231 assets in the company, determine the tax treatment of the asset's salvage value in the following cases.

(a) It is sold in year 3 for $220,000. (6.15.2)

(b) It is sold in year 3 for $160,000. (6.15.2)

(c) It is sold in year 3 for $10,000. (6.15.2)

52. Rework Problem 51 assuming alternate MACRS is used to recover the asset. (6.15.2)

53. A residential property is placed in service in the first month of the tax year for $240,000 and recovered using regular MACRS allowances. Assuming we do not know the status of other Section 1231 assets in the company, determine the tax treatment of the asset's salvage value in the following cases.

(a) It is sold after 3 years, 8 months for $250,000. (6.15.2)

(b) It is sold after 3 years, 8 months for $160,000. (6.15.2)

54. Rework Problem 53 assuming alternate MACRS is used. (6.15.2)

55. A nonresidential real property is placed in service for $700,000 and recovered using regular MACRS allowances. It is purchased in the fourth month of the tax year. Assuming we do not know the status of other Section 1231 assets in the company, determine the tax treatment of the asset's salvage value in the following cases.

(a) It is sold after 5 years, 2 months for $780,000. (6.15.2)

(b) It is sold after 5 years, 2 months for $680,000. (6.15.2)

(c) It is sold after 5 years, 2 months for $250,000. (6.15.2)

56. Rework Problem 55 assuming alternate MACRS is used. (6.15.2)

57. HKE Construction purchases a small crane (5-year property) for $200,000. The crane will be sold during the sixth year for an estimated $20,000. Income less operating and maintenance costs for the crane have been $60,000/year. Using a 34% income tax rate and regular MACRS deductions, compute the after-tax cash flows for the 6-year period. *Assume the investment tax credit has been reinstated, and the full 10% will be taken at the end of year 1.* (6.16)

58. The J. B. Keatstein company proposes to invest $50,000 in an automatic control panel related to research and experimentation (5-year property). It is anticipated that this investment will cause a reduction in annual disbursements of $18,000 over 7 years with a $2000 salvage value at that time. Regular MACRS depreciation will be used and the tax rate is 34%. MARR = 15%. Determine the equivalent present worth of after-tax cash flows for the panel. *Assume the investment tax credit has been reinstated, and the full 10% will be taken at the end of year 1.* (6.16)

59. Mand-O-Man, Inc. is considering a $50,000 heat recovery incinerator (5-year property) which is expected to cause a reduction in net out-of-pocket costs of $20,000/year for 6 years. The incinerator will be depreciated using regular MACRS allowances and no salvage value is expected. The tax rate is 34%, and the Section 179 expense deduction will be taken in year 1. If MARR is 20%, determine the present worth of the after-tax cash flows. (6.17)

60. JES and Associates plan to invest $75,000 in a material-handling system (5-year property) that will reduce annual operating costs by $20,000/year over a 6-year planning horizon, after which the salvage value will be $4000. The tax rate is 34% and MARR = 15% after taxes. Regular MACRS depreciation will be used and the full Section 179 expense deduction will be taken in year 1. What is the equivalent after-tax annual cost of the investment? (6.17)

61. An automatic copier (5-year property) is being considered that will cost $24,000, have a life of 6 years and no prospective salvage value at the end of that time. Because of this venture, there will be an increasing yearly gross income estimated to be $18,000 in the first year, increasing by $2000 each year thereafter. The operating cost during the first year will be $6000, increasing by $500 each year thereafter. The company contemplating the purchase of this asset has an effective tax rate of 34% and uses regular MACRS allowances. Determine the after-tax cash flow for each year and the present worth of this investment if MARR is a combined rate of 15%, including an inflation component of 7%.

    (a) Use a then-current analysis. (6.19)
    (b) Use a constant-worth analysis. (6.19)

62. A construction company invested $70,000 in equipment (5-year property) that will have a useful life of 7 years with a $5000 salvage value at that time. The firm's income will increase according to the estimate $30,000(1.10)' in year *t* over the 7 years. The tax rate is 34%, the combined MARR is 20%, inflation is running at 9%, and regular MACRS allowances are used. Based on the present worth of the after-tax cash flows, should the company have invested?

    (a) Use a then-current analysis. (6.19)
    (b) Use a constant-worth analysis. (6.19)

63. A firm is considering purchasing a minicomputer for $400,000, installed and complete. It is expected to last 7 years with a salvage value of $30,000 at that

time. Additional gross income due to the computer will be $100,000 the first year, increasing by $10,000 each year thereafter. Operating and maintenance costs will be $50,000 in year 1, increasing by $5000 each year thereafter. Regular MACRS allowances are used, the tax rate is 34%, the real discount rate is 10%, and the general inflation rate is 5%. Based on the present worth of the after-tax cash flows, should the company invest?

(a) Use a then-current analysis. (6.19)

(b) Use a constant-worth analysis. (6.19)

64. JES (see Problem 60) is a very smart manager and wishes to leverage his money by financing $50,000 of the $75,000 investment at 13% over 4 years. The principal will be paid back in equal annual amounts each year, plus interest on the unrecovered balance. He still believes cost reductions will be flat at $20,000/year over the 6 years. Inflation is projected to average 6%, and the MARR of 15% is a combined rate. Now, what is the equivalent after-tax annual worth of the investment? (6.19)

(a) Use a then-current analysis. (6.19)

(b) Use a constant-worth analysis. (6.19)

65. A gold mine that is expected to produce 30,000 ounces of gold is purchased for $2,400,000. The gold can be sold for $450/ounce; however, it costs $265/ounce for mining and processing costs. If 3500 ounces are produced this year, what will be the depletion allowance for (a) unit depletion, and (b) percentage depletion? (6.20)

66. A West Virginia coal mine having 6 million tons of coal has a first cost of $24 million. The gross income for this coal is $50/ton. Operating costs are $20/ton. If, during the first 2 years of operation, the mine yields 150,000 tons and 200,000 tons, respectively, determine the after-tax cash flow for each year, assuming that this mine is the owner's only venture. Use the better method of depletion each year. (6.20)

CHAPTER 7

Economic Analysis of Projects in the Public Sector

7.1 INTRODUCTION

Knowing how to evaluate and select projects to be approved, paid for, and operated by the government is at least as important to today's engineer as a similar knowledge relating to the private sector. Fortunately, analysis methods for public and private projects are very similar, even though there are some basic, significant differences between the two. The methods considered in this chapter are those most frequently used in evaluating government (national, state, or local) projects—*benefit-cost* and *cost-effectiveness* analysis. More emphasis will be spent on benefit-cost methods, which require that benefits and costs be evaluated on a monetary basis. Cost effectiveness requires a numerical measure of effectiveness; however, that measure need not be in terms of money.

7.2 THE NATURE OF PUBLIC PROJECTS

There are many types of government projects and many agencies involved. Four classes reasonably cover the spectrum of projects entered into by government. They include cultural development, protection, economic services, and natural resources. *Cultural development* is enhanced through education, recreation, historic, and similar institutions or preservations. *Protection* is achieved through military services, police and fire protection, and the judicial system. *Economic services* include transportation, power generation, and housing loan programs. *Natural resource* projects might entail wildland management, pollution control, and flood control. Although these are obviously incomplete project lists in each class, it is not so obvious that some projects belong in more than one area. For example, flood control is certainly a form of protection for some, as well as being related to natural resources.

Government projects have a number of interesting characteristics that set them apart from projects in the private sector. Many government projects are huge, having first costs of tens of millions of dollars. They tend to have extremely long lives, such as 50 years for a bridge or a dam. The multiple-use concept is common, as in wildland management projects, where economic (timber), wildlife preservation (deer, squirrel), and recreation projects (camping, hiking) are each considered uses of import for the land. The benefits or enjoyment of government projects are often completely out of proportion to the financial support of individuals or groups. Also, there are almost always multiple government agencies that have an interest in a project. Further, public sector projects are not easily evaluated, since it may be many years before their benefits are realized. Finally, there is usually not a clear-cut measure of success or failure in the public sector such as the rate of return or net present worth criteria.

7.3 OBJECTIVES IN PROJECT EVALUATION

If large, complex, lengthy, multiple-use projects of interest to several groups are to be evaluated for their desirability, the criteria for the evaluation must first be agreed on. The setting for modern evaluation of government projects dates back to the River and Harbor Act of 1902, which "required a board of engineers to report on the desirability of Army Corps of Engineers' river and harbor projects, taking into account the amount of commerce *benefited* and the *cost*." Even more applicable to today is the criterion specified in the Flood Control Act of June 22, 1936, which stated ". . . that the Federal Government should improve or participate . . . if the *benefits to whomsoever they may accrue* are in excess of the *estimated costs* . . ." [15]. Obviously, the idea of benefit-cost analysis has been around for a long time, even though only in recent years has there been much written about it.

Prest and Turvey [11] give a short, reasonable definition of benefit-cost analysis: It is "a practical way of assessing the desirability of projects where it is important to take a long view (in the sense of looking at repercussions in the further, as well as the nearer, future) and a wide view (in the sense of allowing for side-effects of many kinds on many persons, industries, regions, etc.); that is, it implies the enumeration and evaluation of all the relevant costs and benefits." The "long view" is nothing more than considering the entire planning horizon which, granted, is generally far longer for projects in the public sector than for those in the private sector. The "wide view" and the notion of evaluating all "relevant costs and benefits" probably spell out the greatest single difference between government and private economic evaluation. That is, government projects often affect many individuals, groups, and things, either directly or indirectly, for better or for worse. In evaluating these projects, the analyst tries to capture these effects on the public, quantify them and, where possible, make the measures in monetary terms. Positive effects are referred to as *benefits*, while negative effects are *disbenefits*. In contrast, the private sector "effects" of primary importance are those that relate to income being returned to the organization. *Costs* of construction, financing, operation, and maintenance are estimated in much the same way in both the public and private sectors.

7.4   BENEFIT-COST ANALYSIS

The notion of benefit-cost analysis is simple in principle. It follows the same systematic approach used in selecting between economic investment alternatives, including the following seven steps adapted from Chapter Five.

1. Define the set of feasible, mutually exclusive, public sector alternatives to be compared.
2. Define the planning horizon to be used in the benefit-cost study.
3. Develop the cost-savings and benefit-disbenefit profiles in monetary terms for each alternative.
4. Specify the *MARR* to be used.
5. Compare the alternatives using a specified measure of worth.
6. Perform supplementary analyses.
7. Select the preferred alternative.

These will be discussed from the viewpoint of a benefit-cost analysis.

The heart of many engineering problems is in identifying all of the *alternatives* available to achieve a particular goal. Some alternatives can be excluded from further consideration immediately. Each remaining alternative should be described thoroughly. This involves specifying all of the good and bad effects that a project will have on the public. This includes effects directly on people, land values, and the environment. In addition, all aspects of project development, operation, maintenance, and eventual salvage must be stated.

Defining the *planning horizon* is essential in order to define the period over which the best project(s) is to be selected. If the planning horizon is longer than the life of a nonrenewable public project, there may be several years during which no benefits or costs are considered. If the planning horizon is shorter than the life of a project, residual values (same concept as salvage values) may be estimated and used.

Quantifying all *costs and savings* refers to governmental expenditures and incomes received relating to a project over the planning horizon. Disbursements include all first and continuing costs of a project, while income may result from tolls, fees, or other charges to the user public. Residual or salvage values may exist if the project is believed to still be in operation at the end of the planning horizon, and these are treated as savings or negative costs at that time. Costs, unlike benefits, are quantified for public projects in a way similar to private projects.

Each *benefit or disbenefit* during the planning horizon must also be quantified monetarily. Benefits, or the positive effects of a government investment, refer to desirable consequences on the public instead of on any governmental body. Disbenefits are the negative effects on the public instead of on any government organization. Unfortunately, placing dollar values on benefits received by a diverse public is often not an easy task. Neither is deciding on the interest rate or *MARR* to be used.

A base measure such as annual equivalent benefits and costs or present worths of benefits and costs is then established, and the *measure of worth* is chosen. Benefit-cost analyses frequently use the benefit-cost ratio (B/C) or, to a lesser extent, a measure of benefits less costs ($B - C$). If

B_{jt} = public benefits associated with project j during year t, $t =$ 1, 2, . . . , n

C_{jt} = governmental costs associated with project j during year t, $t =$ 0, 1, 2, . . . , n

and

i = appropriate interest rate

then the B/C criterion may be expressed mathematically, using a present worth base measure, as

$$B/C_j(i) = \frac{\sum_{t=1}^{n} B_{jt}(1 + i)^{-t}}{\sum_{t=0}^{n} C_{jt}(1 + i)^{-t}} \qquad (7.1)$$

Note that the B/C ratio is an alternative name for the savings/investment ratio treated in Chapter Four. The $B - C$ criterion is expressed as

$$(B - C)_j(i) = \sum_{t=0}^{n} (B_{jt} - C_{jt})(1 + i)^{-t} \qquad (7.2)$$

which is similar to the present worth method described in Chapter Four.

When two or more project *alternatives are being compared* using a B/C ratio, the analysis should be done on an *incremental basis*. That is, first the alternatives should be ordered from lowest to highest cost (present worth, annual equivalent, etc.). Then, the incremental benefits of the second alternative over the first, $\Delta B_{2-1}(i)$, are divided by the incremental costs of the second over the first, $\Delta C_{2-1}(i)$. That is,

$$\Delta B/C_{2-1}(i) = \frac{\Delta B_{2-1}(i)}{\Delta C_{2-1}(i)} = \frac{\sum\limits_{t=1}^{n} (B_{2t} - B_{1t})(1 + i)^{-t}}{\sum\limits_{t=0}^{n} (C_{2t} - C_{1t})(1 + i)^{-t}} \tag{7.3}$$

Note that if the first alternative is to do nothing, the incremental B/C ratio is also the straight B/C ratio for the second alternative. As long as $\Delta B/C_{2-1}(i)$ exceeds 1.0, Alternative 2 is preferable to Alternative 1. Otherwise, Alternative 1 is preferred to Alternative 2. The winner of these is then compared on an incremental basis with the next most costly alternative. These pairwise comparisons continue until all alternatives have been exhausted, and only one "best" project remains. The procedure used is very similar to that specified for the rate-of-return method in Chapter Four.

With the $B - C$ criterion, an incremental basis may be used following the same rules as for the B/C ratio, but preferring Alternative 2 to Alternative 1 as long as the following condition holds:

$$\Delta(B - C)_{2-1}(i) = \Delta B_{2-1}(i) - \Delta C_{2-1}(i)$$

$$= \sum_{t=1}^{n} [B_{2t}(i) - B_{1t}(i)](1 + i)^{-t}$$

$$- \sum_{t=0}^{n} [C_{2t}(i) - C_{1t}(i)](1 + i)^{-t} \geq 0 \tag{7.4}$$

Where benefits and costs are known directly, the value of $(B - C)_j$ for each Alternative j may be calculated and the maximum value selected.

Apparently straightforward, there are a number of potential pitfalls in the evaluation of government projects. That is, economic analyses of public projects can be easily biased unknowingly by a project evaluator. The more significant of these pitfalls are discussed in subsequent sections.

Next, *supplementary analyses* in the form of risk analysis, sensitivity analysis, and break-even analysis may be performed. These are often useful in determining how critical various inputs can be and how close various alternatives are economically, and they otherwise quantify much other information that would likely remain intangible.

The final step is *selecting the preferred alternative*. Not only is it to be selected, but all quantitative and qualitative supporting considerations should be recorded in detail. This is particularly true in the case of public sector projects.

These seven major steps in conducting a benefit-cost analysis, as well as the B/C and $B - C$ relationship to the present worth, annual equivalent, and rate of return criteria, are illustrated in the remainder of this section.

Example 7.1

We are given the task of deciding between three highway alternatives to replace a winding, old, dangerous road. The length of the current route is 26 miles. Alternative A is to overhaul and resurface the old road at a cost of $3 million. Resurfacing will then be required at a cost of $2.5 million at the end of each 10-year period. Annual maintenance for Alternative A will cost $10,000/mile. Alternative B is to cut a new road following the terrain; it will be only 22 miles long. Its first cost will be $10 million, and surface renovation will be required every 10 years at a total cost of $2,250,000. Annual maintenance will be $10,000/mile. Alternative C also involves a new highway which, for practical considerations, will be built along a 20.5-mile straight line. Its first cost, however, will be $18 million, because of the extensive additional excavating necessary along this route. It, too, will require resurfacing every 10 years at a cost of $2,250,000. Annual maintenance will be $18,000/mile. This increase over route B is due to the additional roadside bank retention efforts that will be required.

Our task is to select one of these alternatives, considering a planning horizon of 30 years with negligible residual value for each of the highways at that time. One of these alternatives is required, since the old road has deteriorated below acceptable standards. We can calculate the annual equivalent first cost and maintenance cost of each alternative using an interest rate of 8%.

Construction and resurfacing cost:

Route A:
$$[\$3,000,000 + \$2,500,000(P|F\ 8,10)$$
$$+ \ \$2,500,000(P|F\ 8,20)](A|P\ 8,30)$$
$$= [\$3,000,000 + \$2,500,000(0.4632) + \$2,500,000(0.2145)]$$
$$\times (0.0888)$$
$$= \$416,849/\text{year}$$

Route B:
$$[\$10,000,000 + \$2,250,000(P|F\ 8,10)$$
$$+ \ \$2,250,000(P|F\ 8,20)](A|P\ 8,30)$$
$$= [\$10,000,000 + \$2,250,000(0.4632) + \$2,250,000(0.2145)]$$
$$\times (0.0888)$$
$$= \$1,023,404/\text{year}$$

Route C:
$$[\$18,000,000 + \$2,250,000(P|F\ 8,10)$$
$$+ \ \$2,250,000(P|F\ 8,20)](A|P\ 8,30)$$
$$= [\$18,000,000 + \$2,250,000(0.4632) + \$2,250,000(0.2145)]$$
$$\times (0.0888)$$
$$= \$1,733,804/\text{year}$$

Maintenance cost:

Route A: $\left(10,000 \ \dfrac{\$}{\text{mile} - \text{year}}\right)(26 \text{ miles}) = \$260,000/\text{year}$

Route B: ($10,000)(22) = \$220,000/year

Route C: ($18,000)(20.5) = \$369,000/year

Clearly, route A costs less than route B, which itself costs less than route C. Can we now conclude that route A should be selected based on cost to the government? Absolutely not, according to the criterion that seeks to maximize benefits to the public as a whole, minus costs. In fact, we have analyzed only one side of the problem. Now, we must attempt to quantify the benefits along each of the routes.

Traffic density along each of the three routes will fluctuate widely from day to day, but will average 4000 vehicles/day throughout the year. This volume is composed of 350 light commercial trucks, 250 heavy trucks, 80 motorcycles, and the remainder are automobiles. The average cost per mile of operation for these vehicles is \$0.35, \$0.50, \$0.10, and \$0.20, respectively.

There will be a time savings because of the different distances along each of the routes, as well as the different speeds which each of the routes will sustain. Route A will allow heavy trucks to average 35 miles/hour, while other traffic can maintain an average speed of 45 miles/hour. Routes B and C will allow heavy trucks to average 40 miles/hour, and the rest of the vehicles can average 50 miles/hour. The cost of time for all commercial traffic is valued at \$15/vehicle/hour, and for noncommercial traffic, \$5/vehicle/hour. Twenty-five percent of the automobiles and all of the trucks are considered commercial.

Finally, there is a significant safety factor that should be included. Along the old winding road, there has been an excessive number of accidents per year. Route A will reduce the number of vehicles involved in accidents to 105, and routes B and C are expected to involve only 75 and 70 vehicles in accidents, respectively, per year. The average cost per vehicle in an accident is estimated to be \$7500, considering actual physical property damages, lost wages because of injury, medical expenses, and other relevant costs.

We now set about to analyze the various benefits in monetary terms. We have considered savings in vehicle operation, time, and accident prevention. The costs incurred by the public for these items are calculated in the following steps.

Operational costs

Route A: $\left(350 \ \dfrac{\text{light trucks}}{\text{day}}\right)\left(26 \ \dfrac{\text{miles}}{\text{light truck}}\right)\left(0.35 \ \dfrac{\$}{\text{mile}}\right)\left(365 \ \dfrac{\text{days}}{\text{year}}\right)$

$+ \left(250 \ \dfrac{\text{heavy trucks}}{\text{day}}\right)\left(26 \ \dfrac{\text{miles}}{\text{heavy truck}}\right)\left(0.50 \ \dfrac{\$}{\text{mile}}\right)\left(365 \ \dfrac{\text{days}}{\text{year}}\right)$

$+ \left(80 \ \dfrac{\text{motorcycles}}{\text{day}}\right)\left(26 \ \dfrac{\text{miles}}{\text{motorcycle}}\right)\left(0.10 \ \dfrac{\$}{\text{mile}}\right)\left(365 \ \dfrac{\text{days}}{\text{year}}\right)$

$$+ \left(3320 \frac{\text{automobiles}}{\text{day}}\right)\left(26 \frac{\text{miles}}{\text{automobile}}\right)\left(0.20 \frac{\$}{\text{mile}}\right)\left(365 \frac{\text{days}}{\text{year}}\right)$$

$$= \$8,726,055/\text{year}$$

Route B: $[350(\$0.35) + 250(\$0.50) + 80(\$0.10) + 3320(\$0.20)]$
$\times (22)(365)$
$= \$7,383,585/\text{year}$

Route C: $[350(\$0.35) + 250(\$0.50) + 80(\$0.10) + 3320(\$0.20)]$
$\times (20.5)(365)$
$= \$6,880,159/\text{year}$

Time costs

Route A: $\left(350 \frac{\text{light trucks}}{\text{day}}\right)\left(26 \frac{\text{miles}}{\text{light trucks}}\right)\left(\frac{1 \text{ hour}}{45 \text{ miles}}\right)\left(365 \frac{\text{days}}{\text{year}}\right)$

$\times \left(15 \frac{\$}{\text{hour}}\right) + \left(250 \frac{\text{heavy trucks}}{\text{day}}\right)\left(26 \frac{\text{miles}}{\text{heavy truck}}\right)\left(\frac{1 \text{ hour}}{35 \text{ miles}}\right)$

$\times \left(365 \frac{\text{days}}{\text{year}}\right)\left(15 \frac{\$}{\text{hour}}\right) + \left(80 \frac{\text{motorcycles}}{\text{day}}\right)\left(26 \frac{\text{miles}}{\text{motorcycle}}\right)$

$\times \left(\frac{1 \text{ hour}}{45 \text{ miles}}\right)\left(365 \frac{\text{days}}{\text{year}}\right)\left(5 \frac{\$}{\text{hour}}\right) + \left(3320 \frac{\text{automobiles}}{\text{day}}\right)$

$\times \left(26 \frac{\text{miles}}{\text{automobile}}\right)\left(\frac{1 \text{ hour}}{45 \text{ miles}}\right)\left(365 \frac{\text{days}}{\text{year}}\right)$

$\times \left(0.25 \times 15 \frac{\$}{\text{hour}} + 0.75 \times 5 \frac{\$}{\text{hour}}\right) = \$7,459,441/\text{year}$

Route B: $\left[\frac{350}{50} (\$15) + \frac{250}{40} (\$15) + \frac{80}{50} (\$5)\right.$

$\left. + \frac{3320}{50} (0.25 \times \$15 + 0.75 \times \$5)\right]$

$\times (22)(365) = \$5,659,143/\text{year}$

Route C: $\left[\frac{350}{50} (\$15) + \frac{250}{40} (\$15) + \frac{80}{50} (\$5)\right.$

$\left. + \frac{3320}{50} (0.25 \times \$15 + 0.75 \times \$5)\right] (20.5)(365)$

$= \$5,273,292/\text{year}$

Safety costs

Route A: $\left(105 \frac{\text{vehicles}}{\text{year}}\right)\left(7500 \frac{\$}{\text{vehicle}}\right) = \$787,500/\text{year}$

Route B: $(75)(\$7500) = \$562,500/\text{year}$
Route C: $(70)(\$7500) = \$525,000/\text{year}$

All relevant government and public costs are summarized in Table 7.1.

TABLE 7.1

Summary of Annual Equivalent Government and Public Costs, Example 7.1

	Route A	Route B	Route C
Government first cost of highway	$ 416,849/year	$ 1,023,404/year	$ 1,733,804/year
Government cost of highway maintenance	260,000/year	220,000/year	369,000/year
Public operational costs	8,726,055/year	7,383,585/year	6,880,159/year
Public time costs	7,459,441/year	5,659,143/year	5,273,292/year
Public safety costs	787,500/year	562,500/year	525,000/year
Total government costs	676,849/year	1,243,404/year	2,102,804/year
Total public costs	16,972,996/year	13,605,228/year	12,678,451/year

We desire to compare these three alternative routes using benefit-cost criteria. Let our first criterion be the popular benefit-cost ratio. Since one of these alternatives must be selected, we will assume the lowest government cost alternative, route A, will be selected unless the extra expenditures for routes B or C prove more worthy. Since we have not defined "benefits" per se, user benefits will be taken as the incremental reduction in user costs from the less expensive to the more expensive alternatives. Since we are looking at incremental benefits, it makes sense to compare these against the respective incremental costs needed to achieve these additional benefits.

The incremental benefits and costs for route B as compared to route A for $i = 8\%$ are given as follows:

$$\Delta B_{B-A}(8) = \text{public costs}_A(8) - \text{public costs}_B(8)$$
$$\$16,972,996 - \$13,605,228 = \$3,367,768/\text{year}$$

$$\Delta C_{B-A}(8) = \text{government costs}_B(8) - \text{government costs}_A(8)$$
$$\$1,243,404 - \$676,849 = \$566,555/\text{year}$$

That is, for an incremental expenditure of $566,555/year, the government can provide added benefits of $3,367,768/year for public. The appropriate benefit cost ratio is then

$$\Delta B | C_{B-A}(8) = \frac{\Delta B_{B-A}(8)}{\Delta C_{B-A}(8)} = \frac{\$3,367,768}{\$566,555} = 5.94$$

This clearly indicates that the additional funds for route B are worthwhile, and we desire route B over route A.

Using a similar analysis, we now calculate the benefits, costs, and $\Delta B | C$ ratio to determine whether or not route C is preferable to route B.

$$\Delta B_{C-B}(8) = \$13,605,228 - \$12,678,451 = \$926,777/\text{year}$$

$$\Delta C_{C-B}(8) = \$2,102,804 - \$1,243,404 = \$859,400/\text{year}$$

$$\Delta B / C_{C-B}(8) = \frac{\$926,777}{\$859,400} = 1.08$$

This benefit-cost ratio, being greater than 1.00, indicates that the additional expenditure of $859,400/year to build and maintain route C would provide commensurate benefits to the public. In fact, the user savings would be $926,777/year. Of the three alternative routes, route C is preferred.

The next benefit-cost criterion takes advantage of the fact that if $\Delta B/C > 1$, then $\Delta(B - C) > 0$. That is, the difference in incremental benefits and costs may be used in place of the incremental benefit-cost ratio.

Example 7.2

Applying this measure to the routes in the previous example results in the following calculations for comparing routes A and B. We know that

$$\Delta B_{B-A}(8) = \$3,367,768/\text{year}$$

$$\Delta C_{B-A}(8) = \$\ 566,555/\text{year}$$

This results in

$$\Delta(B - C)_{B-A}(8) = \Delta B_{B-A}(8) - \Delta C_{B-A}(8) = \$3,367,768 - \$566,555$$
$$= \$2,801,213/\text{year}$$

Thus, we again conclude that route B is preferred to route A. Similarly,

$$\Delta(B - C)_{C-B}(8) = \Delta B_{C-B}(8) - \Delta C_{C-B}(8)$$
$$= \$926,777 - \$859,400 = \$67,377/\text{year}$$

which indicates that route C is worthy of the additional expenditure required; hence route C should be constructed.

The benefit-cost criteria are consistent with the methods of alternative evaluation presented in Chapter Five. For example, suppose it is desired to base a decision on the present worth criterion, minimizing the sum of government construction and maintenance costs as well as public user costs.

Example 7.3

The present worth of all costs for each highway alternative is given by the following formula:

$$PW(\text{total}) = PW_{\text{total government costs}}(i) + PW_{\text{total public costs}}(i)$$

Then,

Route A: $PW_A(8) = \$676,849(P|A\ 8,30) + \$16,972,996(P|A\ 8,30)$
$$= \$676,849(11.2578) + \$16,972,996(11.2578)$$
$$= \$198,698,425$$

Route B: $PW_B(8) = \$1{,}243{,}404(P|A\ 8{,}30) + \$13{,}605{,}228(P|A\ 8{,}30)$
$= \$1{,}243{,}404(11.2578) + \$13{,}605{,}228(11.2578)$
$= \$167{,}162{,}930$

Route C: $PW_C(8) = \$2{,}102{,}804(P|A\ 8{,}30) + \$12{,}678{,}451(P|A\ 8{,}30)$
$= \$2{,}102{,}804(11.2578) + \$12{,}678{,}451(11.2578)$
$= \$166{,}404{,}413$

Since the present worth of all costs for route C is smallest, we again see that this alternative is preferred.

The annual equivalent cost is calculated by simply multiplying a constant times the present worth. Thus, the benefit-cost criteria are also comparable to the annual worth measure of merit.

Example 7.4

Calculating the equivalent uniform annual cost using

$$AW_{total}(i) = AW_{total\ government\ costs}(i) + AW_{total\ public\ costs}(i)$$

we have

Route A: $AW_A(8) = \$676{,}849 + \$16{,}972{,}996 = \$17{,}649{,}845/\text{year}$

Route B: $AW_B(8) = \$1{,}243{,}404 + \$13{,}605{,}228 = \$14{,}848{,}632/\text{year}$

Route C: $AW_C(8) = \$2{,}102{,}804 + \$12{,}678{,}451 = \$14{,}781{,}255/\text{year}$

Again route C is preferred.

It is interesting and logical to note that the differences in equivalent uniform annual costs for route A,B and B,C are the same as the difference in incremental benefits minus incremental costs for routes A,B and B,C. That is,

$$AW_A(8) - AW_B(8) = \$17{,}649{,}845 - \$14{,}848{,}632 = \$2{,}801{,}213/\text{year}$$

$$\Delta B_{B-A}(8) - \Delta C_{B-A}(8) = \$3{,}367{,}768 - \$566{,}555 = \$2{,}801{,}213/\text{year}$$

and

$$AW_B(8) - AW_C(8) = \$14{,}848{,}632 - \$14{,}781{,}255 = \$67{,}377/\text{year}$$

$$\Delta B_{C-B}(8) - \Delta C_{C-B}(8) = \$926{,}777 - \$859{,}400 = \$67{,}377/\text{year}$$

Finally, the benefit-cost criteria can even be related to the internal rate-of-return approach discussed in Chapter Four. That is, the interest rate is found that causes the annual equivalent incremental benefits less incremental costs to equal zero.

Example 7.5

The internal rate-of-return approach may be used by finding the value of the interest rate i that equates the incremental benefits of Alternative k over Alternative j to their incremental costs. That is,

$$AW_{k-j}(i) = 0 = \Delta B_{k-j}(i) - \Delta C_{k-j}(i)$$

Using this expression to consider route B versus route A, we have

$$\Delta B_{B-A}(i) = \$16,972,996 - \$13,605,228 = \$3,367,768/\text{year}$$

$$\begin{aligned}
\Delta C_{B-A}(i) = &[\$10,000,000 + \$2,250,000(P|F\ i,10) \\
&+ \$2,250,000(P|F\ i,20)](A|P\ i,30) + \$220,000 \\
&- \{[\$3,000,000 + \$2,500,000(P|F\ i,10) \\
&+ \$2,500,000(P|F\ i,20)][A|P\ i,30] + \$260,000\}
\end{aligned}$$

Searching for the value of i yielding $AW_{B-A}(i) = \Delta B_{B-A}(i) - \Delta C_{B-A}(i) = 0$, we find

$$AW_{B-A}(40) = \$611,348/\text{year}$$

$$AW_{B-A}(50) = -\$90,032/\text{year}$$

The interest rate we are seeking is between 40 and 50%, or approximately 48.72%. In any event, i easily exceeds 8%, and we prefer route B to route A. Using the rate-of-return approach, we can also compare routes B and C.

$$\Delta B_{C-B}(i) = \$13,605,228 - \$12,678,451 = \$926,777/\text{year}$$

$$\begin{aligned}
\Delta C_{C-B}(i) = &[\$18,000,000 + \$2,250,000(P|F\ i,10) \\
&+ \$2,250,000(P|F\ i,20)](A|P\ i,30) + \$369,000 \\
&- \{[\$10,000,000 + \$2,250,000(P|F\ i,10) \\
&+ \$2,250,000(P|F\ i,20)][A|P\ i,30] + \$220,000\}
\end{aligned}$$

The rate of interest causing $AW_{C-B}(i) = \Delta B_{C-B}(i) - \Delta C_{C-B}(i) = 0$ is between 8 and 9%, since

$$AW_{C-B}(8) = \$67,377/\text{year}$$

$$AW_{C-B}(9) = -\$623/\text{year}$$

The solution value for i is approximately 8.99%; hence, we prefer route C to route B.

Several different approaches to the same problem have been illustrated and shown to be consistent project evaluators. Typically, where government projects are involved, one of the benefit-cost criteria (B/C or $B - C$) is used. More often than not, the criterion is the benefit-cost ratio. This is unfortunate because, just as in rate of return analyses in the private sector, the benefit-cost ratio is easy to misuse and misinterpret, and it is very sensitive to the classification of problem elements as "benefits" or "costs." These problems will be discussed subsequently.

The present worth, annual worth, and rate-of-return methods are seldom used in government analyses. Even though the mechanics underlying these approaches are perfectly suitable, their underlying philosophy is more attuned to return on investment, discounted net profits, and minimum attractive rate of return, implying measures of return to *investors* based on capital investment *by the same investors*. Government investments, however, do not necessarily result in *any* monetary incomes to the government and are seldom evaluated solely on the basis of monetary return. Also, government employees and politicians prefer benefit-cost analyses to enable them to speak intelligently of the "benefits" derived by the public as a result of government's expenditures of the public's tax, assessment, and other money.

7.5 IMPORTANT CONSIDERATIONS IN EVALUATING PUBLIC PROJECTS

In the section on benefit-cost analysis, it was stated that there are a number of pitfalls that can affect the analysis of government projects. Actually, opportunities for error pervade benefit-cost analyses, from the very initial philosophy, through to the interpretation of a B/C ratio. It is important to talk about the more significant of these, both to help prevent analysts from erring, and to help those who may be reviewing a biased evaluation.

The major topics to be considered include the following:

1. Point of view (national, state, local, individual).
2. Selection of the interest rate.
3. Assessing benefit-cost factors.
4. Overcounting.
5. Unequal lives.
6. Tolls and fees.
7. Multiple-use projects.
8. Problems with the B/C ratio.

7.5.1 Point of View

The stance taken by the engineer in analyzing a public venture can have an extensive effect on the economic "facts." The analyst may take any of several viewpoints, including those of:

1. An individual who will benefit or lose.
2. A particular governmental organization.
3. A local area such as a city or county.

4. A regional area such as a state.

5. The entire nation.

The first of these viewpoints is not particularly interesting from the standpoint of economic analysis. Nonetheless, all too frequently, an isolated road is paved, a remote stretch of water or sewer line is extended under exceptional circumstances, or a seemingly ideal location for a public works facility is suddenly eliminated from consideration. In these cases, the benefit-cost analysis, its review, and the implementation decision are usually made by a small, select group.

The other four viewpoints are, however, of considerable interest to those involved in public works evaluation. Analyzing projects or project components from viewpoint 2, that of a particular government agency, is analogous to economic comparisons in private enterprise. That is, only the gains and losses to the organization involved are considered. This viewpoint, which seems contrary to benefit-cost optimization to the public as a whole, may be appropriate under certain circumstances.

Example 7.6

Consider a Corps of Engineers construction project in which the water table must be lowered in the immediate area so work can proceed. Any of several water cutoff or dewatering systems may be employed. Water cutoff techniques include driving a sheet pile diaphram or using a bentonite slurry trench to cut off the flow of water to the construction area. Dewatering methods include deep-well turbines, an eductor system, or wellpoints for lowering the water level. It is sometimes appropriate for the Corps to evaluate these different techniques from an "organization" point of view, since each of the feasible methods provides the same service or outcome—a dry construction site. Therefore, the most economical decision from the Corps' point of view is also correct from the view of the public as a whole, since the benefits or contributions to the project are the same regardless of the method chosen.

The third point of view, that of a locality such as a city or county, is popular among local government employees and elected officials. Unfortunately, seemingly localized projects often impact a much wider range of the citizenry than is apparent.

Example 7.7

County officials are to decide whether or not future refuse service should be county owned and operated, or whether a private contractor should be employed. The job requires front-end loader compaction trucks as well as roll on–off container capability. Primarily rural roads are traveled, and from 1 container (a roadside picnic area) to 50 containers (a large rurally located industrial plant) must be collected at each stop. Front-end loader containers range from 2 to 8 cubic yards, while roll on-off containers are sized from 15 to 45 cubic yards. Several trucks and drivers will be required, including a base for operations and maintenance.

The cost in dollars per ton of refuse collected, removed, and disposed of, is given below:

Personnel services	$ 6.08/ton
Materials, supplies, utilities	5.13/ton
Maintenance and repair	5.62/ton
Overhead	3.41/ton
Depreciation	3.61/ton
Five percent interest on half financed by bonds	1.80/ton
TOTAL COUNTY COST	$25.65/ton
Federal taxes foregone	$ 1.87/ton
State taxes foregone	0.20/ton
Property taxes foregone	1.54/ton
Eight percent return on half financed by tax money	2.88/ton
Not necessarily paid by county	$ 6.49/ton

County cost to provide refuse service will be $25.65/ton. The county is not, however, required to pay the additional $6.49/ton for federal and state taxes, *ad valorem* taxes, or a return on appropriated money, as would a private firm.

It is obvious from the example that a "local" county decision can affect a much wider public. Suppose, based on $25.65/ton, the county decided to own and operate the needed refuse service. Federal taxes of $1.87/ton that would have been paid by a private contractor will not be paid. Since the federal government will still have the same revenue requirements, the difference will be made up by passing on an infinitesimally small burden to the national public as a whole. Although this is easily rationalized at the local level—to spread a portion of the cost of county refuse service over the entire country—consider the result if every town, city, and county took this attitude.

An analogous argument follows for state income taxes foregone; however, now the burden is being spread over the people of the state. Even though this is much smaller than the national population, the burden per person to make up the lost state tax income is still very small. Again, providing refuse service at less cost to the local populace, at the expense of the state, is tempting from a parochial point of view.

If the county were to plan on not having to pay the *ad valorem* tax, the slack would be taken up by increasing the property tax rates in the county. Although this approach increases the burden on county property owners, that burden may be entirely disproportionate when compared to the refuse service each requires.

The last two points of view include a regional (e.g., a state) or national perspective. Ideally, a national outlook is preferable for local, regional, and national public works projects. Experience indicates, however, that the primary concern of public works officials and politicians is their particular constituency.

Perhaps the best advice for evaluators and decision makers in the public realm is to examine multiple viewpoints. That is, project evaluators are often not in a position to decide on a single specific point of view. They should instead present a thorough analysis clearly indicating any benefits or costs that depend on the perspective taken. Similarly, decision makers should require multiple points of view so they can be aware of the kind and degree of repercussions resulting from their actions.

7.5.2 Selection of the Interest Rate

The interest rate, discount rate, or minimum attractive rate of return is another premier factor to be decided upon when evaluating public works projects. In the route selection examples, the interest rate was taken at 8% with no question of the appropriateness of such a figure. Clearly, however, the interest rate has a significant effect on the net present worth or annual worth of cash flows in the private sector. Similarly, there is a significant effect on the net present worth of benefits minus costs or the benefit-cost ratio in public sector analyses.

Example 7.8

Let us reconsider the highway route selection Examples 7.1 to 7.5 again, this time using a minimum attractive rate of return of 10%. Since Route B was equivalent to Route A at i = 48.72%, the preference of Route B to Route A continues to hold. However, Route C was equivalent to Route B using a *MARR* of 8.99%. Thus, at a *MARR* of 10%, Route C would not be preferred to Route B and spending the substantial additional funds for Route C is not worthwhile.

One may argue with the above example, saying that $i = 10\%$ is higher than would ever be used. Such an argument is certainly not true for private industry. In public works, however, various knowledgeable individuals have explicitly proposed values on i ranging from 0 to 13%, and rather vague support has been seen ranging from 0 to over 25%. Before considering various schools of thought on the appropriate interest rate, it is helpful to know how public activities are financed.

Financing of Government Projects: There are several different ways that units of government finance public sector projects. The most obvious way is, of course, through taxation such as income tax, property tax, sales tax, and road user tax. Another popular approach is through the issuance of bonds for either specific projects or general use. Other forms of borrowing, such as notes, may be thought of in the same category as bonds. A third type of fund raising includes income generating activities such as a municipally owned power plant, a toll road, or other activity where a charge is made to cover (or partially offset) the cost of the service performed. Although these are the primary sources of government funds, there are a number of ways in which this money may be passed from one government authority to another by way of direct payments, loans, subsidies, and grants.

Federal funds are raised through tax money and federal borrowing. Federal projects may then be financed through direct payment. In this case, no monetary return is expected by the government; however, the "return" is expressed through the benefits incurred by the public. Direct payment financing may be total as in the case of many corps of engineers projects, or partial, for example 90%, as in cost sharing with states for interstate highways.

Financing for projects of national interest and impact may also be available through no-interest or low-interest rate loans. Both are available for long periods of time, say up to 40 years, with terms obviously more favorable than could normally be expected from conventional sources of money. Such loans are available for financing large projects such as the Tennessee Valley Authority where revenues resulting from the projects are used to pay back the loans plus interest, if any. Certain university buildings such as dormitories may also be financed in this way. There are also occasions when principal payment deferment is permitted during the early years of the project.

Other forms of federal financing include subsidies and federal loan insurance. Subsidies are used to encourage projects or services believed to be in the public's best interest, such as in the area of transportation. Loan insurance is used to eliminate the private lending institution's risk, allowing lower interest loans over longer periods than conventionally available. Insured loans began with the Federal Housing Administration.

State and local public projects are financed from taxes or bonds. There are, however, constraints on bond financing. First, bond issues must be approved by the voters. Often, there must be a 60% or even two-thirds vote in favor of the bond. Second, in order to prevent excessive borrowing, states have limited the amount of bond debt that may be undertaken. This is often a fraction of the property valuation assessment in the local area. Finally, there are also restrictions that govern the payback requirements and lives of the bonds. Considering these restrictions, the temperament of the public regarding bond issues, and the future needs that are likely to require bond financing, care is obviously needed in selecting public works projects to be implemented or put before the people.

Considerations in the Selection of the Interest Rate: Many arguments over the correct philosophy to use in selecting the interest rate have surfaced over time. For practical purposes, most of these philosophies are somewhat aligned with one of the following positions:

1. A zero interest rate is appropriate when tax monies are used for financing.
2. The interest value need only reflect society's time preference rate.
3. The interest rate should match that paid by government for borrowed money.
4. The appropriate interest rate is dictated by the opportunity cost of those investments forgone by private investors who pay taxes or purchase bonds.
5. The appropriate interest rate is dictated by the opportunity cost of those investments foregone by government agencies due to budget constraints.

Advocates of a zero interest rate when tax money is used argue that current taxes require no principal or interest payment at all. Hence, current tax monies (e.g., highway user taxes) should be considered "free" money, and a no-interest or discount rate applied. Counterarguments to this stance point out that a zero (or even low) interest rate will allow very marginal projects or marginal "add-on" project enhancements to achieve a B/C ratio greater than 1. This, in turn, takes money away from other projects that are truly deserving. If, in fact, it is not true that more deserving projects are precluded while very marginal projects (say 1 or 2% "return") are being approved, perhaps there is an excess of money available and taxes should be lowered. A final position against the zero interest rate advocates is that if government does not invest funds in high-benefit or "profitable" projects, the people should be able to retain and invest their own money to provide benefits more economically.

The "societal time preference rate" advocates contend that i is merely "a planning parameter reflecting society's feelings about providing for the future as opposed to current consumption" says Howe [6]. Henderson [5] says i is the rate that "reflects the government's judgment about the relative value which the community as a whole is believed to assign, or which the government feels it ought to assign, to present as opposed to future consumption at the margin." As such, Howe [6] makes clear that the societal time preference rate "need bear no relation to the rates of return in the private sector, interest rates, or any other measurable market phenomena." Estimated societal time preference rates tend to be quite low (e.g., 3 to 6%).

Many people back the use of an interest rate that matches that paid by government for borrowed money. This seems reasonable in that government bonds are in direct competition with other investment opportunities available in the private sector. Of course, many government bond coupon rates tend to be lower because of their tax exempt status (the bond interest is not taxable) than their private industry counterparts. One way of obtaining an interest figure is to determine the rate on "safe" long-term federal bonds [6]. Another possibility is to use the rate paid on borrowings by the particular government unit in question. That is, if a project will be financed by borrowing, the rate of interest to be paid is used. If nonborrowed funds are to be used, the appropriate rate is the average rate of interest being paid on long-term (over 15 years) borrowing.

There are also good arguments that the cost of government money may be too low an interest rate. For example, the opportunities forgone by other government agencies or investors in the private sector may have provided a far higher "return" than investments approved using the cost of government money. In addition, using a rate equal to the cost of government borrowings includes no provision for risk, nor does it include the subsidizing effect of the tax exemption. Finally, it is argued that plain good judgment dictates that projects worthy of approval under higher interest rates are more justifiable in cases where benefits and disbenefits received by the public are very nonuniform.

The next philosophy calls for an opportunity cost approach, taking into account many of the factors not considered in the pure cost of government borrowed money. For example, Howe [6] says this philosophy is that "no public project should be undertaken that would generate a rate of return less than the rate of return that would have been experienced on the private uses of funds that would be precluded by the financing of the public project (say, through taxes or bonds)." How are these private use rates determined? Consider a situation in which private investments would yield an average annual return of $i_1\%$. Also consider the rate of return on consumption, which is measured by the rate that consumers are willing to pay to consume now instead of later. This rate is, say, 10 to 13% for a house, 12 to 24% for a new or used car, and so on. Let this rate of return on consumption average out to $i_2\%$. Now, if a fraction α of government financing precludes private investments, while $1 - \alpha$ precludes consumption, the rate of return foregone to finance public projects is given as i.

$$i = \alpha i_1 + (1 - \alpha)i_2 \qquad 0 \le \alpha \le 1 \qquad (7.5)$$

An empirical study conducted by Haveman [4] in 1966 showed the appropriate weighted average to be 7.4% *at that time*. With the many economic changes since that time, a rate of perhaps 12 to 15% is more appropriate for the 1990s.

The last philosophy also requires an opportunity cost approach in which an artificial interest rate reflects the rates of return forgone on government projects by virtue of having insufficient funds. That interest rate is found by continuing to increase the value of i until only the projects remain having $B/C > 1$ or $B - C > 0$ which can be afforded with monies available.

What conclusions can be reached from these philosophical arguments? What guidelines are available for evaluators of public works projects? Clearly, no one answer is universally applicable. But, as a general rule, we recommend using a rate that is at *least* as high as the average effective yield on tax exempt long-term government bonds. That is, projects or enhancements to projects not providing at least this rate of return should not be undertaken. This is not to say, however, just because a project is attractive at the *MARR*, that it should be implemented. Only the most worthy projects, in the environment of limited funds, should be constructed. These projects may be identified by conducting several analyses of prospective projects, continually raising the interest rate ($i = 6, 8, 10, 12, . . . \%$) until only the projects remain that may be accomplished with available or legally borrowable funds.

7.5.3 Assessing Benefit-Cost Factors

The benefit-cost analyst knows, before starting a study, that placing a monetary figure on certain "social benefits" may be difficult. Actually, there is even a more fundamental problem—what factors to assess. Some insight is available by considering the following four types of factors discussed by Cohn [2].

1. *"Internal"* effects are those which accrue directly or indirectly to the individual(s) or organization(s) with which the analyst is primarily concerned. These effects are always included in a benefit-cost analysis.

2. *"External"* *technological* (*or real*) effects are those which cause changes in the physical opportunities for consumption or production. For example, effects on navigation and water sport recreation due to a new hydroelectric plant would fall into this category. These effects should be included in an analysis.

3. *"External"* *pecuniary* effects relate to changes in the distribution of incomes through changes in the prices of goods, services, and production factors. For example, if a firm's expansion is sufficiently large to affect industry prices, an increase in its output is most likely to result in lower prices for output from other firms in the industry as a whole. Many authors agree that these effects can safely be ignored.

4. *Secondary effects* involve changes in the demand for and supply of goods, services, resources, and production factors which *arise from* a particular project. As an example, phosphate mining in Idaho on government lands will bring instant population increases to nearby small towns. Secondary effects include increasing the incomes of various producers such as vehicle repairmen and barbers who provide the vehicle repairmen with haircuts. McKean [9] contends that only incremental income arising from such effects should be included, if reasonable to do so.

Example 7.9

A dam and reservoir are contemplated in an effort to reduce flood damage to homes and crops in a low area of northeastern Oklahoma. Annual damage to property varies from year to year, but averages approximately $230,000/year. The dam and reservoir contemplated should virtually eliminate damage to the area in question. No other benefits (e.g., irrigation, power generation, recreation) will be provided.

In performing a benefit-cost analysis of the flood control project, the engineer notes that the primary benefit to the public will be the $230,000/year damage prevented. However, the engineer argues that this will also cut back on money paid to contractors and servicemen for home and car repair, to health care units, for insurance premiums, and so forth. In other words, the building of a dam, it could be argued, will provide disbenefits to those who would normally receive part of their livelihood from helping flood-damaged families. Should the engineer include in the evaluation only the direct benefits to the flood damage victims, or should the other effects be included as well? That depends on how the analyst chooses to handle secondary effects, as indicated in the following discussion.

The disbenefits to those who would lose income if the dam and reservoir were built are considered *secondary effects*. That is, there would be a decrease in the demand for and supply of post-flood restoration goods and services. It is argued that only the *incremental* incomes or incremental profits (losses or lost profits in this case) should be considered when secondary effects are involved.

Another argument calls upon the "ripple" effect of the economy. That is, every secondary effect disbursement by one person or organization is a receipt to another person or organization. Each receipt then contributes to another disbursement, and so on. If the ripple philosophy were tracked or followed for the secondary effects of a particular alternative, the sum of receipts less disbursements would equal zero, and there would be no economic evaluation.

With either philosophy, the secondary effects of the example's flood control dam and reservoir would be small, if not negligible. This is intuitively reasonable, because the dam's main benefits represent the measure of the direct usefulness of the dam serving its intended purpose, whereas the diseconomies described are, in fact, secondary and diffuse.

Example 7.10

Now, reconsider the previous flood control dam and reservoir of Example 7.9. Suppose the reservoir would cause a loss of agricultural land for grazing and crops. Should this loss be considered in the benefit-cost analysis? Yes, because it is an external real effect causing changes in the physical opportunities for consumption or production.

External technological effects are often well defined and, in practice, are usually included in benefit-cost analyses. External pecuniary effects, however, may not be vividly apparent, as illustrated in the following example.

Example 7.11

A large irrigation project is being considered in the heart of cotton country. The irrigation will provide a significant effect on the quantity and quality of the cotton grown. This additional supply of cotton will, however, depress the price of cotton, lowering the profitability of other cotton growers. Also, the same effect will be felt throughout, say, the clothing industry and manufacturers of cotton substitutable products (products that may be used in place of cotton items) will likely have to reduce prices. At the same time, producers of cotton complementary goods (items that go well with or are used in conjunction with cotton products) will note increased demand, and may increase prices because of an insufficient supply.

Which of these effects would we include in an evaluation of the irrigation project? Of the factors mentioned above, none would be included in the analysis. Each of the effects described relates to changes in the distribution of incomes through changes in the prices of goods, services, and production factors. As such, they are considered external pecuniary effects, which are not "real" benefits or disbenefits and, hence, are not included.

In the above examples, internal and external technological effects were considered to be factors in an analysis, while external pecuniary and secondary effects were not included. There is, of course, no complete agreement on either these classifications or whether or not they should be evaluated. For the practitioner, good practical judgment is probably the best asset in deciding what factors to include.

As a guide, all identifiable effects of a project should be delineated. Some effects will clearly provide direct benefits or disbenefits to the public that should be counted. There may also be some factors that obviously should not be included. The third group—the controversial factors, if any—should be studied in depth and either included for good reason or used implicitly as supporting information. The reason for including or not considering these controversial factors should be stated in writing and become a part of the evaluation for the record.

7.5.4 Overcounting

A common dysfunction of trying to consider a large variety of effects in a benefit-cost analysis is to overcount, or unknowingly count some factors twice.

Example 7.12

Many years ago, Connecticut's Department of Health calculated the following loss from *preventable* disability, hoping to increase interest in health problems.
Individual income lost

Number of wage earners affected	350,000
Average days per year lost due to preventable accidents	4
Average hourly wage	$2.50
TOTAL INDIVIDUAL LOSS PER YEAR	$3,500,000

Industry loss:

It is accepted that the loss to industry in disorganization, idle overhead, and lessened production is 2½ times the wage loss, or	
TOTAL INDUSTRY LOSS PER YEAR	$8,750,000

These figures were totally inadequate because of a classic (but not overly apparent) case of double-counting. If we assume the loss to individuals is, correctly, their forgone earnings, then the loss to industry would be *its* forgone earnings, or its lost profits, and not the entire value of lessened production. As the figures stand above, lost wages are counted once from the viewpoint of the employee, and are again included as part of industry's losses. That is, Weisbrod [13] emphasizes that employee wages are double counted.

Example 7.13

Reconsider Example 7.11 involving a cotton irrigation system. The increased quantity of cotton will require that additional gin and seed mill hands be employed, removing a significant number of persons from the welfare rolls. The amount of their new wage, equal to the sum of their old welfare payments plus some increase, represents an increase in real output and constitutes a legitimate national benefit of the project. To then add the reduction in welfare payments from the taxpayers to the unemployed would be to double-count welfare payments, once from the standpoint of the recipient, and once from the taxpayer's viewpoint.

7.5.5 Unequal Lives

When comparing one-shot public works projects having unequal lives, the planning horizon will commonly coincide with the longest lived alternative. When some projects are expected to have a long life while others have a shorter life, changes in the discount rate could change the attractiveness of some projects with respect to the alternatives.

Example 7.14

Two projects each have first costs of $200,000, with annual operating costs of $30,000. The life of Project A is 15 years, while that of Project B is 30 years, and benefits accrued to the public are estimated at $60,000/year and $54,000/year, respectively. Each is a one-shot project; hence, Project A will have no benefits or costs after year 15. A 30-year planning horizon will be used. If the *MARR* is set at 8%, which project is more attractive?

$$(B - C)_A(8) = \$60,000(P|A\ 8,15) - \$30,000(P|A\ 8,15) - \$200,000$$
$$= \$60,000(8.5595) - \$30,000(8.5595) - \$200,000$$
$$= \$56,785$$

$$(B - C)_B(8) = \$54,000(P|A\ 8,30) - \$30,000(P|A\ 8,30) - \$200,000$$
$$= \$54,000(11.2578) - \$30,000(11.2578) - \$200,000$$
$$= \$70,187$$

Over a 30-year planning horizon, and using an interest rate of 8%, Project B is clearly better. Now suppose that insufficient money is available to implement all of the many projects, B included, that a government agency would like to do. A decision is made to consider all projects using an interest rate that reflects the opportunity cost of investments foregone by government agencies. That is, the interest rate is increased until only those projects that can be afforded remain. Assuming the interest rate is up to 12%, now which project is more desirable?

$$(B - C)_A(12) = \$60,000(P|A\ 12,15) - \$30,000(P|A\ 12,15) - \$200,000$$
$$= \$60,000(6.8109) - \$30,000(6.8109) - \$200,000$$
$$= \$4327$$

$$(B - C)_B(12) = \$54,000(P|A\ 12,30) - \$30,000(P|A\ 12,30) - \$200,000$$
$$= \$54,000(8.0552) - \$30,000(8.0552) - \$200,000$$
$$= -\$6675$$

At the higher interest rate, Project A is more favorable because of the increased emphasis on early year net benefits as opposed to the heavily discounted net benefits during the latter years of the planning horizon.

When a planning horizon shorter than some project durations is selected, a residual value must be estimated for those projects. The residual value is handled in the same way as a salvage value.

Example 7.15

Reconsidering the previous example and selecting a 15-year planning horizon, let us now determine the more favorable alternative, assuming project B has a residual value of 40% of first cost. Let $i = 12\%$.

$$(B - C)_A(12) = \$60,000(P|A\ 12,15) - \$30,000(P|A\ 12,15) - \$200,000$$
$$= \$60,000(6.8109) - \$30,000(6.8109) - \$200,000$$
$$= \$4327$$

$$(B - C)_B(12) = \$54,000(P|A\ 12,15) + 0.4(\$200,000)(P|F\ 12,15)$$
$$- \$30,000(P|A\ 12,15) - \$200,000$$
$$= \$54,000(6.8109) + \$80,000(0.1827) - \$30,000(6.8109)$$
$$- \$200,000 = -\$21,922$$

7.5.6 Tolls, Fees, and User Charges

Tolls, fees, and user charges have an interesting effect on the fiscal aspects of public projects. If a toll, fee, or user charge is regarded as a payment or partial payment for benefits derived, it can be argued that net benefits received are reduced by the amount of the payment. Similarly, the amount of the payment decreases the cost of the project to the government. Thus, the B/C ratio will change, but the $B - C$ measure of merit will remain constant so long as total user benefits remain constant.

Example 7.16

Suppose 10,000 people/year attend a public facility that has an equivalent uniform annual cost of $20,000. The people, on the average, receive recreational benefits in the amount of $3 each. The B/C ratio would be

$$B/C = \frac{3(\$10,000)}{\$20,000} = 1.5$$

and the $B - C$ measure of merit is

$$B - C = 3(\$10,000) - \$20,000 = \$10,000/\text{year}$$

Based on either criterion, the public facility appears worthwhile. Now suppose that a fee of $1.50/season is charged. The net benefits are now $3 − $1.50, or $1.50/person, and the government cost is reduced by $15,000/year. Thus, the B/C and $B - C$ measures are as follows.

$$B/C = \frac{\$30,000 - \$15,000}{\$20,000 - \$15,000} = 3$$

$$B - C = \$30,000 - \$15,000 - (\$20,000 - \$15,000) = \$10,000/\text{year}$$

Note that in the example B/C changed while $B - C$ did not. This phenome-non will be discussed in a subsequent section on "Problems With the B/C Ratio."

It might be concluded that tolls, fees, and user charges are irrelevant, at least with respect to the $B - C$ measure of merit. This, however, *is not* true if the number of users or degree of use is linked to the fee charged, as it almost always will be.

Example 7.17

As an extreme, suppose that the 10,000 users of the public facility in the previous example receive different levels of benefits, but they average out to $3/person. The actual breakout is that 8000 persons perceive $1.45 worth of enjoyment, 1000 persons perceive $3 worth, and 1000 persons expect to derive $15.40 in recreational benefits. With a user fee of $1.50/person, it is assumed only 2000 will patronize the facility. Thus, the B/C and $B - C$ measures would be:

$$B/C = \frac{1000(\$3) + 1000(\$15.40) - 2000(\$1.50)}{\$20,000 - 2000(\$1.50)} = \frac{\$15,400}{\$17,000} = 0.91$$

$$
\begin{aligned}
B - C &= [1000(\$3) + 1000(\$15.40) - 2000(\$1.50)] \\
&\quad - [\$20,000 - 2000(\$1.50)] \\
&= -\$1600
\end{aligned}
$$

In this case, the reaction of demand to a fee would cause the costs to exceed realized benefits. Thus, when tolls, fees, and user charges are ex-pected, their effect on user demand, and hence total user benefits, must be determined and accounted for.

7.5.7 Multiple-Use Projects

Multiple-use projects receive a great deal of attention, both pro and con. Multiple uses, and hence multiple benefits, are often available at slight incre-mental costs over single-use projects. Of course, the incremental capital and net operating costs required for an additional use must provide at least a like worth of benefits.

Example 7.18

A dam and reservoir for irrigation will provide present worth benefits of $25 million over the next 50 years. The present-worth cost of construction, operation, and maintenance of the irrigation facility will be $14,500,000. A single-purpose flood control dam providing present-worth benefits of $6 million would cost a present worth of $9 million. Suitable design modifications can be made to the irrigation dam and reservoir to provide the flood control benefits, too, at a total package present worth cost of $18,500,000. Funds permitting, what should be done?

First, it must be understood that the benefits and costs discounted to the present were discounted at the *MARR* deemed suitable by the decision maker. Assuming this to be true, it is clear that the irrigation project is worthwhile, providing a benefit-cost ratio of

$$B/C_{\text{irrigation}} = \frac{\$25,000,000}{\$14,500,000} = 1.72^1$$

As a single-purpose facility, a flood control dam would not provide benefits commensurate with its costs, yielding a *B/C* ratio of

$$B/C_{\text{flood control}} = \frac{\$6,000,000}{\$9,000,000} = 0.67^1$$

As a multiple-use facility, however, the flood control benefits may be provided at a sufficiently low incremental cost to be justifiable. The incremental *B/C* ratio is

$$\Delta B/\Delta C_{\substack{\text{irrigation plus} \\ \text{flood control}}} = \frac{\$6,000,000}{\$18,500,000 - \$14,500,000} = 1.5$$

Thus, a multiple-use facility should be built.

The example illustrates how multiple uses can draw on each other, providing benefits economically that could never have been provided using a single-purpose facility (e.g., the *B/C* flood control ratio of 0.67). Multiple-purpose projects also have their problems. For example, it is frequently desirable to "allocate" the costs of a project to its various uses.

Example 7.19

A city's refuse is used to fire a power-generation facility owned by a municipality. Not only is electrical power supplied to a segment of the city, but burning of the refuse after processing has virtually eliminated the need for an expensive landfill operation. Since this project is self-supporting, construction, operation, and maintenance costs must be allocated between the disposal and power benefits provided in order to determine the user charge for refuse disposal and electrical energy. Arguments for cost allocation range from (1) no costs should be allocated to refuse disposal because the refuse is

---

[1] This is also the incremental *B/C* ratio of irrigation over doing nothing.

being used in place of fuel oil or coal and, in fact, a credit should be issued, to (2) refuse disposal should receive sufficient cost allocation to raise rates above those for conventional disposal to include the aesthetic benefits of no unsightly public landfill.[2]

Another problem with multiple-purpose projects includes proponents from and opposition by various interest groups, on the basis of politics, environment, and the like. Yet another problem has to do with coordinating project finances, where some aspects may be federally financed with no payback required, other aspects may be self-supporting, and yet others may require state funds. Finally, there may be conflicts between the various purposes involved. That is, an enhancement of one use may detract from another use. Nevertheless, multiple-use projects are here to stay and will be confronted frequently by engineers in the public sector.

7.5.8 Problems with the B/C Ratio

There are two frequent problems with the B/C ratio that require an explanation and warning. Either can give misleading results that may cause an otherwise perfect analysis to point toward the wrong project.

First, it is sometimes difficult to decide whether an item is a benefit to the public or a cost savings to the government. Similarly, there is often uncertainty between disbenefits and costs.

Example 7.20

A project provides annual equivalent benefits of $100,000, disbenefits of $60,000/year, and annual costs of $5000. What is the benefit-cost ratio? Let us first calculate the B/C ratio for the problem as stated.

$$B/C = \frac{\$100,000 - \$60,000}{\$5000} = 8$$

Such a high ratio leads one to believe that the project is outstanding.

Now another analyst notes that the government will reimburse those incurring damages from the project in an amount equivalent to $60,000/year. Thus, the analyst concludes that the public disbenefits have been compensated for by the government and calculates a B/C ratio of

$$B/C = \frac{\$100,000}{\$5000 + \$60,000} = 1.54$$

which is considerably lower.

---

[2] These extremes in arguments have actually been used by public officials of one major city.

Example 7.20 shows that a wide range of B/C ratios may reasonably be obtained on a single project simply by interpreting certain elements of the problem differently. The resolution of this problem is not difficult. The analyst should simply calculate the net benefits less the net costs.

Example 7.21

Let us reconsider Example 7.20 and calculate the annual net benefits less net costs.

The first analyst would have calculated

$$B - C = (\$100,000 - \$60,000) - (\$5000) = \$35,000/\text{year}$$

and the second analyst would have calculated

$$B - C = (\$100,000) - (\$5000 + \$60,000) = \$35,000/\text{year}$$

Calculating $B - C$ eliminates the inherent bias in the B/C ratio and does not require an incremental approach between alternatives where benefits and costs are known directly for each alternative. That is, if mutually exclusive alternatives are involved, over the same time horizon, the one having the highest $B - C$ value should be selected. Unfortunately, the B/C ratio is by far the more popular criterion of the two.

The potentiality for the other B/C problem was illustrated in the dam and reservoir irrigation example. That is, when the B/C ratio is used, it should be based on *incremental benefits* and *incremental costs*.[3] Simply to calculate the B/C ratio of each alternative and take the one with the largest ratio is incorrect and will frequently lead to errors in project selection.

Example 7.22

In the dam and reservoir of Example 7.18 we calculated the B/C irrigation ratio to be 1.72. To compare this value against a total project $B/C_{\text{irrigation}+\text{flood control}}$ ratio of

$$\frac{\$25,000,000 + \$6,000,000}{\$18,500,000} = 1.68$$

would cause us to select irrigation only, in error.

---

[3] Actually, the B/C irrigation ratio in the referenced example is a ratio of incremental benefits to incremental costs, where the pairwise comparison is between irrigation and doing nothing.

A related error is to require the incremental B/C ratio to be above that for the previous incremental B/C ratio. In the dam and reservoir example, had the incremental B/C ratio, $\Delta B/\Delta C_{\text{irrigation plus flood control}} = 1.5$ been compared against the $B/C_{\text{irrigation}} = 1.72$ ratio, again an incorrect conclusion would have resulted. As long as the incremental B/C ratio exceeds 1, the incremental benefits justify the incremental costs. In this regard, the B/C ratio criterion is closely akin to the rate-of-return criterion discussed in Chapter Four.

7.6 COST-EFFECTIVENESS ANALYSIS

To this point it has been assumed that public project effects were measurable, either directly or indirectly, in monetary terms. There are circumstances, however, when project outputs are not measurable monetarily and must be expressed in physical units appropriate to the project. In these cases, *cost-effectiveness analysis* has proven to be a useful technique for deciding between projects or systems for the accomplishment of certain goals. Although cost-effectiveness is most often associated with the economic evaluation of complex defense and space systems, it has also proven useful in the social and economic sectors. In fact, it has roots dating back to Arthur M. Wellington's *The Economic Theory of Railway Location* in 1887 [14].

Cost-effectiveness analyses require that three conditions be met, according to Kazanowski [7]. They are:

1. Common goals or purposes must be identifiable and attainable.
2. There must be alternative means of meeting the goals.
3. There must be perceptible constraints for bounding the problem.

Common goals are required in order to have a basis for comparison. For example, it would not make sense to compare a submarine with a sophisticated single sideband communication network. Obviously, alternative methods of accomplishing the goals must be available in order to have a comparison. Finally, reasonable bounds for constraining the problem by time, cost, and/or effectiveness are necessary to limit and better define the alternatives to be considered.

7.6.1 The Standardized Approach

Kazanowski [7] presents 10 standardized steps that constitute a correct approach to cost-effectiveness analyses. They are, in their usual order, presented below.

Step 1. *Define the goals,* purpose, missions, and so forth, that are to be met. Cost-effectiveness analysis will identify the best alternative way of meeting these goals.

Step 2. *State the requirements* necessary for attainment of the goals. That is, state any requirements that are essential if the goals are to be attained.

Step 3. *Develop alternatives* for achieving the goals. There must be at least two alternative ways of meeting or exceeding the goals.

Step 4. *Establish* evaluation *measures* which relate capabilities of alternatives to requirements. An excellent list of evaluation criteria is contained in Kazanowski [8]. Typical measures are performance, availability, reliability, maintainability, and so forth.

Step 5. *Select* the fixed-effectiveness or fixed-cost *approach.* The fixed-effectiveness criterion is minimum alternative cost to achieve the specified goals or effectiveness levels. Alternatives failing to achieve these levels may either be eliminated or assessed penalty costs. The fixed-cost criterion is the amount of effectiveness achieved at a given cost. "Cost" is usually taken to mean a present worth or annual equivalent of "life-cycle cost," which includes research and development, engineering, construction, operation, maintenance, salvage, and other costs incurred throughout the life cycle of the alternative.

Step 6. *Determine capabilities* of the alternatives in terms of the evaluation measures.

Step 7. *Express* the alternatives and their *capabilities* in a suitable manner.

Step 8. *Analyze* the various *alternatives* based upon the effectiveness criteria and cost considerations. Often, some alternatives are clearly dominated by others and should be removed from consideration.

Step 9. *Conduct a sensitivity analysis* to see if minor changes in assumptions or conditions cause significant changes in alternative preferences.

Step 10. *Document all* considerations, analyses, and decisions from the above nine steps.

Clearly, one cannot become a cost-effectiveness expert by reading these 10 steps. They do, however give an indication of what is involved in a cost-effectiveness study. Excellent detailed presentations on cost effectiveness are available in English [3] and Blanchard [1].

Example-7.23

Three different propulsion systems are under consideration. The life-cycle cost is not to exceed $2.4 million. A single effectiveness measure is decided

TABLE 7.2

Evaluation of Candidate Propulsion
Systems for Example 7.23

Propulsion System	Life Cycle Cost in Millions	Reliability
1	2.4	0.99
2	2.4	0.98
3	2.0	0.98
4	2.0	0.97

on, that being reliability. Reliability would be defined as "the probability that the propulsion system will perform without failure under given conditions for a given period of time." Four contractors submit candidate systems, which are evaluated as shown in Table 7.2. Since there is only one effectiveness measure, the results given in Table 7.2 may be expressed graphically as shown in Figure 7.1.

If we look at these systems in pairs, comparing just 1 and 2 is equivalent to a fixed-cost comparison in which we prefer system 1 because of its higher reliability. Similar reasoning leads us to prefer system 3 over system 4. If we were to compare only systems 2 and 3, this would be a fixed-effectiveness comparison, in which case we prefer system 3 due to its lower cost. Clearly, system 1 dominates 2, 3 dominates 2, and 3 dominates 4. We are left with making the decision between systems 1 and 3. This choice will depend on whether or not an additional percent reliability justifies the expenditure of an additional $400,000.

In this example, it is tempting to go the next step and say that based on the

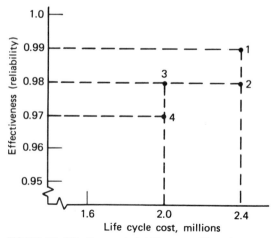

FIGURE 7.1. Effectiveness (reliability versus life-cycle cost.

ratios of reliability to cost, 0.99/\$2.4 million and 0.98/\$2.0 million, system 3 should be selected. Unfortunately, apparent as such a decision may seem, the correct decision may depend on many other considerations, not the least of which are the payload (perhaps human life) and the consequences of system failure (perhaps total destruction with no payroll recovery).

7.7 SUMMARY

Benefit-cost and cost-effectiveness analyses are accepted facts of life in the public sector. Because of the rapid shifting of the majority of employees from private industry to the public and service sectors, no doubt tomorrow's economic evaluators will be more and more closely involved with these techniques over time.

BIBLIOGRAPHY

1. Blanchard, Ben S., "Cost Effectiveness/Life-Cycle Cost Analysis," unpublished report, Virginia Polytechnic Institute and State University, Blacksburg, Virginia, 24061.

2. Cohn, Elchanan, *Public Expenditure Analysis*, D. C. Heath, 1972.

3. English, J. Morley (Editor), *Cost Effectiveness*, Wiley, 1968.

4. Haveman, R. H., "The opportunity cost of displaced private spending and the social discount rate." *Water Resources Research, 5* (5), 1969, pp. 947–57.

5. Henderson, P. D., "The Investment Criteria for Public Enterprises," in *Public Enterprises*, edited by Ralph Turvey, Penguin Books, 1968.

6. Howe, Charles W., *Benefit-Cost Analysis for Water System Planning*, Water Resources Monograph 2, American Geophysical Union, Washington, D.C., 1971.

7. Kazanowski, A. D., "A Standardized Approach to Cost-Effectiveness Evaluations," in *Cost Effectiveness*, edited by J. Morley English, Wiley, 1968.

8. Kazanowski, A. D., "Some Cost-Effectiveness Evaluation Criteria," Appendix B in *Cost Effectiveness*, edited by J. Morley English, Wiley, 1968.

9. McKean, Roland N., *Efficiency in Government Through Systems Analysis*, Wiley, 1958.

10. Park, Chan S., "Probabilistic Benefit-Cost Analysis," *The Engineering Economist, 29* (2) Winter 1984, Institute of Industrial Engineers, pp. 83–100.

11. Prest, A. R., and Turvey, Ralph, "Cost-Benefit Analysis: A Survey," *Economic Journal, 75*, December 1965, pp. 683–735.

12. Smith, Douglas L., "Quantifying Risk for Establishing Rates of Return in Regulated Industries," *The Engineering Economist, 28* (4) Summer 1983, Institute of Industrial Engineers, pp. 267–92.

13. Weisbrod, Burton A., *Economics of Public Health*, University of Pennsylvania Press, 1961, pp. 7–8.

14. Wellington, A. M., *The Economic Theory of Railway Location*, Wiley, 1887.
15. United States Code, 1940 edition, U.S. Government Printing Office, Washington, D.C., p. 2964.

PROBLEMS

1. Identify the benefits, both monetary and intangible, that would accrue to the public for the following projects: (a) a museum, (b) a fence enclosed walkway over a busy four-lane road connecting two city subdivisions, one having a playground and park, (c) a city sanitary system, and (d) a visiting troupe of performers to appear in a downtown city mall for several days. (7.3)

2. Two four-lane roads intersect and traffic is controlled by a standard green, yellow, red stoplight. From each of the four directions a left turn is permitted from the inner lane; however, this impedes the flow of traffic while the person desiring to turn left waits until a safe turn may be accomplished. The light operates on a cycle allowing 30 seconds of green–yellow, followed by 30 seconds of red light for each direction. Approximately an eighth of the 8000 vehicles using the intersection daily are held up for one full minute cycle of the light, and made to perform one extra stop-start operation, solely because of the left turn bottleneck. These delays are only during the 260 working days/year. A stop-start costs 1.0¢/vehicle. A widening of the intersection to accommodate a left-turn lane plus a new lighting system to provide for a left-turn signal will cost $100,000. The cost of time for commercial traffic is three times that of private traffic, and private traffic accounts for 80% of the vehicles. How much must commercial traffic time be worth to justify the intersection changes if the interest rate is 8% and a lifetime of 10 years is expected? (7.4)

3. Three alternatives are available, A, B, and C. Their respective annual benefits, disbenefits, costs, and savings are as follows:

	A	B	C
Benefits	$200,000	$300,000	$400,000
Disbenefits	42,000	75,000	118,000
Costs	150,000	250,000	325,000
Savings	15,000	40,000	55,000

   (a) Calculate the B/C ratios for each project. Can you tell from these calculations which should be selected?

   (b) Determine which should be selected using the incremental B/C ratio.

   (c) Calculate $B - C$ for each alternative. (7.4)

4. The following costs and benefits have been listed for a dam and reservoir which will be used for both flood control and electrical power generation.
   Investment

Dam, including access roads, clearing, and foundation treatment	$31,330,000
Generation equipment and transmission apparatus	16,176,000

Land	2,200,000
Highway relocation	2,770,000
Miscellaneous	180,000

Operating and maintenance costs
Two percent of investment during the first year and increasing by 5% of itself each subsequent year.
Annual benefits

Flood losses prevented	$ 835,000
Property value enhancement	149,000
Power value	3,190,000

For a planning horizon of 50 years and $i = 5\%$, what is the B/C ratio? (7.4)

5. A city is trying to decide between coal, fuel oil 3, low-sulphur fuel oil, and natural gas to power their electrical generators. Fuel forecasts indicate the following needs for the upcoming year, and the gradient increase during each subsequent year.

Fuel	First Year	Gradient Each Subsequent Year
Coal	460,000 tons	23,000 tons
Fuel oil 3	1,760,000 barrels	88,000 barrels
Low-sulphur fuel oil	1,820,000 barrels	91,000 barrels
Natural gas	11,000 10^6 cubic feet	550 10^6 cubic feet

The cost of the fuel, transportation, and various pollution effects have been estimated as follows,

Fuel	Cost	Transportation and Storage	Health	Crops	Uncleanliness
Coal	$5.80/ton	$2.25/ton	$0.70/ton	$1.75/ton	$1.05/ton
Fuel oil 3	2.15/barrel	0.20/barrel	0.10/barrel	0.25/barrel	0.15/barrel
Low-sulphur fuel oil	2.75/barrel	0.20/barrel	0.03/barrel	0.075/barrel	0.045/barrel
Natural gas	0.00058/cubic foot	0.00002/cubic foot	Negligible	Negligible	Negligible

Calculate the annual equivalent benefits and costs of these fuels considering a life of 30 years and (a) $i = 5\%$ and (b) $i = 10\%$. Use an incremental B/C analysis to determine the best fuel to use. (7.4)

6. A highway is to be built connecting Baldhill with Broken Arrow (Baldhill is south of Bixby). Route A follows the old road and costs $4 million initially and $210,000/year thereafter. A new route, B, will cost $7 million initially and $180,000/year thereafter. Route C is simply an enhanced version of route B with wider lanes, shoulders, and so on. It will cost $9 million at first, plus $240,000/year to maintain. Relevant annual user costs considering time, operation, and

safety are $1 million for A, $700,000 for B, and $500,000 for C. Using a *MARR* of 8%, a 15-year study period, and a residual (salvage) value of 50% of first cost, which should be constructed? Use a *B/C* analysis. Which route is preferred if $i = 0$? (7.4)

7. Solve problem 6 using the $B - C$ criterion. (7.4)

8. A municipal zoo is to be enlarged, and the initial cost of the enlargement for physical facilities will be $400,000. Animals for the addition will cost another $100,000. Maintenance, food, and animal care will run $62,000/year. The zoo is expected to be in operation for an indefinite period; however, a study period of 20 years is to be assumed, with a residual (salvage) value of 50% for all physical facilities. Interest is 8%. An estimated 300,000 persons will visit the zoo each year, and they will receive, on the average, an additional $0.50/person in enjoyment when the new area is complete. Should the new area be built? Use a $B - C$ measure of worth. (7.4)

9. A proposed expressway is under study and is found to have an unfortunately high annual cost of $2,200,000 as compared to benefits of only $1,300,000. However, with only inconsequential changes in design, the highway may be eligible for incorporation into the interstate highway system, in which case 90% of the cost would be paid by the federal government. If so, it is argued that the expressway would cost only $220,000/year, yielding a handsome *B/C* ratio of 5.91. Is this reasoning sound? Discuss the reasoning from different points of view. (7.5)

10. Seven projects are available as summarized below. It is desired to select only the projects that government can afford on an available budget of $75,000 for first cost. Operating and maintenance costs are no worry. All have a favorable *B/C* ratio at the cost of money, $i = 6\%$. Raise the interest rate until only those projects remain that continue to have a $B/C > 1$ and that government can afford. Use a 10-year planning horizon. Which projects are selected? What is the opportunity cost of those investments foregone by government? (7.5)

Project	First Cost	Operating and Maintenance	Residual Value After 10 Years	Benefits/Year
A	$33,000	$3,000	$16,000	$14,400
B	27,000	1,500	8,000	7,630
C	41,000	2,500	12,000	10,800
D	38,000	1,600	14,500	8,460
E	30,000	4,200	16,000	8,600
F	34,000	600	8,000	5,400
G	25,000	5,100	3,000	8,980

11. Three projects, each having a first cost of $1 million and annual operating costs of $100,000 are proposed. The lives of Projects A, B, and C are 20, 30, and 40 years, respectively, after which the project will be over, providing no benefits and requiring no costs. The annual benefits provided over the lives of Projects A, B, and C will be $268,000, $250,000, and $243,000, respectively.

Select only one project to be implemented. Use a present worth basis and a $B - C$ measure of worth. First, use an $MARR$ of 5%, and a study period of 40 years. Projects A and B will be void during the last 20 and 10 years of the study period, respectively. Second, continually increase the interest rate, again using a study period of 40 years, until only one project remains with a favorable $B - C$ value. Do the two methods point to the same project? Why or why not? (7.5)

12. Many benefits resulting from public sector projects may be argued to be at the expense of someone else, thus counterbalancing the supposed direct benefits. For example, in Example 7.1, operating costs saved include the cost of fuel, oil, tires, wear and tear, and the like. This is money that will *not* be spent at service stations, tire stores, and new car dealers. The safety costs saved include fees that otherwise would go to lawyers, hospitals, doctors, auto repair shops, and so forth. Do you think the fact that the direct savings represent lost revenues to others should void or nullify these types of benefits in a benefit-cost analysis? Why or why not? (7.5)

13. Comment on the following analysis if you think it is incorrect. Benefits of $1.75 each are now received by 9000 persons and benefits are perceived to be $3 each by another 9000 users. The annual cost of the recreational facility is $36,000. Commissioners have argued that an entrance fee of $2 should be charged to make the facility self-supporting. They argue that $2 is below the average benefit received per person. They do point out, though, that 9000 persons do not perceive but a $1.75 recreational value, and hence a $0.25 disbenefit per person should be noted. The B/C ratio is then

$$B/C = \frac{\$3(9,000) + \$1.75(9,000) - \$0.25(9,000)}{36,000 - \$2(18,000)} = \infty$$

and the user fee of $2 should definitely be implemented immediately. What do you think? (7.5)

14. A recreation area suitable for camping, picnicking, hiking, and water sports is to be developed at an annual equivalent cost of $112,500, including initial cost of construction (clearing, water, restrooms, road, etc.), operation, upkeep, and security. An average of 50 families will camp each night throughout the year. In addition, another 100 persons will be admitted for day use of the recreational facilities. The perceived benefits provided to overnight and day users will vary, due to the subjective nature of people. However, the average family camping at the area is estimated to be willing to pay $5 for the privilege, and the average day user $1. What is the B/C ratio of this recreational area? $B - C$? Should it be built? (7.5)

15. In problem 14, assume that 50% of the potential camping families perceive $5 in benefits, 25% perceive $7.50 in benefits, and 25% expect to receive $10 worth of recreation. If a charge of $7.50 is imposed on campers only, and day users are permitted free, recalculate B/C and $B - C$. Should the facility be built? (7.5)

16. Reconsider problem 8 and assume that half of the people will derive $0.50 of enjoyment per person, while the other half anticipate only $0.25 in benefits. With this information available, and assuming that an incremental $0.25 entrance fee/person will be charged separately for attendance to the addition, do you recommend building the addition? (7.5)

17. Analysts recognize that with a $2 entrance fee as in problem 13, only 9000 persons will utilize the facility, and the benefits derived by 9000 more persons at $1.75 each will be lost. They argue that the *true* benefits are those received only by 9000 persons at $3 each, less the $2 fee, less a disbenefit of $1.75(9000) for the persons who chose not to pay $2 to enter and lose the previous recreational enjoyment from the facility. As such, the B/C ratio is as follows:

$$B/C = \frac{\$3(9,000) - \$2(9,000) - \$1.75(9,000)}{36,000 - \$2(9,000)} = -.38$$

Do you agree with this analysis? Why or why not? (7.5)

18. A government has the following estimates:

Annual benefits	$1,150,000
Annual disbenefits	850,000
Annual costs	765,000
Annual savings	750,000

(a) Calculate the B/C ratio.
(b) Mistakenly treating disbenefits as costs and savings as benefits, calculate the B/C ratio.
(c) Calculate $B - C$.
(d) Which do you prefer, $B - C$ or B/C? Why? (7.5)

19. Six subsystem designs have been proposed for a critical part of a communications network. Each will be "burned in" to eliminate the infant mortality or early failure problem when the subsystem is placed in service. Hence, reliability of the subsystem will have a constant hazard rate. That is, the failure density function will be exponential and reliability may be expressed as

$$R(t) = e^{-\lambda t}$$

where t is the mission time. Mission time for the subsystem will be one year, as a thorough annual preventative maintenance and recalibration procedure is standard, returning the subsystem to "new" condition. The unit will be on call 24 hours/day, seven days/week. The design data are as follows:

System	Annual Equivalent Initial Cost	Annual Maintenance Cost	λ
A	$10,000	$2,000	1.1446886×10^{-5}
B	10,800	1,700	1.0989011×10^{-5}
C	11,200	2,100	1.175824×10^{-5}
D	12,000	1,600	1.070000×10^{-5}
E	12,000	1,800	1.081250×10^{-5}
F	14,100	1,200	0.998000×10^{-5}

Using a cost-effectiveness analysis, plot reliability versus annual cost. Which subsystems can be eliminated from consideration? How might you decide among the rest? (7.6)

20. Four designs are under consideration for a space shuttle designed to carry a particular type of payload. It is estimated that each will have equal lives; however, the first cost, operating cost, and renovation cost, as well as the salvage value, will differ, as will the size of the payload that may be carried. These characteristics are summarized below for shuttles having lives of 3 years, and fired once per year at the end of the year.

Shuttle Design	First Cost	Operating Cost/Firing	Renovation Cost/Firing	Salvage Value	Payload in Units
1	$28,000,000	$8,000,000	$1,400,000	$2,800,000	180
2	32,000,000	6,400,000	2,400,000	3,200,000	155
3	34,000,000	7,200,000	1,900,000	3,400,000	140
4	40,000,000	5,600,000	1,200,000	4,000,000	175

Based on $i = 9\%$, and using a cost-effectiveness analysis, which designs may be eliminated, and which should be considered further? (7.6)

CHAPTER

8

Break-Even, Sensitivity, and Risk Analyses

8.1   INTRODUCTION

In the previous chapters we have assumed that all of the values of the parameters of the economic models were known with certainty. In particular, correct estimates of the values for the length of the planning horizon, the minimum attractive rate of return, and each of the individual cash flows were assumed to be available. In this chapter we consider the consequences of estimating the parameters incorrectly.

The discussion will concentrate on answering a number of "what if . . ." questions concerning the effects of different parameter values on the measure of economic effectiveness of interest. For example, when we are completely *uncertain* of the possible values a parameter can take on, we will be interested in determining the set of values for which an investment alternative is justified economically and the set of values for which an alternative is not justified; this process is called *break-even analysis*. Alternatively, when we are reasonably sure of the possible values a parameter can take on, but *uncertain* of their chances of occurrence, we will be interested in the sensitivity of the measure of merit to various parameter values; this process is referred to as *sensitivity analysis*. Finally, when probabilities can be assigned to the occurrence of the various values of the parameters, we can make probability statements concerning the values of the measure of merit for the various alternatives; this process is referred to as *risk analysis*.

The conditions that lead to break-even and sensitivity analyses are described in the economic analysis literature as conditions under *uncertainty*,

371

since one is completely uncertain of the chances of a parameter taking on a given value. When the conditions are such that probabilities can be assigned to the various parameter values, the decision environment that results is said to be a decision under *risk*; hence, the term risk analysis is used.

One of the most perplexing (and frustrating) experiences for the engineering economic analyst is to encounter an individual who refuses to analyze systematically the economic performances of investment alternatives because "it's not possible to estimate future cash flows exactly." Two responses come to mind. First, failure to analyze the alternatives systematically implies a qualitative judgment will be made, based on even more imperfect information than would be available using a systematic approach. Second, the selection decision might not require precise estimates of the cash flows. In the latter case, the selection decision might be relatively insensitive to a broad range of possible values for the cash flows.

The purpose of break-even and sensitivity analyses is to *reduce* the amount of information needed to make a good decision. Instead of needing a "point" estimate for, say, the salvage value of a piece of equipment, an "interval" estimate might be sufficient. Break-even analysis can reduce the forecasting requirement to determining if the magnitude of a cash flow will exceed a specific value.

It is important to recognize that break-even, sensitivity, and risk analyses are intended to make more realistic the economic comparison of investment alternatives. Without using such approaches, one must make decisions under the assumption of perfect information regarding the future.

In Chapter Nine we will present *prescriptive* or *normative* models under uncertainty and risk conditions. Models will be presented that allow us to *prescribe* the best or optimal course of action using a number of different decision criteria. In this chapter we present *descriptive* models under uncertainty and risk conditions. Instead of attempting to *prescribe* the action to be taken, we attempt to *describe* the behavior of the measure of merit under conditions of uncertainty and risk. Thus, we view break-even, sensitivity, and risk analyses as descriptive processes, not normative processes. Our objective in this chapter will be to gain insight into the behavior of the measure of merit under conditions of uncertainty and risk; the decision of how to respond, given the information from the sensitivity analysis depends on the decision process employed by the decision maker. Hence, the decision models from Chapter Nine may be used to assist the decision maker in making the final selection.

8.2 BREAK-EVEN ANALYSIS

Although we did not label the process as such, in Chapter Three we performed a number of break-even analyses; there we referred to the process as *equivalence*. In particular, in presenting the concept of equivalence, a situation was posed, and you were asked to determine the value of a particular parameter in order for two cash flow profiles to be equivalent. Another way of stating the problem could have been "Determine the value of X that will yield a break-even situation between the two alternatives." In this case, X denotes the parameter whose value is to be determined. Additionally, when the cash flow profiles for two alternatives are equivalent, a break-even situation can be said to exist between the two alternatives.

Another application of break-even analysis occurs in the use of the rate-of-return method. Specifically, the internal rate of return can be interpreted as the break-even value of the reinvestment rate, since such a reinvestment rate will yield a zero future worth for either an individual alternative or the differences in two alternatives.

Break-even analysis is certainly not an unfamiliar concept. Furthermore, the information obtained from a break-even analysis can be of considerable aid to anyone faced with an investment alternative involving a degree of uncertainty concerning the value of some parameter. The term *break-even* is derived from the desire to determine the value of a given parameter that will result in neither a profit nor a loss, as discussed in Chapter Two.

Example 8.1

Suppose a firm is considering manufacturing a new product and the following data have been provided:

Sales price	$12.50/unit
Equipment cost	$200,000
Overhead cost	$50,000/year
Operating and maintenance cost	$25/operating hour
Production time/1000 units	100 hours
Planning horizon	5 years
Minimum attractive rate of return	15%

Assuming a zero salvage value for all equipment at the end of 5 years, and letting X denote the annual sales for the product, the annual worth for the investment alternative can be determined as follows:

$$AW(15\%) = -\$200,000(A|P\ 15,5) - \$50,000 - 0.100(\$25)X + \$12.50X$$
$$= -\$109,660 + \$10.00X$$

Solving for X yields a *break-even* sales value of 10,966/year. If it is felt that annual sales of at least 10,966 units can be achieved each year, then the alternative appears to be worthwhile economically. Even though one does not know with certainty how many units of the new product will be sold annually, it is felt that the information provided by the break-even analysis will assist management in deciding whether or not to undertake the new venture.

A graphical representation of the example is given in Figure 8.1. The chart is referred to as a *break-even* chart, since one can determine graphically the

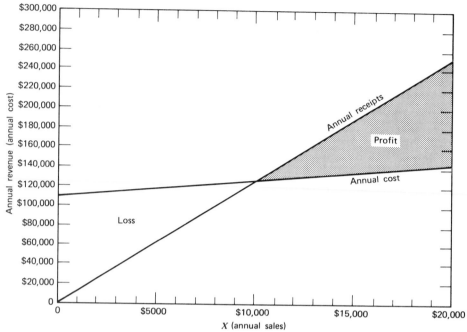

FIGURE 8.1. Break-even chart.

break-even point by observing the value of X when annual revenue equals annual cost.

Example 8.2

Consider a contractor who experiences a seasonal pattern of activity for compressors. The manager currently owns eight compressors and suspects that this number will not be adequate to meet the demand. The contractor realizes that there will arise situations when more than eight compressors will be required, and is considering purchasing an additional compressor for use during heavy demand periods.

A local equipment rental firm will rent compressors at a cost of $50/day. Compressors can be purchased for $6000. The difference in operating and maintenance costs between owned and rented compressors is estimated to be $3000/year.

Letting X denote the number of days a year that more than eight compressors are required, the following break-even analysis is performed. A planning horizon of 5 years, zero salvage values, and 20% minimum attractive rate of return are assumed.

Annual worth (purchasing compressor)

$$AW_1(20\%) = -\$6000(A|P\ 20,5) - \$3000$$
$$= -\$5006.40$$

Annual worth (renting compressor)

$$AW_2(20\%) = -\$50X$$

Setting the annual worths equal for the two alternatives yields a *break-even value* of $X = 100.128$ days/year.

Hence, if the contractor anticipates that a demand will exist for an additional compressor more than 100 days/year over the next 5 years, then an additional compressor should be purchased.

Example 8.3

Let us extend the previous example problem to the question of how many compressors should be purchased. It is interesting to note that compressors should continue to be purchased so long as each one reduces the number of compressor-rental-days by at least 100/year over the 5-year period. To illustrate, if the contractor provides a forecast of daily demand for compressors as given in Table 8.1, it is seen that one additional compressor should be purchased.

To show that one more compressor should be purchased, let us compute

TABLE 8.1

Forecasts of Demand for Compressors, Example 8.3

	X Number of Compressors Demanded On a Given Day	$f(X)$ Number of Days per Year Demand Equals X
	≤ 8	140
	9	40
	10	30
	11	20
	12	20
	13	10
	≥ 14	0

N Number of Compressors Owned by the Contractor	Number of Compressor-Rental-Days if N are Owned	Difference in Number of Compressor-Rental- Days if N versus $N + 1$ Are Owned
8	$40(1) + 30(2) + 20(3) + 20(4) + 10(5) = 290$	—
9	$30(1) + 20(2) + 20(3) + 10(4) = 170$	120
10	$20(1) + 20(2) + 10(3) = 90$	80
11	$20(1) + 10(2) = 40$	50
12	$10(1) = 10$	30
13	0	10

the annual worths for the alternatives of purchasing 0, 1, 2, and 3 additional compressors. From Table 8.1,

$$AW_0(20\%) = -\$50(290)$$
$$= -\$14,500$$

$$AW_1(20\%) = -\$6000(A|P\ 20,5) - \$3000 - \$50(170)$$
$$= -\$13,506.40$$

$$AW_2(20\%) = -\$12,000(A|P\ 20,5) - \$6000 - \$50(90)$$
$$= -\$14,512.80$$

$$AW_3(20\%) = -\$18,000(A|P\ 20,5) - \$9000 - \$50(40)$$
$$= -\$17,019.20$$

Thus, one additional compressor should be purchased in order to provide a total of nine compressors to meet annual demand.

Example 8.4

As a fourth illustration of break-even analysis, consider an office building that is being converted from a coal-burning furnace to a furnace that burns either natural gas or number two fuel oil. The cost of converting to natural gas is estimated to be $50,000 initially; additionally, annual operating and maintenance costs are estimated to be $6000 less than that experienced currently. Approximately 1000 Btu's are produced per cubic foot of natural gas; it is estimated natural gas will cost $0.001/cubic foot. The cost of converting to number two fuel oil is estimated to be $110,000 initially; annual operating costs are estimated to be $4000 less than that experienced using the coal furnace. Approximately 140,000 Btu's are produced per gallon of fuel oil; number two fuel oil is anticipated to cost $0.12/gallon. A planning horizon of 20 years is used; zero salvage values and a 20% minimum attractive rate of return are assumed. It is desired to determine the break-even value for annual Btu requirements for the heating system.

Letting X denote the Btu requirement per year, the equivalent uniform annual cost for the natural gas alternative is found to be

$$EUAC_1(10\%) = \$50,000(A|P\ 20,20) - \$6000 + \frac{\$0.001}{1000}X$$
$$= \$4270 + \$1.0 \times 10^{-6}X$$

Similarly, for the number two fuel oils, the equivalent uniform annual cost will be

$$EUAC_2(10\%) = \$110,000(A|P\ 20,20) - \$4000 + \frac{\$0.12}{140,000}X$$
$$= \$18,594 + \$0.8571 \times 10^{-6}X$$

Equating the annual costs and solving for X yields a break-even value of approximately 10.035×10^9 Btu's/year. Since a consulting engineer has estimated that Btu requirements will be considerably less than the break-even value, it is decided to convert to natural gas.

Example 8.5

A firm is contemplating purchasing versus leasing a computer. If the computer is purchased, a service contract can be entered into with the computer supplier for maintenance and software support. Alternatively, the company can perform its own maintenance and software support. Finally, if the computer is leased, the lease cost will include maintenance and software support.

The following cost estimates are available for the 5-year study period. Outright purchase of the computer will cost $500,000 initially. Purchase of the computer with a 5-year service contract will be $600,000 initially, followed by four annual payments of $55,000. Alternately, the computer can be leased for $165,000/year payable at the beginning of the year.

The firm is uncertain as to the cost of doing its own maintenance and software support. However, they do believe the first year's cost will be approximately $40,000 and that it will increase at an unknown annual compound rate. Additionally, the firm is unsure of the *MARR* to be used; however, they have agreed to use either 10% or 15% in their calculations.

Using break-even analysis to provide a recommendation to the firm reveals the following:

Present Worth of Purchase with Service Contract Alternative

$$PW(10\%) = -\$600,000 - \$55,000(P|A\ 10,4)$$
$$= -\$774,344.50$$

$$PW(15\%) = -\$600,000 - \$55,000(P|A\ 15,4)$$
$$= -\$757,025$$

Present Worth of Lease Alternative

$$PW(10\%) = -\$165,000 - \$165,000(P|A\ 10,4)$$
$$= -\$688,033.50$$

$$PW(15\%) = -\$165,000 - \$165,000(P|A\ 15,4)$$
$$= -\$636,075$$

The lease alternative is preferred to the alternative of purchasing with a service contract regardless of the *MARR* selection. Thus, the problem reduces to a comparison of outright purchasing the computer versus leasing the computer.

Present Worth of Purchase Alternative

$$PW(10\%) = -\$500,000 - \$40,000(P|A\ 10,j,5)$$

$$PW(15\%) = -\$500,000 - \$40,000(P|A\ 15,j,5)$$

Equating the present worths of leasing and purchasing yields the following:

$$PW(10\%) = -\$688,033.50 = -\$500,000 - \$40,000(P|A\ 10,j,5)$$

or

$$4.7008 = (P|A\ 10,j,5)$$

or

$$j \approx 11.9\%$$

$$PW(15\%) = -\$636,075 = -\$500,000 - \$40,000(P|A\ 15,j,5)$$

or

$$3.4019 = (P|A\ 15,j,5)$$

or

$$j \approx 1.0\%$$

From the break-even analysis it is concluded that the computer should be leased if *MARR* is 15%, since at least a 1% annual increase in maintenance and software support is anticipated. If the *MARR* is 10%, then an annual increase of at least 11.9% in maintenance and software support costs is needed before leasing is preferred. Based on the analysis the firm decides to lease the computer. Do you agree with their decision?

Example 8.6

A manufacturing firm has the following annual revenue and cost functions:

$$\text{Annual revenue} = AR(x) = \$1500x - \$0.05x^2$$

$$\text{Annual cost} = AC(x) = \$1,200,000 + \$1000x - \$0.01x^2$$

where x is the annual production volume. It is desired to determine the break-even production volume.

The net profit before taxes is equal to the difference in the revenue and cost functions, or

$$\text{Net profit} = NP(x) = \$500x - \$0.04x^2 - \$1,200,000$$

Setting the net profit to zero and solving the quadratic equation for x yields values of $x = 3240$ and $x = 9260$. Thus, the operation will be profitable so long as annual production is between 3240 and 9260 units.

Example 8.7

A manufacturing firm is performing a make-or-buy analysis for a particular component. They find that the cost to manufacture x units in-house is

$$TC(x) = \$1000 + \$150x - \$0.01x^2$$

A quotation has been received from an outside supplier to provide x units at a cost of $3000 + \$65x$. For what range of production volumes should the component be purchased from the outside supplier?

Equating the two costs and solving for x gives the following:

$$\$1000 + \$150x - \$0.01x^2 = \$3000 + \$65x$$

$$\$0.01x^2 - \$85x + \$2000 = 0$$

$$x = 23.6 \text{ and } 8476$$

Thus, for a production volume between 24 and 8476, the firm should purchase the components. If more than 8476 units (or less than 24 units) are to be produced, then the firm should manufacture the component.

8.3 SENSITIVITY ANALYSIS

Break-even analysis is normally used when an accurate estimate of a parameter cannot be provided, but intelligent judgments can be made as to whether or not the parameter's value is less than or greater than some break-even value. Sensitivity analysis is used to analyze the effects of making errors in estimating parameter values.

Although the analysis techniques employed in break-even analysis and sensitivity analysis are quite similar, there are some subtle differences in the objectives of each. Because of the similarities in the two, it is not uncommon to see the terms used interchangeably.

Example 8.8

To illustrate what we mean by a sensitivity analysis, consider the investment alternative depicted in Figure 8.2. The alternative is to be compared against the do-nothing alternative, which has a zero present worth. If errors are made in estimating the size of the required investment ($10,000), the magnitude of the annual receipt ($3000), the duration of the project (5 years), the minimum attractive rate of return (12%), and/or the form of the series of receipts (uniform versus nonuniform), then the economic desirability of the alternative might be affected. It is anticipated that the future states (possible values) for each parameter will be contained within an interval having a

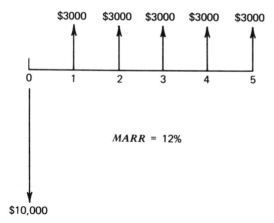

FIGURE 8.2. Cash-flow diagram.

range from −40 to +40% of the initial estimate. Hence, in the uncertain environment we have defined, an infinite number of future states are to be considered in the continuum from −40 to +40% of the initial estimate.

If it is assumed that all estimates are correct except the estimate of annual receipts, the annual worth for the alternative can be given as

$$AW(12\%) = -\$10,000(A|P\ 12,5) + \$3000(1 + X)$$

where X denotes the percent error (decimal equivalent) in estimating the value for annual receipts. Plotting annual worth as a function of the percent error in estimating the value of annual receipts yields the straight line having positive slope shown in Figure 8.3. Performing similar analyses for the initial investment required, the duration of the investment (planning horizon) and the minimum attractive rate of return yields the results given in Figure 8.3.

As shown in Figure 8.3, the net annual worth for the investment is affected differently by errors in estimating the values of the various parameters. The net annual worth is relatively insensitive to changes in the minimum attractive rate of return; in fact, as long as the *MARR* is less than approximately 15.25%, the investment will be recommended. A *break-even* situation exists if either the annual receipts decrease by approximately 7.47% to $2774/year or the required investment increases by approximately 8.14%, or $10,814. If the project life is 4 years or less, then the investment will not be profitable.

The analysis depicted in Figure 8.3 examines the sensitivity of individual parameters one at a time. In practice, estimation errors can occur for more than one parameter. In such a situation, instead of a sensitivity curve, a sensitivity surface is needed.

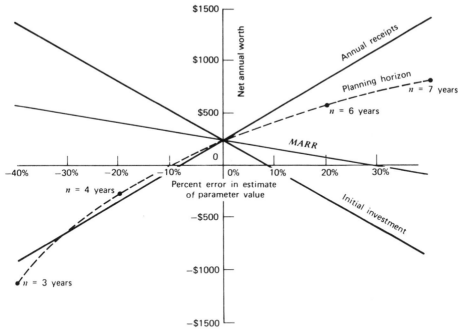

FIGURE 8.3. Deterministic sensitivity analysis.

Example 8.9

Consider an investment alternative involving the modernization of a warehousing operation in which automated storage and retrieval equipment is to be installed in a new warehouse facility. The building has a projected life of 30 years, and the equipment has a projected life of 15 years. A minimum attractive rate of return of 15% is to be used in the economic analysis. It is anticipated that the new warehouse system will require 70 fewer employees than the present system. Each employee costs approximately $18,000/year. The building is estimated to cost $2,500,000, and the equipment is estimated to cost $3,500,000. Annual operating and maintenance costs for the building and equipment are estimated to be $150,000/year more than the current operation. Existing equipment and buildings not included in the new warehouse have terminal salvage values totaling $600,000. Investing in the new warehouse will negate the need to replace existing equipment in the future; the present-worth savings in replacement cost is estimated to be $200,000. The estimated cash flow profile for the investment alternative is given in Table 8.2.

A 30-year planning horizon is to be employed in the analysis. Since equipment life is estimated to be 15 years, it is assumed that identical replacement equipment will be purchased after 15 years. Furthermore, since constant worth dollar estimates are being used in contending with inflationary effects,

TABLE 8.2

Estimated Cash Flows for Warehouse System, Example 8.9

EOY	Building	Equipment	Labor Savings	Operating and Maintenance	Salvage[a]	Total
0	−$2,500,000	−$3,500,000			$800,000	−$5,200,000
1–15			$1,260,000	−$150,000		1,110,000
15		− 3,500,000				− 3,500,000
16–30			1,260,000	− 150,000		1,110,000
30	0	0				0

[a] Includes present worth of savings in equipment replacement.

it is assumed that the replacement equipment will have cash flows that are identical to those that occur during the first 15 years.

The architectural and engineering estimate of $2,500,000 for the building is believed to be quite accurate, as are the estimates of $150,000 for annual operating and maintenance costs, terminal salvage values totaling $600,000, and savings of $200,000 in replacement costs. However, it is felt that the estimate of $3,500,000 for equipment and the labor savings estimate of 70 employees are subject to error. A sensitivity analysis for these two parameters is to be performed on a before-tax basis.

Letting x denote the percent error in the estimate of equipment cost and y denote the percent error in estimating the annual labor savings, it can be seen that the warehouse modernization will be justified economically, if

$$PW = -\$2,500,000 - \$3,500,000(1 + x)[1 + (P|F\ 15,15)]$$
$$-\$150,000(P|A\ 15,30) + \$18,000(70)(1 + y)(P|A\ 15,30)$$
$$+\$800,000 \geq 0$$

or if

$$PW = \$1,658,110 - \$3,930,150x + \$8,273,160y \geq 0$$

Solving for y gives

$$y \geq -0.2004 + 0.47505x$$

Plotting the equation, as shown in Figure 8.4, indicates the favorable region ($PW > 0$) lies above the break-even line, and the unfavorable region ($PW < 0$) lies below the break-even line.

If no errors are made in estimating the equipment cost, that is, $x = 0$, then up to a 20.04% reduction in annual labor savings can be tolerated. Likewise, if no errors are made in estimating the annual labor savings, that is, $y = 0$, then up to a 42.185% increase in equipment cost could occur, and the warehouse modernization would continue to be justified economically.

Since very little of the ±20% estimation error zone results in a negative present worth, it appears that the recommendation to modernize the ware-

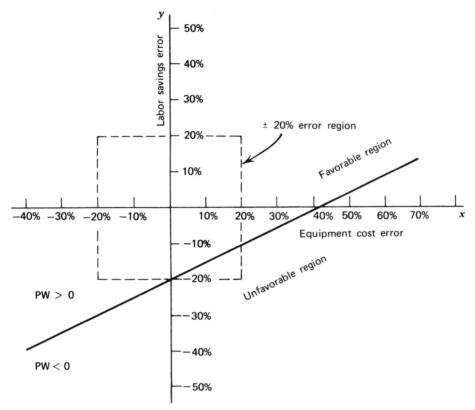

FIGURE 8.4. Multiparameter sensitivity analysis.

house is insensitive to errors in estimating either equipment cost, labor savings, or both. However, the decision is more sensitive to errors in estimating labor savings than in estimating equipment costs.

(This particular example illustrates the difficulties in distinguishing between break-even and sensitivity analyses. However, no matter what you call it, the analysis can be quite beneficial in gaining added understanding of the possible outcomes associated with a new venture involving a large capital investment.)

Example 8.10

To illustrate the use of sensitivity analysis in comparing investment alternatives, consider three mutually exclusive alternatives having cash flow profiles as given in Table 8.3.

Suppose some uncertainty exists concerning the appropriate *MARR* to use in the analysis. As depicted in Figure 8.5, Alternative 3 is preferred if the *MARR* is less than approximately 17%; Alternative 1 is preferred if the *MARR* is greater than 17%; and Alternative 2 is never preferred.

TABLE 8.3

Cash Flow Profiles for Three Mutually
Exclusive Investment Alternatives,
Example 8.10

t	A_{1t}	A_{2t}	A_{3t}
0	$0	−$10,000	−$15,000
1	− 12,000	− 9,000	− 9,000
2	− 12,000	− 9,000	− 8,000
3	− 12,000	− 9,000	− 7,000
4	− 12,000	− 9,000	− 6,000
5	− 12,000	− 9,000	− 5,000

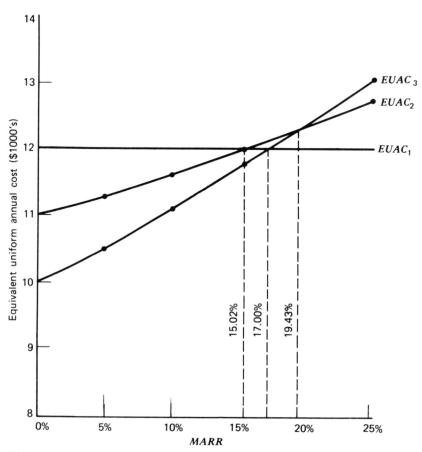

FIGURE 8.5. Sensitivity of equivalent uniform annual cost to the *MARR*.

If the internal rate-of-return method were used to compare the alternatives, the differences in Alternatives 1 and 2 would be analyzed, and a rate of return (break-even value) of 15.02% would be obtained; next, Alternatives 2 and 3 would be compared, and a break-even value of 19.43% would be obtained. The combination of Alternatives 1 and 3 would not be considered when the rate of return method is used.

Example 8.10 illustrates two points: the rates of return obtained from the internal rate-of-return method are actually break-even values for the minimum attractive rate of return for the combinations of alternatives considered; and sensitivity analysis and break-even analysis are closely related.

The final approach we consider in performing sensitivity analyses assumes the possible number of values for the parameters can be reasonably represented by three values for each: pessimistic, most likely, and optimistic estimates will be provided. Given the three estimates of the value of each parameter, the value of the measure of merit is determined for each combination of estimates.

Example 8.11

To illustrate the suggested approach, consider once more the example problem depicted in Figure 8.2 and involving an initial investment of $10,000 followed by annual receipts of $3000 for 5 years. Realizing that any or all of the parameter estimates could be in error, pessimistic, most likely, and optimistic estimates are provided in Table 8.4 for the parameters. As can be seen, there are 3^m combinations to consider, where m is the number of parameters to be included in the sensitivity analysis. Thus, in this case, 81

TABLE 8.4
Pessimistic, Most Likely, and Optimistic Estimates, Example 8.11

	Pessimistic[a]	Most Likely	Optimistic
Investment	−$12,000	−$10,000	−$8,000
Annual receipts	$ 2,500	$ 3,000	$4,000
Discount rate	20%	15%	12%
Planning horizon	4 years	5 years	6 years

[a] Here we interpret the pessimistic value as a value which will lower the present worth; hence, a high discount rate is viewed as a pessimistic value.

combinations are to be considered. Four representative combinations are given below.

$$PW = -\$12,000 + \$4000(P|A\ 15,4) = -\$580$$

$$PW = -\$8000 + \$2500(P|A\ 20,6) = \$313.75$$

$$PW = -\$10,000 + \$3000(P|A\ 15,4) = -\$1435$$

$$PW = -\$12,000 + \$3000(P|A\ 12,5) = -\$1185.60$$

The range on present worth can be obtained by evaluating the totally pessimistic case and the totally optimistic case.

$$PW(\text{pessimistic}) = -\$12,000 + \$2500(P|A\ 20,4) = -\$5528.25$$

$$PW(\text{optimistic}) = -\$8000 + \$4000(P|A\ 12,6) = \$8445.60$$

A tabulation of the present worths for the 81 possible combinations of pessimistic (P), most likely (M), and optimistic (O) estimates is provided in Table

TABLE 8.5
Present Worth Values for Combination of Estimates, Example 8.11

PPPP	−$5,528.25	MPPP	−$3,528.25	OPPP	−$1,528.25
PPPM	− 4,523.50	MPPM	− 2,523.50	OPPM	− 523.50
PPPO	− 3,686.25	MPPO	− 1,686.25	OPPO	313.75
PPMP	− 4,862.50	MPMP	− 2,862.50	OPMP	− 862.50
PPMM	− 3,619.50	MPMM	− 1,619.50	OPMM	380.80
PPMO	− 2,538.75	MPMO	− 538.75	OPMO	1,461.25
PPOP	− 4,406.75	MPOP	− 2,406.75	OPOP	− 406.75
PPOM	− 2,988.00	MPOM	− 988.00	OPOM	1,012.00
PPOO	− 1,721.50	MPOO	278.50	OPOO	2,278.50
PMPP	− 4,233.90	MMPP	− 2,233.90	OMPP	− 233.90
PMPM	− 3,028.20	MMPM	− 1,028.20	OMPM	971.80
PMPO	− 2,023.50	MMPO	− 23.50	OMPO	1,976.50
PMMP	− 3,435.00	MMMP	− 1,435.00	OMMP	565.00
PMMM	− 1,943.40	MMMM	56.60	OMMM	2,056.60
PMMO	− 646.50	MMMO	1,353.50	OMMO	3,353.50
PMOP	− 2,888.10	MMOP	− 888.10	OMOP	1,111.90
PMOM	− 1,185.60	MMOM	814.40	OMOM	2,814.40
PMOO	334.20	MMOO	2,334.20	OMOO	4,334.20
POPP	− 1,645.20	MOPP	354.80	OOPP	2,354.80
POPM	− 37.60	MOPM	1,962.40	OOPM	3,962.40
POPO	1,302.00	MOPO	3,302.00	OOPO	5,302.00
POMP	− 580.00	MOMP	1,420.00	OOMP	3,420.00
POMM	1,408.80	MOMM	3,408.80	OOMM	5,408.80
POMO	3,138.00	MOMO	5,138.00	OOMO	7,138.00
POOP	149.20	MOOP	2,149.20	OOOP	4,149.20
POOM	2,419.20	MOOM	4,419.20	OOOM	6,419.20
POOO	4,445.60	MOOO	6,445.60	OOOO	8,445.60

8.5. The legend PMOM indicates a pessimistic estimate for the investment required, a most likely estimate for annual receipts, an optimistic estimate for the discount rate, and a most likely estimate for the planning horizon were used to determine the present worth of −$1185.60. Plotting the results in the form of a frequency histogram yields the frequencies provided in Table 8.6.

Unfortunately, it is difficult to draw any significant conclusions from the data given in Tables 8.5 and 8.6 concerning the parameter that has the greatest effect on present worth. Instead, the use of the three estimates for each parameter provides us with a feel for the possible values of present worth and their relative frequencies. However, we should not place too much weight on the relative frequencies, since we have not verified that each combination of estimates is equally likely to occur. The assignment of probabilities to combinations of parameter estimates lies in the domain of risk analysis and will be treated next.

8.4 RISK ANALYSIS

Risk analysis will be defined as the process of developing probability distributions for some measure of merit for an investment proposal. Typically, probability distributions are developed for either present worth, annual worth, or the rate of return for an individual investment alternative. Consequently, probability distributions are required for random variables such as

TABLE 8.6

Frequency Tabulation of Present Worths, Example 8.11

Present Worth	Frequency
−$6,000 to −$5,000.01	1
− 5,000 to − 4,000.01	4
− 4,000 to − 3,000.01	5
− 3,000 to − 2,000.01	8
− 2,000 to − 1,000.01	9
− 1,000 to − 0.01	11
0 to 999.99	10
1,000 to 1,999.99	9
2,000 to 2,999.99	7
3,000 to 3,999.99	6
4,000 to 4,999.99	4
5,000 to 5,999.99	3
6,000 to 6,999.99	2
7,000 to 7,999.99	1
8,000 to 8,999.99	1

the cash flows, planning horizon, and the discount rate. The probability distributions are then aggregated analytically or through simulation to obtain the desired probability distribution for the measure of merit.

The cash flow occurring in a given year is often a function of a number of other variables such as selling prices, size of the market, share of the market, market growth rate, investment required, inflation rate, tax rates, operating costs, fixed costs, and salvage values of all assets. The values of a number of these random variables can be correlated with each other, as well as autocorrelated.[1] Consequently, an analytical development of the probability distribution for the measure of merit is not easily achieved in most real-world situations. Thus, simulation is widely used in performing risk analyses.

Risk analysis has gained acceptance in a large number of industries. One industry survey indicated that over 50% of the companies responding to the survey were using risk analysis for operational and/or strategic planning.

By incorporating the concept of utility theory, it is possible to extend risk analysis and obtain a normative or prescriptive model. To be more specific, by combining a manager's utility function and the probability distribution for, say, present worth, it is possible to specify the alternative that maximizes expected utility. Lifson [13] has recommended this approach previously. Hillier [8, 9, 10] incorporates utility theory in applying risk analysis to interrelated investment alternatives.

Our discussion will concentrate on the contributions of risk analysis as a descriptive technique instead of as a normative technique. For those interested in extending our discussion to the normative mode as well as the incorporation of utility theory with risk analysis, see Hillier [8, 9, 10].

8.4.1 Distributions

The risk analysis procedure was developed to take into consideration the imprecision in estimating the values of the inputs required in making economic evaluations. The imprecision is represented in the form of a probability distribution.

Probability distributions for the random variables are usually developed on the basis of subjective probabilities. Typically, the further an event is into the future, the less precise is our estimate of the value of the outcome of the event. Hence, by letting the variance reflect our degree of precision, we would expect the variance of the probability distributions to increase with time.

Among the theoretical probability distributions commonly used in risk analysis are the normal distribution and the beta distribution. Examples of these distributions are depicted in Figures 8.6 and 8.7. For a discussion of a number of distributions and their process generators in the context of simulation, see Schmidt and Taylor [16] and Shannon [17], among others.

In some situations the subjective probability distribution cannot be repre-

---

[1] The term *autocorrelation* means correlated with itself over time.

FIGURE 8.6. Normal distribution.

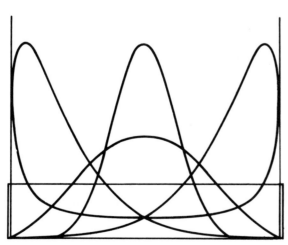

FIGURE 8.7. Sample beta distributions.

sented accurately using a well-known theoretical distribution. Instead, one must estimate directly the probability distribution for the random variable.

One approach that can be used to estimate the subjective probability distribution is to provide optimistic, pessimistic, and most likely estimates for the random variable. The optimistic and pessimistic values should be ones that you do not anticipate will be exceeded with a significant probability (e.g., 1% chance). Given the practical limits on the range of values anticipated for the random variable, an estimate is provided of the chance that the most likely estimate will not be exceeded. Next, a smooth curve is passed through the three points obtained; the resulting cumulative distribution function for the random variable is divided into an appropriate number of intervals and the individual probabilities estimated.

Example 8.12

To illustrate the process, suppose we wish to develop the probability distribution for, say, the anticipated salvage value of a machine in 5 years. We estimate the salvage value will range from $0 to $3000, with the most likely value being $1250. We estimate that there is a 40% chance of salvage value being less than $1250. Since we believe the extreme values of $0 and $3000 are not likely to occur, we use an S-shaped curve to represent the cumulative distribution function, as depicted in Figure 8.8a. Using intervals of $500, the cumulative distribution function is transformed into the probability distribution function shown in Figure 8.8b. Letting the midpoints of the intervals represent the probabilities associated with the intervals yields the probability mass function given in Figure 8.8c. Depending on the use to be made of the probabilities obtained, either of the three representations of the probability distribution could be used.

The process described above is quite subjective, since a number of different curves could be used to describe the cumulative distribution function. However, the probability distribution itself is subjectively based. What is sought is a probability distribution that best describes one's beliefs about the outcomes of the random variable. If the probability distribution obtained above does not reflect your beliefs, the process should be repeated until a satisfactory probability distribution is obtained.

The critics of risk analysis cite the degree of subjectivity involved in developing probability distributions. However, it is argued by those who favor the technique that the only alternative to using subjectively based probabilities is to use the traditional single-estimate approach, which implies that conditions of certainty exist. Interestingly, the final distribution obtained for the measure of merit is often quite insensitive to deviations in the shapes of the distributions for the parameters.

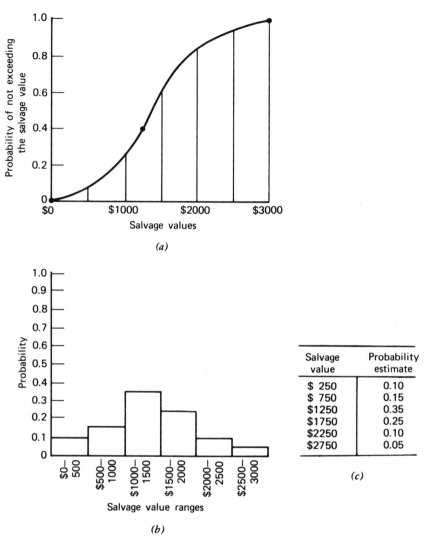

FIGURE 8.8. Developing the subjective probability distribution.

8.4.2 Risk Aggregation

Given the essential factors and their associated probability distributions, we are in a position to aggregate the distributions and obtain the probability distribution for the measure of merit. Three measures of merit have been mentioned: present worth, annual worth, and rate of return. In practice, a combination of the rate of return and either the present-worth or annual-worth measures of merit are often used. Each method of comparing investment alternatives has its limitations. For example, if the rate of return is being determined, the explicit reinvestment rate of return method is recommended, since multiple rates of return can arise when using the internal rate of return method.

Risk aggregation is achieved in basically two ways: by using simulation and analytically. Analytic approaches can be used in a number of simple cases. For more complex situations involving a large number of variables, simulation is used.

Simulation Approaches: Simulation, in the general sense, may be thought of as performing experiments on a model. Basically, simulation is an "if . . . , then . . ." device (i.e., *if* a certain input is specified, *then* the output can be determined). Some of the major reasons for using simulation in risk analysis follow:

1. Analytic solutions are impossible to obtain without great difficulty.
2. Simulation is useful in selling a system modification to management.
3. Simulation can be used as a verification of analytical solutions.
4. Simulation is very versatile.
5. Less background in mathematical analysis and probability theory is generally required.

Some of the major disadvantages of simulation are:

1. Simulations can be quite expensive.
2. Simulations introduce a source of randomness not present in analytic solutions.
3. Simulations do not reproduce the input distributions exactly (especially the tails of the distribution).
4. Validation is easily overlooked in using simulation.
5. Simulation is so easily applied it is often used when analytic solutions can be easily obtained at considerably less cost.

Example 8.13

To illustrate the simulation approach in performing a risk analysis, consider an investment of $10,000 over a 4-year period that returns R_t at the end of year t, with R_t being a statistically independent random variable. Using a *MARR* of 20% it is desired to use simulation to estimate the probability of the investment being unprofitable (or risky!). The following probability distribution is assumed for R_t.

R_t	Probability	Random Number
$2,000	0.10	0
3,000	0.20	1, 2
4,000	0.30	3, 4, 5
5,000	0.40	6, 7, 8, 9

(One-digit random numbers are assigned to the values of the random variables in proportion to their probability of occurrence.)

Consulting the random numbers given in Table 8.6A and beginning with column 1, row 1, the following sequence of four random digits is obtained: 9, 7, 5, and 9. Letting the first random digit (9) represent the return in the first year ($5000), the second random digit (7) represent the return in the second year ($5000), and so on yields simulated returns of $5000, $5000, $4000, and $5000 over the 4-year period for a present worth of $2364.80. Obviously, one simulated investment is not sufficient to estimate the probability of the investment being profitable. Continued simulation yields the results given in Table 8.7. For the 20 simulated investments, 6 had negative present worths; thus, the probability of the investment being unprofitable might be estimated to be 0.30. However, one should be cautioned against drawing conclusions from only 20 simulated investments. In practice, it would be desirable to develop a computer program to perform the simulation and obtain perhaps

TABLE 8.6A
Two-Digit Random Numbers, Example 8.13

90	43	78	83	82	99	54	02
78	31	58	98	68	09	87	80
51	81	42	35	21	42	03	62
93	97	15	95	07	56	60	39
27	37	12	63	31	35	66	93
79	39	44	22	83	96	51	00
89	61	73	29	43	84	91	34
29	38	30	84	90	18	00	10
97	64	33	29	17	48	26	04
07	64	15	02	44	32	92	99
82	13	50	83	35	39	50	51
59	83	21	30	86	90	16	09
04	46	19	63	60	53	33	97
96	54	91	43	44	40	09	02
31	27	71	78	03	65	53	62
03	45	70	42	22	16	67	13
08	35	45	92	79	97	46	02
37	60	80	55	05	35	75	57
90	43	63	17	56	21	69	09
22	07	69	85	38	74	02	58
05	33	79	00	69	29	67	08
48	97	91	14	53	00	03	42
94	68	64	58	97	32	27	80
15	39	85	87	82	38	52	16
09	37	81	73	37	01	66	84

1000 simulated investments before estimating the probability of a negative present worth.

Example 8.14

As a second illustration of risk analysis, consider the following situation. An individual is planning on purchasing a used microcomputer for $1000 and performing certain billing and accounting functions for several retail businesses in the neighborhood. It is anticipated that the business will last only 4 years because of the growth of competition. If income in the third year exceeds expenses by more than $400, operations will continue the fourth year. However, if income is less than or equal to $400 more than expenses in the third year, the computer and software will probably be sold at the end of the third year. For simplicity, a probability of 0.70 is assigned to the possibility of selling out at the end of the third year, given income is not greater than $400 above expenses in the third year. Let

$$E_j = \text{expense for year } j$$
$$I_j = \text{income for year } j$$
$$n = \text{life of the investment}$$
$$S_n = \text{salvage value based on an } n\text{-year life}$$

Assuming a zero discount rate for simplicity of calculations, the present worth of the investment is given as

$$PW = -\$1000 + \sum_{j=1}^{n} (I_j - E_j) + S_n$$

The probability distributions assumed to hold for this example are provided in Table 8.8. As in the previous example, a computer should be used to perform the simulation. However, to illustrate the technique, we will manually perform 10 simulations of the investment. The table of two-digit random numbers given in Table 8.6A will be used. Using the worksheet given in Table 8.9, 10 simulations of the investment yielded an average present worth of $805 for the investment. Of course, 10 simulations is not an adequate number of trials to draw strong conclusions concerning the investment. However, the example does demonstrate the simulation approach.

To illustrate the approach taken, the first random number selected will provide the simulated value for expenses in the first year. A random number of 90 was obtained from row 1, column 1 in Table 8.6A. Consulting Table 8.8, it is seen that a random number of 90 represents an expense of $400; hence $400 is entered appropriately on the worksheet given in Table 8.9. The second random number is selected from row 2, column 1 of Table 8.6A to generate the income for year 1. A random number of 78 is obtained and, from Table 8.8, a simulated income of $600 is obtained for the first year. Continuing through the third year, it is found that income exceeds expenses by $500; hence the business will continue through the fourth year. A random number of 97 is drawn to generate the salvage value of $400 for the investment.

TABLE 8.7

Simulation Results for Example 8.13

Investment Trial	Year 1		Year 2		Year 3		Year 4		Present Worth, PW(20%)
	RN	Value	RN	Value	RN	Value	RN	Value	
1	9	$5,000	7	$5,000	5	$4,000	9	$5,000	$2,364.80
2	2	3,000	7	5,000	8	5,000	2	3,000	312.30
3	9	5,000	0	2,000	8	5,000	5	4,000	378.00
4	0	2,000	9	5,000	3	4,000	0	2,000	− 1,582.00
5	0	2,000	3	4,000	9	5,000	2	3,000	− 1,215.40
6	0	2,000	4	4,000	9	5,000	1	3,000	− 1,215.40
7	0	2,000	0	2,000	8	5,000	1	3,000	− 2,604.20
8	3	4,000	7	5,000	9	5,000	9	5,000	2,110.20
9	9	5,000	7	5,000	7	5,000	2	3,000	1,978.90
10	9	5,000	4	4,000	6	5,000	1	3,000	1,284.50
11	3	4,000	8	5,000	7	5,000	0	2,000	663.30
12	2	3,000	5	4,000	8	5,000	4	4,000	100.20
13	5	4,000	9	5,000	4	4,000	3	4,000	1,049.20
14	8	5,000	9	5,000	3	4,000	3	4,000	1,882.50
15	6	5,000	3	4,000	6	5,000	6	5,000	2,249.10
16	1	3,000	8	5,000	4	4,000	5	4,000	215.90
17	2	3,000	4	4,000	3	4,000	6	5,000	3.80
18	4	4,000	0	2,000	3	4,000	9	5,000	− 551.70
19	6	5,000	3	4,000	3	4,000	3	4,000	1,188.10
20	1	3,000	1	3,000	7	5,000	7	5,000	− 111.90

TABLE 8.8

Data for the Risk Analysis, Example 8.14

E_1	$p(E_1)$	RN	I_1	$p(I_1)$	RN
$200	0.25	00–24	$ 400	0.50	00–49
300	0.50	25–74	600	0.50	50–99
400	0.25	75–99			

E_2	$p(E_2)$	RN	I_2	$p(I_2)$	RN
$300	0.10	00–09	$ 500	0.20	00–19
400	0.40	10–49	750	0.40	20–59
500	0.40	50–89	1,000	0.40	60–99
600	0.10	90–99			

E_3	$p(E_3)$	RN	I_3	$p(I_3)$	RN
$400	0.20	00–19	$ 800	0.30	00–29
500	0.30	20–49	1,000	0.50	30–79
600	0.30	50–79	1,200	0.20	80–99
700	0.20	80–99			

E_4	$p(E_4)$	RN	I_4	$p(I_4)$	RN
$500	0.25	00–24	$ 600	0.25	00–24
600	0.25	25–49	800	0.25	25–49
700	0.25	50–74	1,000	0.25	50–74
800	0.25	75–99	1,200	0.25	75–99

N	$p(N\|I_3 - E_3 \leq \$400)$	RN
3	0.70	00–69
4	0.30	70–99

N	$p(N\|I_3 - E_3 > \$400)$
4	1.00

S_3	$p(S_3)$	RN	S_4	$p(S_4)$	RN
$400	0.50	00–49	$ 300	0.60	00–59
500	0.50	50–99	400	0.40	60–99

Note that in the second trial it was decided that the business should be discontinued after 3 years. Furthermore, income was never less than expenses in any year; this illustrates insufficient observations (trials) have been obtained since, in year 2, expense can exceed income with probability

$$Pr(E_2 = 600 \text{ and } I_2 = 500) = 0.10(0.20) = 0.02$$

and in year 4 expense can exceed income with probability

$$Pr(E_4 = 700 \text{ and } I_4 = 600) + Pr(E_4 = 800 \text{ and } I_4 = 600) = 0.125$$

As an alternative to using the discrete probability distributions given in Table 8.8, we might employ continuous distributions such as the normal, gamma, or beta distributions to represent income and expenses. If such distributions are to be used, appropriate techniques for generating simulated values of the random variables are required. Since a detailed treatment of simulation is beyond the scope of this text, we refer you to texts devoted to simulation for additional discussion [16], [17].

Analytic Approaches: As an illustration of the use of analytic approaches in developing the probability distribution for present worth, consider the following present-worth relation.

$$PW = \sum_{j=0}^{n} C_j(1 + i)^{-j} \tag{8.1}$$

Suppose the cash flows, C_j, are *random variables* with expected values $E(C_j)$ and variances $V(C_j)$. Since the *expected value* of a sum of random variables is given by the sum of the expected values of the random variables, the expected present worth is given by

$$E(PW) = \sum_{j=0}^{n} E[C_j(1 + i)^{-j}] \tag{8.2}$$

Furthermore, since the expected value of the product of a constant and a random variable is given by the product of the constant and the expected value of the random variable,

$$E[C_j(1 + i)^{-j}] = E(C_j)(1 + i)^{-j} \tag{8.3}$$

Substituting Equation 8.3 into Equation 8.2 yields

$$E(PW) = \sum_{j=0}^{n} E(C_j)(1 + i)^{-j}$$

Hence, we see that the expected present worth of a series of cash flows is found by summing the present worths of the expected values of the individual cash flows.

TABLE 8.9

Simulation of the Sample Investment Problem in Example 8.14

Trial	Year	RN(E)	E	RN(I)	I	I − E	RN(N)	RN(S)	S	PW
1	1	90	$400	78	$ 600	$200	—			
	2	51	500	93	1,000	500				
	3	27	500	79	1,000	500		97	$400	$ 600
	4	89	800	29	800	0				
2	1	07	200	82	600	400				
	2	59	500	04	500	0				
	3	96	700	31	1,000	300	03	08	400	100
	4	—	—	—	—	—				
3	1	37	300	90	600	300	—			
	2	22	400	05	500	100				
	3	48	500	94	1,200	700				
	4	15	500	09	600	100		43	300	500
4	1	31	300	81	600	300				
	2	97	600	37	750	150				
	3	39	500	61	1,000	500				
	4	38	600	64	1,000	400	—	64	400	750
5	1	13	200	83	600	400				
	2	46	400	54	750	350				
	3	27	500	45	1,000	500				
	4	35	600	60	1,000	400		43	300	950

TABLE 8.9 Continued

6	1	07	200	33	400	200				
	2	97	600	68	1,000	400				
	3	39	500	37	1,000	500				
	4	78	800	58	1,000	200	—	42	300	600
7	1	15	200	12	400	200				
	2	44	400	73	1,000	600				
	3	30	500	33	1,000	500				
	4	15	500	50	1,000	500	—	21	300	1,100
8	1	19	200	91	600	400				
	2	71	500	70	1,000	500				
	3	45	500	80	1,200	700				
	4	63	700	69	1,000	300	—	79	400	1,300
9	1	91	400	64	600	200				
	2	85	500	81	1,000	500				
	3	83	700	98	1,200	500				
	4	35	600	95	1,200	600	—	63	400	1,200
10	1	22	200	29	400	200				
	2	84	500	29	750	250				
	3	02	400	83	1,200	800				
	4	30	600	63	1,000	400	—	43	300	950
										8,050

Example 8.15

Recall Example 8.13 involving an investment of $10,000 for a 4-year period. The expected value of the return in a given year is given by

$$E[\text{Return}] = \$2000(0.10) + \$3000(0.20)$$
$$+ \$4000(0.30) + \$5000(0.40)$$
$$= \$4000$$

Thus, the expected present worth for the investment situation is

$$E(PW) = \$4000(P|A\ 20,4) - \$10,000$$
$$= \$354.80$$

(From Table 8.7, a computation of the average simulated present worth yields $425.01. This serves to illustrate the need for many simulated trials.)

To determine the variance of the present worth, we first recall that if X_1, X_2, X_3, and Y are random variables related as follows,

$$Y = X_1 + X_2 + X_3 \tag{8.4}$$

then the variance of Y is given by

$$\boxed{\text{Var}(Y) = E(Y^2) - E(Y)^2} \tag{8.5}$$

or

$$\boxed{\text{Var}(Y) = E[(X_1 + X_2 + X_3)^2] - E(X_1 + X_2 + X_3)^2}$$

Expanding and collecting terms yields

$$\text{Var}(Y) = E(X_1^2 + X_2^2 + X_3^2 + 2X_1X_2 + 2X_1X_3 + 2X_2X_3)$$
$$- [E(X_1) + E(X_2) + E(X_3)]^2$$
$$= E(X_1^2) + E(X_2^2) + E(X_3^2) + 2E(X_1X_2) + 2E(X_1X_3) + 2E(X_2X_3)$$
$$- E(X_1)^2 - E(X_2)^2 - E(X_3)^2 - 2E(X_1)E(X_2) - 2E(X_1)E(X_3)$$
$$- 2E(X_2)E(X_3)$$
$$= E(X_1^2) - E(X_1)^2 + E(X_2^2) - E(X_2)^2 + E(X_3^2) - E(X_3)^2$$
$$+ 2[E(X_1X_2) - E(X_1)E(X_2)] + 2[E(X_1X_3) - E(X_1)E(X_3)]$$
$$+ 2[E(X_2X_3) - E(X_2)E(X_3)] \tag{8.6}$$

Since $E(X_k^2) - E(X_k)^2$ defines the variance of the random variable X_k and $E(X_pX_k) - E(X_p)E(X_k)$ defines the covariance of the random variables X_p and X_k, we see that

$$\text{Var}(Y) = \text{Var}(X_1) + \text{Var}(X_2) + \text{Var}(X_3) + 2 \text{ Cov}(X_1 X_2) \\ + 2 \text{ Cov}(X_1 X_3) + 2 \text{ Cov}(X_2 X_3)$$

(8.7)

where $\text{Cov}(X_p X_k)$ denotes the covariance of X_p and X_k. If the random variables X_p and X_k are statistically independent, then $\text{Cov}(X_p X_k) = 0$.

Generalizing to the sum of $(n + 1)$ random variables, if

$$Y = \sum_{j=0}^{n} X_j$$

(8.8)

then the expected value and variance of Y are given by

$$E(Y) = \sum_{j=0}^{n} E(X_j)$$

$$\text{Var}(Y) = \sum_{j=0}^{n} \text{Var}(X_j) + 2 \sum_{j=0}^{n-1} \sum_{k=j+1}^{n} \text{Cov}(X_j X_k)$$

(8.9)

Additionally, if a_1 and a_2 are constants, and X_1, X_2, and Y are random variables related by

$$Y = a_1 X_1 + a_2 X_2$$

(8.10)

recall that

$$\text{Var}(Y) = a_1^2 \text{ Var}(X_1) + a_2^2 \text{ Var}(X_2) + 2a_1 a_2 \text{ Cov}(X_1 X_2)$$

(8.11)

Combining the relationships given in Equations 8.9 and 8.11, the variance of present worth is found to be

$$\text{Var}(PW) = \sum_{j=0}^{n} \text{Var}(C_j)(1 + i)^{-2j} \\ + 2 \sum_{j=0}^{n-1} \sum_{k=j+1}^{n} \text{Cov}(C_j C_k)(1 + i)^{-(j+k)}$$

(8.12)

Example 8.16

Continuing Example 8.15 to compute the variance of present worth, recall from Example 8.13 that the random variables are statistically independent. Thus, $Cov(C_j C_k) = 0$ for all j and k. The variance of an annual return is determined as follows:

$$Var[Return] = E[Return^2] - E[Return]^2$$

and

$$
\begin{aligned}
E[Return^2] &= (\$2000)^2(0.10) + (\$3000)^2(0.20) \\
&\quad + (\$4000)^2(0.30) + (\$5000)^2(0.40) \\
&= 17 \times 10^6
\end{aligned}
$$

Thus

$$
\begin{aligned}
Var[Return] &= (17 \times 10^6) - (4000)^2 \\
&= 1 \times 10^6
\end{aligned}
$$

The standard deviation is the square root of the variance and equals

$$SD[Return] = \$1000$$

To compute the variance of present worth, from Equation 8.12

$$Var(PW) = Var[Investment] + \sum_{t=1}^{4} Var(R_t)(1.20)^{-2t}$$

$$= 0 + \sum_{t=1}^{4} (1 \times 10^6)(1.20)^{-2t}$$

$$= (1 \times 10^6) \sum_{t=1}^{4} (P|F\ 20,2t)$$

$$= (1 \times 10^6)(0.6944 + 0.4823 + 0.3349 + 0.2326)$$

$$= 1.7442 \times 10^6$$

Hillier [6] argues that it is probably unrealistic to expect investment analysts to develop accurate estimates for covariances in Equation 8.12. Consequently, he suggests that the net cash flow in any year be divided into those components of cash flow that are reasonably independent from year to year and those that are correlated over time. Specifically, it is assumed that

$$\boxed{C_j = X_j + Y_{j1} + Y_{j2} + \cdots + Y_{jm}} \qquad (8.13)$$

where the X_j values are mutually independent over j but, for a given value of h, $Y_{0h}, Y_{1h}, \ldots, Y_{nh}$ are *perfectly* correlated.

When two random variables X and Y are perfectly correlated, one can be expressed as a linear function of the other. Hence, if

$$Y = a + bX \qquad (8.14)$$

where a and b are constants, then the covariance of X and Y is given by

$$\text{Cov}(XY) = E(XY) - E(X)E(Y) \qquad (8.15)$$

However, from Equation 8.14, we note that

$$E(Y) = a + bE(X) \qquad (8.16)$$

and

$$\text{Var}(Y) = b^2 \,\text{Var}(X) \qquad (8.17)$$

Substituting Equations 8.14 and 8.16 into Equation 8.15 gives

$$\begin{aligned}
\text{Cov}(XY) &= E(aX + bX^2) - E(X)[a + bE(X)] \\
&= aE(X) + bE(X^2) - aE(X) - bE(X)^2 \\
&= b\,\text{Var}(X)
\end{aligned} \qquad (8.18)$$

Recalling that the *standard deviation* of a random variable is defined as the square root of the variance of the random variable, we see that Equation 8.18 can be expressed as

$$\text{Cov}(XY) = bSD(X)SD(X) \qquad (8.19)$$

where $SD(X)$ denotes the standard deviation of X. However, from Equation 8.17, we see that $SD(Y)$ equals $bSD(X)$. Hence,

$$\text{Cov}(XY) = SD(X)SD(Y) \qquad (8.20)$$

when X and Y are perfectly correlated.

The model suggested by Hillier yields the following expressions for the expected present worth and variance of present worth:

$$E(PW) = \sum_{j=0}^{n} E(X_j)(1 + i)^{-j} + \sum_{j=0}^{n} \sum_{h=1}^{m} E(Y_{jh})(1 + i)^{-j} \qquad (8.21)$$

and

$$\text{Var}(PW) = \sum_{j=0}^{n} \text{Var}(X_j)(1 + i)^{-2j} + \sum_{h=1}^{m} \left[\sum_{j=0}^{n} SD(Y_{jh})(1 + i)^{-j} \right]^2 \qquad (8.22)$$

where $SD(Y_{jh})$ denotes the standard deviation of the random variable Y_{jh}.

A very important theorem from probability theory, the *central limit theorem*, is usually invoked at this point. The central limit theorem establishes under very general conditions that the sum of independently distributed, random variables tends to be distributed normally as the number of terms in the summation approaches infinity. Hence, it is argued that present worth, as defined by Equation 8.1, is normally distributed with mean and variance, as given by Equations 8.21 and 8.22. Of course, this assumes that the C_j are statistically independent. However, the view is usually taken that the normal distribution is a reasonable approximation to the distribution of present worth.

Given that present worth is assumed to be normally distributed, one can compute for each investment alternative the probability of achieving a given aspiration level. For example, one is usually interested in knowing the probability that present worth is less than zero. Some analysts interpret this to be a measure of the risk associated with an investment alternative.

In performing an analysis of the risk associated with an investment alternative, we have treated the simplest situation; only the cash flows were considered to be random variables, and the measure of merit employed was present worth, not rate of return. When either the discount rate or the planning horizon are treated as random variables and when the probability distribution for rate of return is desired, simulation approaches are usually employed.

Example 8.17

As an illustration of the analytic approach in developing the probability distribution for *PW*, consider the following example problem based on one given by Giffin [4].

A flight school operator is considering the alternatives of purchasing a utility category training aircraft versus purchasing an acrobatic version of the same aircraft. Having been in the flight training business for a number of years, the operator has reason to believe that income and expense from a utility category aircraft will be nearly independent from year to year. The pertinent data are summarized in Table 8.10.

Investment in the acrobatic aircraft is a more risky but promising investment. The acrobatic aircraft is also expected to have a life of 5 years. The

TABLE 8.10

Estimated Net Cash Flows for a Utility Model Aircraft,[a] Example 8.17

Year	Source	Symbol	Expected Value	Range	Standard Deviation
0	Purchase	C_0	−$11,000	$9,500–12,500	$500
1	Income-expense	C_1	2,200	2,050– 2,350	50
2	Income-expense	C_2	2,200	1,900– 2,500	100
3	Income-expense	C_3	2,200	1,900– 2,500	100
4	Income-expense	C_4	2,000	1,700– 2,300	100
5	Income-expense	C_{51}	1,000	700– 1,300	100
5	Salvage	C_{52}	6,000	4,800– 7,200	400

[a] The data have been modified slightly from those given by Giffin [4].

operator feels that maintenance costs will be nearly independent year to year with this aircraft. However, since the flight school has never offered an acrobatic course before, there is some uncertainty regarding the demand for time in such an aircraft. It is felt that the net cash flow from the sale of flight time for each of the 5 years will be perfectly correlated. The pertinent data are summarized in Table 8.11. An interest rate of 10% is to be used in the analysis.

From Equations 8.21 and 8.22, we obtain the following:

$$E(PW_1) = \$190 \qquad E(PW_2) = \$355$$
$$\mathrm{Var}(PW_1) = 334{,}741 \qquad \mathrm{Var}(PW_2) = 11{,}631{,}930$$

TABLE 8.11

Estimated Net Cash Flows for an Acrobatic Model Aircraft,[a] Example 8.17

Year	Source	Symbol	Expected Value	Range	Standard Deviation
0	Purchase	X_0	−$14,000	$13,100–14,900	$ 300
1	Expense	X_1	− 10,000	9,100–10,900	300
2	Expense	X_2	− 10,000	9,100–10,900	300
3	Expense	X_3	− 11,000	10,100–11,900	300
4	Expense	X_4	− 12,000	10,800–13,200	400
5	Expense	X_{51}	− 12,000	10,800–13,200	400
1	Income	Y_{11}	12,500	9,500–15,500	1,000
2	Income	Y_{21}	13,500	10,500–16,500	1,000
3	Income	Y_{31}	13,500	10,800–16,200	900
4	Income	Y_{41}	14,500	12,100–16,900	800
5	Income	Y_{51}	13,500	11,400–15,600	700
5	Salvage	X_{52}	7,500	6,000– 9,000	500

[a] The data have been modified slightly from those given by Giffin [4].

Assuming normally distributed PW, the probability of an equivalent present worth less than zero is found to be

$$Pr(PW_1 < 0 | i = 10\%) = 0.37$$

for the utility model aircraft and

$$Pr(PW_2 < 0 | i = 10\%) = 0.46$$

for the acrobatic model aircraft. Here we are faced with a choice between the alternative having the greatest expected value and the alternative having the smallest variance. The choice will depend on the owner's attitudes toward risk. At this point we leave the flight school operator "up in the air" relative to a choice between the investment alternatives. If the decision were yours to make, which model aircraft would you choose? Why?

Risk analysis offers a number of important advantages over traditional deterministic approaches. Klausner [12] summarizes some of the most significant advantages as follows:

1. Uncertainty Made Explicit. The uncertainty that an estimator feels about his estimate of an element value is brought out into the open and incorporated into the investment analysis. The analysis technique permits maximum information utilization by providing a vehicle for the inclusion of "less likely" estimates in the analysis.

2. More Comprehensive Analysis. This technique permits a determination of the effect of simultaneous variation of all the element values on the outcome of an investment. This approximates the "real world" conditions under which an actual investment's outcome will be determined. The Probabilistic Cash Flow Simulation generates an overall indication of potential variation in outcome and project risk. This indicator, in the form of a probability distribution, accounts statistically for element interaction.

3. Variability of Outcome Measured. One of the most significant advantages of this analysis technique is that it gives a measure of the dispersion around the investment outcome based on the expected cash flow. This dispersion, or variability, is an important consideration in the comparison of alternative investments. Other things being equal, lower variability for the same return is usually desirable. The probability distribution associated with each investment's outcome gives a clear picture of this important evaluation consideration.

4. Promotes More Reasoned Estimating Procedures. By requiring that element values be given as probability distributions rather than as single values, more reasoned consideration is given to the estimating procedure. Judgment is applied to the individual element values rather than to the investment's outcome, which is jointly determined by all of the elements. Thinking through the uncertainties in a project and recognizing what is known and unknown will go far toward ensuring the best investment decision. Understanding and dealing effectively with uncertainty and risk is the key to rational decision making.

Risk analysis is a technique that has been used by a number of firms to improve the decision-making process in a risk environment. When applied properly, risk analysis can enhance significantly the manager's understanding of the risks associated with an investment alternative. The major premise underlying risk analysis is the belief that a manager can make better decisions when he or she has a fuller understanding of the implications of the decision.

8.5 SUMMARY

Descriptive approaches were presented in this chapter for coping with risk and uncertainty conditions in performing economic analyses. In particular, break-even, sensitivity, and risk analyses were discussed. In Chapter Nine, we employ normative or prescriptive approaches for dealing with risk and uncertainty.

The extent to which supplementary analyses are justified depends largely on the magnitude of the initial investment. In particular, if an analysis has been performed using the techniques described in Chapters Four through Seven and one of the alternatives is not clearly the superior alternative, a risk analysis will probably yield additional insight concerning the decision to be made. If either small amounts of money are involved, or if the choice is not close using the techniques described in Chapters Four through Seven under assumed certainty, then supplementary analyses are not justified. In those cases where supplementary analysis is justified and a sensitivity analysis is not sufficient to allow a decision, a risk analysis is performed.

BIBLIOGRAPHY

1. Buck, J. R., and Askin, R. G., "Partial Means in the Economic Risk Analysis of Projects," *The Engineering Economist, 31* (3) Spring 1986, Institute of Industrial Engineers, pp. 189–212.

2. Canada, J. R., and White, J. A., Jr., *Capital Investment Decision Analysis for Management and Engineering,* Prentice-Hall, 1980.

3. Fleischer, G. A., *Risk and Uncertainty: Non-Deterministic Decision Making in Engineering Economy,* AIIE Monograph Series, American Institute of Industrial Engineers, 1975.

4. Giffin, W. C., *Introduction to Operations Engineering,* R. D. Irwin, 1971.

5. Hertz, D. B., "Risk Analysis in Capital Investments," *Harvard Business Review, 42* (1), January–February 1964.

6. Hillier, F. S., "The Derivation of Probabilistic Information for the Evaluation of Risky Investments," *Management Science, 9* (3), April 1963.

7. Hillier, F. S., "Supplement to 'The Derivation of Probabilistic Information Evaluation of Risky Investments," *Management Science, 11* (3), January 1965.

8. Hillier, F. S., *The Evaluation of Risky Interrelated Investments,* North-Holland Publishing, 1969.

9. Hillier, F. S., "A Basic Approach to the Evaluation of Risky Interrelated Investments," Chapter 1 in *Studies in Budgeting,* edited by R. F. Byrne, W. W. Cooper, A. Charnes, O. A. Davis, and D. A. Gilford, North-Holland Publishing, 1971.

10. Hillier, F. S., "A Basic Model for Capital Budgeting of Risky Interrelated Projects," *The Engineering Economist, 17* (1), 1971.

11. Kaplan, S., and Barish, N., "Decision Making Allowing for Uncertainty of Future Investment Opportunities," *Management Science, 13* (10), June 1967.

12. Klausner, R. F., "The Evaluation of Risk in Marine Capital Investments," *The Engineering Economist, 14* (4), Summer 1969.

13. Lifson, M., "Evaluation Modeling in the Context of the Systems Decision Process," *Decision and Risk Analysis: Powerful New Tools for Management,* Proceedings of the Sixth Triennial Symposium, published by *The Engineering Economist,* 1971.

14. Miller, E. C., *Advanced Techniques for Strategic Planning,* AMA Research Study 104, American Management Association, 1971.

15. Saxena, U., "Investment Analysis Under Uncertainty," *The Engineering Economist, 29* (1) Fall 1983, Institute of Industrial Engineers, pp. 33–40.

16. Schmidt, J. W., and Taylor, R. E., *Simulation and Analysis of Industrial Systems,* R. D. Irwin, 1970.

17. Shannon, R. E., *Systems Simulation: The Art and Science,* Prentice-Hall, 1975.

18. Shashua, L., and Goldschmidt, Y., "Break-Even Analysis under Inflation," *The Engineering Economist, 32* (2) Winter 1987, Institute of Industrial Engineers, pp. 79–88.

PROBLEMS

1. Two 100-horsepower motors are under consideration by the Mighty Machinery Company. Motor Q costs $5000 and operates at 90% efficiency. Motor R costs $3500 and is 88% efficient. Annual operating and maintenance costs are estimated to be 15% of the initial purchase price. Power costs 3.2¢/kilowatt-hour. How many hours of full-load operation are necessary each year in order to justify the purchase of motor Q? Use a 15-year planning horizon; assume that salvage values will equal 20% of the initial purchase price; and let the *MARR* be 15%. (*Note.* 0.746 kilowatts = 1 horsepower.)(8.2)

2. An aluminum extrusion plant manufactures a particular product at a variable cost of $0.04 per unit, including material cost. The fixed costs associated with manufacturing the product equal $30,000/year. Determine the break-even value for annual sales if the selling price per unit is (a) $0.40, (b) $0.30, and (c) $0.20. (8.2)

3. Owners of a nationwide motel chain are considering locating a new motel in Duluth, Georgia. The complete cost of building a 150-unit motel (excluding

furnishings) is $5 million; the firm estimates that the furnishings in the motel must be replaced at a cost of $1,875,000 every 5 years. Annual operating and maintenance cost for the facility is estimated to be $125,000. The average rate for a unit is anticipated to be $45/day. A 15-year planning horizon is used by the firm in evaluating new ventures of this type; a terminal salvage value of 20% of the original building cost is anticipated; furnishings are estimated to have no salvage value at the end of each 5-year replacement interval; land cost is not to be included. Determine the break-even value for the daily occupancy percentage based on a *MARR* of (a) 0%, (b) 10%, (c) 20%, and (d) 30%. (Assume that the motel will operate 365 days/year.) (8.2)

4. A consulting engineer is considering two pumps to meet a demand of 15,000 gallons/minute at 12 feet total dynamic head. The specific gravity of the liquid being pumped is 1.50. Pump A operates at 70% efficiency and costs $12,000; pump B operates at 75% efficiency and costs $18,000. Power costs $0.015/kilo-watt-hour. Continuous pumping for 365 days/year is required (i.e., 24 hours/day). Using a *MARR* of 20% and assuming equal salvage values for both pumps, how many years of service are required for pump B to be justified economically? (*Note:* Dynamic head times gallon/minute times specific gravity divided by 3960 equals horsepower required. Horsepower times 0.746 equals kilowatts required.) (8.2)

5. A business firm is contemplating the installation of an improved material-han-dling system between the packaging department and the finished goods ware-house. Two designs are being considered. The first consists of an automated guided vehicle system (AGVS) involving three vehicles on the loop. The second design consists of a pallet conveyor installed between packaging and the ware-house. The AGVS will have an initial equipment cost of $280,000 and annual operating and maintenance costs of $25,000. The pallet conveyor has an initial cost of $360,000 and annual operating and maintenance costs of $2000. The firm is not sure what planning horizon to use in the analysis; however, the salvage value estimates given in the following table have been developed for various planning horizons. Using a *MARR* of 20%, determine the break-even value for *N*, the planning horizon. (8.2)

Salvage Value Estimates		
N	AGVS	Pallet Conveyor
1	$230,000	$300,000
2	185,000	245,000
3	145,000	200,000
4	110,000	160,000
5	80,000	125,000
6	55,000	95,000
7	35,000	70,000
8	20,000	50,000
9	10,000	35,000
10	5,000	25,000
11	—	20,000
12	—	20,000

6. A manufacturing plant in Michigan has been contracting snow removal at a cost of $325/day. The past 3 years have produced heavy snowfalls, resulting in the cost of snow removal being of concern to the plant manager. The plant engineer has found that a snow-removal machine can be purchased for $65,000; it is estimated to have a useful life of 10 years, and a zero salvage value at that time. Annual costs for operating and maintaining the equipment are estimated to be $15,000. Determine the break-even value for the number of days per year that snow removal is required in order to justify the equipment, based on a *MARR* of (a) 0%, (b) 10%, and (c) 30%. (8.2)

7. The motor on a gas-fired furnace in a small foundry is to be replaced. Three different 15-horsepower electric motors are being considered. Motor X sells for $1500 and has an efficiency rating of 90%; motor Y sells for $1000 and has a rating of 85%; motor Z sells for $750 and is rated to be 80% efficient. The cost of electricity is $0.05/kilowatt-hour. An 8-year planning horizon is used, and zero salvage values are assumed for all three motors. A *MARR* of 12% is to be used. Assume that the motor selected will be loaded to capacity. Determine the range of values for annual usage of the motor (in hours) that will lead to the preference of each motor. (*Note:* 0.746 kilowatts = 1 horsepower.) (8.2)

8. A machine can be purchased at $t = 0$ for $20,000. The estimated life is 5 years, with an estimated salvage value of zero at that time. The average annual operating and maintenance expenses are expected to be $4000. If *MARR* = 20%, what must the average annual revenues be in order to be indifferent between (a) purchasing the machine, or (b) do nothing? (8.2)

9. A firm has decided to manufacture widgets. There are two production processes available for consideration. Process A involves the purchase of a $22,000 machine that will last for 10 years and have a $2000 salvage value at that time. Annual operating and maintenance costs amount to $15,000/year for the machine. In addition, using Process A requires additional costs amounting to $0.15/ widget produced. The second process, Process B, requires an investment of $15,000 in a machine that will last for 10 years and have a $2000 salvage value at that time. Annual operating and maintenance costs amount to $13,000/year for the machine. Additional costs of $0.20/widget produced result when Process B is used.

With a 10% interest rate, for what annual production volume is Process A preferred? (8.2)

10. Two condensers are being considered by the Ajax Company. A copper condenser can be purchased for $3000; annual operating and maintenance costs are estimated to be $500. Alternatively, a ferrous condenser can be purchased for $2500; since the Ajax Company has not had previous experience with ferrous condensers, they are not sure what annual operating and maintenance cost estimate is appropriate. A 5-year planning horizon is to be used, salvage values are estimated to be 25% of the original purchase price, and a *MARR* of 20% is to be used. Determine the break-even value for the annual operating and maintenance cost for the ferrous condenser. (8.2)

11. The ABC Company is faced with three proposed methods for making one of their products. Method A involves the purchase of a machine for $25,000. It will have a 7-year life, with a zero salvage value at that time. Using Method A involves additional costs of $0.20/unit of product produced per year. Method B involves the purchase of a machine for $35,000. It will have a $2000 salvage value when disposed of in 7 years. Using Method B involves additional costs of $0.15/unit of product produced per year. Method C involves buying a machine for $30,000. It will have a $2000 salvage value when disposed of in 7 years. Additional costs of $0.25/unit of product per year arise when Method C is used. A 15% interest rate is used by the ABC Company in evaluating investment alternatives. For what range of values for annual production volume is each method preferred? (8.2)

12. Two manufacturing methods are being considered. Method A has a fixed cost of $5000 and a variable cost of $50. Method B has a fixed cost of $2500 and a variable cost of $150. For what production volume would one prefer (a) Method A, and (b) Method B? (8.2)

13. In performing a make-or-buy analysis, the following cost functions were obtained:

$$M(x) = \$2500 + \$125x - \$0.025x^2$$

$$B_1(x) = \$5000 + \$75x$$

For what range of production volumes (x) should the component be purchased rather than made? (8.2)

14. Consider the following annual revenue and cost functions for a firm.

$$AR(x) = \$2250x - \$0.075x^2$$

$$AC(x) = \$1,500,000 + \$1500x - \$0.025x^2$$

Where x is the annual production volume. Determine the break-even production volume(s). (8.2)

15. In problem 13 suppose the firm is approached by a second supplier with the following quotation,

$$B_2(x) = \$2000 + \$100x + 0.005x^2$$

What would be your recommendation to the firm for various values of production volumes? (8.2)

16. In problem 5 suppose a 10-year planning horizon is specified. Perform a sensitivity analysis comparable to that given in Figure 8.3 to determine the effect of errors in estimating the *differences* in initial investment, the *differences* in annual operating and maintenance costs, and the *differences* in salvage values for the two investment alternatives. (8.3)

17. In problem 3 suppose the following pessimistic, most likely, and optimistic estimates are given for building cost, furnishings cost, annual operating and maintenance costs and the average rate per occupied unit.

	Pessimistic	Most Likely	Optimistic
Building cost	$7,500,000	$5,000,000	$4,000,000
Furnishings cost	3,000,000	1,875,000	1,000,000
Annual operating and maintenance costs	200,000	125,000	75,000
Average rate	25/day	45/day	55/day

Determine the pessimistic and optimistic limits on the break-even value for the daily occupancy percentage based on a *MARR* of 15%. Assume the motel will operate 365 days/year. (8.3)

18. A warehouse modernization plan requires an investment of $3 million in equipment; at the end of the 10-year planning horizon, it is anticipated the equipment will have a salvage value of $600,000. Annual savings in operating and maintenance costs due to the modernization are anticipated to total $1,400,000/year. A *MARR* of 20% is used by the firm. Perform a sensitivity analysis to determine the effects on the economic feasibility of the plan due to errors in estimating the initial investment required, and the annual savings. (8.3)

19. An investment of $10,000 is to be made into a savings account. The interest rate to be paid each year is uncertain; however, it is estimated that it is twice as likely to be 8% as it is to be 6%, and it is equally likely to be either 6% or 10%. Determine the probability distribution for the amount in the fund after 3 years, assuming the interest rate is not autocorrelated. (8.4)

20. Assume an initial investment of $10,000, annual receipts of $3500, and an uncertain life for the investment. Use a 25% *MARR*. Let the probability distribution for the life of the investment be given as follows:

N	p(N)
1	0.10
2	0.15
3	0.20
4	0.25
5	0.15
6	0.10
7	0.05

(a) Perform 20 simulations and estimate the probability of the investment being profitable.
(b) Analytically determine the probability of the investment being profitable. (8.4)

21. In problem 20 suppose the *MARR* is not known with certainty and the following probability distribution is anticipated to hold:

i	p(i)
0.10	0.20
0.15	0.60
0.20	0.20

Use simulation and analytical methods to estimate or determine the probability of the investment being profitable. (8.4)

22. In problem 20 suppose the magnitude of the annual receipts (R) is subject to random variation. Assume that each annual receipt will be identical in value and the annual receipt has the following probability distribution:

R	$p(R)$
$3,000	0.20
3,500	0.60
4,000	0.20

Use simulation and analytical methods to estimate or determine the probability of the investment's being profitable. (8.4)

23. In problem 20 suppose the minimum attractive rate of return is distributed as given in problem 21 and suppose the annual receipts are distributed as given in problem 22. Analytically determine that the investment will be profitable. (8.4)

24. In problem 23 suppose the initial investment is equally likely to be either $9,000 or $11,000. Analytically determine the probability that the investment will be profitable. (8.4)

25. Suppose $n = 4$, $i = 0\%$, a $11,000 investment is made, and the receipt in year j, $j = 1, \ldots, 4$, is statistically independent and distributed as in problem 22. Determine the probability distribution for present worth. (8.4)

26. Recall Example 8.13. Continue the simulation for 10 additional investment trials, estimate the probability of a negative present worth, and compute the average present worth over the total of 30 trials. (8.4)

27. Recall Example 8.14. Continue the simulation for 10 additional trials and compute the cumulative average present worth. (8.4)

28. In Example 8.14, suppose the individual had used a minimum attractive rate of return of 10% in the calculations. Would the first 10 simulation trials have yielded a positive average present worth? (8.4)

29. Consider an investment alternative having a 6-year planning horizon and expected values and variances for statistically independent cash flows as given below:

j	$E(C_j)$	$V(C_j)$
0	−$25,000	625×10^4
1	4,000	16×10^4
2	5,000	25×10^4
3	6,000	36×10^4
4	7,000	49×10^4
5	8,000	64×10^4
6	9,000	81×10^4

Using a discount rate of 20%, determine the expected values and variances for both present worth and annual worth. Based on the central limit theorem, compute the probability of a positive present worth; compute the probability of a positive annual worth. (8.4)

30. Solve problem 29 using a discount rate of (a) 0%, (b) 15%. (8.4)

31. Solve problem 29 when the $C_j, j = 1, \ldots, 6$, are perfectly correlated. (8.4)

32. Two investment alternatives are being considered. Alternative A requires an initial investment of $13,000 in equipment; annual operating and maintenance costs are anticipated to be normally distributed, with a mean of $5000 and a standard deviation of $500; the terminal salvage value at the end of the 8-year planning horizon is anticipated to be normally distributed, with a mean of $2000 and a standard deviation of $800. Alternative B requires end-of-year annual expenditures over the planning horizon. The annual expenditure will be normally distributed, with a mean of $7500 and a standard deviation of $750. Using a *MARR* of 25%, what is the probability that Alternative A is the most economic alternative? (8.4)

33. In problem 32, suppose the *MARR* were 10% with probability 0.25, 15% with probability 0.50, and 20% with probability 0.25, what is the probability that Alternative A is the most economic alternative? (8.4)

34. Company W is considering investing $10,000 in a machine. The machine will last N years, at which time it will be sold for L. Maintenance costs for this machine are estimated to increase by 10%/year over its life. The maintenance cost for the first year is estimated to be $1000. The company has a 20% *MARR*. Based on the probability distributions given below for N and L, what is the expected equivalent uniform annual cost for the machine? (8.4)

N	L	p(N)
6	$3,000	0.2
8	2,000	0.4
10	1,000	0.4

35. Two investment alternatives are under consideration. The data for the alternatives are given below. It is assumed the cash flows are not autocorrelated.

EOY	E[CF(A)]	SD[CF(A)]	E[CF(B)]	SD[CF(B)]
0	−$10,000	$1,000	−$15,000	$1,500
1	4,000	400	8,000	800
2	5,000	500	8,000	800
3	6,000	600	8,000	800
4	7,000	700	8,000	800
5	8,000	800	8,000	800
6	9,000	900	8,000	800

(a) For each alternative, determine the mean and standard deviation for present worth and annual worth using a *MARR* of 15%.

(b) For each alternative, based on the central limit theorem, compute the probability of a positive present worth using a *MARR* of 15%.

(c) Develop the mean and standard deviation for the *incremental* present worth using a *MARR* of 15%.

(d) Based on the central limit theorem, compute the probability of a positive incremental present worth using a *MARR* of 15%. (8.4)

CHAPTER 9

Decision Models

9.1 INTRODUCTION

This chapter presents basic models for displaying decision situations involving risk and uncertainty and suggests criteria for choosing among alternatives. These models will be *prescriptive* or *normative*, in that they prescribe a decision action to be taken. The risk-analysis discussion and models presented in Chapter Eight were descriptive; they described the behavior of the measure of effectiveness under conditions of uncertainty and risk. This chapter is concerned with specifying the preferred choice from a set of feasible alternatives and represents the final step in the seven-step procedure for comparing alternatives presented in Chapter Five.

The first section of this chapter discusses a matrix model for decisions under risk and uncertainty and suggests criteria for choosing among alternatives. The second section presents a *network* or *tree model* for decision situations involving a sequence of decisions over a time horizon where the outcomes of alternatives are uncertain. The final section of the chapter offers an introductory discussion on methodology for choosing among alternatives when multiple measures of effectiveness are involved. The matrix model for decisions under risk and uncertainty is now introduced by an example concerning material handling.

Example 9.1

For purposes of introducing a matrix model for decision making, consider the following manufacturing situation. A new product group is to be added by the ABC Company, with initial production scheduled 2 years hence. The new product group, which will consist of a single product in varying sizes and finishes, requires that a new and separate facility be added to the existing plant. The company is presently completing plans for the new facility, and plant construction will begin in the near future. Excess floor space is planned in order to increase production if demand increases. This increase is expected, with demand reaching stability after 5 years of production.

The manufacturing system consists, essentially, of several machining centers that operate on a job-shop basis instead of on a production-line basis. The manufacturing sequence is fixed except for slight variations, depending on product size and finish required. Thus, a subproblem of the total manufacturing system is to determine the material-handling requirements. After considerable study and discussion, it has been decided that industrial lift trucks provide the flexibility of handling required in the job-shop environment. Now, questions arise as to the type of lift truck, the number required, the attachments needed, whether they should be leased or purchased and, if leased, which of several leasing plans should be chosen, and so forth. The answers to these and other relevant questions depend primarily on the demand for the new product during the first 5-year production period. This demand is uncertain, and ABC Company management can influence demand through advertising only to a limited degree.

Management could make a decision on material-handling requirements based on the demand forecast for the first year of production, then evaluate the situation at the end of the first year, make a new decision for the next year, reevaluate, and so on, in sequential fashion. However, let us assume that it has been decided to choose a material-handling system now for the first 3-year production period and then reevaluate at the end of that period.

Given the above managerial decision, the analysts investigating the material-handling requirements make the judgment that demand for the product group during the 3-year planning horizon will be such that either four, five, or six lift trucks are required. They further reason that the feasible alternatives are:

A_1, A_2, A_3—lease four, five, or six trucks, respectively

A_4, A_5, A_6—purchase four, five, or six trucks, respectively

These lease plans do not include a maintenance contract, since the ABC Company presently employs skilled maintenance personnel. Furthermore, the analysts have not considered any lease plan with an "option to purchase" clause.

The analysts select before-tax, equivalent uniform annual cost as the single measure of effectiveness for a given "A_j versus number of trucks required" combination. It is necessary to make gross estimates of these annual costs, but essentially, for leasing, there is a fixed monthly charge per truck plus a judgmental penalty cost associated with production delays when the number of trucks required exceeds the number of trucks available. On the other hand, the company can obtain a more favorable leasing charge as the number of trucks leased increases. For the purchase alternatives, similar fixed (actually fixed and variable) annual operating and maintenance costs, penalty costs, and quantity discounts are involved. After much study, somewhat gross estimations, and a liberal application of judgment, the analysts present the model of Table 9.1 to management for a final decision. The values in the table are equivalent, uniform, and annual costs.

On the basis of the single, equivalent uniform annual cost measure of effectiveness, any *purchase* alternative is preferable to any *lease* alternative. For example, the purchase alternative A_4 has a lower equivalent annual cost than any lease alternative whether four, five, or six trucks are required. Thus, Alternative A_4 dominates the A_1, A_2, and A_3 alternatives, and they may therefore be deleted from the decision matrix. (It is noted that this might not have been the case if after-tax annual cost values had been calculated. Furthermore, a consideration of intangible factors by management could result in a lease alternative being chosen over a purchase alternative even though annual costs are higher.) The purchase alternatives A_5 and A_6 also dominate the three lease alternatives, but no one purchase alternative dominates any other purchase alternative. That is, no single purchase alternative has lower annual costs than any other purchase alternative when viewed across all "number of trucks required" conditions, anyone of which it is assumed can occur. Dominance will be discussed in greater detail subsequently.

A solution to the lift truck problem will not be presented here. The example has been cited to illustrate the matrix format of Table 9.1 and to suggest the model as an effective portrayal of many decision situations.

The elements in the decision model of Table 9.1 are now summarized as:

States = the number of lift trucks required in the 3-year planning horizon.

TABLE 9.1
Decision Model for the Lift Truck Example

Alternatives	Number of Trucks Required		
	4	5	6
A_1—Lease four	$18,000	$20,000	$24,000
A_2—Lease five	20,000	20,000	22,000
A_3—Lease six	21,000	21,000	21,000
A_4—Purchase four	12,500	14,500	18,500
A_5—Purchase five	14,000	14,000	16,000
A_6—Purchase six	15,000	15,000	15,000

Feasible alternatives	= the number of lift trucks available, whether leased or purchased.
Outcomes	= the results of particular "state versus alternative" combinations. In this case, the general outcomes are that the number of trucks required is less than, equal to, or greater than the number of trucks available.
Value of the outcomes	= the measure of effectiveness for a particular "state versus alternative" combination. In this case, equivalent uniform annual costs.

A matrix decision model with general symbolism, adapted from Morris [25], is given in Figure 9.1 and further discussed in the next section.

9.2 THE MATRIX DECISION MODEL

The symbolism employed is defined as follows:

A_j = an *alternative* or *strategy* under the decision maker's control,

where $\quad j = 1, 2, \ldots, n$.

S_k = a *state* or *possible future* that can occur given that alternative A_j is chosen, where $k = 1, 2, \ldots, m$.

θ_{jk} = the *outcome* of choosing alternative A_j and having state S_k occur.

		p_1	p_2	$\cdots$	$\cdots$	p_k	$\cdots$	$\cdots$	p_m
A_j	S_k	S_1	S_2	$\cdots$	$\cdots$	S_k	$\cdots$	$\cdots$	S_m
A_1		$V(\theta_{11})$	$V(\theta_{12})$	$\cdots$	$\cdots$	$V(\theta_{1k})$	$\cdots$	$\cdots$	$V(\theta_{1m})$
A_2		$V(\theta_{21})$	$V(\theta_{22})$	$\cdots$	$\cdots$	$V(\theta_{2k})$	$\cdots$	$\cdots$	$V(\theta_{2m})$
$\vdots$		$\vdots$	$\vdots$			$\vdots$			$\vdots$
$\vdots$		$\vdots$	$\vdots$			$\vdots$			$\vdots$
A_j		$V(\theta_{j1})$	$V(\theta_{j2})$	$\cdots$	$\cdots$	$V(\theta_{jk})$	$\cdots$	$\cdots$	$V(\theta_{jm})$
$\vdots$		$\vdots$	$\vdots$			$\vdots$			$\vdots$
$\vdots$		$\vdots$	$\vdots$			$\vdots$			$\vdots$
A_n		$V(\theta_{n1})$	$V(\theta_{n2})$	$\cdots$	$\cdots$	$V(\theta_{nk})$	$\cdots$	$\cdots$	$V(\theta_{nm})$

FIGURE 9.1. A matrix decision model with general symbolism.

$V(\theta_{jk})$ = the *value* of outcome θ_{jk}, which may be in terms of dollars, time, distance, utility, etc.

p_k = the *probability* that state S_k will occur.[1]

The term p_k was not previously introduced in the lift truck decision problem but will subsequently serve as a means by which decisions may be categorized into decisions under assumed certainty, decisions under risk, and decisions under uncertainty. This categorization has essentially been made in Chapter Eight but will be repeated in order to present decision rules for the normative models of this chapter.

The matrix model, in general, describes a set of alternatives available where a single alternative is to be selected at the present time. It is implied that the outcomes possible for a given alternative do not result in the necessity for subsequent decisions at future times. If a sequence of decisions were involved, the decision tree model presented later would probably be more appropriate.

It is also assumed in the matrix model that the alternatives are mutually exclusive, feasible, and represent the total set of alternatives the decision maker may wish to consider. Recall from the discussion in Chapter Five that mutually exclusive alternatives cannot occur together, and thus a single alternative will be chosen from the feasible set. The question now arises, "What constitutes feasibility?" The answer to the question lies in the original statement of objectives that the alternative is to accomplish. For example, if one were considering the replacement of a production machine because of increasing maintenance costs and inadequate capacity, an objective might be that the new machine must be capable of producing 1000 units of Product A per day. Those machines incapable of satisfying this constraint of 1000 units/day are infeasible and rejected from further consideration. Those machines capable of meeting or exceeding the requirement of 1000 units/day are feasible, and the "best" replacement machine is then selected from the feasible set according to a more discriminating criterion, such as minimum estimated equivalent annual cost or maximum estimated annual profit.

Determining the *total set* of feasible alternatives is also a difficulty deserving additional comment. Rarely does the analyst have the time or even the insight to define all the possible alternatives available in a given decision situation. How to execute a search procedure and when to terminate the search are interesting and important questions, but they will not be pursued here. Suffice it to say that experience plays a major role in the search procedure, and when to terminate the search is primarily a matter of judgment in the practical world of business decisions. The decision to terminate the search for alternatives is a trade-off of the cost of finding an additional alternative versus the benefits resulting from the discovery of the additional alternative. Admittedly, the benefits cannot actually be evaluated until after

---

[1] It is assumed the probability that a particular state will occur does not depend on the alternative chosen by the decision maker.

the alternative is discovered and thus, this guideline for terminating search is largely conceptual in nature.

Experience and judgment also play an important role in defining the relevant states for a given decision situation and the set of feasible alternatives involved. To convey further the meaning of a *state,* other terms in the literature are *possible futures, states of nature, external conditions,* and *future events.*

In the discussion on risk analysis in Chapter Eight, states were values for the random variables of purchase price, annual operating expenses, annual revenues, service lives, and salvage values. However, in other decision situations, the states may be defined more grossly; for instance, the market demand for product A will be low, medium, or high. Furthermore, the number of states may be finite or infinite. For the examples of this text, the number of states will be assumed finite and few in number. This seems a realistic assumption for most business and engineering decisions because, although the states may actually be a very large finite number or even infinite, the decision maker wishes to limit the scope of the problem and uses a sample of states. Using the industrial lift truck example presented earlier, knowledge of the machining processes involved and a recorded history of machining times can provide the basis for estimating values for the states in terms of the number of trucks required to service different levels of production output. This same experience can also provide the range of state values to consider. In any case, it will be assumed here that once the states are defined, they are collectively exhaustive (i.e., the total set of possible states). Furthermore,

1. The occurrence of one state precludes the occurrence of the others (i.e., they are mutually exclusive).
2. The occurrence of a state is not influenced by the alternative selected by the decision maker.
3. The occurrence of a state is not known with certainty by the decision maker.

For each alternative-versus-state combination, there is an outcome and a value associated with the outcome. In the previous industrial lift truck example, a given "number of trucks available versus number of trucks required" combination resulted in the trucks available being less than, greater than, or equal to the number of trucks required. For each outcome, it was assumed that an annual equivalent dollar cost could be calculated and was the preferred decision criterion. Other value measurements may of course be appropriate in other decision situations. The matrix values may be one of two general kinds—objective or subjective. Objective values represent physical quantities such as dollars, hours, liters, kilograms, and so forth. With the exception of the section on multiple goals in this chapter, the subsequent examples are concerned with dollars or objective values.[2] Subjective values,

---

[2] A linear utility function for the decision maker is also assumed. That is, minimizing costs or maximizing profits are both the same as maximizing the decision maker's utility.

on the other hand, are numbers that represent the decision maker's relative preferences; for example, a score of 90 on an examination is preferred to a score of 60. The values of 90 and 60 are subjective values assigned on an arbitrary scale of, say, 0 to 100. An example of a subjective value matrix will be given in the multiple-goals section of this chapter. In any case, as a general statement, the problems of measurement and value determination are often very substantial in real-world situations.

Let the matrix model of Figure 9.1 now be recalled in order to classify decisions on the basis of the amount of information available to the decision maker in regard to the probability of occurrence for a given state.

9.3 DECISIONS UNDER ASSUMED CERTAINTY

It is reasonable to assume in many decision situations that only one state is relevant. Then, the decision maker assumes this single state will occur with certainty (i.e., with probability = 1.0). This kind of case is termed a *decision under assumed certainty*. For example, it may be a matter of company policy to depreciate new equipment purchases fully in 5 years. If equivalent annual cost is chosen as the measure of effectiveness to compare several new equipment candidates, the determination of these annual costs would be based on a 5-year life assumed for each candidate. The outcome of a 5-year life and the associated annual operating costs for each candidate are assumed certain. The actual functional life of an equipment candidate might be 4, 5, 6, 7, 8, . . . , years. However, a judgment is made by the decision maker to suppress uncertainty because of convenience or practicality and assume only the single state of a 5-year life. All of the analysis in Chapters Three to Seven was based on such judgments, and single-value estimates for operating costs, service life, and the like were examples of assumed certainty.

In terms of the matrix decision model, a decision under assumed certainty would appear as follows:

	S
A_1	$V(\theta_1)$
A_2	$V(\theta_2)$
$\vdots$	$\vdots$
A_j	$V(\theta_j)$
$\vdots$	$\vdots$
A_{n-1}	$V(\theta_{n-1})$
A_n	$V(\theta_n)$

If, in the above model, the values for each alternative were profits (or gains), the principle for choice might be "choose the alternative that maximizes profit." Conversely, if the values were costs (or losses), one might "choose the alternative that minimizes costs."

Example 9.2

Suppose that an investor is considering a $10,000 purchase of United States government securities. Different government agencies offer such securities, and they are available on the open market. Assuming that the various securities have common maturity dates, the investor judges that each agency is financially secure and payoffs are certain. A logical criterion for choice among the securities would be the effective annual yield on the $10,000 purchase. (It is assumed there is no reason for allocating the $10,000 to different securities.) Suppose the investor considers five government securities and, for a maturity date of 5 years hence, has calculated the effective annual yield on each security to be

$$
\begin{array}{ll}
A_1 & 8.0\% \\
A_2 & 7.3\% \\
A_3 & 8.7\% \\
A_4 & 6.0\% \\
A_5 & 6.5\%
\end{array}
$$

Since the investor would logically desire to maximize the yield on the investment, A_3 would be chosen. Recall that equal risk and maturity dates were assumed for each security.

Example 9.3

The cash flow profiles for three mutually exclusive investment alternatives are given in Table 9.2. Considering the minimum attractive rate of return of 10% as the single state that will occur with certainty and the present worth of each alternative as the value of interest, the matrix model of this decision appears in Table 9.3. If the principle for choice is "select the alternative that maximizes the present worth," A_3 should be chosen.

9.4 DECISIONS UNDER RISK

A decision situation is called a *decision under risk* when the decision maker elects to consider several states and the probabilities of their occurrence are explicitly stated. In some decision problems, the probability values may be objectively known from historical records or objectively determined from analytical calculations. For example, if an unbiased die is rolled once, the six

TABLE 9.2

Cash Flow Profiles for Three Mutually Exclusive
Investment Alternatives ($MARR = 10\%$)

t	A_1	A_2	A_3
0	$ 0	-$10,000	-$15,000
1	4,000	7,000	7,000
2	4,000	7,000	8,000
3	4,000	7,000	9,000
4	4,000	7,000	10,000
5	4,000	7,000	11,000
Present worth	$15,163.20	$16,535.60	$18,397.33

possible states are that one, two, three, four, five, or six dots will occur on the top face exposed. The objective probability that each state will occur prior to the roll is $\frac{1}{6}$. The sum of the individual probabilities over all states equals the value of 1.0. This property of $\Sigma p_k = 1.0$ agrees with an earlier statement that, once the set of possible states is decided on, the states are treated as if they constituted a set of mutually exclusive and collectively exhaustive events. Indeed, the property of $\Sigma p_k = 1.0$ is a necessary restriction if probability theory is to be used as a guide to decision making.

The decision maker may not have past records available to arrive at objective probability values. This was the case in the industrial truck example earlier. However, if the decision maker feels that experience and judgment are sufficient to assign probability values subjectively to the occurrence of each state, the decision is still treated as one under risk. The restriction of $\Sigma p_k = 1.0$ still applies in this case. Principles of choice for selecting an alternative in a decision under risk, which commonly appear in the literature on decision theory, are illustrated in Example 9.4.

TABLE 9.3

Present Worths of
Three Mutually
Exclusive Investment
Alternatives
($MARR = 10\%$)

	S_1
A_j	$i = 10\%$
A_1	$15,163.20
A_2	16,535.60
A_3	18,397.33

Example 9.4

The three mutually exclusive investment alternatives presented in Table 9.2 will again be used to illustrate guidelines for a decision under risk. The single state previously defined was a minimum attractive rate of return equal to 10%. For the purpose of this example, the two additional states of $MARR = 20\%$ and $MARR = 30\%$ will be assumed. The realism of considering these states as probabilistic in nature may be questionable, but an argument supporting this claim can be posed. In periods of an unstable economy, the analyst may perceive the $MARR$ as uncertain and be willing to define different values with associated probabilities. The matrix model of Table 9.4 assumes that this is the case, and cell values are present worths (rounded) for a 5-year planning horizon.

The states of $S_1 = 10\%$, $S_2 = 20\%$, and $S_3 = 30\%$ may have been defined by judging these values to be optimistic, most likely, and pessimistic estimates, as discussed in Chapter Eight. A $MARR$ that lowers the present worth of the investment alternative is considered a pessimistic $MARR$. The chance of occurrence for the states may have been reasoned as follows: S_3 is twice as likely to occur as S_2, which, in turn is three times as likely to occur as S_1. This reasoning can be stated as the equation $p_1 + 3p_1 + 6p_1 = 1.0$. From this, $p_1 = 0.10$, $p_2 = 0.30$, and $p_3 = 0.60$ can be determined.

9.4.1 Dominance

The *dominance* principle is described as follows. Given two alternatives, if one would always be preferred no matter which state occurs, this preferred alternative is said to dominate the other, and the dominated alternative can be deleted from further consideration. Applying this principle will not necessarily solve the decision problem by selecting a unique alternative, but it may reduce the number of alternatives for further consideration.

The dominance principle is a way of deciding which alternatives not to select. If the values are in terms of *costs*, a more formal statement of the

TABLE 9.4

Matrix Model with States of $MARR = 10\%$, 20%, and 30%, Example 9.4

A_j	$MARR$	$p_1 = 0.10$ S_1 10%	$p_2 = 0.30$ S_2 20%	$p_3 = 0.60$ S_3 30%
A_1		$15,163	$11,962	$9,742
A_2		$16,536	$10,934	$7,049
A_3		$18,397	$10,840	$5,679

dominance relation is: *If there exists a pair of alternatives A_j and A_l such that $V(\theta_{jk}) \leq V(\theta_{lk})$ for all k, A_j is said to dominate A_l. Alternative A_l may then be discarded from the decision problem.* If the values of a decision matrix are in terms of *gains*, the condition for A_j to dominate A_l is that $V(\theta_{jk}) \geq V(\theta_{lk})$ for all k.

In Table 9.4, no alternative has a present worth value greater than the present worth value of another alternative for all states. Thus, there is no dominant alternative for this example, but recall that there were dominated alternatives in Table 9.1.

9.4.2 Expectation-Variance Principle

If X is a discrete random variable that is defined for a finite number of values, and $p(x)$ denotes the probability of a particular value occurring, then the expected or mean value of the random variable is defined as

$$E(X) = \sum_{\text{all } x} xp(x) \tag{9.1}$$

The variance of the random variable is defined as $\text{Var}(X)$, where

$$\text{Var}(X) = \sum_{\text{all } x} [x - E(X)]^2 p(x) \tag{9.2}$$

For the matrix of Figure 9.1, the random variable is $V(\theta_{jk})$, and the expected value of an alternative A_j is

$$E(A_j) = \sum_{\text{all } k} V(\theta_{jk})p_k \tag{9.3}$$

where p_k denotes the probability of state k occurring.

The expected values for each of the alternatives of Table 9.4 are calculated as

$$E(A_1) = \$15{,}163(0.10) + \$11{,}962(0.30) + \$9742(0.60)$$
$$= \$10{,}950.10$$

$$E(A_2) = \$16{,}536(0.10) + \$10{,}934(0.30) + \$7049(0.60)$$
$$= \$9163.20$$

$$E(A_3) = \$18{,}397(0.10) + \$10{,}840(0.30) + \$5679(0.60)$$
$$= \$8499.10$$

If the principle of maximizing (minimizing) the expected gain (loss) is followed, alternative A_1 would be chosen, since the expected values are positive present worths. A corollary of the expected-value criterion is that, in the event of a tie in expected value for two or more alternatives, the alternative having minimum variance should be chosen. The argument for this secondary criterion is, essentially, that a larger variance is indicative of greater uncertainty, and since most decision makers have an aversion to uncertainty, the alternative with smallest variance should be chosen. Conrath [7], in an article on the implementation of statistical decision theory in practical decisions under risk, states that ". . . as yet we have little guidance as to what the trade-offs between mean and variance might be or ought to be." That the expected value and the variance are two independent decision criteria instead of dependent criteria is an interesting speculation (or perhaps observation) that deserves further research.

9.4.3 Most Probable Future Principle

If, in a decision under risk, one state has a probability of occurrence considerably greater than any other, the *most probable future principle* is to consider this state as certain and all other states as having a zero chance of occurrence. The decision is thereby reduced to a decision under assumed certainty. Then an alternative is chosen that maximizes (or minimizes) the measure of effectiveness being used. For Table 9.4, S_3 has a probability of 0.60 assigned and is therefore the most probable state. Assuming certainty for S_3 or $p_3 = 1.0$, the reduced matrix is

$$S_3$$

	S_3
A_1	\$9742
A_2	\$7049
A_3	\$5679

To obtain the maximum present worth, A_1 is selected as the preferred alternative.

Another example of the most probable future principle is the case of a person commuting to work. Each morning that the person leaves for work there is a nonzero probability that an accident could occur and serious injury result. However, the most probable future is that the commuter will reach work safely. The commuter considers this future certain, or else he (she) would probably not go to work!

As a final remark on this particular principle of choice, it seems appealing only if one state has a significantly higher probability of occurring than any other state and the values in the total matrix do not differ significantly. One would not wish to suppress the consideration of a state that, albeit with low probability, would inflict a severe loss on the decision maker. An example of this is the matrix below where the positive values are profits.

	$p_1 = 0.95$	$p_2 = 0.05$
A_1	\$10	\$40
A_2	\$50	-\$5,000

Following the most probable future principle, A_2 would be chosen. However, the suppression of S_2 does not alter the fact that S_2 could occur, by virtue of $p_2 = 0.05$; in that event the decision maker would suffer a loss of \$5000.

9.4.4 Aspiration-Level Principle

In most real-world decisions, the complexity of the decision prevents the discovery and selection of an alternative that will yield the single "best" result, and decision makers set their goals in terms of outcomes that are good enough. That is, decision makers set aspiration levels and then evaluate alternatives against them. An interpretation of this philosophy in terms of a decision under risk is to select an alternative that maximizes the probability of achieving the desired aspiration level. Typical aspiration-level objectives might be to choose the alternative that maximizes the probability of either (1) a 25% internal rate of return, (2) annual costs less than $8000, or (3) annual profits greater than $9500. In terms of the example of Table 9.4, let the objective be to choose the alternative that maximizes the probability that the present worth value will be equal to or greater than $8000. This objective can be symbolized as $P(PW \geq \$8000)$. From Table 9.4 it can be seen that, for A_1, the present worth will be greater than $8000 when S_1, S_2, or S_3 occurs. Thus, a present worth greater than $8000 is certain if A_1 is chosen. Because the states are mutually exclusive, the additive-probability law permits the sum of the individual state probabilities to be the probability of the joint event, "$PW \geq \$8000$" for A_1. The calculations for each alternative are

$$\text{For } A_1, P(PW \geq \$8000) = P(S_1) + P(S_2) + P(S_3)$$
$$= 0.10 + 0.30 + 0.60 = 1.0$$

$$\text{For } A_2, P(PW \geq \$8000) = P(S_1) + P(S_2)$$
$$= 0.10 + 0.30 = 0.40$$

$$\text{For } A_3, P(PW \geq \$8000) = P(S_1) + P(S_2)$$
$$= 0.10 + 0.30 = 0.40$$

and A_1 maximizes $P(PW \geq \$8000)$.

A variety of aspiration-level objectives are, of course, possible for even a single decision under risk. The relevant aspiration-level objective is a decision maker's preference.

Do not interpret any of the principles for choice above or those to follow in the next section as the way an alternative *should* be selected. Instead, the principles for choice presented result from observing how people *seem* to decide on alternatives; therefore, they are offered only as possible guidelines to the decision maker.

9.5 DECISIONS UNDER UNCERTAINTY

A decision situation where several states are possible and sufficient information is not available to assign probability values to their occurrence is termed a *decision under uncertainty*. It can be argued that any decision maker, if able to define the states, should have sufficient knowledge to make at least gross subjective probability estimates for these states. On the other hand, in situations such as (1) installing new safety devices on machinery where the states are the number of injuries expected per year, or (2) the introduction of a new product where the states are units demanded per year, one may feel quite inadequate to assign probabilities to these states. This may be the case even though the analyst is not totally lacking in knowledge concerning the number of injuries expected per year or the units of new product demanded per year. The principles for choice subsequently presented are based on the premise that state probabilities cannot be assigned with any degree of confidence.

9.5.1 The Laplace Principle

The philosophy of the *Laplace principle,* named after an early nineteenth-century mathematician [7], is simply that if one cannot assign probabilities to the states, the states should be considered as equally probable. Then, consider the decision as one under risk. Applying this principle to the previous example of Table 9.4, the probability value assigned to each of the three states is 1/3. Then,

$$E(A_1) = \$15{,}163(1/3) + \$11{,}962(1/3) + \$9742(1/3)$$
$$= \$12{,}289$$

$$E(A_2) = \$16{,}536(1/3) + \$10{,}934(1/3) + \$7049(1/3)$$
$$= \$11{,}506.33$$

$$E(A_3) = \$18{,}397(1/3) + \$10{,}840(1/3) + \$5679(1/3)$$
$$= \$11{,}638.67$$

and A_1 would be chosen to maximize expected present worth.

9.5.2 Maximin and Minimax Principles

The *maximin* and *minimax* principles for choice represent a single philosophy, depending on whether the matrix values are *gains* or *losses,* respectively. These principles hold considerable appeal for the conservative or pessimistic decision maker. In applying the principles, if the matrix values are *gains,* the minimum gain associated with each alternative (when viewed over all states) is determined, and the maximum value in the set of minimum values designates the alternative to be chosen. In other words, determine the worst value possible for *each* alternative and then choose the best value from the set of worst values. Applying the maximin principle to the example problem of Table 9.4 gives

	Minimum Present Worth Value	Maximum of These
A_1	$9742	$9742
A_2	7049	
A_3	5679	

Thus, alternative A_1 is selected as the alternative that will maximize the minimum present-worth value that could occur. More formally stated, the maximin principle is to

Select the alternative, j, associated with the $\max_{j} \min_{k} V(\theta_{jk})$.

In the case of a decision dealing with *losses* or *costs,* the minimax philosophy is also one of extreme pessimism. In applying the minimax principle to a matrix of costs, the maximum cost associated with each alternative (when viewed over all states) is determined, and the minimum value in the set of maximum values designates the alternative to be chosen. Again, this is choosing the best (or minimum) cost from the set of worst (or maximum) costs. An example to illustrate the application of the minimax principle is given in Table 9.5, where the cell values are equivalent uniform annual costs for the alternatives. Thus, from Table 9.5, alternative A_1 would be chosen. Formally stated, the minimax principle is to

Select the alternative, j, associated with the $\min_{j} \max_{k} V(\theta_{jk})$.

9.5.3 Maximax and Minimin Principles

As illustrated in the previous section, the maximin principle for gains and the minimax principle for losses represent a philosophy of extreme pessimism on behalf of the decision maker. For the person who is an extreme optimist in a given decision situation, an *optimistic* rule for choice among alternatives that involve *gains* is the *maximax principle.* That is, the decision maker desires to select the alternative that affords the opportunity to obtain the largest value given in the matrix. A formal statement for the maximax principle is to

Select the alternative, j, associated with the $\max_{j} \max_{k} V(\theta_{jk})$.

Then, if the matrix cell values were losses, an optimistic philosophy of choice is to select the alternative that affords the opportunity to obtain the

TABLE 9.5

Matrix of Equivalent Uniform Annual Costs

Alternatives / States	S_1	S_2	S_3	S_4	Maximum Cost	Minimum of These
A_1	$7,000	$10,000	$ 6,500	$ 5,000	$10,000	$10,000
A_2	9,000	4,000	9,000	15,000	15,000	
A_3	4,000	6,000	12,000	8,500	12,000	
A_4	2,000	8,000	3,000	11,000	11,000	

minimum loss value given in the matrix. A formal statement for the *minimin principle* involving *loss* values is

> **Select the alternative, *j*, associated with the min min $V(\theta_{jk})$.**
> $\qquad\qquad\qquad\qquad\qquad\qquad\qquad\qquad j \quad k$

Applying the maximax principle to the "gains" example of Table 9.4 selects alternative A_3 ($18,397), and applying the minimin principle to the cost example of Table 9.5 selects alternative A_4 ($2000).

9.5.4 Hurwicz Principle

The *Hurwicz principle* considers that a decision maker's view may, in the case of gains, fall between the extreme pessimism of the maximin principle and the extreme optimism of the maximax principle and offers a method by which various levels of optimism–pessimism may be incorporated into the decision. The Hurwicz principle defines an index of optimism, α, on a scale from 0 to 1. A value of $\alpha = 0$ thus indicates zero optimism or extreme pessimism and, conversely, a value of $\alpha = 1$ indicates extreme optimism.

Assuming that a decision maker is able to reflect a degree of optimism by assigning a specific value to α, and again emphasizing that the decision is in terms of *gains*, the maximum gain value for each alternative is multiplied by α, and the minimum gain value for each alternative is multiplied by $(1 - \alpha)$. The linear sum of these two products for each alternative j is called the Hurwicz value, H_j, and the alternative that maximizes this value is selected. That is,

> **Select an index of optimism, α, such that $0 \leq \alpha \leq 1$. For each alternative j, compute $H_j = \alpha\,[\max_k V(\theta_{jk})] + (1 - \alpha)\,[\min_k V(\theta_{jk})]$ and select the alternative that maximizes this quantity.**

Note that if $\alpha = 0$ (extreme pessimism), only the minimum gain for each alternative will be examined, and the maximum of these values would then be selected—the maximin principle of choice. If $\alpha = 1$ (extreme optimism), the maximax principle would be executed.

Suppose the decision maker concerned with the example problem of Table 9.4 is a middle-of-the-road type of person and assigns $\alpha = 0.5$. Then

$$H_1 \text{ for } A_1 = (0.5)(\$15,163) + (0.5)(\$9742) = \$12,452.50$$

$$H_2 \text{ for } A_2 = (0.5)(\$16,536) + (0.5)(\$7049) = \$11,792.50$$

$$H_3 \text{ for } A_3 = (0.5)(\$18,397) + (0.5)(\$5679) = \$12,038.00$$

Choosing the maximum of these values is to select alternative A_1.

Shortcomings of the Hurwicz principle include (1) ignoring intermediate values for each alternative, (2) the inability to select a particular alternative when two or more alternatives have the same Hurwicz value, and (3) the

TABLE 9.6

Data for the Hurwicz Example

	S_1	S_2	S_3
A_1	$ 2	$10	$30
A_2	18	18	18
A_3	24	12	10

practical difficulty of assigning a specific value to α. This latter objection can be circumvented by the graphic solution discussed below.

It will be noted from the expression for calculating the Hurwicz value that the equation is linear in α, which ranges in value from 0 to 1.0. Any alternative thus yields a linear function that can be plotted, and preferred alternatives can be determined for various ranges of α. An example to illustrate the procedure is given in Table 9.6, where the cell values are gains.

Applying the principle for the data of Table 9.6, the Hurwicz values are

$$H_1 \text{ for } A_1 = \alpha(\$30) + (1 - \alpha)(\$2) = 28\alpha + 2$$

$$H_2 \text{ for } A_2 = \alpha(\$18) + (1 - \alpha)(\$18) = 18$$

$$H_3 \text{ for } A_3 = \alpha(\$24) + (1 - \alpha)(\$10) = 14\alpha + 10$$

These equations are plotted in Figure 9.2.

From Figure 9.2 it can be seen that if a decision maker's index of optimism is less than α_1, alternative A_2 would be preferred; if it is greater than α_1, alternative A_1 would be preferred. A specific value for α_1 can be determined graphically or analytically by, in this case, equating the Hurwicz value functions for any pair of alternatives (since there is a single intersection for the three linear functions). For example,

$$H_1 = H_2$$

$$28\alpha_1 + 2 = 18 \quad \text{and} \quad \alpha_1 = 4/7$$

If the decision problem involves costs or losses, a reinterpretation of the Hurwicz principle is necessary:

Select an index of optimism, α, such that $0 \leq \alpha \leq 1$. For each alternative A_j, compute $H_j = \alpha[\min_k V(\theta_{jk})] + (1 - \alpha)[\max_k V(\theta_{jk})]$ and select the alternative that minimizes this quantity.

Note that if $\alpha = 0$, only the maximum costs would be examined for each alternative and the minimum of these would be selected—the pessimistic minimax criterion. If $\alpha = 1$, the optimistic minimin principle would be executed.

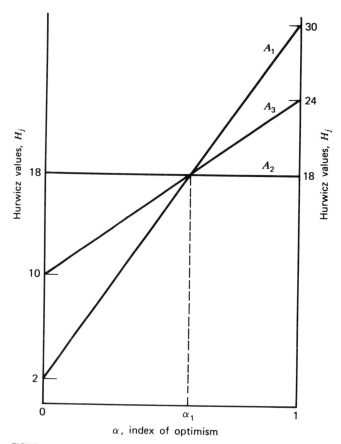

FIGURE 9.2. Graphical interpretation of the Hurwicz principle.

9.5.5 Savage Principle (Minimax Regret)

This principle, proposed by L. J. Savage [22], introduces and defines a quantity termed *regret*. A matrix consisting of regret values is first developed. Then the maximum regret value for each alternative A_j is determined, and the alternative associated with the minimum regret value is chosen from this set of maximum regret values.

If the original decision matrix values are *gains*, the procedure for determining the regret matrix is as follows.

1. For a given state S_l, search the matrix column values over all alternatives and determine the largest gain. Assign this gain a zero regret value.

2. For all other gain values *under* the given state S_l, subtract these from the largest gain value from step 1. The difference is interpreted as units of regret for a particular alternative A_j, given that state S_l occurs.

TABLE 9.7

Regret Matrix for the Minimax
Regret Example

	S_1	S_2	S_3
A_1	3,234	0	0
A_2	1,861	1,028	2,693
A_3	0	1,122	4,063

3. Repeat steps 1 and 2 for each state S_k, $k \neq l$, until the regret matrix is completed.

Once the regret matrix is developed, examine each alternative over all states and select the maximum regret value. Then *choose the alternative that minimizes the maximum regret*. This principle of choice is conservative and is very similar to the maximin principle. It is emphasized, however, that the Savage principle deals in units of regret, not gain values in the original matrix; therefore, if both the maximin principle and the minimax regret principle are applied to a common decision problem, different alternatives may be selected.

Applying the Savage principle to the present-worth example of Table 9.4, the regret matrix given in Table 9.7 is obtained.

The maximum regret values are 3234, 2693, and 4063 for alternatives A_1, A_2, and A_3, respectively. Thus, the minimum of these is 2693, and alternative A_2 would be preferred.

In a decision problem involving *costs*, the "logic" for creating the regret matrix is the same as the problem involving gains. That is,

1. For a given state S_l, search the matrix column values over all alternatives and determine the smallest cost. Assign this cost a zero regret value.

2. For all other cost values *under* the given state S_l, subtract the smallest cost value determined in step 1 from these to determine regret values.

3. Repeat steps 1 and 2 for each state, S_k, $k \neq l$, until the regret matrix is completed.

After the regret matrix is obtained, the procedural steps for selecting an alternative are the same as before.

9.6 SEQUENTIAL DECISIONS

In the previous sections, the decision situations happened once. We now wish to consider situations that may require multiple decisions in sequential fashion. The reader can no doubt recall from personal experience or imagine real-world decisions that are sequential. For example, medical decisions are typically sequential. Based on the results of a routine blood analysis, a physician may require additional medical tests. The physician may then perform exploratory surgery and, based on these findings, recommend additional surgery for the patient. Following surgery, the recuperative procedure may also be conditional and involve further sequential decisions on behalf of the physician.

Sequential decisions in the business and industrial world are also common. For example, the number and type of material-handling units to purchase may depend on the forecast of product demand in each of the next 5 years. In the field of quality control, sampling is concerned with sequential decisions. For example, if a sample of five units is taken from a production machine, subsequent actions taken in regard to the production machine might very well depend on whether 0, 1, 2, 3, 4, or 5 defective units were found in the sample of five units.

In this section, sequential decisions under risk will be discussed, the technique of *decision tree* representation will be used, and the logic of Bayes's theorem will be applied in the solution of the decision-tree problem.

9.6.1 Decision Trees

Magee [17, 18] is generally credited with representing sequential decisions in decision trees and advocating the expectation criterion for solving the problem. Hespos and Strassman [13] developed stochastic decision trees incorporating risk-analysis methods and simulation techniques into the basic decision tree methodology originated by Magee. Before presenting the graphic technique and a solution procedure, it is necessary to define certain symbolism.

Δ = a decision point, which will be labeled D_i for the ith decision point in the tree

Δ—— = a branch emanating from a decision point; represents an alternative that can be chosen at this point

0 = a fork (or node) in the tree where chance events influence the outcomes of an alternative choice

0—— = a branch representing a probabilistic outcome for a given alternative. The notation $(\theta; p)$ represents an outcome θ, having an associated value $v(\theta; p)$, which occurs with probability p. It is assumed that branches emanating from a fork in the tree represent mutually exclusive and collectively exhaustive outcomes such that their probabilities sum to 1.0

V = a value associated with a particular outcome (branch)

In Figure 9.3, a symbolic decision tree for a simple sequential decision problem is presented. The sequential decision problem depicted involves two decision points, D_1 and D_2. If alternative A_1 is chosen, it can have two

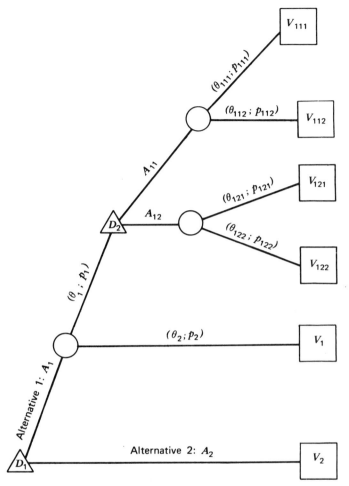

FIGURE 9.3. A symbolic decision tree.

outcomes, θ_1 and θ_2. If θ_1 occurs, then a decision, D_2, is required. If alternative A_{11} is chosen, there can be the two outcomes θ_{111} and θ_{112}. If alternative A_{12} is chosen at D_2, there can be the two outcomes θ_{121} and θ_{122}. For each branch in the tree there is an associated value. For example, if alternative A_1 is chosen *and* outcome θ_1 occurs *and* alternative A_{11} is chosen *and* outcome θ_{111} occurs, than value V_{111} results with the conditional probability p_{111}. The first decision to be made is at decision point D_1, and if alternative A_1 is chosen, then a second decision is required at decision point D_2. The ultimate question to answer is which alternative to choose at D_1?

The principle of *maximizing expected gain* or *minimizing expected loss* is adopted as the principle of choice at each decision point. The solution procedure advocated is to reach the best decision at the decision point (or points) most distant from the base (the first decision) of the tree. Then, replace this most distant decision point with the best expected value and work backward through decision points until the best decision is made at the initial decision point, D_1. To illustrate the solution procedure, consider the decision tree of Figure 9.3 and assume all the values are *gains*. The procedure follows.

1. At decision point D_2, calculate the expected gains for A_{11} and A_{12}. That is,

$$E(A_{11}) = p_{111}V_{111} + p_{112}V_{112}$$

and

$$E(A_{12}) = p_{121}V_{121} + p_{122}V_{122}$$

If $E(A_{11}) > E(A_{12})$, then alternative A_{11} is chosen as best at D_2. Assume this is the case.

2. Replace D_2 with $E(A_{11})$, and the new reduced decision tree becomes as shown in Figure 9.4.

3. Calculate the expected gains for alternatives A_1 and A_2. That is,

$$E(A_1) = p_1 E(A_{11}) + p_2 V_1$$

and

$$E(A_2) = V_2 (\text{a certain event with } p = 1.0).$$

If $E(A_1) > E(A_2)$, alternative A_1 is chosen as the best at decision point D_1. If $E(A_2) > E(A_1)$, then A_2 would, of course, be chosen, and if $E(A_2) = E(A_1)$, then one would be indifferent between the choice of A_1 and A_2. Given that $E(A_1) > E(A_2)$ from step 3, the optimal sequence of decisions is to choose A_1 at D_1 and then choose A_{11} later on if outcome θ_1 does occur.

Example 9.5

The Jax Tool and Engineering Company is a medium-size, job-order machine shop and has been in business for 10 years. After surviving two na-

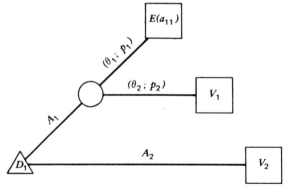

FIGURE 9.4. The reduced decision tree for Figure 9.3.

tional economic recessions with shrewd and industrious management, the company has built a solid reputation for dependability and high-quality work. During the past 2 years, sales have sharply increased, and the company is operating near capacity. A market test has recently been made for a new product—electroplating tanks—and the response has been very favorable. However, production floor space to establish the new product line is limited in the present building. The people in management have three alternatives: (1) do nothing, (2) rearrange an area of the existing floor space, or (3) build an addition to the present plant. Also, there is the consideration of competition from one other job-order machine shop in the local area. If the electroplating tank venture is successful and the market demand cannot be fully met by the Jax Company, there is a good chance that competition will occur.

Management assigns representatives from production engineering and marketing to make a detailed study of the alternatives. Because management desires a report as soon as possible, it is necessary to make rather gross estimates on many items in the analysis. The analysts decide to use a 5-year planning horizon, a minimum attractive rate of return of 30%, and an incremental present-worth value for each alternative relative to the do-nothing alternative. The results of the analysis now follow.

Rearrangement does not require any new construction, but machinery must be moved, some new machinery purchased, and new storage areas created. Production would be limited to 10 tanks/week on a two-shift operation. If there is competition in the first year from the other local firm, the Jax Company would probably take no further action, and a present-worth value of $50,000 is estimated for this outcome. On the other hand, if there is no competition in the first year, the Jax Company will not be able to meet the expected demand. Subjective probabilities of 0.4 and 0.6 are assigned to the outcomes of (1) no competition in the first year, and (2) competition in the first year, respectively. If no competition, Jax management would then face the decision of either expanding the product line by building an addition or not expanding.

Expansion at this point would be expensive; a building addition plus a layout of the existing plant would be involved. However, no competition is expected if the expansion is made now and demand can be met with increased capacity. A present value of $30,000 is estimated for this outcome. If there is no expansion, two outcomes are possible. One possibility is that the other local machine shop can still enter the competition. However, a 1-year advantage in the market will have been gained, and the present worth of this outcome is estimated to be $10,000. It is very unlikely that competition will occur after a 2-year period, and the probability of such an event is assumed to be zero. A subjective probability value of 0.7 is assigned to the outcome of competition in the second year. The other possible outcome is, of course, no competition in the second year, with a probability of 0.3 and estimated present worth of $50,000.

The "build addition initially" alternative has higher first and recurring costs than the "new layout" alternative, but production capacity is estimated at 25 tanks/week, which should satisfy average annual demand over the next 5 years. With adequate production capacity initially, the chance of competition is low—about 20%. The present-worth estimate in the event of no competition is $90,000. If there is competition, then the reduced sales and high annual fixed costs for the new building would result in a present worth of −$100,000. A decision tree representation of this example is given in Figure 9.5 (values are in thousands of dollars).

Using the principle of choice of maximizing expected present worth, a best decision is sought at D_2, the decision point most distant from the base of the tree, D_1. The expected values at D_2 are

$$E(\text{expansion}) = \$30,000$$

and

$$E(\text{no expansion}) = \$10,000(0.7) + \$50,000(0.3)$$
$$= \$22,000$$

Thus, the alternative of expansion is chosen to maximize expected present worth, and this *best* value of $30,000 replaces D_2 in the tree, as shown in Figure 9.6.

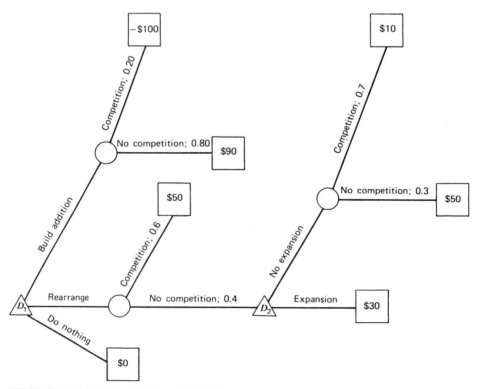

FIGURE 9.5. Jax Company example—original tree.

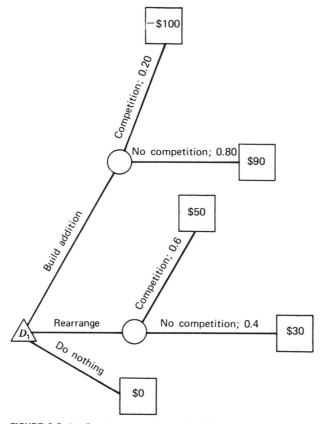

FIGURE 9.6. Jax Company example—reduced tree.

From Figure 9.6, the expected values at D_1 are calculated as:

$$E(\text{build addition}) = -\$100,000(0.20) + \$90,000(0.80)$$
$$= \$52,000$$

$$E(\text{rearrange}) = \$50,000(0.6) + \$30,000(0.4)$$
$$= \$42,000$$

$$E(\text{do nothing}) = \$0$$

Thus, the alternative of building an addition should be chosen in order to maximize expected present worth. It is interesting to note from Figure 9.6 that if this alternative is indeed chosen, the Jax Company is exposed to the possibility of a −$100,000 present worth. This event is "smoothed" by the expectation principle, on which most decision tree analysis is based. The −$100,000 figure is not necessarily a dollar loss, since a $MARR = 30\%$ was assumed in the calculation of the values. However, management may wish to choose the "rearrange" alternative instead because of the large negative figure; if so, they would be exercising a maximin philosophy.

Example 9.6

For further illustration of the decision tree approach to a problem and exposure to fundamental laws of probability, consider an urn containing five white (W) balls and seven black (B) balls. Three consecutive, independent, random (without bias in the selection) draws are made from the urn without replacing any ball drawn. The probability of selecting a white or black ball on the first draw is unconditional; that is, it does not depend on a previous draw (event). The probability of selecting a white or black ball on the second draw is, however, conditional on the color of ball selected on the first draw, and the third draw is conditional on both the first and second draws. In this example, we are interested in all the possible sequences of colors for the three draws and their probabilities of occurrence. A decision tree representation of this problem is helpful and is given in Figure 9.7.

All the eight possible sequences of colors are indicated in Figure 9.7. The objective probabilities of occurrence for each sequence will be determined

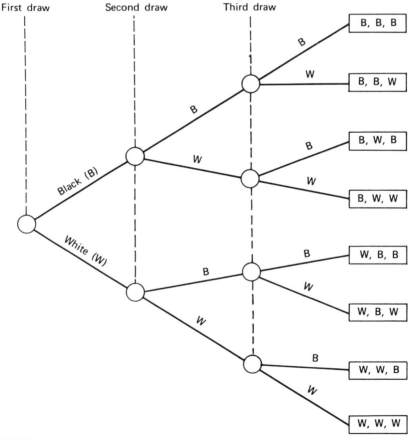

FIGURE 9.7. Tree representation of urn example.

by a relative-frequency argument. That is, for a general event A, the probability of A, $P(A)$, is calculated as the ratio

$$P(A) = \frac{\text{Number of ways favorable for the event}}{\text{Total number of ways possible}}$$

In the urn example, the probability of the unconditional event "a white ball on the first draw," W_1, is given by

$$P(W_1) = \frac{5 \text{ white balls (favorable number)}}{12 \text{ balls (total number)}}$$

Similarly, the probability of the unconditional event "a black ball on the first draw," B_1, is

$$P(B_1) = \frac{7}{12}$$

It is noted that there are only two possible outcomes (collectively exhaustive) for each of the three draws; either a white or black ball will be drawn. Since these two outcomes are mutually exclusive (cannot occur together) and collectively exhaustive, their individual probabilities of occurrence are additive and sum to 1.0. Thus,

$$P(W_1) + P(B_1) = \frac{5}{12} + \frac{7}{12} = 1.0$$

In general, for two mutually exclusive events A and B, the *additive law of probability* states

$$P \text{ (either A or B)} = P(A \cup B) = P(A) + P(B) \qquad (9.4)$$

where the symbol ($\cup$) stands for the union of the events.

For the second draw, the probability of the event "a white ball on the second draw," W_2, is conditional on whether the first draw results in a white or black ball [that is, $P(W_2|W_1)$ or $P(W_2|B_1)$]. The probability that a white ball is selected on the second draw given that a white ball is selected on the first draw is reasoned to be

$$P(W_2|W_1) = \frac{4 \text{ (white balls remaining after first draw)}}{11 \text{ (total balls remaining after first draw)}}$$

Similarly,

$$P(W_2|B_1) = \frac{5}{11}$$

The probability of the event "a black ball on the second draw," B_2, is also conditional on whether the first draw results in a white or black ball, and

$$P(B_2|W_1) = \frac{7}{11}$$

$$P(B_2|B_1) = \frac{6}{11}$$

We again note that the additive law of probability holds from

$$P(W_2|W_1 \cup B_2|W_1) = P(W_2|W_1) + P(B_2|W_1)$$
$$= \frac{4}{11} + \frac{7}{11} = 1.0$$

and

$$P(W_2|B_1 \cup B_2|B_1) = P(W_2|B_1) + P(B_2|B_1)$$
$$= \frac{5}{11} + \frac{6}{11} = 1.0$$

The additive law is easily expanded to include multiple, mutually exclusive events.

For the third draw, if the ball selected on each of the first and second draws is white, then the probability of selecting a white ball on the third draw is denoted $P(W_3|W_1, W_2)$ and reasoned, by the relative-frequency logic, to be

$$P(W_3|W_1, W_2) = \frac{3 \text{ (white balls remaining after first and second draw)}}{10 \text{ (total balls remaining after first and second draw)}}$$

The symbolism $(W_3|W_1, W_2)$ is interpreted to mean the event of drawing a white ball on the third draw given that a white ball was selected on the first draw *and* a white ball was selected on the second draw.

Similarly,

$$P(B_3|W_1, W_2) = \frac{7}{10}$$

$$P(W_3|W_1, B_2) = \frac{4}{10}$$

$$P(B_3|W_1, B_2) = \frac{6}{10}$$

$$P(W_3|B_1, W_2) = \frac{4}{10}$$

$$P(B_3|B_1, W_2) = \frac{6}{10}$$

$$P(W_3|B_1, B_2) = \frac{5}{10}$$

$$P(B_3|B_1, B_2) = \frac{5}{10}$$

A summary of all these events and their probabilities is shown in Figure 9.8.

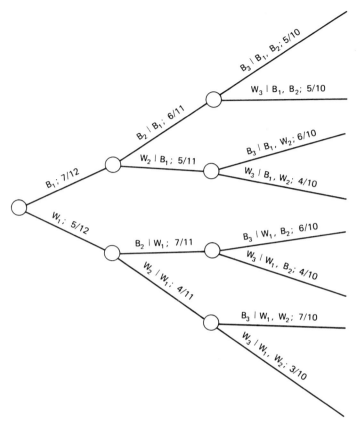

FIGURE 9.8. Probability tree for urn example.

Up to this point the probabilities of each individual outcome (event) of the example have been determined. However, the probability of each sequence (branch of the tree) occurring is yet to be determined (see Figure 9.7 to recall the possible sequences of colors). For instance, it is desired to know the probability of the joint event of a white ball being selected with each of the three draws or $P(W_1, W_2, W_3)$, which may also be written with set notation as $P(W_1 \cap W_2 \cap W_3)$ to mean the joint occurrence of these three events (conditional events in this particular case). This probability may be calculated as

$$P(W_1, W_2, W_3) = P(W_1) \cdot P(W_2|W_1) \cdot P(W_3|W_1, W_2)$$

$$= \left(\frac{5}{12}\right)\left(\frac{4}{11}\right)\left(\frac{3}{10}\right) = \frac{60}{1320}$$

A full justification of the above multiplicative calculation requires a probability theory development that is outside the scope of this textbook. However, the calculation of the joint occurrence of three conditional events is a direct extension of the conditional probability theorem presented below in section 9.6.2. Here are the other possible sequences of this example:

$$P(W_1, W_2, B_3) = P(W_1) \cdot P(W_2|W_1) \cdot P(B_3|W_1, W_2)$$
$$= \left(\frac{5}{12}\right)\left(\frac{4}{11}\right)\left(\frac{7}{10}\right) = \frac{140}{1320}$$

$$P(W_1, B_2, W_3) = P(W_1) \cdot P(B_2|W_1) \cdot P(W_3|W_1, B_2)$$
$$= \left(\frac{5}{12}\right)\left(\frac{7}{11}\right)\left(\frac{4}{10}\right) = \frac{140}{1320}$$

$$P(W_1, B_2, B_3) = P(W_1) \cdot P(B_2|W_1) \cdot P(B_3|W_1, B_2)$$
$$= \left(\frac{5}{12}\right)\left(\frac{7}{11}\right)\left(\frac{6}{10}\right) = \frac{210}{1320}$$

$$P(B_1, B_2, B_3) = P(B_1) \cdot P(B_2|B_1) \cdot P(B_3|B_1, B_2)$$
$$= \left(\frac{7}{12}\right)\left(\frac{6}{11}\right)\left(\frac{5}{10}\right) = \frac{210}{1320}$$

$$P(B_1, B_2, W_3) = P(B_1) \cdot P(B_2|B_1) \cdot P(W_3|B_1, B_2)$$
$$= \left(\frac{7}{12}\right)\left(\frac{6}{11}\right)\left(\frac{5}{10}\right) = \frac{210}{1320}$$

$$P(B_1, W_2, B_3) = P(B_1) \cdot P(W_2|B_1) \cdot P(B_3|B_1, W_2)$$
$$= \left(\frac{7}{12}\right)\left(\frac{5}{11}\right)\left(\frac{6}{10}\right) = \frac{210}{1320}$$

$$P(B_1, W_2, W_3) = P(B_1) \cdot P(W_2|B_1) \cdot P(W_3|B_1, W_2)$$
$$= \left(\frac{7}{12}\right)\left(\frac{5}{11}\right)\left(\frac{4}{10}\right) = \frac{140}{1320}$$

The sum of the probabilities for each of these eight possible sequences is 1320/1320 = 1.0, and the complete probability tree is shown in Figure 9.9.

9.6.2 The Conditional Probability Theorem

The probability of the joint occurrence of two dependent events, A and B, is given by

$$P(A \text{ and } B) = P(A \cap B) = P(A) \cdot P(B|A) \qquad (9.5)$$

where $P(A)$ is an unconditional probability and $P(B|A)$ is the conditional probability of event B given that event A has occurred or will occur. If event B is not conditional on event A, then events A and B are *independent* events and Equation 9.5 may be written as

$$P(A \cap B) = P(A) \cdot P(B) \qquad (9.6)$$

An extension of Equation 9.5 for three dependent events is given by

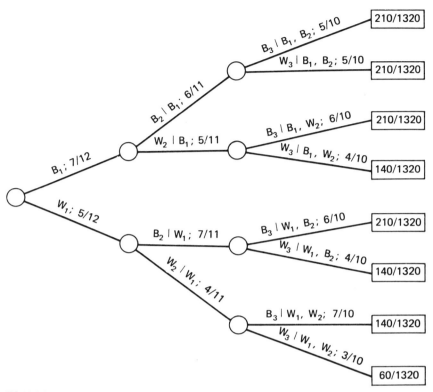

FIGURE 9.9. Complete probability tree for urn example.

$$P(A \cap B \cap C) = P(A) \cdot P(B|A) \cdot P(C|A, B) \qquad (9.7)$$

which was the relationship used in the previous urn example.

For the purpose of the development to follow, Equation 9.5 is now written in its complete form as

$$P(A \cap B) = P(A) \cdot P(B|A) = P(A|B) \cdot P(B) \qquad (9.8)$$

from which it follows that

$$P(B|A) = \frac{P(A|B)P(B)}{P(A)} \qquad (9.9)$$

or

$$P(A|B) = \frac{P(B|A)P(A)}{P(B)} \qquad (9.10)$$

Example 9.7

Either Equation 9.9 or 9.10 is a form of Bayes's theorem [23], which may provide a logical guide to sequential decision making under risk. To illustrate this claim, let us consider another urn example where one urn, U_1, contains two white and eight black balls; the other urn, U_2, contains five white and five black balls. Suppose the urn contents are known to you, but the urns themselves are hidden from your view. Now, another person selects a single ball from one of the urns. Without knowing the color of the ball drawn, which urn would you guess was selected? Then, if the person reported a black ball had been drawn, which urn would you guess had been selected?

Without benefit of the sample information concerning the color of the ball drawn, most persons would feel that it was equally likely that U_1 or U_2 had been chosen. That is, since $P(U_1) = P(U_2) = \frac{1}{2}$, guess either urn. However, after the sample information of a black ball is received, intuition suggests that U_1 was the urn selected because the probability of a black ball in U_1 is greater than the probability of a black ball in U_2. We now wish to formalize this example and show that Bayes's theorem can be used to support the intuitive guess of U_1 after receiving the report.

Before receiving the report of a black ball, the prior belief that U_1 or U_2 was selected is subjectively stated as

$$P(U_1) = \text{the } prior \text{ probability of selecting } U_1$$
$$= \frac{1}{2}$$

and

$$P(U_2) = \text{the } prior \text{ probability of selecting } U_2$$
$$= \frac{1}{2}$$

Note that

$$W = \text{the event of a white ball being drawn}$$

and

$$B = \text{the event of a black ball being drawn}$$

Also, before receiving the sample information, the likelihood of a particular report (W or B), given that U_1 or U_2 was selected, can be determined, based on knowledge of the contents of each urn. That is,

$$P(W|U_1) = \text{the conditional probability of a white ball being drawn}$$
$$\text{given that } U_1 \text{ was selected for the draw}$$
$$= \frac{2}{10}$$

Similarly,

$$P(B|U_1) = \frac{8}{10}$$

$$P(W|U_2) = \frac{5}{10}$$

$$P(B|U_2) = \frac{5}{10}$$

In the literature on sequential decision making, these conditional probabilities, either objectively known as above or subjectively assigned, are termed *likelihood statements*.

Ultimately, it is desired to assess the probability of whether U_1 or U_2 was chosen based on the sample information received (the report of a black ball in this case). Thus, the *posterior probabilities* of $P(U_1|B, P(U_2|B), P(U_1|W)$, and $P(U_2|W)$, need to be determined.

In the calculation of these posterior probabilities, it is convenient to calculate the probability of event W and event B before the ball is drawn. It is reasoned that event W could occur if either U_1 or U_2 were chosen. The same is true for the event B. Thus, B could occur if (1) U_1 were selected and B occurred, or (2) U_2 were selected and B occurred. That is,

$$
\begin{aligned}
P(B) &= P(U_1 \text{ and } B \text{ or } U_2 \text{ and } B) \\
&= P(U_1 \cap B \cup U_2 \cap B) \\
&= P(U_1 \cap B) + P(U_2 \cap B)
\end{aligned}
$$

Then, using Equation 9.8,

$$
P(B) = P(U_1)P(B|U_1) + P(U_2)P(B|U_2)
$$

and, from the previous data,

$$
P(B) = \left(\frac{1}{2}\right)\left(\frac{8}{10}\right) + \left(\frac{1}{2}\right)\left(\frac{5}{10}\right) = \frac{13}{20}
$$

Similarly,

$$
P(W) = P(U_1)P(W|U_1) + P(U_2)P(W|U_2)
$$
$$
= \left(\frac{1}{2}\right)\left(\frac{2}{10}\right) + \left(\frac{1}{2}\right)\left(\frac{5}{10}\right) = \frac{7}{20}
$$

With these results, the posterior probabilities may be calculated from Bayes's theorem as

$$
P(U_1|B) = \frac{P(B|U_1)P(U_1)}{P(B)} = \frac{(8/10)(1/2)}{13/20} = \frac{8}{13}
$$

$$
P(U_2|B) = \frac{P(B|U_2)P(U_2)}{P(B)} = \frac{(5/10)(1/2)}{13/20} = \frac{5}{13}
$$

$$
P(U_1|W) = \frac{P(W|U_1)P(U_1)}{P(W)} = \frac{(2/10)(1/2)}{7/20} = \frac{2}{7}
$$

$$
P(U_2|W) = \frac{P(W|U_2)P(U_2)}{P(W)} = \frac{(5/10)(1/2)}{7/20} = \frac{5}{7}
$$

It is noted from the results of $P(U_1|B) = 8/13$ and $P(U_2|B) = 5/13$ that the event of B should therefore increase the belief that U_1 was selected for taking the sample if one is to be consistent in a probabilistic sense. That is, the prior probability of U_1 was 1/2 and the posterior probability, given the event B, of U_1 is 8/13. If, after receiving the information of B, the guess had been U_1, then the above calculations would support the intuitive guess. If a second sample of a ball is taken from an urn, then the posterior probabilities $P(U_1|B)$ and $P(U_2|B)$ become the prior probabilities $P(U_1)$ and $P(U_2)$ for the second trial.

9.6.3 The Value of Perfect Information

In order to use the logic of Bayes's theorem in a more practical way to aid decision making, it is of interest to evaluate the value of the sample information in a monetary sense. Recalling the previous example, suppose the person who drew the ball from the urn had offered to pay a $10 prize if the correct urn were guessed. However, the person will charge for the sample information of whether a white or black ball is drawn. The question then is how much should be paid for the *sample information*. Before answering this particular question, a rationale for determining the *value of perfect information* is considered (i.e., the knowledge of exactly which urn was selected for the draw).

The *prior* probabilities of $P(U_1) = P(U_2) = 1/2$ are now recalled and written with the more suggestive notation of $PR(U_1) = PR(U_2) = 1/2$ [23]. Without any sample information, one should be indifferent as to which urn is guessed and the *prior expected profit* (*PEP*) would be $5, as determined from the matrix of Table 9.8.

The prior expected profit is $5, as determined from either of the two calculations below.

$$E(\text{profit}|\text{guess } U_1) = \left(\frac{1}{2}\right) (\$10) + \left(\frac{1}{2}\right) (\$0) = \$5$$

$$E(\text{profit}|\text{guess } U_2) = \left(\frac{1}{2}\right) (\$0) + \left(\frac{1}{2}\right) (\$10) = \$5$$

If these expected profits were different, the alternative with the larger expected value would be chosen, and the prior expected profit is the larger value.

Now, given that the person drawing the ball states that U_1 was selected

TABLE 9.8

Matrix for the Urn Example

			$PR(U_1) = 1/2$	$PR(U_2) = 1/2$
Urn Guessed	Actual Urn Selected		U_1	U_2
	U_1		$10	$ 0
	U_2		$ 0	$10

and the person is perfectly reliable, U_1 would obviously be guessed, and a $10 payoff received. The same is true, of course, for the report of U_2. The *expected profit given perfect information* ($EP|PI$) is thus $10 in this example. It is now reasoned that the maximum amount one should pay for perfect information is $10 − $5 (expected profit without any information) = $5. We can formalize the *expected value of perfect information* ($EVPI$) as the difference between the *expected profit given perfect information* ($EP|PI$) and the *prior expected profit* (PEP), or

$$EVPI = EP|PI − PEP \qquad (9.11)$$

9.6.4 The Value of Imperfect Information

In most real-world decision situations, information is incomplete and not perfectly reliable, as in the urn example just discussed. The report that a black ball had been drawn gave additional information to the decision maker, but still did not provide the absolute answer as to which urn was selected. In order to define the *expected value of sample information* ($EVSI$), the posterior probability (with new notation) results are repeated below.

For the event of B, the posterior probabilities were

$$PO(U_1|B) = \frac{8}{13}$$

and

$$PO(U_2|B) = \frac{5}{13}$$

Thus, since $8/13 > 5/13$, the best guess is that U_1 was selected. The actual urn chosen could have been U_1 or U_2, and the expected profit given a guess of U_1 is

$$E(\text{profit}|\text{guess } U_1) = (\text{expected profit}|\text{guess } U_1 \text{ and is } U_1)PO(U_1)$$
$$+ (\text{expected profit}|\text{guess } U_1 \text{ and is } U_2)PO(U_2)$$

$$= (\$10) \left(\frac{8}{13}\right) + (\$0) \left(\frac{5}{13}\right)$$

$$= \frac{\$80}{13}$$

For the event of W, the posterior probabilities were

$$PO(U_1|W) = \frac{2}{7}$$

$$PO(U_2|W) = \frac{5}{7}$$

Thus, if the report had been W, and since $5/7 > 2/7$, the best guess would be U_2 with

$$E(\text{profit}|\text{guess } U_2) = (\text{expected profit}|\text{guess } U_2 \text{ and is } U_1)PO(U_1)$$
$$+ (\text{expected profit}|\text{guess } U_2 \text{ and is } U_2)PO(U_2)$$

$$= (\$0)\left(\frac{2}{7}\right) + (\$10)\left(\frac{5}{7}\right)$$

$$= \frac{\$50}{7}$$

However, *before* the sample is taken and the report is given, both events B and W are possible, with $P(B) = 13/20$ and $P(W) = 7/20$. The *expected profit given sample information* $(EP|SI)$ is

$$EP|SI = (\text{expected profit}|B)P(B) + (\text{expected profit}|W)P(W)$$

$$= \left(\frac{\$80}{13}\right)\left(\frac{13}{20}\right) + \left(\frac{\$50}{7}\right)\left(\frac{7}{20}\right)$$

$$= \$6.50$$

In the above calculations, both expected profit values (expected profit $|B$) and (expected profit $|W$) assume that the best guess would, in fact, be made for the reports of B and W, respectively.

Finally, it is reasoned that the *expected value of sample information* is the difference between the *expected value given sample information* and the *prior expected profit*, or

$$EVSI = EP|SI - PEP$$
$$= \$6.50 - \$5 \qquad\qquad (9.12)$$
$$= \$1.50$$

which is the maximum amount one should pay for the sample information of this example. This value is, of course, less than the $5 value for perfect information.

A decision tree representation of this example is given in Figure 9.10 for a *sample information charge* of $1.50.

From Figure 9.10, the appropriate calculations at D_2 are

$$E(\text{guess } U_1) = (\$8.50)\left(\frac{8}{13}\right) + (-\$1.50)\left(\frac{5}{13}\right)$$

$$= \frac{\$60.5}{13}$$

$$E(\text{guess } U_2) = (-\$1.50)\left(\frac{8}{13}\right) + (\$8.50)\left(\frac{5}{13}\right)$$

$$= \frac{\$30.5}{13}$$

Since $60.5/13 > $30.5/13, the best guess given a report of B would be U_1.

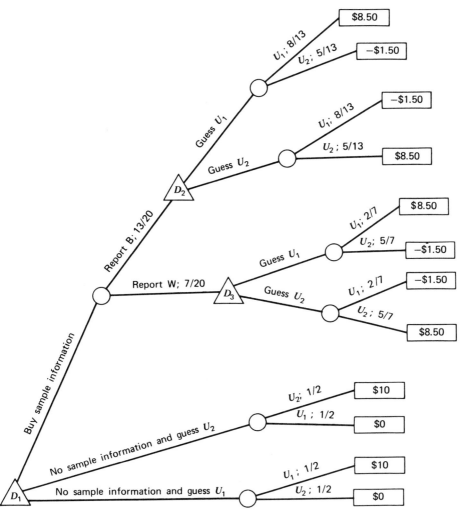

FIGURE 9.10. Decision tree for urn example.

The appropriate calculations at D_3 are

$$E(\text{guess } U_1) = (\$8.50) \left(\frac{2}{7}\right) + (-\$1.50) \left(\frac{5}{7}\right)$$

$$= \frac{\$6.5}{7}$$

$$E(\text{guess } U_2) = (-\$1.50) \left(\frac{2}{7}\right) + (\$8.50) \left(\frac{5}{7}\right)$$

$$= \frac{\$39.5}{7}$$

Since $\$39.5/7 > \$6.5/7$, the best guess given a report of W would be U_2.

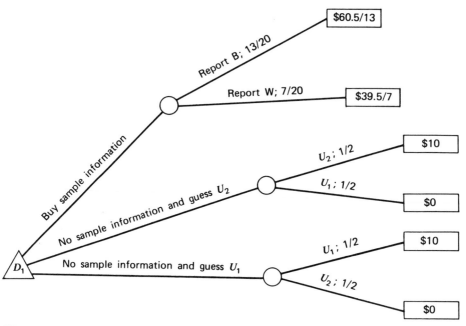

FIGURE 9.11. Reduced decision tree for urn example.

Replacing D_2 and D_3 with these best expected values yields the reduced decision tree of Figure 9.11. From Figure 9.11, the appropriate calculations at D_1 are

$$E(\text{buy sample information}) = \left(\frac{\$60.5}{13}\right)\left(\frac{13}{20}\right) + \left(\frac{\$39.5}{7}\right)\left(\frac{7}{20}\right)$$
$$= \$5$$

$E(\text{no sample information and guess } U_2)$
$$= E(\text{no sample information and guess } U_1)$$
$$= \$5$$

Thus, one is indifferent between the three alternatives at D_1 given that the charge for sample information is \$1.50. If the charge is less than this amount, the decision maker should choose the "buy sample information" alternative in order to maximize expected profits.

For a similar sequential decision problem involving only costs, the basic relationships to determine expected value of information become

$EVPI$ = prior expected cost − expected cost given perfect information
$= PEC - EC|PI$ (9.13)

$EVSI$ = prior expected cost − expected cost given sample information
$= PEC - EC|SI$ (9.14)

TABLE 9.9

Sixty Manuscript Decisions/Outcomes, Example 9.3

Decision/Outcome When Marketed	No Outside Expert Review	Outside Reviewer's Evaluation			Total
		Good	Fair	Poor	
Not published	5	0	14	12	31
Low market demand	2	0	4	3	9
Best-seller	1	2	4	0	7
Best-seller + movie	1	4	3	0	8
Low market + B movie	1	0	3	1	5
Total	10	6	28	16	60

Example 9.8[3]

The editor of the American Book Company is considering a manuscript for a historical novel. The editor estimates the book can be marketed for $5, with the company receiving $4 and the author $1. Since other publishers are interested in the manuscript, the author desires a commitment from the American Book Company in the very near future. American has already invested $300 in a preliminary review of the manuscript and a quick market survey. The editor considers this manuscript competitive with other recently published historical novels, and the author, although relatively new, has other published works that have received good reviews.

The editor identifies the immediate alternatives of (1) reject the manuscript, (2) accept the manuscript, or (3) obtain an additional review of the manuscript by an expert. An additional review will cost $400, and the expert's opinion will not perfectly predict the historical novel's success in the market. The editor reasons that at best the expert can only rate the manuscript good, fair, or poor.

If the manuscript is accepted for publication, the editor speculates on the possible market outcomes; there could be (1) a low market demand, or (2) the novel could be a best-seller. Additionally, there is the question of movie rights. There may be no interest in filming the novel, but, on the other hand, a best-seller could yield $30,000 for the movie rights. Even if there is low market demand, a film company might pay $4000 for the rights to make a low-budget, grade B film.

The American Book Company has available historical data on 60 recent manuscripts of similar books, given in Table 9.9.

---

[3] This example was presented at the Sixth Triennial Symposium, Engineering Economy Division, ASEE, June 19–20, 1971, in a paper entitled "Introduction to Decision Theory" by Barnard E. Smith. The article appears in the publication, *Decision and Risk Analysis: Powerful New Tools For Management,* The Engineering Economist, Stevens Institute of Technology, Hoboken, N.J., and is presented here by the permission of the publisher.

From available sales figures, the editor can estimate reasonably well the revenues given a particular outcome for a published book. He considers the present worth of after-tax revenues over the effective life of a book using a 10% expected rate of return. For a best-seller, a present worth of $70,000 is estimated, and a low-demand publication yields an estimated $5000 present worth. Offers for movie rights, if any, are usually made 1 year after the manuscript is under contract, and these after-tax revenues should be added to the book sales (assume that $4000 for a grade B film and $30,000 for a best-seller film are appropriate figures). It costs $15,000 to publish a book.

If it is assumed the editor has a linear utility function for dollars and uses a "maximize present worth" principle for choice among alternatives, what decision should be made at the present time?

The editor faces the immediate alternatives of

A_1 = reject the manuscript without any additional review by an expert
A_2 = accept the manuscript without obtaining an expert's review
A_3 = obtain an expert's review

The value of alternative A_1 is the $300 cost, which has already been spent on a preliminary review [that is, $V(A_1) = \$300$]. Indeed, the $300 for a preliminary review is a sunk cost and applies whether alternative A_1, A_2, or A_3 is chosen at the present time. This $300 cost can therefore be dropped from consideration, and $V(A_1) = 0$ with this adjustment.

For alternative A_2 there could be four possible market outcomes, defined as

θ_1 = low market demand
θ_2 = a best-seller
θ_3 = best-seller plus movie rights
θ_4 = low market demand plus grade B movie

Thus, the value associated with each of these outcomes is determined as

$$V(\theta_1|A_2) = \text{revenues} - \text{publication costs}$$
$$= \$5000 - \$15,000 = -\$10,000$$

$$V(\theta_2|A_2) = \$70,000 - \$15,000 = \$55,000$$

$$V(\theta_3|A_2) = \$70,000 + \$30,000 - \$15,000 = \$85,000$$

$$V(\theta_4|A_2) = \$5000 + \$4000 - \$15,000 = -\$6000$$

For alternative A_3, the expert's evaluations could be either

$$Z_1 = \text{Good}$$
$$Z_2 = \text{Fair}$$
$$Z_3 = \text{Poor}$$

For *each* of these evaluations, the editor must make one of two possible decisions, defined as

$$X_1 = \text{reject the manuscript}$$
$$X_2 = \text{accept the manuscript}$$

Given the decision of X_1 for *each* of the three possible evaluations (Z_j), the cost to the American Book Company would be the same—the additional cost of an expert's review. That is,

$$V(X_1|A_3, Z_j) = -\$400 \qquad \text{for } j = 1, 2, 3$$

Then, for the decision of X_2 associated with each of the reviewer's evaluations, there are the four possible market outcomes defined previously. The value for any *given* market outcome *and* decision X_2 is the same for all Z_j evaluations. This value is equal to the values for alternative A_2 minus \$400 for the expert's review. That is,

$$V(\theta_1|A_3, Z, X_2) = -\$10,000 - \$400 = -\$10,400$$
$$V(\theta_2|A_3, Z, X_2) = \$55,000 - \$400 = \$54,600$$
$$V(\theta_3|A_3, Z, X_2) = \$85,000 - \$400 = \$84,600$$
$$V(\theta_4|A_3, Z, X_2) = -\$6000 - \$400 = -\$6400$$

A decision tree representation of this example, including probability values for the various outcomes, is given by Figure 9.12, where the values are in thousands of dollars.

The probability values for the various outcomes of the decision tree given in Figure 9.12 are computed using the data of Table 9.9. First, consider alternative A_2 (accept for publication without further review). It is noted from Table 9.9 that out of the ten manuscripts that were not reviewed by an outside expert, five were published. Two of the five had low market demand, one was a bestseller, one was a best-seller plus movie rights, and another had low market demand plus the rights sold for a grade B movie. Thus, by a relative frequency logic, the following probabilities can be determined:

$$P(\theta_1|A_2) = \frac{2}{5}$$

$$P(\theta_2|A_2) = \frac{1}{5}$$

$$P(\theta_3|A_2) = \frac{1}{5}$$

$$P(\theta_4|A_2) = \frac{1}{5}$$

For the 50 manuscripts that were reviewed by an expert, 6 received a good evaluation, 28 received a fair evaluation, and 16 received a poor evaluation. Thus,

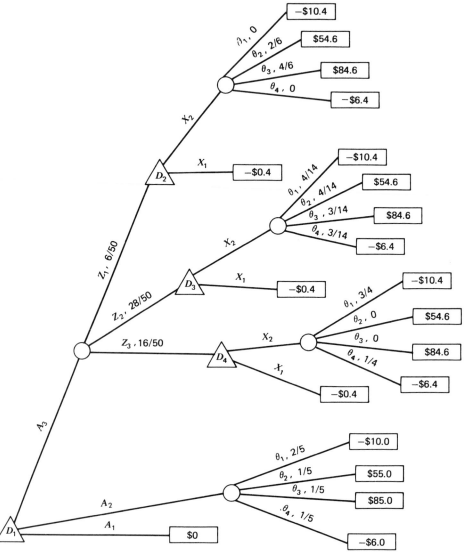

FIGURE 9.12. Decision tree for American Book Company example.

$$P(Z_1) = \frac{6}{50}$$

$$P(Z_2) = \frac{28}{50}$$

$$P(Z_3) = \frac{16}{50}$$

Similarly, the following conditional probabilities may be determined from the data of Table 9.9.

$$P(\theta_1|Z_1) = 0, \qquad P(\theta_1|Z_2) = \frac{4}{14}, \qquad P(\theta_1|Z_3) = \frac{3}{4}$$

$$P(\theta_2|Z_1) = \frac{2}{6}, \qquad P(\theta_2|Z_2) = \frac{4}{14}, \qquad P(\theta_2|Z_3) = 0$$

$$P(\theta_3|Z_1) = \frac{4}{6}, \qquad P(\theta_3|Z_2) = \frac{3}{14}, \qquad P(\theta_3|Z_3) = 0$$

$$P(\theta_4|Z_1) = 0, \qquad P(\theta_4|Z_2) = \frac{3}{14}, \qquad P(\theta_4|Z_3) = \frac{1}{4}$$

It is recognized that these probability values are estimates but would seem to be less gross than purely subjective evaluations made without any historical data as a basis.

The solution to the decision tree of Figure 9.12 follows. The appropriate calculations for D_2 are

$$E(X_1) = -0.4 = -\$400$$

$$E(X_2) = (-\$10.4)(0) + (\$54.6)\left(\frac{2}{6}\right) + (\$84.6)\left(\frac{4}{6}\right) + (-\$6.4)(0)$$

$$= \$74.6 = \$74,600$$

and the best decision at D_2 is therefore X_2 (accept the manuscript). The calculations for D_3 are

$$E(X_1) = -\$400$$

$$E(X_2) = (-\$10.4)\left(\frac{4}{14}\right) + (\$54.6)\left(\frac{4}{14}\right) + (\$84.6)\left(\frac{3}{14}\right) + (-\$6.4)\left(\frac{3}{14}\right)$$

$$= \$29.386 = \$29,386$$

and the best decision at D_3 is therefore X_2 (accept the manuscript). The calculations for D_4 are

$$E(X_1) = -\$400$$

$$E(X_2) = (-\$10.4)\left(\frac{3}{4}\right) + (\$54.6)(0) + (\$84.6)(0) + (-\$6.4)\left(\frac{1}{4}\right)$$

$$= -\$9.4 = -\$9400$$

and the best decision at D_4 is therefore X_1 (reject the manuscript).

The reduced decision tree is given by Figure 9.13, and the appropriate calculations for D_1 are

$$E(A_3) = (\$74,600)\left(\frac{6}{50}\right) + (\$29,386)\left(\frac{28}{50}\right) + (-\$400)\left(\frac{16}{50}\right)$$

$$= \$25,280$$

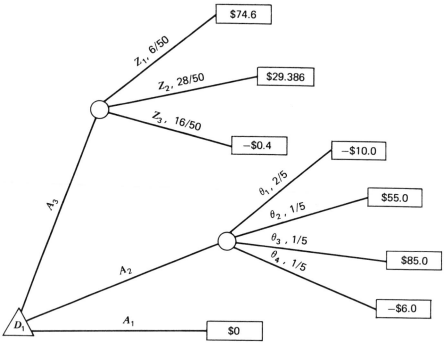

FIGURE 9.13. Reduced decision tree for American Book Company example (value in thousands of dollars).

$$E(A_2) = (-\$10,000)\left(\frac{2}{5}\right) + (\$55,000)\left(\frac{1}{5}\right) + (\$85,000)\left(\frac{1}{5}\right) + (-\$6000)\left(\frac{1}{5}\right)$$

$$= \$22,800$$

$$E(A_1) = \$0$$

and the best decision at D_1 in order to maximize the expected present worth is alternative A_3 (obtain an expert's opinion). Then, if the evaluation is either good or fair, accept the manuscript for publication; but if the evaluation is poor, reject the manuscript. The expected value of sample information in this example is the difference between $E(A_3)$ and $E(A_2)$ plus the $400 paid the expert, or $2880.

$$EVSI = EV|SI - PEP + \$400$$
$$= \$25,280 - \$22,800 + \$400$$
$$= \$2880$$

Clearly the expert's opinion is well worth the $400 fee in this example.

9.6.5 Sequential Decisions—Summary Comments

The decision tree representation of a sequential-decision situation is, in our opinion, an effective device for visualization. By virtue of developing the tree, serious attention is given to explicitly identifying and describing alternatives, to assessing the outcomes that might occur as a result of the alternative choice, and to making any necessary future decisions. Even if the probabilities for outcomes cannot be adequately assessed or values for the "branches" accurately estimated, considerable insight into the decision situation may be achieved and decision making aided. On the other hand, whether the expectation principle of choice among alternatives is an appropriate criterion or not is a matter of personal preference, and any real-world sequential-decision situation may readily become too large for effective description or convenient solution. As the number of alternatives, the number of outcomes per alternative, and the number of decision points increase, so do the uncertainty of data and the burden of calculation. Furthermore, the expense in time and money to obtain the information required may become prohibitive when compared with the benefits expected from the analysis. An evaluation of such a trade-off is largely, if not primarily, a matter of judgment.

9.7 MULTIPLE OBJECTIVES

In order to initiate a discussion of multiple objectives and the confounding effect they have on the decision-making process, we will paraphrase the work of Miller and Starr [19], quoted in Chapter One: Being unable to describe goals satisfactorily in terms of one objective, people customarily maintain various objectives, and multiple objectives are frequently in conflict with each other. Because organizations are made up of people this statement would seem to apply to an organizational unit as well as an individual. Accepting this premise then, managers, who are responsible for the organization's behavior and performance, usually expect a decision to achieve multiple objectives and evaluate recommendations made to them on this basis.

Establishing the relevant objectives is an important problem for a given decision maker, and the problem is further complicated by conflicting objectives and the difficulty (if not impossibility) of obtaining a measurement of these objectives.

At the outset, we acknowledge that the discussion to follow is somewhat superficial, and we have drawn heavily from Churchman, Ackoff, and Arnoff [6, Chapters 5 and 6], Miller and Starr [19, Chapters 1–3 and pp. 236–241], Kepner and Tregoe [15], and Morris [22, Chapter 7]. The literature on the subject of multiple objectives and the theory of measurement is very extensive, since these factors are involved in all areas of human endeavor. The primary reference sources are in the general areas of psychology, statistics, mathematics, and management. The previous references have been cited because they are concerned with managerial decision making instead of the psychological motivation stimulating the multiple objectives or the mathematical theory of measurement. For the reader particularly interested in the subject of multiple objectives from a managerial decision-making point of view, the cited references are recommended readings.

Kepner and Tregoe [15] state that "making the best decision will involve a sequence of procedures based on the following seven concepts:

1. The objectives of a decision must be established first.
2. The objectives are classified as to importance.
3. Alternative actions are developed.
4. The alternatives are evaluated against the established objectives.
5. The choice of the alternative best able to achieve all the objectives represents the tentative decision.
6. The tentative decision is explored for future possible adverse consequences.
7. The effects of the final decision are controlled by taking other actions to prevent possible adverse consequences from becoming problems, and by making sure the actions decided on are carried out."

For a single objective, Chapters Three to Nine of this text have been concerned primarily with steps 3 to 6 of the above procedure. This particular section of Chapter Nine will be concerned mainly with steps 2 and 4 of the above procedure. Churchman, Ackoff, and Arnoff [6] present an interesting discussion on establishing the objectives of a decision. Morse [26] states that objectives are derived from two general areas: the results expected to come from a decision, and the resources available for expenditure in carrying out a decision. The results expected may be to achieve certain things, but it may also be desired to maintain certain things. For example, over the next 5 years, management may wish to increase sales by 40% relative to the past year and render better customer service, retaining good employee morale and the company's good financial credit rating. In any case, for the subsequent discussion, it will be assumed that objectives can be precisely stated and are relevant to the decision.

9.7.1 Classifying Objectives According to Importance

Assessing the relative importance of multiple objectives, measuring the outcomes of alternatives in terms of these objectives, obtaining a common measure for multiple outcomes having different measures, and combining all these into a single measure of merit form the heart of the multiple-objective problem and the source of much difficulty and controversy. We cannot offer a solution to the problem, but can only discuss approaches to it, citing cautions at appropriate points.

When objectives can be sorted by classes such that each objective in the class is equivalent, then this is nothing more than a classification system or *nominal* system of measurement. However, when objectives can be sorted into classes that can be ordered, this is termed an *ordinal* system of measurement.

An illustration of a nominal classification of objectives is the concept advanced by Kepner and Tregoe according to which objectives are first categorized on a "must" or "desirable" basis. For example, a university library must serve the faculty and student population, but perhaps it is only desirable to serve residents of the local community. Once objectives are classed on a "must" versus "desirable" basis, the desirable objectives may be considered only subjectively thereafter, whereas the must objectives are deserving of greater attention and, if possible, quantification. Some of the approaches taken to quantify objectives will now be treated; in the discussion, the term *goal* is used synonymously with *objective,* and the symbolism adopted to denote the kth goal or objective of interest is G_k.

9.7.2 Ranking

If a decision maker is capable of stating preferences among objectives, objectives may be ranked in terms of relative importance. A technique to assist the decision maker in making consistent preference statements is the method of paired comparisons. In order to illustrate the method, assume that four goals (objectives) are relevant: G_1, G_2, G_3, and G_4. The method of paired comparisons enumerates all possible pairs $[n(n - 1)/2]$, makes preference statements, and deduces a ranking for each goal. In this example, the possible pairs are:

$$G_1 \text{ versus } G_2 \qquad G_2 \text{ versus } G_3 \qquad G_3 \text{ versus } G_4$$
$$G_1 \text{ versus } G_3 \qquad G_2 \text{ versus } G_4$$
$$G_1 \text{ versus } G_4$$

If the symbol $>$ represents "preferred to" and the symbol $<$ represents "not preferred to," suppose the results of the decision about the pairs were:

$$G_1 > G_2 \qquad G_2 < G_3 \qquad G_3 > G_4$$
$$G_1 < G_3 \qquad G_2 > G_4$$
$$G_1 > G_4$$

Rewriting so that all comparisons are of the "preferred to" variety yields:

$$G_3 > G_1 \qquad G_1 > G_2 \qquad G_2 > G_4$$
$$G_3 > G_2 \qquad G_1 > G_4$$
$$G_3 > G_4$$

It can then be concluded that G_3 is preferred to all others, G_1 is preferred to two others, G_2 is preferred to one other, and G_4 is preferred to zero others, and the following ranks should be assigned: $G_3 = \text{I}$, $G_1 = \text{II}$, $G_2 = \text{III}$, and $G_4 = \text{IV}$.

If, in the above procedure, the preference statements had been $G_1 > G_2$, $G_2 > G_4$, and $G_1 < G_4$, the decision maker should reconsider the inconsistent judgments. These results are "intransitive" in a mathematical sense.

The ranking method of classifying objectives according to relative importance is certainly useful; indeed, ranking was implicitly used in the comparison of alternative investments under conditions of assumed certainty in the earlier chapters of this textbook. There are, however, obvious deficiencies in the ranking method. For instance, the ranks do not indicate the *extent* to which one objective is preferred over another. Furthermore, because rankings are an *ordinal scale of measurement,* the mathematical operations of addition, subtraction, multiplication, division, and so forth, cannot be performed. Therefore a higher-order scale of measurement (interval or ratio scale) is generally required to determine choice among alternatives under conditions of risk and uncertainty [22]. Unfortunately, accepted methods for measuring objectives on a ratio scale do not currently exist.

9.7.3 Weighting Objectives

Before proceeding with a discussion of certain methods to assign relative weights to multiple objectives, a distinction between objectives and outcomes should be made. This is perhaps most easily done by means of an example where a machine-replacement situation is of interest, and the following two objectives are defined:

G_1: minimize the equivalent annual cost over a 5-year planning horizon
G_2: minimize the number of production personnel transferred to other jobs within the firm

Let

θ_{jk} = the outcome resulting from taking the jth alternative course of action in terms of the kth objective or goal
$V(\theta_{jk})$ = the value of the outcome resulting from selecting the jth alternative in terms of the kth objective

Suppose, in this example, that

A_1 = the alternative of keeping the present machining center
A_2 = the alternative of replacing the present center with a new machining center

The outcomes for each of these alternatives in terms of the two objectives, G_1 and G_2, are:

$$\theta_{11} = \text{an annual cost of \$80,000}$$
$$\theta_{12} = \text{no persons transferred}$$
$$\theta_{21} = \text{an annual cost of \$60,000}$$
$$\theta_{22} = \text{two persons transferred}$$

In order to compare the two alternatives in a quantitative fashion, values must be assigned to each outcome; the values must be weighted in some way to reflect the relative importance of the two objectives, and some type of aggregate number must be determined for each alternative to enable a comparison and final choice. For example, the end result (without regard for validity at this point) might appear as

$$V(A_1) = w_1 V(\theta_{11}) + w_2 V(\theta_{12})$$

$$V(A_2) = w_1 V(\theta_{21}) + w_2 V(\theta_{22})$$

where $V(A_2) > V(A_1)$.

The w_k (or w_1, w_2) values are weighting factors identifying the relative importance of the objectives G_k (or G_1, G_2) and hence are applied to the appropriate outcomes (θ_{jk}) of the jth alternative. Determining the weighting factors is thus one issue in quantifying multiple objectives. The values $V(\theta_{jk})$ may be in different dimensions and, if so, may have to be transformed to another common dimension. Determining the transformation function is another issue in quantifying multiple objectives. Finally, the values for each alternative, $V(A_j)$, were determined by a linear function. Determining the mathematical functional form of the model is yet another issue in quantifying multiple objectives. Again, all these issues will not be resolved in this chapter, but examples of approaches that have been taken will be presented.

One method for assigning relative weights to multiple objectives in order to indicate the extent by which one objective is preferred to another is simply to assign weights by judgment. That is, in the case of four objectives (G_1, G_2, G_3, G_4), the decision maker may reason that G_3 is twice as important as G_1, three times as important as G_2 and five times as important as G_4. Assigning a weighting value of 1.00 to G_3, it can be reasoned that the set of weighting values is

For G_3: $w_3 = 1.000$
For G_1: $w_1 = 0.500$
For G_2: $w_2 = 0.333$
For G_4: $w_4 = 0.200$

These weights may be transformed to a scale from 0 to 1.0, or "normalized," by dividing each weight by the sum of all weights; that is,

$$w_3' = \frac{1.000}{2.033} = 0.4918$$

$$w_1' = \frac{0.500}{2.033} = 0.2459$$

$$w_2' = \frac{0.333}{2.033} = 0.1639$$

$$w_4' = \frac{0.200}{2.033} = 0.0984$$

Another method of assigning weights to objectives by judgment, but still keeping consistency, is adapted from Churchman, Ackoff, and Arnoff [6], and the commentary on this method given by Morris [22]. The references state that the method is applicable to weight objectives or outcomes. For a decision involving multiple objectives $G_1, \ldots, G_m$ to be evaluated, this method requires two major assumptions. Paraphrasing Morris, these are:

1. It must be possible for the decision maker to think about and judge the value of any combination of objectives. That is, it must be possible to consider not only the importance or weight value for, say, G_1 but also the weighting value for sums of objectives.

2. Values are assumed to be additive. Given the individual weighting values for, say, G_1 and G_2, it is assumed that the weighting value for both objectives is the sum of their individual weighting values.

The general procedure for the method now follows.

1. Rank the objectives in order of importance, where G_1 indicates the most important, G_2 the next most important, and so forth, and G_m is the least important.

2. Assign the weighting value of 1.00 to G_1 (that is, $w_1 = 1.00$) and weighting values to the other objectives to reflect their importance relative to G_1. These two steps actually complete the judgmental process, and the following steps serve to refine the initial judgments and aid consistency.

3. Compare G_1 to the linear combination of all other objectives. That is, compare G_1 versus $(G_2 + G_3 + \cdots + G_m)$.

  a. If $G_1 > (G_2 + G_3 + \cdots + G_m)$, adjust (if necessary) the value of w_1 such that $w_1 > (w_2 + w_3 + \cdots + w_m)$. Attempt, in all adjustments of the procedural steps, to keep the relative values within the objective *group* invariant. Proceed to step 4.

  b. If $G_1 = (G_2 + G_3 + \cdots + G_m)$, adjust w_1 so that $w_1 = (w_2 + \cdots + w_m)$ and proceed to step 4.

  c. If $G_1 < (G_2 + G_3 + \cdots + G_m)$, adjust (if necessary) the value of w_1 so that $w_1 < (w_2 + \cdots + w_m)$.

    (1) Now compare G_1 versus $(G_2 + G_3 + \cdots + G_{m-1})$.

      (a) If $G_1 > (G_2 + G_3 + \cdots + G_{m-1})$, adjust (if necessary) the values so that $w_1 > (G_2 + G_3 + \cdots + G_{m-1})$ and proceed to step 4.

      (b) If $G_1 = (G_2 + G_3 + \cdots + G_{m-1})$, adjust (if necessary) the values such that $w_1 = (w_2 + w_3 + \cdots + w_{m-1})$ and proceed to step 4.

      (c) If $G_1 < (G_2 + G_3 + \cdots + G_{m-1})$, adjust (if necessary) the values such that $w_1 < (w_2 + w_3 + \cdots + w_{m-1}$. Then, compare G_1 versus $(G_2 + G_3 + \cdots + G_{m-2})$, and so forth, either until G_1 is preferred or equal to the rest, then proceed to step 4, or until the comparison of G_1 versus $(G_2 + G_3)$ is completed, then proceed to step 4.

4. Compare G_2 versus $(G_3 + G_4 + \cdots + G_m)$ and proceed as in step 3.

5. Continue until the comparison of G_{m-2} versus $(G_{m-1} + G_m)$ is completed.

6. If desired, convert each w_k into a normalized value, dividing w_k by the sum $\sum_{k=1}^{m} w_k$.

Example 9.9

Assume that it is desired to determine weighting values for four objectives, where these have been ranked in order of importance such that G_1 is most important, and G_4 is the least important. Step 1 of the procedure is thus accomplished.

STEP 2. The objectives are tentatively assigned the weights $w_1 = 1.00$, $w_2 = 0.80$, $w_3 = 0.50$, and $w_4 = 0.20$.

STEP 3. Assume that G_1 is preferred to the linear combination of the other objectives. That is, $G_1 > (G_2 + G_3 + G_4)$. Since $w_1 < (w_2 + w_2 + w_3)$, it is necessary to adjust w_1 to be greater than the sum of 1.50—say, $w_1 = 1.75$. Then proceed to step 4.

STEP 4. Comparing G_2 versus $(G_3 + G_4)$, assume that $G_2 < (G_3 + G_4)$, and the step 3 procedure is repeated. Since the weighting values do not reflect this [that is, $w_2 = 0.80$ is not less than $(0.50 + 0.20)$], w_2 is adjusted to be 0.65. [The adjustment to decrease w_2 does not violate the previous preference of $G_1 > (G_2 + G_3 + G_4)$ with the new value of $w_1 = 1.75$ to reflect this preference.] Since, in this example of only four objectives, the comparison of G_{m-2} versus $(G_{m-1} + G_m)$ or G_2 versus $(G_3 + G_4)$ has been completed at this point, step 5 of the procedure has been done.

If it is desired to normalize the results, new weighting values are calculated as

$$
\begin{aligned}
w_1' &= 1.75/(1.75 + 0.65 + 0.50 + 0.20) = 0.5645 \\
w_2' &= 0.65/3.10 & = 0.2097 \\
w_3' &= 0.50/3.10 & = 0.1613 \\
w_4' &= 0.20/3.10 & = \underline{0.0645} \\
& \text{TOTAL} & 1.0000
\end{aligned}
$$

For a small number of objectives, this method is relatively easy to use, but it becomes cumbersome as the number of objectives increases. However, it is believed that many real-world problems reduce to a few primary objectives. It should also again be emphasized that the method still depends on the decision maker's judgment in assigning the relative weights.

9.7.4 Determining the Value of Multiple Objectives

In the previous section on weighting objectives, an example of a machine-replacement situation was hypothesized to introduce the discussion. This example is now recalled, where the objectives were:

G_1: minimize the equivalent annual cost over a 5-year planning horizon
G_2: minimize the number of production personnel transferred to other jobs within the firm

For the two alternatives of A_1 (keep present machining center) and A_2 (replace with new machining center), the outcomes were

A_1: \$80,000 annual cost (θ_{11}); no persons transferred (θ_{12})
A_2: \$60,000 annual cost (θ_{21}); two persons transferred (θ_{22})

Let it be assumed that, by some method, the objectives have been weighted as $w_1 = 0.60$ and $w_2 = 0.40$.

One method to assign values to the outcomes of the two alternatives is to make a second series of judgments as to the degree to which each outcome succeeds in meeting the objective. Suppose these judgments are made in the form of numbers on an arbitrary scale from 0 to 1, where the higher the scale value, the closer to meeting the desired objective. Assume the results of such judgments are as given below.

Outcome	Value
θ_{11}	0.50
θ_{12}	1.00
θ_{21}	0.85
θ_{22}	0.60

Then, assuming that a linear model is appropriate, the values for each alternative are calculated as

$$V(A_1) = w_1 V(\theta_{11}) + w_2 V(\theta_{12})$$
$$= (0.60)(0.50) + (0.40)(1.00) = 0.70$$

$$V(A_2) = w_1 V(\theta_{21}) + w_2 V(\theta_{22})$$
$$= (0.60)(0.85) + (0.40)(0.60) = 0.75$$

Thus, since $V(A_2) > V(A_1)$, alternative A_2 would be chosen, and the present machining center would be replaced.

Clearly, these results are sensitive to the weighting values assigned by judgment. Furthermore, the magnitude of the difference in values between $V(A_1)$ and $V(A_2)$ is a function of the arbitrary scale (from 0 to 1) chosen. The linear model assumed is also questionable [19]. All these points are criticisms of this method of quantifying multiple objectives. In the literature on the theory of utility functions, it is argued that if a decision maker's utility function can be established for each objective, then values of outcomes (e.g., \$80,000 annual cost) can be converted to a utility value (e.g., \$80,000 annual cost equals 0.75) from the decision maker's utility curve. Such a utility value could be determined for each objective, and our model for the alternatives above would become

$$U(A_1) = w_1 U(\theta_{11}) + w_2 U(\theta_{12})$$

$$U(A_2) = w_1 U(\theta_{21}) + w_2 U(\theta_{22})$$

Then, the decision maker should choose to maximize the weighted utility value. Whether such utility functions for each objective can be readily obtained for a single decision maker is still questionable in our opinion, and even if this could be accomplished, there is yet another question of whether interaction between multiple objectives would be totally missed. As stated in Chapter One, the study of value measurement, and utility theory in particular, is interesting, but questions remain on the practical application and implementation of utility theory to business and engineering problems. In any event, it is judged that a treatment of utility theory is outside the scope of this textbook.

9.8 SUMMARY

In this chapter, some prescriptive or normative models for decisions under risk and uncertainty have been presented. For decisions under risk and uncertainty where the matrix model is appropriate, criteria that are commonly cited in the literature for choosing among mutually exclusive alternatives were discussed. None of these criteria were cited as the best or optimal basis for selecting an alternative; they merely represent observations on how decision makers seem to decide.

The decision tree model for a sequence of decisions where the outcomes of alternatives are chance events was also presented, and the expectation-variance criterion was suggested as a guideline for choice among alternatives. Some fundamental laws of probability were reviewed, and Bayes's theorem of conditional probability was illustrated as a guideline for revising one's opinion in sequential decisions under risk when only imperfect information is available to the decision maker.

The last section of the chapter was an introductory presentation of some approaches in the literature for quantifying multiple objectives. The discussion offered and examples used were restricted to decisions under assumed certainty. However, the methods illustrated are easily extended to decisions under risk and uncertainty and the matrix model used for analysis.

BIBLIOGRAPHY

1. Agee, M. H., Taylor, R. E., and Torgersen, P. E., *Quantitative Analysis for Management Decision*, Prentice-Hall, 1976.
2. Bunn, D. W., *Applied Decision Analysis,* McGraw-Hill, 1984.
3. Byrd, J., Jr., and Moore, L. T., *Decision Models for Management,* McGraw-Hill, 1982.
4. Chankong, V., and Haimes, Y. Y., *Multiobjective Decision Making: Theory and Methodology,* (Volume 8 of the North-Holland Series in System Science and Engineering), Elsevier, 1983.
5. Chernoff, H., and Moses, L. E., *Elementary Decision Theory,* Wiley, 1959.
6. Churchman, C. W., Ackoff, R. L., and Arnoff, E. L., *Introduction to Operations Research,* Wiley, 1957.

7. Conrath, David W., "From Statistical Decision Theory to Practice: Some Problems with the Transition," *Management Science, 19* (8), April 1973. Reprinted in EE Monograph Series No. 2, entitled *Risk and Uncertainty: Non-Deterministic Decision Making in Engineering Economy,* American Institute of Industrial Engineers, 1975, pp. 32–37.

8. de Neufville, R., and Stafford, J. H., *Systems Analysis for Engineers and Managers,* McGraw-Hill, 1958.

9. Fishburn, P. C., "Decision Under Uncertainty: An Introductory Exposition," in EE Monograph Series No. 2, entitled *Risk and Uncertainty: Non-Deterministic Decision Making in Engineering Economy,* American Institute of Industrial Engineers, 1975, pp. 19–31.

10. Goetz, B. E., "Perplexing Problems in Decision Theory," *The Engineering Economist, 14* (3), Spring 1969.

11. Goicoechea, A, Hansen, D. R., and Duckstein, L., *Multiobjective Decision Analysis with Engineering and Business Applications,* Wiley, 1982.

12. Harpaz, G., and Thomadakis, S. B., "Project Valuation with Imperfect Information," *The Engineering Economist, 29* (2) Institute of Industrial Engineers, pp. 101–112.

13. Hespos, R. F., and Strassman, P. A., "Stochastic Decision Trees for the Analysis of Investment Decisions," *Management Science, 11* (10), August 1965. Reprinted in EE Monograph Series No. 2, entitled *Risk and Uncertainty: Non-Deterministic Decision Making in Engineering Economy,* American Institute of Industrial Engineers, 1975, pp. 48–55.

14. Keeney, R. L., and Raiffa, H., *Decisions with Multiple Objectives: Preferences and Value Tradeoffs,* Wiley, 1976.

15. Kepner, C. H., and Tregoe, B. B., *The Rational Manager,* McGraw-Hill, 1965.

16. Luce, R. D., and Raiffa, H., *Games and Decisions: Introduction and Critical Survey,* Wiley, 1958.

17. Magee, J. F., "Decision Trees for Decision Making," *Harvard Business Review, 42* (4), July–August 1964.

18. Magee, J. F., "How to Use Decision Trees in Capital Investment," *Harvard Business Review, 42* (5), September–October 1964.

19. Miller, D. W., and Starr, M. K., *Executive Decisions and Operations Research,* Second Edition, Prentice-Hall, 1969.

20. Miller, E. C., *Advanced Techniques for Strategic Planning,* AMA Research Study 104, American Management Association, 1971.

21. Moore, W. T., and Chen, S., "The Value of Perfect Information in Capital Budgeting Decisions with Unknown Cash Flow Parameters," *The Engineering Economist, 29* (1) Fall 1983, Institute of Industrial Engineers, pp. 41–51.

22. Morris, W. T., *The Analysis of Management Decisions,* Revised Edition, R. D. Irwin, 1964.

23. Morris, W. T., *Management Science: A Bayesian Introduction,* Prentice-Hall, 1968.

24. Morris, W. T., *Engineering Economic Analysis,* Reston Publishing Company, Inc., A Prentice-Hall Company, 1976.

25. Morris, W. T., *Decision Analysis,* Grid, Inc., 1977.

26. Morse, P. M., *Library Effectiveness: A Systems Approach,* M.I.T. Press, 1968.

27. Raiffa, H., *Decision Analysis: Introductory Lectures on Choices Under Uncertainty,* Second Printing, Addison-Wesley, 1970.

28. Spiegel, M. R., *Schaum's Outline of Theory and Problems of Statistics,* Schaum Publishing, 1961.

29. Szidarovsky, F., Gershon, M. E., and Duckstein, L., *Techniques for Multiobjective Decision Making in Systems Management,* (Volume 2 of the series entitled, Advances in Industrial Engineering), Elsevier, 1986.

PROBLEMS

1. Apply the various decision-under-uncertainty principles of choice to the matrix below. The values in the matrix are cost units. (9.4.1, 9.5)

	S_1	S_2	S_3	S_4
A_1	\$13	\$13	\$5	\$9
A_2	8	8	8	8
A_3	0	21	5	5
A_4	9	17	5	5
A_5	5	7	7	5

2. Assume the following decision matrix, where the cell values are units of gain.

	S_1	S_2	S_3	S_4
A_1	\$4	\$ 2	\$0	\$1
A_2	− 4	10	3	7
A_3	4	8	2	3
A_4	2	4	9	5

If a person's index of optimism is estimated as $\alpha = 0.25$, predict the person's choice of alternative by applying the various principles of choice for a decision under uncertainty. (9.4.1, 9.5)

3. Analyze the cost matrix below by the use of the Hurwicz principle. (9.5.4)

	S_1	S_2	S_3
A_1	$ 6	$6	$6
A_2	12	8	0
A_3	4	7	3

4. Assume the values in the matrix of Problem 3 are units of gain. Analyze the alternatives by using the Hurwicz principle. (9.5.4).

5. For the matrix given in Problem 3, if the probability of S_1 occurring is 0.10, what are the probabilities for S_2 and S_3 which would make the decision maker indifferent between the choice of A_2 and A_3? Assume the expectation principle of choice. (9.4.2).

6. Assume the following decision matrix.

	p_1 S_1	p_2 S_2	p_3 S_3
A_1	−$ 20	$100	$800
A_2	218	218	218
A_3	650	120	40

(a) Treat the decision as one under uncertainty and the values in the matrix as costs. Then,
  1. If α = an index of optimism, for what value of α is one indifferent between A_1 and A_2? (9.5.4).
  2. Which alternative would be chosen by the Savage principle? (9.5.5).

(b) Treat the decision as one under risk where $p_1 = 0.20$, $p_2 = 0.70$, and $p_3 = 0.10$, and the values in the matrix are profits. Which alternative would be chosen by the expectation-variance principle? (9.4.2).

7. A particular production department supervisor is known to be very conservative in business matters. For the decision model below, where values are equivalent annual profits (in thousands of dollars) for a $MARR$ = 15% and a 5-year planning horizon, predict the supervisor's choice of an alternative by each of the principles for a decision under uncertainty. (9.4.1, 9.5)

	S_1	S_2	S_3	S_4
A_1	$16	$16	$ 8	$12
A_2	9	20	10	6
A_3	15	10	8	9
A_4	3	14	6	11
A_5	8	16	9	12

8. Given the decision matrix shown below (with *cost* elements), determine the preferred alternative using the following principles of choice. (9.5)
   (a) Minimax.
   (b) Minimin.
   (c) Minimax regret.
   (d) Expected cost, if each future state is expected to occur with equal probability.
   (e) Hurwicz (with $\alpha = 0.80$).

	S_1	S_2	S_3
a_1	$120	$50	$ 10
a_2	60	60	60
a_3	70	50	60
a_4	20	− 50	250

9. Shown below is a matrix of costs for three investment alternatives under three future states. Determine the preferred alternative using (a) the Laplace principle, (b) the minimax principle, and (c) the minimax regret principle. (9.5.1, 9.5.2, 9.5.5).

	S_1	S_2	S_3
A_1	$300	$200	$100
A_2	145	180	200
A_3	210	140	175

10. Given the decision matrix shown below, determine the recommended alternative under the following principles of choice: expectation, most probable future,

maximax, maximin, and minimax regret. Entries in the matrix are profits. (9.4, 9.5)

	$p_1 = 0.1$ S_1	$p_2 = 0.3$ S_2	$p_3 = 0.5$ S_3	$p_4 = 0.1$ S_4
A_1	$-\$\ 50$	$\$100$	$\$200$	$\$400$
A_2	$-\ 20$	50	500	100
A_3	100	$-\ 100$	50	100
A_4	200	$-\ 50$	50	200

11. An electronics firm has recently received a United States government contract to produce a certain quantity of expensive electronic guidance systems. The contracted number of units can be produced in 2 years, and the contract terminates then. The firm won the contract with cost estimates based on using present manufacturing equipment. If the firm had more specialized equipment (with almost zero salvage value immediately after purchase), the unit cost of production would be reduced, and profit per unit would thus be increased. However, this would be true only if the contract were for at least 4 years instead of 2 years. The firm feels there is a 30% chance for an additional 2-year contract, a 30% chance for an additional 4-year contract, and a 40% chance of no contract renewal.

  Using only present equipment, total profit on this job for a 2-year, 4-year, and 6-year contract is estimated as $40,000, $80,000 and $120,000, respectively. If the specialized equipment is purchased, total profit on this job for a 2-year, 4-year, and 6-year contract is estimated as $-$100,000, $90,000, and $180,000, respectively. If an expectation principle is used, should the specialized equipment be purchased? Does the most-probable-future criterion seem reasonable in this decision? (9.4.1, 9.4.2, 9.4.3).

12. An analysis yields a decision under risk given in the matrix below, where the matrix values are annual profits in thousands of dollars.

	$p_k =$ 0.1 S_1	0.1 S_2	0.4 S_3	0.2 S_4	0.1 S_5	0.1 S_6
A_1	$\$12$	$\$\ 5$	$-\$\ 8$	$-\$3$	$\$\ 6$	$\$\ 9$
A_2	7	0	1	5	20	7
A_3	3	3	7	9	$-\ 5$	5
A_4	0	12	15	2	8	$-\ 200$
A_5	$-\ 10$	22	9	0	4	12

a. Which alternative should be chosen in order to minimize the probability of a loss? (9.4.4)

b. Which alternative should be chosen in order to maximize the probability of an annual profit of at least $9000? (9.4.4)

c. Which alternative should be chosen in order to maximize the probability that annual profits will be between $3000 and $10,000? (9.4.4)

d. Would the most-probable-future principle be a reasonable one to follow in selecting an alternative? (9.4.3)

13. A highway construction firm is considering the purchase of a used mobile crane. Two such cranes are available, A_1 and A_2. If either is purchased, the firm will use a 5-year depreciation schedule. The cranes differ in capacity, age, and mechanical condition, but both are presently in operating condition, and each has the capacity to do the handling jobs expected. The firm expects that a major overhaul of each crane will eventually be required, but when the overhaul will be necessary is uncertain. Estimates of the net annual operating expenses (excluding labor) for each crane are given below:

	A_1		A_2	
Year	Expense Schedule 1	Expense Schedule 2	Expense Schedule 1	Expense Schedule 2
1	$ 6,500	$ 6,500	$15,000	$ 8,000
2	7,500	12,500	8,000	8,500
3	8,500	5,500	8,000	18,000
4	14,000	6,000	8,000	6,000
5	5,500	6,500	8,000	6,500

Other data for the two used cranes are

	A_1	A_2
First cost	$40,000	$27,500
Salvage value at the end of 5 years	12,000	8,500

If the firm uses a before-tax $MARR$ = 15% and a present-worth method of comparing investment alternatives, which crane would be purchased if the Laplace principle were used? (9.4.1, 9.5.1).

14. Suppose a coin is biased such that, when the coin is tossed, the probability of a head (H) showing a 0.60. Now assume that the coin is tossed three consecutive times.

a. Sketch a probability tree for the three tosses and determine the probability values for each of the eight possible outcomes: HHH, HHT, and so on. For the three tosses, the following states are now defined:

S_1 = the event of three heads (3H)
S_2 = the event of two heads, one tail—occurring in any order (2H, T)
S_3 = the event of two tails, one head—occurring in any order, (2T, H)
S_4 = the event of three tails (3T) (9.6.1, 9.6.2)

b. Create a decision matrix where four alternatives are to guess the events defined above. Now assume that before tossing the coin three times, a person offers a $4 prize if the (3H) event is guessed correctly, a $3 prize if the (2H, T), event is guessed correctly, a $4 prize if the (2T, H) event is guessed correctly, and a $5 prize if the (3T) event is guessed correctly. However, there is a $1 charge to play the game. If the game is played, would the expectation principle and the most-probable-future principle select the same alternative? (9.2, 9.4.2, 9.4.3)

15.* A vehicle manufacturer plans to develop and operate a public bus system for a community. The manufacturer's purpose is to demonstrate the profitability of the bus system and then sell the system to the community. The manufacturer's most serious concern is whether the Public Utilities Commission will authorize a competing bus system. It is rationalized that there is a small chance that such an event will happen; therefore, the manufacturer subjectively assigns a probability of 0.2 to the event of competition.

A detailed analysis by the manufacturer results in an estimate of 50 vehicles needed in the event of no competition, and 25 vehicles needed if competition results. Furthermore, it is reasoned that operating either a 50-vehicle or a 25-vehicle system will not affect the probability of the Public Utilities Commission authorizing or not authorizing competition.

If the manufacturer initially develops a 50-vehicle system and there is no competition, it is estimated the system will yield a $250,000 profit. In the event of competition, a $120,000 loss is projected.

If a 25-vehicle system is initially developed and competition occurs, the manufacturer plans to sell the system as quickly as possible and estimates a $25,000 profit in this instance. On the other hand, if there is no competition, the 25-vehicle system will be inadequate. Poor service would therefore result, and the value of the demonstration would be reduced, thereby adversely affecting the sales price. Thus, in the event of no competition, the manufacturer reasons that another decision must be faced: whether to expand from 25 vehicles to 50 vehicles or not to expand. If there is no expansion, the manufacturer would sell soon to avoid the bad publicity for expected poor service, estimating a $35,000 profit in this case. If, on the other hand, the system is expanded, and the total system is sold as soon as possible, two outcomes are predicted. One outcome is that poor service could result during the expansion, and a net $10,000 loss would result (a probability of 0.10 is subjectively assigned to this event). The second outcome is that poor service would not result, and a net $140,000 profit is expected in this instance. Create a decision tree model of this situation and solve by use of the expectation principle. (9.6.1, 9.6.2, 9.4.2).

16. Let it be supposed that, out of your sight, an experimenter rolls three dice: one green, one red, and one white. The experimenter covers two of these. Data concerning the dice are that (1) the green die has five surfaces marked H, one

---

* Problem 15 is taken from deNeufville and Stafford [8] by permission of the publisher.

surface marked T, (2) the red die has two surfaces marked H, four surfaces marked T, and (3) the white die has three surfaces marked H, three surfaces marked T. Also

A = the event that the green die is uncovered.
B = the event that the red die is uncovered.
C = the event that the white die is uncovered.
h = the event that a surface marked H is showing on the uppermost surface of the uncovered die.
t = the event that a surface marked T is showing.

Now, the experimenter reports that the letter H is showing on the uppermost surface of the uncovered die.

a. Using the logic of Bayes's theorem, what is the best guess concerning the color of the uncovered die? (9.6.1, 9.6.2)

b. Assume that the experimenter, after the report but before the guess, offers to pay $6 if the green die is guessed correctly, $15 if the red die is guessed correctly, and $15 if the white die is guessed correctly. However, there is a $3 charge to play the game. Given that the game is played, model this situation as a decision-under-risk matrix and determine the best guess by the expectation principle. (9.2, 9.4.2)

17. Work problem 16 above, with the following two modifications:

a. The experimenter reports that the letter T is showing on the uppermost surface of the uncovered die.

b. The experimenter offers to pay $9, $12, and $15 if the green, red, and white die, respectively, is guessed correctly. There is a $5 charge to play the game.

18. The Promax Company purchases a steel casting from four suppliers (A, B, C, and D) on a random basis throughout a production year. The steel casting from each supplier may be of either low (L) or high (H) nickel alloy content. Historical records reveal that the Promax Company has purchased castings from the four suppliers in the proportions shown in the following table. (9.6.1, 9.6.2)

Supplier	Proportion of Total Castings	Proportion by Alloy Content	
		L	H
A	0.45	0.20	0.80
B	0.10	0.65	0.35
C	0.30	0.55	0.45
D	0.15	0.40	0.60

a. If a steel casting is selected at random from the shop floor of the Promax Company, estimate the probability that the casting has a high nickel content.

b. If a casting that is selected from the shop floor of the Promax Company has a low nickel content, estimate the probability that it came from supplier C.

*19. After receiving somewhat unreliable information concerning enemy troop movement along a supply route, a military commander of an artillery battalion faces the following combat decision (assume the matrix values are loss units, arbitrarily chosen).

	$p_1 = 0.7$ $S_1 = $ Troop Movement	$p_2 = 0.3$ $S_2 = $ No Troop Movement
$A_1 = $ bombard supply route	0	4
$A_2 = $ not bombard supply route	12	0

The commander can choose between A_1 and A_2 without additional information, or he can send out a reconnaissance plane for additional information. If the plane is sent out, he reasons the mutually exclusive and collectively exhaustive outcomes of the flight are:

Outcome O_1: the plane is shot down, with a loss of 3 units.
Outcome O_2: the plane returns safely with negative information about troop movement, with a loss of 0.2 units.
Outcome O_3: the plane returns safely with a report of suspicious activity, with a loss of 0.2 units.

The commander assigns the following subjective conditional probabilities to these outcomes:

$$P(O_1|S_1) = 0.4 \qquad P(O_1|S_2) = 0.2$$
$$P(O_2|S_1) = 0.1 \qquad P(O_2|S_2) = 0.8$$
$$P(O_3|S_1) = 0.5 \qquad P(O_3|S_2) = 0.0$$

(a) Determine the posterior probabilities. (9.6.2)
(b) Create a decision tree for this problem. (9.6.1)
(c) Using the minimizing-expected-loss principle of choice, what action should the commander take? (9.4.2)
(d) What is the expected value of sample information in this problem? (9.6.3, 9.6.4)

20. A manufacturer is considering the possibility of introducing a new product and the advisability of a test marketing prior to making the final decision. The alternatives are

$$a_1 = \text{market the product}$$
$$a_2 = \text{do not market the product}$$

For simplicity, only three possible futures are considered; they are shown below, together with the prior probabilities associated with them.

---

* Problem 19 taken from Agee, Taylor, and Torgerson [1] by permission of the publisher."

	Profit	Prior Probability
S_1: the product captures 10% of the market	$10,000,000	0.70
S_2: the product captures 3% of the market	1,000,000	0.10
S_3: the product captures less than 1% of the market	$-5,000,000$	0.20

If the test marketing is made, three possible results are considered:

Z_1: test sales of more than 10% of the market.
Z_2: test sales of 5% to 10% of the market.
Z_3: test sales of less than 5% of the market.

The conditional probabilities of the test results are

	Z_1	Z_2	Z_3
$P(Z\|S_1)$	0.6	0.3	0.1
$P(Z\|S_2)$	0.3	0.6	0.1
$P(Z\|S_3)$	0.1	0.1	0.8

(a) Determine the prior expected profit. (9.6.1, 9.6.2, 9.6.3)
(b) Determine *EVPI*. (9.6.3)
(c) Determine *EVSI*. (9.6.4)
(d) If the test marketing costs $250,000, what would be the expected net gain from the sample information? (9.6.4)

21. Work problem 20 above, with the following two modifications:

(a) The prior probabilities of states S_1, S_2, and S_3 are 0.30, 0.60, and 0.10, respectively.
(b) Test marketing costs are $200,000.

22. A textile firm in a rural town owns two trucks that transport raw materials from and finished goods to a nearby city. One of the trucks (tractor and trailer) is 8 years old, and annual maintenance expenses are increasing significantly. Management is considering the purchase of a new truck but, because of a mild national economic recession, textile sales have declined during the past 2 years, and management is reluctant to make a large capital expenditure at present. Although in the midst of a recession, inflation is also occurring; if the decision to purchase is delayed for a year, the cost of a new truck will probably increase 10% or more.

Adopting a 2-year planning horizon, the following estimates and judgments are made.

Present Truck: The present market value is approximately $9000. If kept 1 more year, operating (excluding labor and depreciation) and maintenance ex-

penses are estimated as either $6000, $10,000, or $12,000 with subjective proba-
bilities of 0.25, 0.50, and 0.25, respectively. The salvage value at the end of the
year is about $8000. If kept 1 more year, a new decision to keep an additional
year or buy a new truck will be made.

If the truck is kept 2 additional years, operating and maintenance expenses
(excluding labor and depreciation) for the second year are estimated to be either
$7000, $10,000, or $15,000 with subjective probabilities of 0.10, 0.70, and 0.20,
respectively. The salvage value at the end of the second year is about $7500.

New Truck: A new truck can be purchased now for $30,000. Average annual
operating, maintenance, and depreciation expenses for each of the first 2 years
are estimated to be $15,000 with reasonable certainty.

If the purchase of the truck is delayed for 1 year, an increase in purchase price
is virtually certain. It is judged there is a 50–50 chance that the purchase price
will be $32,000 or $34,000. If it is $32,000, total operating, maintenance, and
depreciation expenses will be $16,000; if it is $34,000, total expenses will be
$17,500.

For simplicity, assume $MARR = 0\%$ and **(a)** create a decision tree model for
this situation, and then **(b)** determine the best present decision by the expectation principle. (9.6.1, 9.4.2)

23. Suppose a decision may be modeled as

	$p_1 = 0.6$ S_1 Contract Renewed	$p_2 = 0.4$ S_2 Contract Not Renewed
A_1: use present equipment	$V(\theta_{11})$	$V(\theta_{12})$
A_2: purchase new equipment	$V(\theta_{21})$	$V(\theta_{22})$

Assume that analysis produces the following descriptions of the outcomes:

θ_{11} = no new capital required, 20 people required for at least 4 years, average
annual profit units are 85.

θ_{12} = no new capital required, 20 people required for 2 years and then laid off,
average annual profit units are 67.

θ_{21} = two units of new capital required, 8 people required for at least 4 years,
average annual profit units are 150.

θ_{22} = two units of new capital required, 8 people required for 2 years and then
laid off, average annual profit units are 400.

Presume the relevant values for objectives are

V(profit units) $= +0.003$ (profit units) $+ 1.0$
V(capital units required) $= e^{-u}$, where $u =$ the number of capital units required
V(work-years required) $= -0.0145$ (work-years required) $+ 1.0$

Furthermore, the goals of "capital units required" and "annual profit units" are
considered equally important, but the labor consideration is only half as important as either of the other two.

On the basis of this information, which alternative would you recommend?
(9.2, 9.7, 9.7.2)

24. Work Problem 23 above, with the following two modifications:

    (a) The probabilities of states S_1 and S_2 are 0.3 and 0.7, respectively.

    (b) The goal of "capital units required" is double the importance of the "annual profit units" goal and four times the importance of the labor factor. The labor factor is twice the importance of the annual profit units.

25. The Jax Tool and Engineering Company is experiencing considerable problems with in-process material handling and storage of finished goods. A new layout of machinery is possible, but this alternative is subjectively judged by management to be prohibitively expensive, and the alternative is discarded. They wish, therefore, to consider a new material-handling system and identify the following objectives.

G_1: minimize annual costs over a 10-year planning horizon.

G_2: minimize the disruption of production during the installation of the new system.

G_3: install material-handling equipment that has flexibility to meet different handling requirements.

G_4: the material-handling equipment should have a high index of repairability.

G_5: material-handling personnel should require little training in order to operate the equipment.

Management ranks these goals in the following order of importance: $G_1 = $ I, $G_3 = $ II, $G_2 = $ III, $G_5 = $ IV, and $G_4 = $ V. Using the Churchman, Ackoff, and Arnoff method of weighting objectives, assign weights to these five objectives. (9.7.4)

APPENDIX A

DISCRETE COMPOUNDING

SECTION I -- DISCRETE COMPOUND INTEREST FACTORS

SECTION II -- GEOMETRIC SERIES FACTORS

Table A.1 Discrete Compounding: $i = 1/2\%$

	Single payment		Uniform Series				Gradient series	
	Compound amount factor	Present worth factor	Compound amount factor	Sinking fund factor	Present worth factor	Capital recovery factor	Uniform series factor	Present worth factor
	To find F Given P	To find P given F	To find F given A	To find A given F	To find P given A	To find A given P	To find A given G	To find P given G
n	$F\|P\ i,n$	$P\|F\ i,n$	$F\|A\ i,n$	$A\|F\ i,n$	$P\|A\ i,n$	$A\|P\ i,n$	$A\|G\ i,n$	$P\|G\ i,n$
1	1.0050	0.9950	1.0000	1.0000	0.9950	1.0050	0.0000	0.0000
2	1.0100	0.9901	2.0050	0.4988	1.9851	0.5038	0.4988	0.9901
3	1.0151	0.9851	3.0150	0.3317	2.9702	0.3367	0.9967	2.9604
4	1.0202	0.9802	4.0301	0.2481	3.9505	0.2531	1.4938	5.9011
5	1.0253	0.9754	5.0503	0.1980	4.9259	0.2030	1.9900	9.8026
6	1.0304	0.9705	6.0755	0.1646	5.8964	0.1696	2.4855	14.6552
7	1.0355	0.9657	7.1059	0.1407	6.8621	0.1457	2.9801	20.4493
8	1.0407	0.9609	8.1414	0.1228	7.8230	0.1278	3.4738	27.1755
9	1.0459	0.9561	9.1821	0.1089	8.7791	0.1139	3.9668	34.8244
10	1.0511	0.9513	10.2280	0.0978	9.7304	0.1028	4.4589	43.3865
11	1.0564	0.9466	11.2792	0.0887	10.6770	0.0937	4.9501	52.8526
12	1.0617	0.9419	12.3356	0.0811	11.6189	0.0861	5.4406	63.2136
13	1.0670	0.9372	13.3972	0.0746	12.5562	0.0796	5.9302	74.4602
14	1.0723	0.9326	14.4642	0.0691	13.4887	0.0741	6.4190	86.5835
15	1.0777	0.9279	15.5365	0.0644	14.4166	0.0694	6.9069	99.5743
16	1.0831	0.9233	16.6142	0.0602	15.3399	0.0652	7.3940	113.4238
17	1.0885	0.9187	17.6973	0.0565	16.2586	0.0615	7.8803	128.1231
18	1.0939	0.9141	18.7858	0.0532	17.1728	0.0582	8.3658	143.6634
19	1.0994	0.9096	19.8797	0.0503	18.0824	0.0553	8.8504	160.0360
20	1.1049	0.9051	20.9791	0.0477	18.9874	0.0527	9.3342	177.2322
21	1.1104	0.9006	22.0840	0.0453	19.8880	0.0503	9.8172	195.2434
22	1.1160	0.8961	23.1944	0.0431	20.7841	0.0481	10.2993	214.0611
23	1.1216	0.8916	24.3104	0.0411	21.6757	0.0461	10.7806	233.6768
24	1.1272	0.8872	25.4320	0.0393	22.5629	0.0443	11.2611	254.0820
25	1.1328	0.8828	26.5591	0.0377	23.4456	0.0427	11.7407	275.2686
26	1.1385	0.8784	27.6919	0.0361	24.3240	0.0411	12.2195	297.2281
27	1.1442	0.8740	28.8304	0.0347	25.1980	0.0397	12.6975	319.9523
28	1.1499	0.8697	29.9745	0.0334	26.0677	0.0384	13.1747	343.4332
29	1.1556	0.8653	31.1244	0.0321	26.9330	0.0371	13.6510	367.6625
30	1.1614	0.8610	32.2800	0.0310	27.7941	0.0360	14.1265	392.6324
31	1.1672	0.8567	33.4414	0.0299	28.6508	0.0349	14.6012	418.3348
32	1.1730	0.8525	34.6086	0.0289	29.5033	0.0339	15.0750	444.7618
33	1.1789	0.8482	35.7817	0.0279	30.3515	0.0329	15.5480	471.9055
34	1.1848	0.8440	36.9606	0.0271	31.1955	0.0321	16.0202	499.7583
35	1.1907	0.8398	38.1454	0.0262	32.0354	0.0312	16.4915	528.3123
36	1.1967	0.8356	39.3361	0.0254	32.8710	0.0304	16.9621	557.5598
37	1.2027	0.8315	40.5328	0.0247	33.7025	0.0297	17.4317	587.4934
38	1.2087	0.8274	41.7354	0.0240	34.5299	0.0290	17.9006	618.1054
39	1.2147	0.8232	42.9441	0.0233	35.3531	0.0283	18.3686	649.3883
40	1.2208	0.8191	44.1588	0.0226	36.1722	0.0276	18.8359	681.3347
45	1.2516	0.7990	50.3242	0.0199	40.2072	0.0249	21.1595	850.7631
48	1.2705	0.7871	54.0978	0.0185	42.5803	0.0235	22.5437	959.9188
50	1.2832	0.7793	56.6452	0.0177	44.1428	0.0227	23.4624	1035.6966
54	1.3091	0.7639	61.8167	0.0162	47.2214	0.0212	25.2899	1194.2236
60	1.3489	0.7414	69.7700	0.0143	51.7256	0.0193	28.0064	1448.6458
65	1.3829	0.7231	76.5821	0.0131	55.3775	0.0181	30.2475	1675.0272
70	1.4178	0.7053	83.5661	0.0120	58.9394	0.0170	32.4680	1913.6427
75	1.4536	0.6879	90.7265	0.0110	62.4136	0.0160	34.6679	2163.7525
80	1.4903	0.6710	98.0677	0.0102	65.8023	0.0152	36.8474	2424.6455
90	1.5666	0.6383	113.3109	0.0088	72.3313	0.0138	41.1451	2976.0769
100	1.6467	0.6073	129.3337	0.0077	78.5426	0.0127	45.3613	3562.7934
120	1.8194	0.5496	163.8793	0.0061	90.0735	0.0111	53.5508	4823.5051

Table A.2 Discrete Compounding: $i = 7/12\%$

	Single payment		Uniform Series				Gradient series	
	Compound amount factor	Present worth factor	Compound amount factor	Sinking fund factor	Present worth factor	Capital recovery factor	Uniform series factor	Present worth factor
	To find F Given P $F\|P$ i,n	To find P given F $P\|F$ i,n	To find F given A $F\|A$ i,n	To find A given F $A\|F$ i,n	To find P given A $P\|A$ i,n	To find A given P $A\|P$ i,n	To find A given G $A\|G$ i,n	To find P given G $P\|G$ i,n
n								
1	1.0058	0.9942	1.0000	1.0000	0.9942	1.0058	0.0000	0.0000
2	1.0117	0.9884	2.0058	0.4985	1.9826	0.5044	0.4985	0.9884
3	1.0176	0.9827	3.0175	0.3314	2.9653	0.3372	0.9961	2.9538
4	1.0235	0.9770	4.0351	0.2478	3.9423	0.2537	1.4927	5.8848
5	1.0295	0.9713	5.0587	0.1977	4.9137	0.2035	1.9884	9.7702
6	1.0355	0.9657	6.0882	0.1643	5.8794	0.1701	2.4830	14.5987
7	1.0416	0.9601	7.1237	0.1404	6.8395	0.1462	2.9767	20.3593
8	1.0476	0.9545	8.1653	0.1225	7.7940	0.1283	3.4695	27.0411
9	1.0537	0.9490	9.2129	0.1085	8.7430	0.1144	3.9612	34.6331
10	1.0599	0.9435	10.2666	0.0974	9.6865	0.1032	4.4520	43.1245
11	1.0661	0.9380	11.3265	0.0883	10.6245	0.0941	4.9418	52.5048
12	1.0723	0.9326	12.3926	0.0807	11.5571	0.0865	5.4307	62.7632
13	1.0785	0.9272	13.4649	0.0743	12.4843	0.0801	5.9186	73.8893
14	1.0848	0.9218	14.5434	0.0688	13.4061	0.0746	6.4055	85.8727
15	1.0912	0.9165	15.6283	0.0640	14.3225	0.0698	6.8914	98.7030
16	1.0975	0.9111	16.7194	0.0598	15.2337	0.0656	7.3764	112.3700
17	1.1039	0.9059	17.8170	0.0561	16.1395	0.0620	7.8604	126.8637
18	1.1104	0.9006	18.9209	0.0529	17.0401	0.0587	8.3435	142.1739
19	1.1168	0.8954	20.0313	0.0499	17.9355	0.0558	8.8255	158.2906
20	1.1234	0.8902	21.1481	0.0473	18.8257	0.0531	9.3066	175.2041
21	1.1299	0.8850	22.2715	0.0449	19.7107	0.0507	9.7868	192.9045
22	1.1365	0.8799	23.4014	0.0427	20.5906	0.0486	10.2660	211.3822
23	1.1431	0.8748	24.5379	0.0408	21.4654	0.0466	10.7442	230.6275
24	1.1498	0.8697	25.6810	0.0389	22.3351	0.0448	11.2214	250.6308
25	1.1565	0.8647	26.8308	0.0373	23.1998	0.0431	11.6977	271.3829
26	1.1633	0.8597	27.9874	0.0357	24.0594	0.0416	12.1730	292.8742
27	1.1700	0.8547	29.1506	0.0343	24.9141	0.0401	12.6473	315.0956
28	1.1769	0.8497	30.3207	0.0330	25.7638	0.0388	13.1206	338.0378
29	1.1837	0.8448	31.4975	0.0317	26.6086	0.0376	13.5930	361.6917
30	1.1906	0.8399	32.6813	0.0306	27.4485	0.0364	14.0645	386.0483
31	1.1976	0.8350	33.8719	0.0295	28.2835	0.0354	14.5349	411.0987
32	1.2046	0.8302	35.0695	0.0285	29.1137	0.0343	15.0044	436.8340
33	1.2116	0.8254	36.2741	0.0276	29.9390	0.0334	15.4730	463.2454
34	1.2187	0.8206	37.4857	0.0267	30.7596	0.0325	15.9405	490.3242
35	1.2258	0.8158	38.7043	0.0258	31.5754	0.0317	16.4071	518.0617
36	1.2329	0.8111	39.9301	0.0250	32.3865	0.0309	16.8728	546.4495
37	1.2401	0.8064	41.1630	0.0243	33.1928	0.0301	17.3374	575.4790
38	1.2474	0.8017	42.4031	0.0236	33.9945	0.0294	17.8011	605.1418
39	1.2546	0.7970	43.6505	0.0229	34.7916	0.0287	18.2639	635.4297
40	1.2619	0.7924	44.9051	0.0223	35.5840	0.0281	18.7257	666.3343
45	1.2992	0.7697	51.2889	0.0195	39.4777	0.0253	21.0201	829.8256
48	1.3221	0.7564	55.2092	0.0181	41.7602	0.0239	22.3852	934.8099
50	1.3375	0.7477	57.8611	0.0173	43.2599	0.0231	23.2904	1007.5414
54	1.3690	0.7305	63.2581	0.0158	46.2074	0.0216	25.0894	1159.3159
60	1.4176	0.7054	71.5929	0.0140	50.5020	0.0198	27.7591	1401.8899
65	1.4595	0.6852	78.7642	0.0127	53.9681	0.0185	29.9575	1616.7498
70	1.5025	0.6655	86.1471	0.0116	57.3349	0.0174	32.1320	1842.2851
75	1.5469	0.6465	93.7479	0.0107	60.6052	0.0165	34.2827	2077.7072
80	1.5925	0.6279	101.5730	0.0098	63.7817	0.0157	36.4095	2322.2641
90	1.6879	0.5925	117.9227	0.0085	69.8643	0.0143	40.5922	2835.9476
100	1.7890	0.5590	135.2515	0.0074	75.6031	0.0132	44.6806	3377.9941
120	2.0097	0.4976	173.0848	0.0058	86.1264	0.0116	52.5768	4528.2518

Table A.3 Discrete Compounding: $i = 2/3\%$

	Single payment		Uniform Series				Gradient series	
	Compound amount factor	Present worth factor	Compound amount factor	Sinking fund factor	Present worth factor	Capital recovery factor	Uniform series factor	Present worth factor
	To find F Given P	To find P given F	To find F given A	To find A given F	To find P given A	To find A given P	To find A given G	To find P given G
n	$F\|P\ i,n$	$P\|F\ i,n$	$F\|A\ i,n$	$A\|F\ i,n$	$P\|A\ i,n$	$A\|P\ i,n$	$A\|G\ i,n$	$P\|G\ i,n$
1	1.0067	0.9934	1.0000	1.0000	0.9934	1.0067	0.0000	0.0000
2	1.0134	0.9868	2.0067	0.4983	1.9802	0.5050	0.4983	0.9868
3	1.0201	0.9803	3.0200	0.3311	2.9604	0.3378	0.9956	2.9473
4	1.0269	0.9738	4.0402	0.2475	3.9342	0.2542	1.4917	5.8686
5	1.0338	0.9673	5.0671	0.1974	4.9015	0.2040	1.9867	9.7379
6	1.0407	0.9609	6.1009	0.1639	5.8625	0.1706	2.4806	14.5425
7	1.0476	0.9546	7.1416	0.1400	6.8170	0.1467	2.9734	20.2698
8	1.0546	0.9482	8.1892	0.1221	7.7652	0.1288	3.4651	26.9075
9	1.0616	0.9420	9.2438	0.1082	8.7072	0.1148	3.9557	34.4431
10	1.0687	0.9357	10.3054	0.0970	9.6429	0.1037	4.4452	42.8645
11	1.0758	0.9295	11.3741	0.0879	10.5724	0.0946	4.9336	52.1597
12	1.0830	0.9234	12.4499	0.0803	11.4958	0.0870	5.4208	62.3167
13	1.0902	0.9172	13.5329	0.0739	12.4130	0.0806	5.9070	73.3236
14	1.0975	0.9112	14.6231	0.0684	13.3242	0.0751	6.3920	85.1688
15	1.1048	0.9051	15.7206	0.0636	14.2293	0.0703	6.8760	97.8408
16	1.1122	0.8991	16.8254	0.0594	15.1285	0.0661	7.3588	111.3279
17	1.1196	0.8932	17.9376	0.0557	16.0217	0.0624	7.8406	125.6189
18	1.1270	0.8873	19.0572	0.0525	16.9089	0.0591	8.3212	140.7026
19	1.1346	0.8814	20.1842	0.0495	17.7903	0.0562	8.8007	156.5678
20	1.1421	0.8756	21.3188	0.0469	18.6659	0.0536	9.2791	173.2034
21	1.1497	0.8698	22.4609	0.0445	19.5357	0.0512	9.7564	190.5987
22	1.1574	0.8640	23.6107	0.0424	20.3997	0.0490	10.2327	208.7427
23	1.1651	0.8583	24.7681	0.0404	21.2579	0.0470	10.7078	227.6249
24	1.1729	0.8526	25.9332	0.0386	22.1105	0.0452	11.1818	247.2346
25	1.1807	0.8470	27.1061	0.0369	22.9575	0.0436	11.6546	267.5614
26	1.1886	0.8413	28.2868	0.0354	23.7988	0.0420	12.1264	288.5949
27	1.1965	0.8358	29.4754	0.0339	24.6346	0.0406	12.5971	310.3249
28	1.2045	0.8302	30.6719	0.0326	25.4648	0.0393	13.0667	332.7412
29	1.2125	0.8247	31.8763	0.0314	26.2896	0.0380	13.5352	355.8339
30	1.2206	0.8193	33.0889	0.0302	27.1088	0.0369	14.0025	379.5928
31	1.2287	0.8138	34.3094	0.0291	27.9227	0.0358	14.4688	404.0083
32	1.2369	0.8085	35.5382	0.0281	28.7312	0.0348	14.9340	429.0705
33	1.2452	0.8031	36.7751	0.0272	29.5343	0.0339	15.3980	454.7699
34	1.2535	0.7978	38.0203	0.0263	30.3320	0.0330	15.8610	481.0968
35	1.2618	0.7925	39.2737	0.0255	31.1246	0.0321	16.3229	508.0419
36	1.2702	0.7873	40.5356	0.0247	31.9118	0.0313	16.7836	535.5958
37	1.2787	0.7820	41.8058	0.0239	32.6938	0.0306	17.2433	563.7493
38	1.2872	0.7769	43.0845	0.0232	33.4707	0.0299	17.7018	592.4932
39	1.2958	0.7717	44.3717	0.0225	34.2424	0.0292	18.1593	621.8184
40	1.3045	0.7666	45.6675	0.0219	35.0090	0.0286	18.6157	651.7161
45	1.3485	0.7416	52.2773	0.0191	38.7666	0.0258	20.8810	809.4832
48	1.3757	0.7269	56.3499	0.0177	40.9619	0.0244	22.2270	910.4588
50	1.3941	0.7173	59.1104	0.0169	42.4013	0.0236	23.1188	980.2688
54	1.4316	0.6985	64.7427	0.0154	45.2235	0.0221	24.8894	1125.5838
60	1.4898	0.6712	73.4769	0.0136	49.3184	0.0203	27.5125	1356.8710
65	1.5402	0.6493	81.0261	0.0123	52.6084	0.0190	29.6684	1560.8049
70	1.5922	0.6281	88.8303	0.0113	55.7909	0.0179	31.7971	1773.9871
75	1.6460	0.6075	96.8982	0.0103	58.8693	0.0170	33.8987	1995.5956
80	1.7016	0.5877	105.2386	0.0095	61.8472	0.0162	35.9734	2224.8519
90	1.8185	0.5499	122.7741	0.0081	67.5142	0.0148	40.0420	2703.4022
100	1.9434	0.5146	141.5144	0.0071	72.8169	0.0137	44.0638	3204.2153
120	2.2196	0.4505	182.9460	0.0055	82.4215	0.0121	51.6103	4253.7998

Table A.4 Discrete Compounding: $i = 3/4\%$

	Single payment		Uniform Series				Gradient series	
	Compound amount factor	Present worth factor	Compound amount factor	Sinking fund factor	Present worth factor	Capital recovery factor	Uniform series factor	Present worth factor
	To find F Given P	To find P given F	To find F given A	To find A given F	To find P given A	To find A given P	To find A given G	To find P given G
n	F|P i,n	P|F i,n	F|A i,n	A|F i,n	P|A i,n	A|P i,n	A|G i,n	P|G i,n
1	1.0075	0.9926	1.0000	1.0000	0.9926	1.0075	0.0000	0.0000
2	1.0151	0.9852	2.0075	0.4981	1.9777	0.5056	0.4981	0.9852
3	1.0227	0.9778	3.0226	0.3308	2.9556	0.3383	0.9950	2.9408
4	1.0303	0.9706	4.0452	0.2472	3.9261	0.2547	1.4907	5.8525
5	1.0381	0.9633	5.0756	0.1970	4.8894	0.2045	1.9851	9.7058
6	1.0459	0.9562	6.1136	0.1636	5.8456	0.1711	2.4782	14.4866
7	1.0537	0.9490	7.1595	0.1397	6.7946	0.1472	2.9701	20.1808
8	1.0616	0.9420	8.2132	0.1218	7.7366	0.1293	3.4608	26.7747
9	1.0696	0.9350	9.2748	0.1078	8.6716	0.1153	3.9502	34.2544
10	1.0776	0.9280	10.3443	0.0967	9.5996	0.1042	4.4384	42.6064
11	1.0857	0.9211	11.4219	0.0876	10.5207	0.0951	4.9253	51.8174
12	1.0938	0.9142	12.5076	0.0800	11.4349	0.0875	5.4110	61.8740
13	1.1020	0.9074	13.6014	0.0735	12.3423	0.0810	5.8954	72.7632
14	1.1103	0.9007	14.7034	0.0680	13.2430	0.0755	6.3786	84.4720
15	1.1186	0.8940	15.8137	0.0632	14.1370	0.0707	6.8606	96.9876
16	1.1270	0.8873	16.9323	0.0591	15.0243	0.0666	7.3413	110.2973
17	1.1354	0.8807	18.0593	0.0554	15.9050	0.0629	7.8207	124.3887
18	1.1440	0.8742	19.1947	0.0521	16.7792	0.0596	8.2989	139.2494
19	1.1525	0.8676	20.3387	0.0492	17.6468	0.0567	8.7759	154.8671
20	1.1612	0.8612	21.4912	0.0465	18.5080	0.0540	9.2516	171.2297
21	1.1699	0.8548	22.6524	0.0441	19.3628	0.0516	9.7261	188.3253
22	1.1787	0.8484	23.8223	0.0420	20.2112	0.0495	10.1994	206.1420
23	1.1875	0.8421	25.0010	0.0400	21.0533	0.0475	10.6714	224.6682
24	1.1964	0.8358	26.1885	0.0382	21.8891	0.0457	11.1422	243.8923
25	1.2054	0.8296	27.3849	0.0365	22.7188	0.0440	11.6117	263.8029
26	1.2144	0.8234	28.5903	0.0350	23.5422	0.0425	12.0800	284.3888
27	1.2235	0.8173	29.8047	0.0336	24.3595	0.0411	12.5470	305.6387
28	1.2327	0.8112	31.0282	0.0322	25.1707	0.0397	13.0128	327.5416
29	1.2420	0.8052	32.2609	0.0310	25.9759	0.0385	13.4774	350.0867
30	1.2513	0.7992	33.5029	0.0298	26.7751	0.0373	13.9407	373.2631
31	1.2607	0.7932	34.7542	0.0288	27.5683	0.0363	14.4028	397.0602
32	1.2701	0.7873	36.0148	0.0278	28.3557	0.0353	14.8636	421.4675
33	1.2796	0.7815	37.2849	0.0268	29.1371	0.0343	15.3232	446.4746
34	1.2892	0.7757	38.5646	0.0259	29.9128	0.0334	15.7816	472.0712
35	1.2989	0.7699	39.8538	0.0251	30.6827	0.0326	16.2387	498.2471
36	1.3086	0.7641	41.1527	0.0243	31.4468	0.0318	16.6946	524.9924
37	1.3185	0.7585	42.4614	0.0236	32.2053	0.0311	17.1493	552.2969
38	1.3283	0.7528	43.7798	0.0228	32.9581	0.0303	17.6027	580.1511
39	1.3383	0.7472	45.1082	0.0222	33.7053	0.0297	18.0549	608.5451
40	1.3483	0.7416	46.4465	0.0215	34.4469	0.0290	18.5058	637.4693
45	1.3997	0.7145	53.2901	0.0188	38.0732	0.0263	20.7421	789.7173
48	1.4314	0.6986	57.5207	0.0174	40.1848	0.0249	22.0691	886.8404
50	1.4530	0.6883	60.3943	0.0166	41.5664	0.0241	22.9476	953.8486
54	1.4970	0.6680	66.2718	0.0151	44.2686	0.0226	24.6898	1092.9842
60	1.5657	0.6387	75.4241	0.0133	48.1734	0.0208	27.2665	1313.5189
65	1.6253	0.6153	83.3709	0.0120	51.2963	0.0195	29.3801	1507.0910
70	1.6872	0.5927	91.6201	0.0109	54.3046	0.0184	31.4634	1708.6065
75	1.7514	0.5710	100.1833	0.0100	57.2027	0.0175	33.5163	1917.2225
80	1.8180	0.5500	109.0725	0.0092	59.9944	0.0167	35.5391	2132.1472
90	1.9591	0.5104	127.8790	0.0078	65.2746	0.0153	39.4946	2577.9961
100	2.1111	0.4737	148.1445	0.0068	70.1746	0.0143	43.3311	3040.7453
120	2.4514	0.4079	193.5143	0.0052	78.9417	0.0127	50.6521	3998.5621

Table A.5 Discrete Compounding: $i = 1\%$

n	Single payment Compound amount factor To find F Given P $F\vert P\ i,n$	Present worth factor To find P given F $P\vert F\ i,n$	Uniform Series Compound amount factor To find F given A $F\vert A\ i,n$	Sinking fund factor To find A given F $A\vert F\ i,n$	Present worth factor To find P given A $P\vert A\ i,n$	Capital recovery factor To find A given P $A\vert P\ i,n$	Gradient series Uniform series factor To find A given G $A\vert G\ i,n$	Present worth factor To find P given G $P\vert G\ i,n$
1	1.0100	0.9901	1.0000	1.0000	0.9901	1.0100	0.0000	0.0000
2	1.0201	0.9803	2.0100	0.4975	1.9704	0.5075	0.4975	0.9803
3	1.0303	0.9706	3.0301	0.3300	2.9410	0.3400	0.9934	2.9215
4	1.0406	0.9610	4.0604	0.2463	3.9020	0.2563	1.4876	5.8044
5	1.0510	0.9515	5.1010	0.1960	4.8534	0.2060	1.9801	9.6103
6	1.0615	0.9420	6.1520	0.1625	5.7955	0.1725	2.4710	14.3205
7	1.0721	0.9327	7.2135	0.1386	6.7282	0.1486	2.9602	19.9168
8	1.0829	0.9235	8.2857	0.1207	7.6517	0.1307	3.4478	26.3812
9	1.0937	0.9143	9.3685	0.1067	8.5660	0.1167	3.9337	33.6959
10	1.1046	0.9053	10.4622	0.0956	9.4713	0.1056	4.4179	41.8435
11	1.1157	0.8963	11.5668	0.0865	10.3676	0.0965	4.9005	50.8067
12	1.1268	0.8874	12.6825	0.0788	11.2551	0.0888	5.3815	60.5687
13	1.1381	0.8787	13.8093	0.0724	12.1337	0.0824	5.8607	71.1126
14	1.1495	0.8700	14.9474	0.0669	13.0037	0.0769	6.3384	82.4221
15	1.1610	0.8613	16.0969	0.0621	13.8651	0.0721	6.8143	94.4810
16	1.1726	0.8528	17.2579	0.0579	14.7179	0.0679	7.2886	107.2734
17	1.1843	0.8444	18.4304	0.0543	15.5623	0.0643	7.7613	120.7834
18	1.1961	0.8360	19.6147	0.0510	16.3983	0.0610	8.2323	134.9957
19	1.2081	0.8277	20.8109	0.0481	17.2260	0.0581	8.7017	149.8950
20	1.2202	0.8195	22.0190	0.0454	18.0456	0.0554	9.1694	165.4664
21	1.2324	0.8114	23.2392	0.0430	18.8570	0.0530	9.6354	181.6950
22	1.2447	0.8034	24.4716	0.0409	19.6604	0.0509	10.0998	198.5663
23	1.2572	0.7954	25.7163	0.0389	20.4558	0.0489	10.5626	216.0660
24	1.2697	0.7876	26.9735	0.0371	21.2434	0.0471	11.0237	234.1800
25	1.2824	0.7798	28.2432	0.0354	22.0232	0.0454	11.4831	252.8945
26	1.2953	0.7720	29.5256	0.0339	22.7952	0.0439	11.9409	272.1957
27	1.3082	0.7644	30.8209	0.0324	23.5596	0.0424	12.3971	292.0702
28	1.3213	0.7568	32.1291	0.0311	24.3164	0.0411	12.8516	312.5047
29	1.3345	0.7493	33.4504	0.0299	25.0658	0.0399	13.3044	333.4863
30	1.3478	0.7419	34.7849	0.0287	25.8077	0.0387	13.7557	355.0021
31	1.3613	0.7346	36.1327	0.0277	26.5423	0.0377	14.2052	377.0394
32	1.3749	0.7273	37.4941	0.0267	27.2696	0.0367	14.6532	399.5858
33	1.3887	0.7201	38.8690	0.0257	27.9897	0.0357	15.0995	422.6291
34	1.4026	0.7130	40.2577	0.0248	28.7027	0.0348	15.5441	446.1572
35	1.4166	0.7059	41.6603	0.0240	29.4086	0.0340	15.9871	470.1583
36	1.4308	0.6989	43.0769	0.0232	30.1075	0.0332	16.4285	494.6207
37	1.4451	0.6920	44.5076	0.0225	30.7995	0.0325	16.8682	519.5329
38	1.4595	0.6852	45.9527	0.0218	31.4847	0.0318	17.3063	544.8835
39	1.4741	0.6784	47.4123	0.0211	32.1630	0.0311	17.7428	570.6616
40	1.4889	0.6717	48.8864	0.0205	32.8347	0.0305	18.1776	596.8561
45	1.5648	0.6391	56.4811	0.0177	36.0945	0.0277	20.3273	733.7037
48	1.6122	0.6203	61.2226	0.0163	37.9740	0.0263	21.5976	820.1460
50	1.6446	0.6080	64.4632	0.0155	39.1961	0.0255	22.4363	879.4176
54	1.7114	0.5843	71.1410	0.0141	41.5687	0.0241	24.0945	1001.5743
60	1.8167	0.5504	81.6697	0.0122	44.9550	0.0222	26.5333	1192.8061
65	1.9094	0.5237	90.9366	0.0110	47.6266	0.0210	28.5217	1358.3903
70	2.0068	0.4983	100.6763	0.0099	50.1685	0.0199	30.4703	1528.6474
75	2.1091	0.4741	110.9128	0.0090	52.5871	0.0190	32.3793	1702.7340
80	2.2167	0.4511	121.6715	0.0082	54.8882	0.0182	34.2492	1879.8771
90	2.4486	0.4084	144.8633	0.0069	59.1609	0.0169	37.8724	2240.5675
100	2.7048	0.3697	170.4814	0.0059	63.0289	0.0159	41.3426	2605.7758
120	3.3004	0.3030	230.0387	0.0043	69.7005	0.0143	47.8349	3334.1148

Table A.6 Discrete Compounding: $i = 3/2\%$

	Single payment		Uniform Series				Gradient series	
	Compound amount factor	Present worth factor	Compound amount factor	Sinking fund factor	Present worth factor	Capital recovery factor	Uniform series factor	Present worth factor
	To find F Given P $F\|P$ i,n	To find P given F $P\|F$ i,n	To find F given A $F\|A$ i,n	To find A given F $A\|F$ i,n	To find P given A $P\|A$ i,n	To find A given P $A\|P$ i,n	To find A given G $A\|G$ i,n	To find P given G $P\|G$ i,n
n								
1	1.0150	0.9852	1.0000	1.0000	0.9852	1.0150	0.0000	0.0000
2	1.0302	0.9707	2.0150	0.4963	1.9559	0.5113	0.4963	0.9707
3	1.0457	0.9563	3.0452	0.3284	2.9122	0.3434	0.9901	2.8833
4	1.0614	0.9422	4.0909	0.2444	3.8544	0.2594	1.4814	5.7098
5	1.0773	0.9283	5.1523	0.1941	4.7826	0.2091	1.9702	9.4229
6	1.0934	0.9145	6.2296	0.1605	5.6972	0.1755	2.4566	13.9956
7	1.1098	0.9010	7.3230	0.1366	6.5982	0.1516	2.9405	19.4018
8	1.1265	0.8877	8.4328	0.1186	7.4859	0.1336	3.4219	25.6157
9	1.1434	0.8746	9.5593	0.1046	8.3605	0.1196	3.9008	32.6125
10	1.1605	0.8617	10.7027	0.0934	9.2222	0.1084	4.3772	40.3675
11	1.1779	0.8489	11.8633	0.0843	10.0711	0.0993	4.8512	48.8568
12	1.1956	0.8364	13.0412	0.0767	10.9075	0.0917	5.3227	58.0571
13	1.2136	0.8240	14.2368	0.0702	11.7315	0.0852	5.7917	67.9454
14	1.2318	0.8118	15.4504	0.0647	12.5434	0.0797	6.2582	78.4994
15	1.2502	0.7999	16.6821	0.0599	13.3432	0.0749	6.7223	89.6974
16	1.2690	0.7880	17.9324	0.0558	14.1313	0.0708	7.1839	101.5178
17	1.2880	0.7764	19.2014	0.0521	14.9076	0.0671	7.6431	113.9400
18	1.3073	0.7649	20.4894	0.0488	15.6726	0.0638	8.0997	126.9435
19	1.3270	0.7536	21.7967	0.0459	16.4262	0.0609	8.5539	140.5084
20	1.3469	0.7425	23.1237	0.0432	17.1686	0.0582	9.0057	154.6154
21	1.3671	0.7315	24.4705	0.0409	17.9001	0.0559	9.4550	169.2453
22	1.3876	0.7207	25.8376	0.0387	18.6208	0.0537	9.9018	184.3798
23	1.4084	0.7100	27.2251	0.0367	19.3309	0.0517	10.3462	200.0006
24	1.4295	0.6995	28.6335	0.0349	20.0304	0.0499	10.7881	216.0901
25	1.4509	0.6892	30.0630	0.0333	20.7196	0.0483	11.2276	232.6310
26	1.4727	0.6790	31.5140	0.0317	21.3986	0.0467	11.6646	249.6065
27	1.4948	0.6690	32.9867	0.0303	22.0676	0.0453	12.0992	267.0002
28	1.5172	0.6591	34.4815	0.0290	22.7267	0.0440	12.5313	284.7958
29	1.5400	0.6494	35.9987	0.0278	23.3761	0.0428	12.9610	302.9779
30	1.5631	0.6398	37.5387	0.0266	24.0158	0.0416	13.3883	321.5310
31	1.5865	0.6303	39.1018	0.0256	24.6461	0.0406	13.8131	340.4402
32	1.6103	0.6210	40.6883	0.0246	25.2671	0.0396	14.2355	359.6910
33	1.6345	0.6118	42.2986	0.0236	25.8790	0.0386	14.6555	379.2691
34	1.6590	0.6028	43.9331	0.0228	26.4817	0.0378	15.0731	399.1607
35	1.6839	0.5939	45.5921	0.0219	27.0756	0.0369	15.4882	419.3521
36	1.7091	0.5851	47.2760	0.0212	27.6607	0.0362	15.9009	439.8303
37	1.7348	0.5764	48.9851	0.0204	28.2371	0.0354	16.3112	460.5822
38	1.7608	0.5679	50.7199	0.0197	28.8051	0.0347	16.7191	481.5954
39	1.7872	0.5595	52.4807	0.0191	29.3646	0.0341	17.1246	502.8576
40	1.8140	0.5513	54.2679	0.0184	29.9158	0.0334	17.5277	524.3568
45	1.9542	0.5117	63.6142	0.0157	32.5523	0.0307	19.5074	635.0110
48	2.0435	0.4894	69.5652	0.0144	34.0426	0.0294	20.6667	703.5462
50	2.1052	0.4750	73.6828	0.0136	34.9997	0.0286	21.4277	749.9636
54	2.2344	0.4475	82.2952	0.0122	36.8305	0.0272	22.9217	844.2184
60	2.4432	0.4093	96.2147	0.0104	39.3803	0.0254	25.0930	988.1674
65	2.6320	0.3799	108.8028	0.0092	41.3378	0.0242	26.8393	1109.4752
70	2.8355	0.3527	122.3638	0.0082	43.1549	0.0232	28.5290	1231.1658
75	3.0546	0.3274	136.9728	0.0073	44.8416	0.0223	30.1631	1352.5600
80	3.2907	0.3039	152.7109	0.0065	46.4073	0.0215	31.7423	1473.0741
90	3.8189	0.2619	187.9299	0.0053	49.2099	0.0203	34.7399	1709.5439
100	4.4320	0.2256	228.8030	0.0044	51.6247	0.0194	37.5295	1937.4506
120	5.9693	0.1675	331.2882	0.0030	55.4985	0.0180	42.5185	2359.7114

Table A.7 Discrete Compounding: $i= 2\%$

	Single payment		Uniform Series				Gradient series	
	Compound amount factor	Present worth factor	Compound amount factor	Sinking fund factor	Present worth factor	Capital recovery factor	Uniform series factor	Present worth factor
	To find F Given P $F|P\ i,n$	To find P given F $P|F\ i,n$	To find F given A $F|A\ i,n$	To find A given F $A|F\ i,n$	To find P given A $P|A\ i,n$	To find A given P $A|P\ i,n$	To find A given G $A|G\ i,n$	To find P given G $P|G\ i,n$
n								
1	1.0200	0.9804	1.0000	1.0000	0.9804	1.0200	0.0000	0.0000
2	1.0404	0.9612	2.0200	0.4950	1.9416	0.5150	0.4950	0.9612
3	1.0612	0.9423	3.0604	0.3268	2.8839	0.3468	0.9868	2.8458
4	1.0824	0.9238	4.1216	0.2426	3.8077	0.2626	1.4752	5.6173
5	1.1041	0.9057	5.2040	0.1922	4.7135	0.2122	1.9604	9.2403
6	1.1262	0.8880	6.3081	0.1585	5.6014	0.1785	2.4423	13.6801
7	1.1487	0.8706	7.4343	0.1345	6.4720	0.1545	2.9208	18.9035
8	1.1717	0.8535	8.5830	0.1165	7.3255	0.1365	3.3961	24.8779
9	1.1951	0.8368	9.7546	0.1025	8.1622	0.1225	3.8681	31.5720
10	1.2190	0.8203	10.9497	0.0913	8.9826	0.1113	4.3367	38.9551
11	1.2434	0.8043	12.1687	0.0822	9.7868	0.1022	4.8021	46.9977
12	1.2682	0.7885	13.4121	0.0746	10.5753	0.0946	5.2642	55.6712
13	1.2936	0.7730	14.6803	0.0681	11.3484	0.0881	5.7231	64.9475
14	1.3195	0.7579	15.9739	0.0626	12.1062	0.0826	6.1786	74.7999
15	1.3459	0.7430	17.2934	0.0578	12.8493	0.0778	6.6309	85.2021
16	1.3728	0.7284	18.6393	0.0537	13.5777	0.0737	7.0799	96.1288
17	1.4002	0.7142	20.0121	0.0500	14.2919	0.0700	7.5256	107.5554
18	1.4282	0.7002	21.4123	0.0467	14.9920	0.0667	7.9681	119.4581
19	1.4568	0.6864	22.8406	0.0438	15.6785	0.0638	8.4073	131.8139
20	1.4859	0.6730	24.2974	0.0412	16.3514	0.0612	8.8433	144.6003
21	1.5157	0.6598	25.7833	0.0388	17.0112	0.0588	9.2760	157.7959
22	1.5460	0.6468	27.2990	0.0366	17.6580	0.0566	9.7055	171.3795
23	1.5769	0.6342	28.8450	0.0347	18.2922	0.0547	10.1317	185.3309
24	1.6084	0.6217	30.4219	0.0329	18.9139	0.0529	10.5547	199.6305
25	1.6406	0.6095	32.0303	0.0312	19.5235	0.0512	10.9745	214.2592
26	1.6734	0.5976	33.6709	0.0297	20.1210	0.0497	11.3910	229.1987
27	1.7069	0.5859	35.3443	0.0283	20.7069	0.0483	11.8043	244.4311
28	1.7410	0.5744	37.0512	0.0270	21.2813	0.0470	12.2145	259.9392
29	1.7758	0.5631	38.7922	0.0258	21.8444	0.0458	12.6214	275.7064
30	1.8114	0.5521	40.5681	0.0246	22.3965	0.0446	13.0251	291.7164
31	1.8476	0.5412	42.3794	0.0236	22.9377	0.0436	13.4257	307.9538
32	1.8845	0.5306	44.2270	0.0226	23.4683	0.0426	13.8230	324.4035
33	1.9222	0.5202	46.1116	0.0217	23.9886	0.0417	14.2172	341.0508
34	1.9607	0.5100	48.0338	0.0208	24.4986	0.0408	14.6083	357.8817
35	1.9999	0.5000	49.9945	0.0200	24.9986	0.0400	14.9961	374.8826
36	2.0399	0.4902	51.9944	0.0192	25.4888	0.0392	15.3809	392.0405
37	2.0807	0.4806	54.0343	0.0185	25.9695	0.0385	15.7625	409.3424
38	2.1223	0.4712	56.1149	0.0178	26.4406	0.0378	16.1409	426.7764
39	2.1647	0.4619	58.2372	0.0172	26.9026	0.0372	16.5163	444.3304
40	2.2080	0.4529	60.4020	0.0166	27.3555	0.0366	16.8885	461.9931
45	2.4379	0.4102	71.8927	0.0139	29.4902	0.0339	18.7034	551.5652
48	2.5871	0.3865	79.3535	0.0126	30.6731	0.0326	19.7556	605.9657
50	2.6916	0.3715	84.5794	0.0118	31.4236	0.0318	20.4420	642.3606
54	2.9135	0.3432	95.6731	0.0105	32.8383	0.0305	21.7789	715.1815
60	3.2810	0.3048	114.0515	0.0088	34.7609	0.0288	23.6961	823.6975
65	3.6225	0.2761	131.1262	0.0076	36.1975	0.0276	25.2147	912.7085
70	3.9996	0.2500	149.9779	0.0067	37.4986	0.0267	26.6632	999.8343
75	4.4158	0.2265	170.7918	0.0059	38.6771	0.0259	28.0434	1084.6393
80	4.8754	0.2051	193.7720	0.0052	39.7445	0.0252	29.3572	1166.7868
90	5.9431	0.1683	247.1567	0.0040	41.5869	0.0240	31.7929	1322.1701
100	7.2446	0.1380	312.2323	0.0032	43.0984	0.0232	33.9863	1464.7527
120	10.7652	0.0929	488.2582	0.0020	45.3554	0.0220	37.7114	1710.4160

Table A.8 Discrete Compounding: i = 5/2%

	Single payment		Uniform Series				Gradient series	
	Compound amount factor	Present worth factor	Compound amount factor	Sinking fund factor	Present worth factor	Capital recovery factor	Uniform series factor	Present worth factor
	To find F Given P	To find P given F	To find F given A	To find A given F	To find P given A	To find A given P	To find A given G	To find P given G
n	F|P i,n	P|F i,n	F|A i,n	A|F i,n	P|A i,n	A|P i,n	A|G i,n	P|G i,n
1	1.0250	0.9756	1.0000	1.0000	0.9756	1.0250	0.0000	0.0000
2	1.0506	0.9518	2.0250	0.4938	1.9274	0.5188	0.4938	0.9518
3	1.0769	0.9286	3.0756	0.3251	2.8560	0.3501	0.9835	2.8090
4	1.1038	0.9060	4.1525	0.2408	3.7620	0.2658	1.4691	5.5269
5	1.1314	0.8839	5.2563	0.1902	4.6458	0.2152	1.9506	9.0623
6	1.1597	0.8623	6.3877	0.1565	5.5081	0.1815	2.4280	13.3738
7	1.1887	0.8413	7.5474	0.1325	6.3494	0.1575	2.9013	18.4214
8	1.2184	0.8207	8.7361	0.1145	7.1701	0.1395	3.3704	24.1666
9	1.2489	0.8007	9.9545	0.1005	7.9709	0.1255	3.8355	30.5724
10	1.2801	0.7812	11.2034	0.0893	8.7521	0.1143	4.2965	37.6032
11	1.3121	0.7621	12.4835	0.0801	9.5142	0.1051	4.7534	45.2246
12	1.3449	0.7436	13.7956	0.0725	10.2578	0.0975	5.2062	53.4038
13	1.3785	0.7254	15.1404	0.0660	10.9832	0.0910	5.6549	62.1088
14	1.4130	0.7077	16.5190	0.0605	11.6909	0.0855	6.0995	71.3093
15	1.4483	0.6905	17.9319	0.0558	12.3814	0.0808	6.5401	80.9758
16	1.4845	0.6736	19.3802	0.0516	13.0550	0.0766	6.9766	91.0801
17	1.5216	0.6572	20.8647	0.0479	13.7122	0.0729	7.4091	101.5953
18	1.5597	0.6412	22.3863	0.0447	14.3534	0.0697	7.8375	112.4951
19	1.5987	0.6255	23.9460	0.0418	14.9789	0.0668	8.2619	123.7546
20	1.6386	0.6103	25.5447	0.0391	15.5892	0.0641	8.6823	135.3497
21	1.6796	0.5954	27.1833	0.0368	16.1845	0.0618	9.0986	147.2575
22	1.7216	0.5809	28.8629	0.0346	16.7654	0.0596	9.5110	159.4556
23	1.7646	0.5667	30.5844	0.0327	17.3321	0.0577	9.9193	171.9230
24	1.8087	0.5529	32.3490	0.0309	17.8850	0.0559	10.3237	184.6391
25	1.8539	0.5394	34.1578	0.0293	18.4244	0.0543	10.7241	197.5845
26	1.9003	0.5262	36.0117	0.0278	18.9506	0.0528	11.1205	210.7403
27	1.9478	0.5134	37.9120	0.0264	19.4640	0.0514	11.5130	224.0887
28	1.9965	0.5009	39.8598	0.0251	19.9649	0.0501	11.9015	237.6124
29	2.0464	0.4887	41.8563	0.0239	20.4535	0.0489	12.2861	251.2949
30	2.0976	0.4767	43.9027	0.0228	20.9303	0.0478	12.6668	265.1205
31	2.1500	0.4651	46.0003	0.0217	21.3954	0.0467	13.0436	279.0739
32	2.2038	0.4538	48.1503	0.0208	21.8492	0.0458	13.4166	293.1408
33	2.2589	0.4427	50.3540	0.0199	22.2919	0.0449	13.7856	307.3073
34	2.3153	0.4319	52.6129	0.0190	22.7238	0.0440	14.1508	321.5602
35	2.3732	0.4214	54.9282	0.0182	23.1452	0.0432	14.5122	335.8868
36	2.4325	0.4111	57.3014	0.0175	23.5563	0.0425	14.8697	350.2751
37	2.4933	0.4011	59.7339	0.0167	23.9573	0.0417	15.2235	364.7135
38	2.5557	0.3913	62.2273	0.0161	24.3486	0.0411	15.5734	379.1910
39	2.6196	0.3817	64.7830	0.0154	24.7303	0.0404	15.9196	393.6972
40	2.6851	0.3724	67.4026	0.0148	25.1028	0.0398	16.2620	408.2220
45	3.0379	0.3292	81.5161	0.0123	26.8330	0.0373	17.9185	480.8070
48	3.2715	0.3057	90.8596	0.0110	27.7732	0.0360	18.8665	524.0375
50	3.4371	0.2909	97.4843	0.0103	28.3623	0.0353	19.4839	552.6081
54	3.7939	0.2636	111.7570	0.0089	29.4568	0.0339	20.6724	608.9419
60	4.3998	0.2273	135.9916	0.0074	30.9087	0.0324	22.3518	690.8656
65	4.9780	0.2009	159.1183	0.0063	31.9646	0.0313	23.6600	756.2806
70	5.6321	0.1776	185.2841	0.0054	32.8979	0.0304	24.8881	818.7643
75	6.3722	0.1569	214.8883	0.0047	33.7227	0.0297	26.0393	878.1152
80	7.2096	0.1387	248.3827	0.0040	34.4518	0.0290	27.1167	934.2181
90	9.2289	0.1084	329.1543	0.0030	35.6658	0.0280	29.0629	1036.5499
100	11.8137	0.0846	432.5487	0.0023	36.6141	0.0273	30.7525	1125.9747
120	19.3581	0.0517	734.3260	0.0014	37.9337	0.0264	33.4634	1269.3899

Table A.9 Discrete Compounding: $i= 3\%$

	Single payment		Uniform Series				Gradient series	
	Compound amount factor	Present worth factor	Compound amount factor	Sinking fund factor	Present worth factor	Capital recovery factor	Uniform series factor	Present worth factor
	To find F Given P	To find P given F	To find F given A	To find A given F	To find P given A	To find A given P	To find A given G	To find P given G
n	$F\|P\ i,n$	$P\|F\ i,n$	$F\|A\ i,n$	$A\|F\ i,n$	$P\|A\ i,n$	$A\|P\ i,n$	$A\|G\ i,n$	$P\|G\ i,n$
1	1.0300	0.9709	1.0000	1.0000	0.9709	1.0300	0.0000	0.0000
2	1.0609	0.9426	2.0300	0.4926	1.9135	0.5226	0.4926	0.9426
3	1.0927	0.9151	3.0909	0.3235	2.8286	0.3535	0.9803	2.7729
4	1.1255	0.8885	4.1836	0.2390	3.7171	0.2690	1.4631	5.4383
5	1.1593	0.8626	5.3091	0.1884	4.5797	0.2184	1.9409	8.8888
6	1.1941	0.8375	6.4684	0.1546	5.4172	0.1846	2.4138	13.0762
7	1.2299	0.8131	7.6625	0.1305	6.2303	0.1605	2.8819	17.9547
8	1.2668	0.7894	8.8923	0.1125	7.0197	0.1425	3.3450	23.4806
9	1.3048	0.7664	10.1591	0.0984	7.7861	0.1284	3.8032	29.6119
10	1.3439	0.7441	11.4639	0.0872	8.5302	0.1172	4.2565	36.3088
11	1.3842	0.7224	12.8078	0.0781	9.2526	0.1081	4.7049	43.5330
12	1.4258	0.7014	14.1920	0.0705	9.9540	0.1005	5.1485	51.2482
13	1.4685	0.6810	15.6178	0.0640	10.6350	0.0940	5.5872	59.4196
14	1.5126	0.6611	17.0863	0.0585	11.2961	0.0885	6.0210	68.0141
15	1.5580	0.6419	18.5989	0.0538	11.9379	0.0838	6.4500	77.0002
16	1.6047	0.6232	20.1569	0.0496	12.5611	0.0796	6.8742	86.3477
17	1.6528	0.6050	21.7616	0.0460	13.1661	0.0760	7.2936	96.0280
18	1.7024	0.5874	23.4144	0.0427	13.7535	0.0727	7.7081	106.0137
19	1.7535	0.5703	25.1169	0.0398	14.3238	0.0698	8.1179	116.2788
20	1.8061	0.5537	26.8704	0.0372	14.8775	0.0672	8.5229	126.7987
21	1.8603	0.5375	28.6765	0.0349	15.4150	0.0649	8.9231	137.5496
22	1.9161	0.5219	30.5368	0.0327	15.9369	0.0627	9.3186	148.5094
23	1.9736	0.5067	32.4529	0.0308	16.4436	0.0608	9.7093	159.6566
24	2.0328	0.4919	34.4265	0.0290	16.9355	0.0590	10.0954	170.9711
25	2.0938	0.4776	36.4593	0.0274	17.4131	0.0574	10.4768	182.4336
26	2.1566	0.4637	38.5530	0.0259	17.8768	0.0559	10.8535	194.0260
27	2.2213	0.4502	40.7096	0.0246	18.3270	0.0546	11.2255	205.7309
28	2.2879	0.4371	42.9309	0.0233	18.7641	0.0533	11.5930	217.5320
29	2.3566	0.4243	45.2189	0.0221	19.1885	0.0521	11.9558	229.4137
30	2.4273	0.4120	47.5754	0.0210	19.6004	0.0510	12.3141	241.3613
31	2.5001	0.4000	50.0027	0.0200	20.0004	0.0500	12.6678	253.3609
32	2.5751	0.3883	52.5028	0.0190	20.3888	0.0490	13.0169	265.3993
33	2.6523	0.3770	55.0778	0.0182	20.7658	0.0482	13.3616	277.4642
34	2.7319	0.3660	57.7302	0.0173	21.1318	0.0473	13.7018	289.5437
35	2.8139	0.3554	60.4621	0.0165	21.4872	0.0465	14.0375	301.6267
36	2.8983	0.3450	63.2759	0.0158	21.8323	0.0458	14.3688	313.7028
37	2.9852	0.3350	66.1742	0.0151	22.1672	0.0451	14.6957	325.7622
38	3.0748	0.3252	69.1594	0.0145	22.4925	0.0445	15.0182	337.7956
39	3.1670	0.3158	72.2342	0.0138	22.8082	0.0438	15.3363	349.7942
40	3.2620	0.3066	75.4013	0.0133	23.1148	0.0433	15.6502	361.7499
45	3.7816	0.2644	92.7199	0.0108	24.5187	0.0408	17.1556	420.6325
48	4.1323	0.2420	104.4084	0.0096	25.2667	0.0396	18.0089	455.0255
50	4.3839	0.2281	112.7969	0.0089	25.7298	0.0389	18.5575	477.4803
54	4.9341	0.2027	131.1375	0.0076	26.5777	0.0376	19.6073	521.1157
60	5.8916	0.1697	163.0534	0.0061	27.6756	0.0361	21.0674	583.0526
65	6.8300	0.1464	194.3328	0.0051	28.4529	0.0351	22.1841	631.2010
70	7.9178	0.1263	230.5941	0.0043	29.1234	0.0343	23.2145	676.0869
75	9.1789	0.1089	272.6309	0.0037	29.7018	0.0337	24.1634	717.6978
80	10.6409	0.0940	321.3630	0.0031	30.2008	0.0331	25.0353	756.0865
90	14.3005	0.0699	443.3489	0.0023	31.0024	0.0323	26.5667	823.6302
100	19.2186	0.0520	607.2877	0.0016	31.5989	0.0316	27.8444	879.8540
120	34.7110	0.0288	1123.6996	0.0009	32.3730	0.0309	29.7737	963.8635

Table A.10   Discrete Compounding: $i=$ 4%

	Single payment		Uniform Series				Gradient series	
	Compound amount factor	Present worth factor	Compound amount factor	Sinking fund factor	Present worth factor	Capital recovery factor	Uniform series factor	Present worth factor
	To find F Given P	To find P given F	To find F given A	To find A given F	To find P given A	To find A given P	To find A given G	To find P given G
n	F|P i,n	P|F i,n	F|A i,n	A|F i,n	P|A i,n	A|P i,n	A|G i,n	P|G i,n
1	1.0400	0.9615	1.0000	1.0000	0.9615	1.0400	0.0000	0.0000
2	1.0816	0.9246	2.0400	0.4902	1.8861	0.5302	0.4902	0.9246
3	1.1249	0.8890	3.1216	0.3203	2.7751	0.3603	0.9739	2.7025
4	1.1699	0.8548	4.2465	0.2355	3.6299	0.2755	1.4510	5.2670
5	1.2167	0.8219	5.4163	0.1846	4.4518	0.2246	1.9216	8.5547
6	1.2653	0.7903	6.6330	0.1508	5.2421	0.1908	2.3857	12.5062
7	1.3159	0.7599	7.8983	0.1266	6.0021	0.1666	2.8433	17.0657
8	1.3686	0.7307	9.2142	0.1085	6.7327	0.1485	3.2944	22.1806
9	1.4233	0.7026	10.5828	0.0945	7.4353	0.1345	3.7391	27.8013
10	1.4802	0.6756	12.0061	0.0833	8.1109	0.1233	4.1773	33.8814
11	1.5395	0.6496	13.4864	0.0741	8.7605	0.1141	4.6090	40.3772
12	1.6010	0.6246	15.0258	0.0666	9.3851	0.1066	5.0343	47.2477
13	1.6651	0.6006	16.6268	0.0601	9.9856	0.1001	5.4533	54.4546
14	1.7317	0.5775	18.2919	0.0547	10.5631	0.0947	5.8659	61.9618
15	1.8009	0.5553	20.0236	0.0499	11.1184	0.0899	6.2721	69.7355
16	1.8730	0.5339	21.8245	0.0458	11.6523	0.0858	6.6720	77.7441
17	1.9479	0.5134	23.6975	0.0422	12.1657	0.0822	7.0656	85.9581
18	2.0258	0.4936	25.6454	0.0390	12.6593	0.0790	7.4530	94.3498
19	2.1068	0.4746	27.6712	0.0361	13.1339	0.0761	7.8342	102.8933
20	2.1911	0.4564	29.7781	0.0336	13.5903	0.0736	8.2091	111.5647
21	2.2788	0.4388	31.9692	0.0313	14.0292	0.0713	8.5779	120.3414
22	2.3699	0.4220	34.2480	0.0292	14.4511	0.0692	8.9407	129.2024
23	2.4647	0.4057	36.6179	0.0273	14.8568	0.0673	9.2973	138.1284
24	2.5633	0.3901	39.0826	0.0256	15.2470	0.0656	9.6479	147.1012
25	2.6658	0.3751	41.6459	0.0240	15.6221	0.0640	9.9925	156.1040
26	2.7725	0.3607	44.3117	0.0226	15.9828	0.0626	10.3312	165.1212
27	2.8834	0.3468	47.0842	0.0212	16.3296	0.0612	10.6640	174.1385
28	2.9987	0.3335	49.9676	0.0200	16.6631	0.0600	10.9909	183.1424
29	3.1187	0.3207	52.9663	0.0189	16.9837	0.0589	11.3120	192.1206
30	3.2434	0.3083	56.0849	0.0178	17.2920	0.0578	11.6274	201.0618
31	3.3731	0.2965	59.3283	0.0169	17.5885	0.0569	11.9371	209.9556
32	3.5081	0.2851	62.7015	0.0159	17.8736	0.0559	12.2411	218.7924
33	3.6484	0.2741	66.2095	0.0151	18.1476	0.0551	12.5396	227.5634
34	3.7943	0.2636	69.8579	0.0143	18.4112	0.0543	12.8324	236.2607
35	3.9461	0.2534	73.6522	0.0136	18.6646	0.0536	13.1198	244.8768
36	4.1039	0.2437	77.5983	0.0129	18.9083	0.0529	13.4018	253.4052
37	4.2681	0.2343	81.7022	0.0122	19.1426	0.0522	13.6784	261.8399
38	4.4388	0.2253	85.9703	0.0116	19.3679	0.0516	13.9497	270.1754
39	4.6164	0.2166	90.4091	0.0111	19.5845	0.0511	14.2157	278.4070
40	4.8010	0.2083	95.0255	0.0105	19.7928	0.0505	14.4765	286.5303
45	5.8412	0.1712	121.0294	0.0083	20.7200	0.0483	15.7047	325.4028
48	6.5705	0.1522	139.2632	0.0072	21.1951	0.0472	16.3832	347.2446
50	7.1067	0.1407	152.6671	0.0066	21.4822	0.0466	16.8122	361.1638
54	8.3138	0.1203	182.8454	0.0055	21.9930	0.0455	17.6167	387.4436
60	10.5196	0.0951	237.9907	0.0042	22.6235	0.0442	18.6972	422.9966
65	12.7987	0.0781	294.9684	0.0034	23.0467	0.0434	19.4909	449.2014
70	15.5716	0.0642	364.2905	0.0027	23.3945	0.0427	20.1961	472.4789
75	18.9453	0.0528	448.6314	0.0022	23.6804	0.0422	20.8206	493.0408
80	23.0498	0.0434	551.2450	0.0018	23.9154	0.0418	21.3718	511.1161
90	34.1193	0.0293	827.9833	0.0012	24.2673	0.0412	22.2826	540.7369
100	50.5049	0.0198	1237.6237	0.0008	24.5050	0.0408	22.9800	563.1249
120	110.6626	0.0090	2741.5640	0.0004	24.7741	0.0404	23.9057	592.2428

Table A.11   Discrete Compounding: $i = 5\%$

	Single payment		Uniform Series				Gradient series	
	Compound amount factor	Present worth factor	Compound amount factor	Sinking fund factor	Present worth factor	Capital recovery factor	Uniform series factor	Present worth factor
	To find F Given P	To find P given F	To find F given A	To find A given F	To find P given A	To find A given P	To find A given G	To find P given G
n	$F\|P\ i,n$	$P\|F\ i,n$	$F\|A\ i,n$	$A\|F\ i,n$	$P\|A\ i,n$	$A\|P\ i,n$	$A\|G\ i,n$	$P\|G\ i,n$
1	1.0500	0.9524	1.0000	1.0000	0.9524	1.0500	0.0000	0.0000
2	1.1025	0.9070	2.0500	0.4878	1.8594	0.5378	0.4878	0.9070
3	1.1576	0.8638	3.1525	0.3172	2.7232	0.3672	0.9675	2.6347
4	1.2155	0.8227	4.3101	0.2320	3.5460	0.2820	1.4391	5.1028
5	1.2763	0.7835	5.5256	0.1810	4.3295	0.2310	1.9025	8.2369
6	1.3401	0.7462	6.8019	0.1470	5.0757	0.1970	2.3579	11.9680
7	1.4071	0.7107	8.1420	0.1228	5.7864	0.1728	2.8052	16.2321
8	1.4775	0.6768	9.5491	0.1047	6.4632	0.1547	3.2445	20.9700
9	1.5513	0.6446	11.0266	0.0907	7.1078	0.1407	3.6758	26.1268
10	1.6289	0.6139	12.5779	0.0795	7.7217	0.1295	4.0991	31.6520
11	1.7103	0.5847	14.2068	0.0704	8.3064	0.1204	4.5144	37.4988
12	1.7959	0.5568	15.9171	0.0628	8.8633	0.1128	4.9219	43.6241
13	1.8856	0.5303	17.7130	0.0565	9.3936	0.1065	5.3215	49.9879
14	1.9799	0.5051	19.5986	0.0510	9.8986	0.1010	5.7133	56.5538
15	2.0789	0.4810	21.5786	0.0463	10.3797	0.0963	6.0973	63.2880
16	2.1829	0.4581	23.6575	0.0423	10.8378	0.0923	6.4736	70.1597
17	2.2920	0.4363	25.8404	0.0387	11.2741	0.0887	6.8423	77.1405
18	2.4066	0.4155	28.1324	0.0355	11.6896	0.0855	7.2034	84.2043
19	2.5270	0.3957	30.5390	0.0327	12.0853	0.0827	7.5569	91.3275
20	2.6533	0.3769	33.0660	0.0302	12.4622	0.0802	7.9030	98.4884
21	2.7860	0.3589	35.7193	0.0280	12.8212	0.0780	8.2416	105.6673
22	2.9253	0.3418	38.5052	0.0260	13.1630	0.0760	8.5730	112.8461
23	3.0715	0.3256	41.4305	0.0241	13.4886	0.0741	8.8971	120.0087
24	3.2251	0.3101	44.5020	0.0225	13.7986	0.0725	9.2140	127.1402
25	3.3864	0.2953	47.7271	0.0210	14.0939	0.0710	9.5238	134.2275
26	3.5557	0.2812	51.1135	0.0196	14.3752	0.0696	9.8266	141.2585
27	3.7335	0.2678	54.6691	0.0183	14.6430	0.0683	10.1224	148.2226
28	3.9201	0.2551	58.4026	0.0171	14.8981	0.0671	10.4114	155.1101
29	4.1161	0.2429	62.3227	0.0160	15.1411	0.0660	10.6936	161.9126
30	4.3219	0.2314	66.4388	0.0151	15.3725	0.0651	10.9691	168.6226
31	4.5380	0.2204	70.7608	0.0141	15.5928	0.0641	11.2381	175.2333
32	4.7649	0.2099	75.2988	0.0133	15.8027	0.0633	11.5005	181.7392
33	5.0032	0.1999	80.0638	0.0125	16.0025	0.0625	11.7566	188.1351
34	5.2533	0.1904	85.0670	0.0118	16.1929	0.0618	12.0063	194.4168
35	5.5160	0.1813	90.3203	0.0111	16.3742	0.0611	12.2498	200.5807
36	5.7918	0.1727	95.8363	0.0104	16.5469	0.0604	12.4872	206.6237
37	6.0814	0.1644	101.6281	0.0098	16.7113	0.0598	12.7186	212.5434
38	6.3855	0.1566	107.7095	0.0093	16.8679	0.0593	12.9440	218.3378
39	6.7048	0.1491	114.0950	0.0088	17.0170	0.0588	13.1636	224.0054
40	7.0400	0.1420	120.7998	0.0083	17.1591	0.0583	13.3775	229.5452
45	8.9850	0.1113	159.7002	0.0063	17.7741	0.0563	14.3644	255.3145
48	10.4013	0.0961	188.0254	0.0053	18.0772	0.0553	14.8943	269.2467
50	11.4674	0.0872	209.3480	0.0048	18.2559	0.0548	15.2233	277.9148
54	13.9387	0.0717	258.7739	0.0039	18.5651	0.0539	15.8265	293.8208
60	18.6792	0.0535	353.5837	0.0028	18.9293	0.0528	16.6062	314.3432
65	23.8399	0.0419	456.7980	0.0022	19.1611	0.0522	17.1541	328.6910
70	30.4264	0.0329	588.5285	0.0017	19.3427	0.0517	17.6212	340.8409
75	38.8327	0.0258	756.6537	0.0013	19.4850	0.0513	18.0176	351.0721
80	49.5614	0.0202	971.2288	0.0010	19.5965	0.0510	18.3526	359.6460
90	80.7304	0.0124	1594.6073	0.0006	19.7523	0.0506	18.8712	372.7488
100	131.5013	0.0076	2610.0252	0.0004	19.8479	0.0504	19.2337	381.7492
120	348.9120	0.0029	6958.2397	0.0001	19.9427	0.0501	19.6551	391.9751

Table A.12 Discrete Compounding: $i = 6\%$

	Single payment		Uniform Series				Gradient series	
	Compound amount factor	Present worth factor	Compound amount factor	Sinking fund factor	Present worth factor	Capital recovery factor	Uniform series factor	Present worth factor
	To find F Given P	To find P given F	To find F given A	To find A given F	To find P given A	To find A given P	To find A given G	To find P given G
n	$F\|P\ i,n$	$P\|F\ i,n$	$F\|A\ i,n$	$A\|F\ i,n$	$P\|A\ i,n$	$A\|P\ i,n$	$A\|G\ i,n$	$P\|G\ i,n$
1	1.0600	0.9434	1.0000	1.0000	0.9434	1.0600	0.0000	0.0000
2	1.1236	0.8900	2.0600	0.4854	1.8334	0.5454	0.4854	0.8900
3	1.1910	0.8396	3.1836	0.3141	2.6730	0.3741	0.9612	2.5692
4	1.2625	0.7921	4.3746	0.2286	3.4651	0.2886	1.4272	4.9455
5	1.3382	0.7473	5.6371	0.1774	4.2124	0.2374	1.8836	7.9345
6	1.4185	0.7050	6.9753	0.1434	4.9173	0.2034	2.3304	11.4594
7	1.5036	0.6651	8.3938	0.1191	5.5824	0.1791	2.7676	15.4497
8	1.5938	0.6274	9.8975	0.1010	6.2098	0.1610	3.1952	19.8416
9	1.6895	0.5919	11.4913	0.0870	6.8017	0.1470	3.6133	24.5768
10	1.7908	0.5584	13.1808	0.0759	7.3601	0.1359	4.0220	29.6023
11	1.8983	0.5268	14.9716	0.0668	7.8869	0.1268	4.4213	34.8702
12	2.0122	0.4970	16.8699	0.0593	8.3838	0.1193	4.8113	40.3369
13	2.1329	0.4688	18.8821	0.0530	8.8527	0.1130	5.1920	45.9629
14	2.2609	0.4423	21.0151	0.0476	9.2950	0.1076	5.5635	51.7128
15	2.3966	0.4173	23.2760	0.0430	9.7122	0.1030	5.9260	57.5546
16	2.5404	0.3936	25.6725	0.0390	10.1059	0.0990	6.2794	63.4592
17	2.6928	0.3714	28.2129	0.0354	10.4773	0.0954	6.6240	69.4011
18	2.8543	0.3503	30.9057	0.0324	10.8276	0.0924	6.9597	75.3569
19	3.0256	0.3305	33.7600	0.0296	11.1581	0.0896	7.2867	81.3062
20	3.2071	0.3118	36.7856	0.0272	11.4699	0.0872	7.6051	87.2304
21	3.3996	0.2942	39.9927	0.0250	11.7641	0.0850	7.9151	93.1136
22	3.6035	0.2775	43.3923	0.0230	12.0416	0.0830	8.2166	98.9412
23	3.8197	0.2618	46.9958	0.0213	12.3034	0.0813	8.5099	104.7007
24	4.0489	0.2470	50.8156	0.0197	12.5504	0.0797	8.7951	110.3812
25	4.2919	0.2330	54.8645	0.0182	12.7834	0.0782	9.0722	115.9732
26	4.5494	0.2198	59.1564	0.0169	13.0032	0.0769	9.3414	121.4684
27	4.8223	0.2074	63.7058	0.0157	13.2105	0.0757	9.6029	126.8600
28	5.1117	0.1956	68.5281	0.0146	13.4062	0.0746	9.8568	132.1420
29	5.4184	0.1846	73.6398	0.0136	13.5907	0.0736	10.1032	137.3096
30	5.7435	0.1741	79.0582	0.0126	13.7648	0.0726	10.3422	142.3588
31	6.0881	0.1643	84.8017	0.0118	13.9291	0.0718	10.5740	147.2864
32	6.4534	0.1550	90.8898	0.0110	14.0840	0.0710	10.7988	152.0901
33	6.8406	0.1462	97.3432	0.0103	14.2302	0.0703	11.0166	156.7681
34	7.2510	0.1379	104.1838	0.0096	14.3681	0.0696	11.2276	161.3192
35	7.6861	0.1301	111.4348	0.0090	14.4982	0.0690	11.4319	165.7427
36	8.1473	0.1227	119.1209	0.0084	14.6210	0.0684	11.6298	170.0387
37	8.6361	0.1158	127.2681	0.0079	14.7368	0.0679	11.8213	174.2072
38	9.1543	0.1092	135.9042	0.0074	14.8460	0.0674	12.0065	178.2490
39	9.7035	0.1031	145.0585	0.0069	14.9491	0.0669	12.1857	182.1652
40	10.2857	0.0972	154.7620	0.0065	15.0463	0.0665	12.3590	185.9568
45	13.7646	0.0727	212.7435	0.0047	15.4558	0.0647	13.1413	203.1096
48	16.3939	0.0610	256.5645	0.0039	15.6500	0.0639	13.5485	212.0351
50	18.4202	0.0543	290.3359	0.0034	15.7619	0.0634	13.7964	217.4574
54	23.2550	0.0430	370.9170	0.0027	15.9500	0.0627	14.2402	227.1316
60	32.9877	0.0303	533.1282	0.0019	16.1614	0.0619	14.7909	239.0428
65	44.1450	0.0227	719.0829	0.0014	16.2891	0.0614	15.1601	246.9450
70	59.0759	0.0169	967.9322	0.0010	16.3845	0.0610	15.4613	253.3271
75	79.0569	0.0126	1300.9487	0.0008	16.4558	0.0608	15.7058	258.4527
80	105.7960	0.0095	1746.5999	0.0006	16.5091	0.0606	15.9033	262.5493
90	189.4645	0.0053	3141.0752	0.0003	16.5787	0.0603	16.1891	268.3946
100	339.3021	0.0029	5638.3681	0.0002	16.6175	0.0602	16.3711	272.0471
120	1088.1877	0.0009	18119.7958	0.0001	16.6514	0.0601	16.5563	275.6846

Table A.13 Discrete Compounding: $i = 7\%$

	Single payment		Uniform Series				Gradient series	
	Compound amount factor	Present worth factor	Compound amount factor	Sinking fund factor	Present worth factor	Capital recovery factor	Uniform series factor	Present worth factor
	To find F Given P	To find P given F	To find F given A	To find A given F	To find P given A	To find A given P	To find A given G	To find P given G
n	F|P i,n	P|F i,n	F|A i,n	A|F i,n	P|A i,n	A|P i,n	A|G i,n	P|G i,n
1	1.0700	0.9346	1.0000	1.0000	0.9346	1.0700	0.0000	0.0000
2	1.1449	0.8734	2.0700	0.4831	1.8080	0.5531	0.4831	0.8734
3	1.2250	0.8163	3.2149	0.3111	2.6243	0.3811	0.9549	2.5060
4	1.3108	0.7629	4.4399	0.2252	3.3872	0.2952	1.4155	4.7947
5	1.4026	0.7130	5.7507	0.1739	4.1002	0.2439	1.8650	7.6467
6	1.5007	0.6663	7.1533	0.1398	4.7665	0.2098	2.3032	10.9784
7	1.6058	0.6227	8.6540	0.1156	5.3893	0.1856	2.7304	14.7149
8	1.7182	0.5820	10.2598	0.0975	5.9713	0.1675	3.1465	18.7889
9	1.8385	0.5439	11.9780	0.0835	6.5152	0.1535	3.5517	23.1404
10	1.9672	0.5083	13.8164	0.0724	7.0236	0.1424	3.9461	27.7156
11	2.1049	0.4751	15.7836	0.0634	7.4987	0.1334	4.3296	32.4665
12	2.2522	0.4440	17.8885	0.0559	7.9427	0.1259	4.7025	37.3506
13	2.4098	0.4150	20.1406	0.0497	8.3577	0.1197	5.0648	42.3302
14	2.5785	0.3878	22.5505	0.0443	8.7455	0.1143	5.4167	47.3718
15	2.7590	0.3624	25.1290	0.0398	9.1079	0.1098	5.7583	52.4461
16	2.9522	0.3387	27.8881	0.0359	9.4466	0.1059	6.0897	57.5271
17	3.1588	0.3166	30.8402	0.0324	9.7632	0.1024	6.4110	62.5923
18	3.3799	0.2959	33.9990	0.0294	10.0591	0.0994	6.7225	67.6219
19	3.6165	0.2765	37.3790	0.0268	10.3356	0.0968	7.0242	72.5991
20	3.8697	0.2584	40.9955	0.0244	10.5940	0.0944	7.3163	77.5091
21	4.1406	0.2415	44.8652	0.0223	10.8355	0.0923	7.5990	82.3393
22	4.4304	0.2257	49.0057	0.0204	11.0612	0.0904	7.8725	87.0793
23	4.7405	0.2109	53.4361	0.0187	11.2722	0.0887	8.1369	91.7201
24	5.0724	0.1971	58.1767	0.0172	11.4693	0.0872	8.3923	96.2545
25	5.4274	0.1842	63.2490	0.0158	11.6536	0.0858	8.6391	100.6765
26	5.8074	0.1722	68.6765	0.0146	11.8258	0.0846	8.8773	104.9814
27	6.2139	0.1609	74.4838	0.0134	11.9867	0.0834	9.1072	109.1656
28	6.6488	0.1504	80.6977	0.0124	12.1371	0.0824	9.3289	113.2264
29	7.1143	0.1406	87.3465	0.0114	12.2777	0.0814	9.5427	117.1622
30	7.6123	0.1314	94.4608	0.0106	12.4090	0.0806	9.7487	120.9718
31	8.1451	0.1228	102.0730	0.0098	12.5318	0.0798	9.9471	124.6550
32	8.7153	0.1147	110.2182	0.0091	12.6466	0.0791	10.1381	128.2120
33	9.3253	0.1072	118.9334	0.0084	12.7538	0.0784	10.3219	131.6435
34	9.9781	0.1002	128.2588	0.0078	12.8540	0.0778	10.4987	134.9507
35	10.6766	0.0937	138.2369	0.0072	12.9477	0.0772	10.6687	138.1353
36	11.4239	0.0875	148.9135	0.0067	13.0352	0.0767	10.8321	141.1990
37	12.2236	0.0818	160.3374	0.0062	13.1170	0.0762	10.9891	144.1441
38	13.0793	0.0765	172.5610	0.0058	13.1935	0.0758	11.1398	146.9730
39	13.9948	0.0715	185.6403	0.0054	13.2649	0.0754	11.2845	149.6883
40	14.9745	0.0668	199.6351	0.0050	13.3317	0.0750	11.4233	152.2928
45	21.0025	0.0476	285.7493	0.0035	13.6055	0.0735	12.0360	163.7559
48	25.7289	0.0389	353.2701	0.0028	13.7305	0.0728	12.3447	169.4981
50	29.4570	0.0339	406.5289	0.0025	13.8007	0.0725	12.5287	172.9051
54	38.6122	0.0259	537.3164	0.0019	13.9157	0.0719	12.8500	178.8173
60	57.9464	0.0173	813.5204	0.0012	14.0392	0.0712	13.2321	185.7677
65	81.2729	0.0123	1146.7552	0.0009	14.1099	0.0709	13.4760	190.1452
70	113.9894	0.0088	1614.1342	0.0006	14.1604	0.0706	13.6662	193.5185
75	159.8760	0.0063	2269.6574	0.0004	14.1964	0.0704	13.8136	196.1035
80	224.2344	0.0045	3189.0627	0.0003	14.2220	0.0703	13.9273	198.0748
90	441.1030	0.0023	6287.1854	0.0002	14.2533	0.0702	14.0812	200.7042
100	867.7163	0.0012	12381.6618	0.0001	14.2693	0.0701	14.1703	202.2001
120	3357.7884	0.0003	47954.1198	0.0000	14.2815	0.0700	14.2500	203.5103

Table A.14 Discrete Compounding: $i = 8\%$

	Single payment		Uniform Series				Gradient series	
	Compound amount factor	Present worth factor	Compound amount factor	Sinking fund factor	Present worth factor	Capital recovery factor	Uniform series factor	Present worth factor
	To find F Given P	To find P given F	To find F given A	To find A given F	To find P given A	To find A given P	To find A given G	To find P given G
n	$F\|P\ i,n$	$P\|F\ i,n$	$F\|A\ i,n$	$A\|F\ i,n$	$P\|A\ i,n$	$A\|P\ i,n$	$A\|G\ i,n$	$P\|G\ i,n$
1	1.0800	0.9259	1.0000	1.0000	0.9259	1.0800	0.0000	0.0000
2	1.1664	0.8573	2.0800	0.4808	1.7833	0.5608	0.4808	0.8573
3	1.2597	0.7938	3.2464	0.3080	2.5771	0.3880	0.9487	2.4450
4	1.3605	0.7350	4.5061	0.2219	3.3121	0.3019	1.4040	4.6501
5	1.4693	0.6806	5.8666	0.1705	3.9927	0.2505	1.8465	7.3724
6	1.5869	0.6302	7.3359	0.1363	4.6229	0.2163	2.2763	10.5233
7	1.7138	0.5835	8.9228	0.1121	5.2064	0.1921	2.6937	14.0242
8	1.8509	0.5403	10.6366	0.0940	5.7466	0.1740	3.0985	17.8061
9	1.9990	0.5002	12.4876	0.0801	6.2469	0.1601	3.4910	21.8081
10	2.1589	0.4632	14.4866	0.0690	6.7101	0.1490	3.8713	25.9768
11	2.3316	0.4289	16.6455	0.0601	7.1390	0.1401	4.2395	30.2657
12	2.5182	0.3971	18.9771	0.0527	7.5361	0.1327	4.5957	34.6339
13	2.7196	0.3677	21.4953	0.0465	7.9038	0.1265	4.9402	39.0463
14	2.9372	0.3405	24.2149	0.0413	8.2442	0.1213	5.2731	43.4723
15	3.1722	0.3152	27.1521	0.0368	8.5595	0.1168	5.5945	47.8857
16	3.4259	0.2919	30.3243	0.0330	8.8514	0.1130	5.9046	52.2640
17	3.7000	0.2703	33.7502	0.0296	9.1216	0.1096	6.2037	56.5883
18	3.9960	0.2502	37.4502	0.0267	9.3719	0.1067	6.4920	60.8426
19	4.3157	0.2317	41.4463	0.0241	9.6036	0.1041	6.7697	65.0134
20	4.6610	0.2145	45.7620	0.0219	9.8181	0.1019	7.0369	69.0898
21	5.0338	0.1987	50.4229	0.0198	10.0168	0.0998	7.2940	73.0629
22	5.4365	0.1839	55.4568	0.0180	10.2007	0.0980	7.5412	76.9257
23	5.8715	0.1703	60.8933	0.0164	10.3711	0.0964	7.7786	80.6726
24	6.3412	0.1577	66.7648	0.0150	10.5288	0.0950	8.0066	84.2997
25	6.8485	0.1460	73.1059	0.0137	10.6748	0.0937	8.2254	87.8041
26	7.3964	0.1352	79.9544	0.0125	10.8100	0.0925	8.4352	91.1842
27	7.9881	0.1252	87.3508	0.0114	10.9352	0.0914	8.6363	94.4390
28	8.6271	0.1159	95.3388	0.0105	11.0511	0.0905	8.8289	97.5687
29	9.3173	0.1073	103.9659	0.0096	11.1584	0.0896	9.0133	100.5738
30	10.0627	0.0994	113.2832	0.0088	11.2578	0.0888	9.1897	103.4558
31	10.8677	0.0920	123.3459	0.0081	11.3498	0.0881	9.3584	106.2163
32	11.7371	0.0852	134.2135	0.0075	11.4350	0.0875	9.5197	108.8575
33	12.6760	0.0789	145.9506	0.0069	11.5139	0.0869	9.6737	111.3819
34	13.6901	0.0730	158.6267	0.0063	11.5869	0.0863	9.8208	113.7924
35	14.7853	0.0676	172.3168	0.0058	11.6546	0.0858	9.9611	116.0920
36	15.9682	0.0626	187.1021	0.0053	11.7172	0.0853	10.0949	118.2839
37	17.2456	0.0580	203.0703	0.0049	11.7752	0.0849	10.2225	120.3713
38	18.6253	0.0537	220.3159	0.0045	11.8289	0.0845	10.3440	122.3579
39	20.1153	0.0497	238.9412	0.0042	11.8786	0.0842	10.4597	124.2470
40	21.7245	0.0460	259.0565	0.0039	11.9246	0.0839	10.5699	126.0422
45	31.9204	0.0313	386.5056	0.0026	12.1084	0.0826	11.0447	133.7331
48	40.2106	0.0249	490.1322	0.0020	12.1891	0.0820	11.2758	137.4428
50	46.9016	0.0213	573.7702	0.0017	12.2335	0.0817	11.4107	139.5928
54	63.8091	0.0157	785.1141	0.0013	12.3041	0.0813	11.6403	143.2229
60	101.2571	0.0099	1253.2133	0.0008	12.3766	0.0808	11.9015	147.3000
65	148.7798	0.0067	1847.2481	0.0005	12.4160	0.0805	12.0602	149.7387
70	218.6064	0.0046	2720.0801	0.0004	12.4428	0.0804	12.1783	151.5326
75	321.2045	0.0031	4002.5566	0.0002	12.4611	0.0802	12.2658	152.8448
80	471.9548	0.0021	5886.9354	0.0002	12.4735	0.0802	12.3301	153.8001
90	1018.9151	0.0010	12723.9386	0.0001	12.4877	0.0801	12.4116	154.9925
100	2199.7613	0.0005	27484.5157	0.0000	12.4943	0.0800	12.4545	155.6107
120	10252.9929	0.0001	128149.9118	0.0000	12.4988	0.0800	12.4883	156.0885

Table A.15 Discrete Compounding: $i = 9\%$

	Single payment		Uniform Series				Gradient series	
	Compound amount factor	Present worth factor	Compound amount factor	Sinking fund factor	Present worth factor	Capital recovery factor	Uniform series factor	Present worth factor
	To find F Given P	To find P given F	To find F given A	To find A given F	To find P given A	To find A given P	To find A given G	To find P given G
n	$F\|P\ i,n$	$P\|F\ i,n$	$F\|A\ i,n$	$A\|F\ i,n$	$P\|A\ i,n$	$A\|P\ i,n$	$A\|G\ i,n$	$P\|G\ i,n$
1	1.0900	0.9174	1.0000	1.0000	0.9174	1.0900	0.0000	0.0000
2	1.1881	0.8417	2.0900	0.4785	1.7591	0.5685	0.4785	0.8417
3	1.2950	0.7722	3.2781	0.3051	2.5313	0.3951	0.9426	2.3860
4	1.4116	0.7084	4.5731	0.2187	3.2397	0.3087	1.3925	4.5113
5	1.5386	0.6499	5.9847	0.1671	3.8897	0.2571	1.8282	7.1110
6	1.6771	0.5963	7.5233	0.1329	4.4859	0.2229	2.2498	10.0924
7	1.8280	0.5470	9.2004	0.1087	5.0330	0.1987	2.6574	13.3746
8	1.9926	0.5019	11.0285	0.0907	5.5348	0.1807	3.0512	16.8877
9	2.1719	0.4604	13.0210	0.0768	5.9952	0.1668	3.4312	20.5711
10	2.3674	0.4224	15.1929	0.0658	6.4177	0.1558	3.7978	24.3728
11	2.5804	0.3875	17.5603	0.0569	6.8052	0.1469	4.1510	28.2481
12	2.8127	0.3555	20.1407	0.0497	7.1607	0.1397	4.4910	32.1590
13	3.0658	0.3262	22.9534	0.0436	7.4869	0.1336	4.8182	36.0731
14	3.3417	0.2992	26.0192	0.0384	7.7862	0.1284	5.1326	39.9633
15	3.6425	0.2745	29.3609	0.0341	8.0607	0.1241	5.4346	43.8069
16	3.9703	0.2519	33.0034	0.0303	8.3126	0.1203	5.7245	47.5849
17	4.3276	0.2311	36.9737	0.0270	8.5436	0.1170	6.0024	51.2821
18	4.7171	0.2120	41.3013	0.0242	8.7556	0.1142	6.2687	54.8860
19	5.1417	0.1945	46.0185	0.0217	8.9501	0.1117	6.5236	58.3868
20	5.6044	0.1784	51.1601	0.0195	9.1285	0.1095	6.7674	61.7770
21	6.1088	0.1637	56.7645	0.0176	9.2922	0.1076	7.0006	65.0509
22	6.6586	0.1502	62.8733	0.0159	9.4424	0.1059	7.2232	68.2048
23	7.2579	0.1378	69.5319	0.0144	9.5802	0.1044	7.4357	71.2359
24	7.9111	0.1264	76.7898	0.0130	9.7066	0.1030	7.6384	74.1433
25	8.6231	0.1160	84.7009	0.0118	9.8226	0.1018	7.8316	76.9265
26	9.3992	0.1064	93.3240	0.0107	9.9290	0.1007	8.0156	79.5863
27	10.2451	0.0976	102.7231	0.0097	10.0266	0.0997	8.1906	82.1241
28	11.1671	0.0895	112.9682	0.0089	10.1161	0.0989	8.3571	84.5419
29	12.1722	0.0822	124.1354	0.0081	10.1983	0.0981	8.5154	86.8422
30	13.2677	0.0754	136.3075	0.0073	10.2737	0.0973	8.6657	89.0280
31	14.4618	0.0691	149.5752	0.0067	10.3428	0.0967	8.8083	91.1024
32	15.7633	0.0634	164.0370	0.0061	10.4062	0.0961	8.9436	93.0690
33	17.1820	0.0582	179.8003	0.0056	10.4644	0.0956	9.0718	94.9314
34	18.7284	0.0534	196.9823	0.0051	10.5178	0.0951	9.1933	96.6935
35	20.4140	0.0490	215.7108	0.0046	10.5668	0.0946	9.3083	98.3590
36	22.2512	0.0449	236.1247	0.0042	10.6118	0.0942	9.4171	99.9319
37	24.2538	0.0412	258.3759	0.0039	10.6530	0.0939	9.5200	101.4162
38	26.4367	0.0378	282.6298	0.0035	10.6908	0.0935	9.6172	102.8158
39	28.8160	0.0347	309.0665	0.0032	10.7255	0.0932	9.7090	104.1345
40	31.4094	0.0318	337.8824	0.0030	10.7574	0.0930	9.7957	105.3762
45	48.3273	0.0207	525.8587	0.0019	10.8812	0.0919	10.1603	110.5561
48	62.5852	0.0160	684.2804	0.0015	10.9336	0.0915	10.3317	112.9625
50	74.3575	0.0134	815.0836	0.0012	10.9617	0.0912	10.4295	114.3251
54	104.9617	0.0095	1155.1301	0.0009	11.0053	0.0909	10.5917	116.5642
60	176.0313	0.0057	1944.7921	0.0005	11.0480	0.0905	10.7683	118.9683
65	270.8460	0.0037	2998.2885	0.0003	11.0701	0.0903	10.8702	120.3344
70	416.7301	0.0024	4619.2232	0.0002	11.0844	0.0902	10.9427	121.2942
75	641.1909	0.0016	7113.2321	0.0001	11.0938	0.0901	10.9940	121.9646
80	986.5517	0.0010	10950.5741	0.0001	11.0998	0.0901	11.0299	122.4306
90	2335.5266	0.0004	25939.1842	0.0000	11.1064	0.0900	11.0726	122.9758
100	5529.0408	0.0002	61422.6755	0.0000	11.1091	0.0900	11.0930	123.2335
120	30987.0157	0.0000	344289.0639	0.0000	11.1108	0.0900	11.1072	123.4098

Table A.16 Discrete Compounding: $i = 10\%$

	Single payment		Uniform Series				Gradient series	
	Compound amount factor	Present worth factor	Compound amount factor	Sinking fund factor	Present worth factor	Capital recovery factor	Uniform series factor	Present worth factor
	To find F Given P	To find P given F	To find F given A	To find A given F	To find P given A	To find A given P	To find A given G	To find P given G
n	$F\|P\ i,n$	$P\|F\ i,n$	$F\|A\ i,n$	$A\|F\ i,n$	$P\|A\ i,n$	$A\|P\ i,n$	$A\|G\ i,n$	$P\|G\ i,n$
1	1.1000	0.9091	1.0000	1.0000	0.9091	1.1000	0.0000	0.0000
2	1.2100	0.8264	2.1000	0.4762	1.7355	0.5762	0.4762	0.8264
3	1.3310	0.7513	3.3100	0.3021	2.4869	0.4021	0.9366	2.3291
4	1.4641	0.6830	4.6410	0.2155	3.1699	0.3155	1.3812	4.3781
5	1.6105	0.6209	6.1051	0.1638	3.7908	0.2638	1.8101	6.8618
6	1.7716	0.5645	7.7156	0.1296	4.3553	0.2296	2.2236	9.6842
7	1.9487	0.5132	9.4872	0.1054	4.8684	0.2054	2.6216	12.7631
8	2.1436	0.4665	11.4359	0.0874	5.3349	0.1874	3.0045	16.0287
9	2.3579	0.4241	13.5795	0.0736	5.7590	0.1736	3.3724	19.4215
10	2.5937	0.3855	15.9374	0.0627	6.1446	0.1627	3.7255	22.8913
11	2.8531	0.3505	18.5312	0.0540	6.4951	0.1540	4.0641	26.3963
12	3.1384	0.3186	21.3843	0.0468	6.8137	0.1468	4.3884	29.9012
13	3.4523	0.2897	24.5227	0.0408	7.1034	0.1408	4.6988	33.3772
14	3.7975	0.2633	27.9750	0.0357	7.3667	0.1357	4.9955	36.8005
15	4.1772	0.2394	31.7725	0.0315	7.6061	0.1315	5.2789	40.1520
16	4.5950	0.2176	35.9497	0.0278	7.8237	0.1278	5.5493	43.4164
17	5.0545	0.1978	40.5447	0.0247	8.0216	0.1247	5.8071	46.5819
18	5.5599	0.1799	45.5992	0.0219	8.2014	0.1219	6.0526	49.6395
19	6.1159	0.1635	51.1591	0.0195	8.3649	0.1195	6.2861	52.5827
20	6.7275	0.1486	57.2750	0.0175	8.5136	0.1175	6.5081	55.4069
21	7.4002	0.1351	64.0025	0.0156	8.6487	0.1156	6.7189	58.1095
22	8.1403	0.1228	71.4027	0.0140	8.7715	0.1140	6.9189	60.6893
23	8.9543	0.1117	79.5430	0.0126	8.8832	0.1126	7.1085	63.1462
24	9.8497	0.1015	88.4973	0.0113	8.9847	0.1113	7.2881	65.4813
25	10.8347	0.0923	98.3471	0.0102	9.0770	0.1102	7.4580	67.6964
26	11.9182	0.0839	109.1818	0.0092	9.1609	0.1092	7.6186	69.7940
27	13.1100	0.0763	121.0999	0.0083	9.2372	0.1083	7.7704	71.7773
28	14.4210	0.0693	134.2099	0.0075	9.3066	0.1075	7.9137	73.6495
29	15.8631	0.0630	148.6309	0.0067	9.3696	0.1067	8.0489	75.4146
30	17.4494	0.0573	164.4940	0.0061	9.4269	0.1061	8.1762	77.0766
31	19.1943	0.0521	181.9434	0.0055	9.4790	0.1055	8.2962	78.6395
32	21.1138	0.0474	201.1378	0.0050	9.5264	0.1050	8.4091	80.1078
33	23.2252	0.0431	222.2515	0.0045	9.5694	0.1045	8.5152	81.4856
34	25.5477	0.0391	245.4767	0.0041	9.6086	0.1041	8.6149	82.7773
35	28.1024	0.0356	271.0244	0.0037	9.6442	0.1037	8.7086	83.9872
36	30.9127	0.0323	299.1268	0.0033	9.6765	0.1033	8.7965	85.1194
37	34.0039	0.0294	330.0395	0.0030	9.7059	0.1030	8.8789	86.1781
38	37.4043	0.0267	364.0434	0.0027	9.7327	0.1027	8.9562	87.1673
39	41.1448	0.0243	401.4478	0.0025	9.7570	0.1025	9.0285	88.0908
40	45.2593	0.0221	442.5926	0.0023	9.7791	0.1023	9.0962	88.9525
45	72.8905	0.0137	718.9048	0.0014	9.8628	0.1014	9.3740	92.4544
48	97.0172	0.0103	960.1723	0.0010	9.8969	0.1010	9.5001	94.0217
50	117.3909	0.0085	1163.9085	0.0009	9.9148	0.1009	9.5704	94.8889
54	171.8719	0.0058	1708.7195	0.0006	9.9418	0.1006	9.6840	96.2763
60	304.4816	0.0033	3034.8164	0.0003	9.9672	0.1003	9.8023	97.7010
65	490.3707	0.0020	4893.7073	0.0002	9.9796	0.1002	9.8672	98.4705
70	789.7470	0.0013	7887.4696	0.0001	9.9873	0.1001	9.9113	98.9870
75	1271.8954	0.0008	12708.9537	0.0001	9.9921	0.1001	9.9410	99.3317
80	2048.4002	0.0005	20474.0021	0.0000	9.9951	0.1000	9.9609	99.5606
90	5313.0226	0.0002	53120.2261	0.0000	9.9981	0.1000	9.9831	99.8118
100	13780.6123	0.0001	137796.1234	0.0000	9.9993	0.1000	9.9927	99.9202
120	92709.0688	0.0000	927080.6882	0.0000	9.9999	0.1000	9.9987	99.9860

Table A.17   Discrete Compounding: $i = 12\%$

	Single payment		Uniform Series				Gradient series	
	Compound amount factor	Present worth factor	Compound amount factor	Sinking fund factor	Present worth factor	Capital recovery factor	Uniform series factor	Present worth factor
	To find F Given P	To find P given F	To find F given A	To find A given F	To find P given A	To find A given P	To find A given G	To find P given G
n	$F/P\ i,n$	$P/F\ i,n$	$F/A\ i,n$	$A/F\ i,n$	$P/A\ i,n$	$A/P\ i,n$	$A/G\ i,n$	$P/G\ i,n$
1	1.1200	0.8929	1.0000	1.0000	0.8929	1.1200	0.0000	0.0000
2	1.2544	0.7972	2.1200	0.4717	1.6901	0.5917	0.4717	0.7972
3	1.4049	0.7118	3.3744	0.2963	2.4018	0.4163	0.9246	2.2208
4	1.5735	0.6355	4.7793	0.2092	3.0373	0.3292	1.3589	4.1273
5	1.7623	0.5674	6.3528	0.1574	3.6048	0.2774	1.7746	6.3970
6	1.9738	0.5066	8.1152	0.1232	4.1114	0.2432	2.1720	8.9302
7	2.2107	0.4523	10.0890	0.0991	4.5638	0.2191	2.5515	11.6443
8	2.4760	0.4039	12.2997	0.0813	4.9676	0.2013	2.9131	14.4714
9	2.7731	0.3606	14.7757	0.0677	5.3282	0.1877	3.2574	17.3563
10	3.1058	0.3220	17.5487	0.0570	5.6502	0.1770	3.5847	20.2541
11	3.4785	0.2875	20.6546	0.0484	5.9377	0.1684	3.8953	23.1288
12	3.8960	0.2567	24.1331	0.0414	6.1944	0.1614	4.1897	25.9523
13	4.3635	0.2292	28.0291	0.0357	6.4235	0.1557	4.4683	28.7024
14	4.8871	0.2046	32.3926	0.0309	6.6282	0.1509	4.7317	31.3624
15	5.4736	0.1827	37.2797	0.0268	6.8109	0.1468	4.9803	33.9202
16	6.1304	0.1631	42.7533	0.0234	6.9740	0.1434	5.2147	36.3670
17	6.8660	0.1456	48.8837	0.0205	7.1196	0.1405	5.4353	38.6973
18	7.6900	0.1300	55.7497	0.0179	7.2497	0.1379	5.6427	40.9080
19	8.6128	0.1161	63.4397	0.0158	7.3658	0.1358	5.8375	42.9979
20	9.6463	0.1037	72.0524	0.0139	7.4694	0.1339	6.0202	44.9676
21	10.8038	0.0926	81.6987	0.0122	7.5620	0.1322	6.1913	46.8188
22	12.1003	0.0826	92.5026	0.0108	7.6446	0.1308	6.3514	48.5543
23	13.5523	0.0738	104.6029	0.0096	7.7184	0.1296	6.5010	50.1776
24	15.1786	0.0659	118.1552	0.0085	7.7843	0.1285	6.6406	51.6929
25	17.0001	0.0588	133.3339	0.0075	7.8431	0.1275	6.7708	53.1046
26	19.0401	0.0525	150.3339	0.0067	7.8957	0.1267	6.8921	54.4177
27	21.3249	0.0469	169.3740	0.0059	7.9426	0.1259	7.0049	55.6369
28	23.8839	0.0419	190.6989	0.0052	7.9844	0.1252	7.1098	56.7674
29	26.7499	0.0374	214.5828	0.0047	8.0218	0.1247	7.2071	57.8141
30	29.9599	0.0334	241.3327	0.0041	8.0552	0.1241	7.2974	58.7821
31	33.5551	0.0298	271.2926	0.0037	8.0850	0.1237	7.3811	59.6761
32	37.5817	0.0266	304.8477	0.0033	8.1116	0.1233	7.4586	60.5010
33	42.0915	0.0238	342.4294	0.0029	8.1354	0.1229	7.5302	61.2612
34	47.1425	0.0212	384.5210	0.0026	8.1566	0.1226	7.5965	61.9612
35	52.7996	0.0189	431.6635	0.0023	8.1755	0.1223	7.6577	62.6052
36	59.1356	0.0169	484.4631	0.0021	8.1924	0.1221	7.7141	63.1970
37	66.2318	0.0151	543.5987	0.0018	8.2075	0.1218	7.7661	63.7406
38	74.1797	0.0135	609.8305	0.0016	8.2210	0.1216	7.8141	64.2394
39	83.0812	0.0120	684.0102	0.0015	8.2330	0.1215	7.8582	64.6967
40	93.0510	0.0107	767.0914	0.0013	8.2438	0.1213	7.8988	65.1159
45	163.9876	0.0061	1358.2300	0.0007	8.2825	0.1207	8.0572	66.7342
48	230.3908	0.0043	1911.5898	0.0005	8.2972	0.1205	8.1241	67.4068
50	289.0022	0.0035	2400.0182	0.0004	8.3045	0.1204	8.1597	67.7624
54	454.7505	0.0022	3781.2545	0.0003	8.3150	0.1203	8.2143	68.3022
60	897.5969	0.0011	7471.6411	0.0001	8.3240	0.1201	8.2664	68.8100

Table A.18 Discrete Compounding: $i = 15\%$

	Single payment		Uniform Series				Gradient series	
	Compound amount factor	Present worth factor	Compound amount factor	Sinking fund factor	Present worth factor	Capital recovery factor	Uniform series factor	Present worth factor
	To find F Given P	To find P given F	To find F given A	To find A given F	To find P given A	To find A given P	To find A given G	To find P given G
n	$F\vert P\ i,n$	$P\vert F\ i,n$	$F\vert A\ i,n$	$A\vert F\ i,n$	$P\vert A\ i,n$	$A\vert P\ i,n$	$A\vert G\ i,n$	$P\vert G\ i,n$
1	1.1500	0.8696	1.0000	1.0000	0.8696	1.1500	0.0000	0.0000
2	1.3225	0.7561	2.1500	0.4651	1.6257	0.6151	0.4651	0.7561
3	1.5209	0.6575	3.4725	0.2880	2.2832	0.4380	0.9071	2.0712
4	1.7490	0.5718	4.9934	0.2003	2.8550	0.3503	1.3263	3.7864
5	2.0114	0.4972	6.7424	0.1483	3.3522	0.2983	1.7228	5.7751
6	2.3131	0.4323	8.7537	0.1142	3.7845	0.2642	2.0972	7.9368
7	2.6600	0.3759	11.0668	0.0904	4.1604	0.2404	2.4498	10.1924
8	3.0590	0.3269	13.7268	0.0729	4.4873	0.2229	2.7813	12.4807
9	3.5179	0.2843	16.7858	0.0596	4.7716	0.2096	3.0922	14.7548
10	4.0456	0.2472	20.3037	0.0493	5.0188	0.1993	3.3832	16.9795
11	4.6524	0.2149	24.3493	0.0411	5.2337	0.1911	3.6549	19.1289
12	5.3503	0.1869	29.0017	0.0345	5.4206	0.1845	3.9082	21.1849
13	6.1528	0.1625	34.3519	0.0291	5.5831	0.1791	4.1438	23.1352
14	7.0757	0.1413	40.5047	0.0247	5.7245	0.1747	4.3624	24.9725
15	8.1371	0.1229	47.5804	0.0210	5.8474	0.1710	4.5650	26.6930
16	9.3576	0.1069	55.7175	0.0179	5.9542	0.1679	4.7522	28.2960
17	10.7613	0.0929	65.0751	0.0154	6.0472	0.1654	4.9251	29.7828
18	12.3755	0.0808	75.8364	0.0132	6.1280	0.1632	5.0843	31.1565
19	14.2318	0.0703	88.2118	0.0113	6.1982	0.1613	5.2307	32.4213
20	16.3665	0.0611	102.4436	0.0098	6.2593	0.1598	5.3651	33.5822
21	18.8215	0.0531	118.8101	0.0084	6.3125	0.1584	5.4883	34.6448
22	21.6447	0.0462	137.6316	0.0073	6.3587	0.1573	5.6010	35.6150
23	24.8915	0.0402	159.2764	0.0063	6.3988	0.1563	5.7040	36.4988
24	28.6252	0.0349	184.1678	0.0054	6.4338	0.1554	5.7979	37.3023
25	32.9190	0.0304	212.7930	0.0047	6.4641	0.1547	5.8834	38.0314
26	37.8568	0.0264	245.7120	0.0041	6.4906	0.1541	5.9612	38.6918
27	43.5353	0.0230	283.5688	0.0035	6.5135	0.1535	6.0319	39.2890
28	50.0656	0.0200	327.1041	0.0031	6.5335	0.1531	6.0960	39.8283
29	57.5755	0.0174	377.1697	0.0027	6.5509	0.1527	6.1541	40.3146
30	66.2118	0.0151	434.7451	0.0023	6.5660	0.1523	6.2066	40.7526
31	76.1435	0.0131	500.9569	0.0020	6.5791	0.1520	6.2541	41.1466
32	87.5651	0.0114	577.1005	0.0017	6.5905	0.1517	6.2970	41.5006
33	100.6998	0.0099	664.6655	0.0015	6.6005	0.1515	6.3357	41.8184
34	115.8048	0.0086	765.3654	0.0013	6.6091	0.1513	6.3705	42.1033
35	133.1755	0.0075	881.1702	0.0011	6.6166	0.1511	6.4019	42.3586
36	153.1519	0.0065	1014.3457	0.0010	6.6231	0.1510	6.4301	42.5872
37	176.1246	0.0057	1167.4975	0.0009	6.6288	0.1509	6.4554	42.7916
38	202.5433	0.0049	1343.6222	0.0007	6.6338	0.1507	6.4781	42.9743
39	232.9248	0.0043	1546.1655	0.0006	6.6380	0.1506	6.4985	43.1374
40	267.8635	0.0037	1779.0903	0.0006	6.6418	0.1506	6.5168	43.2830
45	538.7693	0.0019	3585.1285	0.0003	6.6543	0.1503	6.5830	43.8051
48	819.4007	0.0012	5456.0047	0.0002	6.6585	0.1502	6.6080	43.9997
50	1083.6574	0.0009	7217.7163	0.0001	6.6605	0.1501	6.6205	44.0958
54	1895.3236	0.0005	12628.8243	0.0001	6.6631	0.1501	6.6382	44.2311
60	4383.9987	0.0002	29219.9916	0.0000	6.6651	0.1500	6.6530	44.3431

Table A.19 Discrete Compounding: $i = 18\%$

	Single payment		Uniform Series				Gradient series	
	Compound amount factor	Present worth factor	Compound amount factor	Sinking fund factor	Present worth factor	Capital recovery factor	Uniform series factor	Present worth factor
	To find F Given P	To find P given F	To find F given A	To find A given F	To find P given A	To find A given P	To find A given G	To find P given G
n	$F\|P\ i,n$	$P\|F\ i,n$	$F\|A\ i,n$	$A\|F\ i,n$	$P\|A\ i,n$	$A\|P\ i,n$	$A\|G\ i,n$	$P\|G\ i,n$
1	1.1800	0.8475	1.0000	1.0000	0.8475	1.1800	0.0000	0.0000
2	1.3924	0.7182	2.1800	0.4587	1.5656	0.6387	0.4587	0.7182
3	1.6430	0.6086	3.5724	0.2799	2.1743	0.4599	0.8902	1.9354
4	1.9388	0.5158	5.2154	0.1917	2.6901	0.3717	1.2947	3.4828
5	2.2878	0.4371	7.1542	0.1398	3.1272	0.3198	1.6728	5.2312
6	2.6996	0.3704	9.4420	0.1059	3.4976	0.2859	2.0252	7.0834
7	3.1855	0.3139	12.1415	0.0824	3.8115	0.2624	2.3526	8.9670
8	3.7589	0.2660	15.3270	0.0652	4.0776	0.2452	2.6558	10.8292
9	4.4355	0.2255	19.0859	0.0524	4.3030	0.2324	2.9358	12.6329
10	5.2338	0.1911	23.5213	0.0425	4.4941	0.2225	3.1936	14.3525
11	6.1759	0.1619	28.7551	0.0348	4.6560	0.2148	3.4303	15.9716
12	7.2876	0.1372	34.9311	0.0286	4.7932	0.2086	3.6470	17.4811
13	8.5994	0.1163	42.2187	0.0237	4.9095	0.2037	3.8449	18.8765
14	10.1472	0.0985	50.8180	0.0197	5.0081	0.1997	4.0250	20.1576
15	11.9737	0.0835	60.9653	0.0164	5.0916	0.1964	4.1887	21.3269
16	14.1290	0.0708	72.9390	0.0137	5.1624	0.1937	4.3369	22.3885
17	16.6722	0.0600	87.0680	0.0115	5.2223	0.1915	4.4708	23.3482
18	19.6733	0.0508	103.7403	0.0096	5.2732	0.1896	4.5916	24.2123
19	23.2144	0.0431	123.4135	0.0081	5.3162	0.1881	4.7003	24.9877
20	27.3930	0.0365	146.6280	0.0068	5.3527	0.1868	4.7978	25.6813
21	32.3238	0.0309	174.0210	0.0057	5.3837	0.1857	4.8851	26.3000
22	38.1421	0.0262	206.3448	0.0048	5.4099	0.1848	4.9632	26.8506
23	45.0076	0.0222	244.4868	0.0041	5.4321	0.1841	5.0329	27.3394
24	53.1090	0.0188	289.4945	0.0035	5.4509	0.1835	5.0950	27.7725
25	62.6686	0.0160	342.6035	0.0029	5.4669	0.1829	5.1502	28.1555
26	73.9490	0.0135	405.2721	0.0025	5.4804	0.1825	5.1991	28.4935
27	87.2598	0.0115	479.2211	0.0021	5.4919	0.1821	5.2425	28.7915
28	102.9666	0.0097	566.4809	0.0018	5.5016	0.1818	5.2810	29.0537
29	121.5005	0.0082	669.4475	0.0015	5.5098	0.1815	5.3149	29.2842
30	143.3706	0.0070	790.9480	0.0013	5.5168	0.1813	5.3448	29.4864
31	169.1774	0.0059	934.3186	0.0011	5.5227	0.1811	5.3712	29.6638
32	199.6293	0.0050	1103.4960	0.0009	5.5277	0.1809	5.3945	29.8191
33	235.5625	0.0042	1303.1253	0.0008	5.5320	0.1808	5.4149	29.9549
34	277.9638	0.0036	1538.6878	0.0006	5.5356	0.1806	5.4328	30.0736
35	327.9973	0.0030	1816.6516	0.0006	5.5386	0.1806	5.4485	30.1773
36	387.0368	0.0026	2144.6489	0.0005	5.5412	0.1805	5.4623	30.2677
37	456.7034	0.0022	2531.6857	0.0004	5.5434	0.1804	5.4744	30.3465
38	538.9100	0.0019	2988.3891	0.0003	5.5452	0.1803	5.4849	30.4152
39	635.9139	0.0016	3527.2992	0.0003	5.5468	0.1803	5.4941	30.4749
40	750.3783	0.0013	4163.2130	0.0002	5.5482	0.1802	5.5022	30.5269
45	1716.6839	0.0006	9531.5771	0.0001	5.5523	0.1801	5.5293	30.7006
48	2820.5665	0.0004	15664.2586	0.0001	5.5536	0.1801	5.5385	30.7587
50	3927.3569	0.0003	21813.0937	0.0000	5.5541	0.1800	5.5428	30.7856
54	7614.2721	0.0001	42295.9563	0.0000	5.5548	0.1800	5.5485	30.8207
60	20555.1400	0.0000	114189.6665	0.0000	5.5553	0.1800	5.5526	30.8465

Table A.20   Discrete Compounding: $i = 20\%$

	Single payment		Uniform Series				Gradient series	
	Compound amount factor	Present worth factor	Compound amount factor	Sinking fund factor	Present worth factor	Capital recovery factor	Uniform series factor	Present worth factor
	To find F Given P	To find P given F	To find F given A	To find A given F	To find P given A	To find A given P	To find A given G	To find P given G
n	F|P i,n	P|F i,n	F|A i,n	A|F i,n	P|A i,n	A|P i,n	A|G i,n	P|G i,n
1	1.2000	0.8333	1.0000	1.0000	0.8333	1.2000	0.0000	0.0000
2	1.4400	0.6944	2.2000	0.4545	1.5278	0.6545	0.4545	0.6944
3	1.7280	0.5787	3.6400	0.2747	2.1065	0.4747	0.8791	1.8519
4	2.0736	0.4823	5.3680	0.1863	2.5887	0.3863	1.2742	3.2986
5	2.4883	0.4019	7.4416	0.1344	2.9906	0.3344	1.6405	4.9061
6	2.9860	0.3349	9.9299	0.1007	3.3255	0.3007	1.9788	6.5806
7	3.5832	0.2791	12.9159	0.0774	3.6046	0.2774	2.2902	8.2551
8	4.2998	0.2326	16.4991	0.0606	3.8372	0.2606	2.5756	9.8831
9	5.1598	0.1938	20.7989	0.0481	4.0310	0.2481	2.8364	11.4335
10	6.1917	0.1615	25.9587	0.0385	4.1925	0.2385	3.0739	12.8871
11	7.4301	0.1346	32.1504	0.0311	4.3271	0.2311	3.2893	14.2330
12	8.9161	0.1122	39.5805	0.0253	4.4392	0.2253	3.4841	15.4667
13	10.6993	0.0935	48.4966	0.0206	4.5327	0.2206	3.6597	16.5883
14	12.8392	0.0779	59.1959	0.0169	4.6106	0.2169	3.8175	17.6008
15	15.4070	0.0649	72.0351	0.0139	4.6755	0.2139	3.9588	18.5095
16	18.4884	0.0541	87.4421	0.0114	4.7296	0.2114	4.0851	19.3208
17	22.1861	0.0451	105.9306	0.0094	4.7746	0.2094	4.1976	20.0419
18	26.6233	0.0376	128.1167	0.0078	4.8122	0.2078	4.2975	20.6805
19	31.9480	0.0313	154.7400	0.0065	4.8435	0.2065	4.3861	21.2439
20	38.3376	0.0261	186.6880	0.0054	4.8696	0.2054	4.4643	21.7395
21	46.0051	0.0217	225.0256	0.0044	4.8913	0.2044	4.5334	22.1742
22	55.2061	0.0181	271.0307	0.0037	4.9094	0.2037	4.5941	22.5546
23	66.2474	0.0151	326.2369	0.0031	4.9245	0.2031	4.6475	22.8867
24	79.4968	0.0126	392.4842	0.0025	4.9371	0.2025	4.6943	23.1760
25	95.3962	0.0105	471.9811	0.0021	4.9476	0.2021	4.7352	23.4276
26	114.4755	0.0087	567.3773	0.0018	4.9563	0.2018	4.7709	23.6460
27	137.3706	0.0073	681.8528	0.0015	4.9636	0.2015	4.8020	23.8353
28	164.8447	0.0061	819.2233	0.0012	4.9697	0.2012	4.8291	23.9991
29	197.8136	0.0051	984.0680	0.0010	4.9747	0.2010	4.8527	24.1406
30	237.3763	0.0042	1181.8816	0.0008	4.9789	0.2008	4.8731	24.2628
31	284.8516	0.0035	1419.2579	0.0007	4.9824	0.2007	4.8908	24.3681
32	341.8219	0.0029	1704.1095	0.0006	4.9854	0.2006	4.9061	24.4588
33	410.1863	0.0024	2045.9314	0.0005	4.9878	0.2005	4.9194	24.5368
34	492.2235	0.0020	2456.1176	0.0004	4.9898	0.2004	4.9308	24.6038
35	590.6682	0.0017	2948.3411	0.0003	4.9915	0.2003	4.9406	24.6614
36	708.8019	0.0014	3539.0094	0.0003	4.9929	0.2003	4.9491	24.7108
37	850.5622	0.0012	4247.8112	0.0002	4.9941	0.2002	4.9564	24.7531
38	1020.6747	0.0010	5098.3735	0.0002	4.9951	0.2002	4.9627	24.7894
39	1224.8096	0.0008	6119.0482	0.0002	4.9959	0.2002	4.9681	24.8204
40	1469.7716	0.0007	7343.8578	0.0001	4.9966	0.2001	4.9728	24.8469
45	3657.2620	0.0003	18281.3099	0.0001	4.9986	0.2001	4.9877	24.9316
48	6319.7487	0.0002	31593.7436	0.0000	4.9992	0.2000	4.9924	24.9581
50	9100.4382	0.0001	45497.1908	0.0000	4.9995	0.2000	4.9945	24.9698
54	18870.6685	0.0001	94348.3427	0.0000	4.9997	0.2000	4.9971	24.9844
60	56347.5144	0.0000	281732.5718	0.0000	4.9999	0.2000	4.9989	24.9942

Table A.21 Discrete Compounding: *i* = 25%

	Single payment		Uniform Series				Gradient series	
	Compound amount factor	Present worth factor	Compound amount factor	Sinking fund factor	Present worth factor	Capital recovery factor	Uniform series factor	Present worth factor
	To find F Given P	To find P given F	To find F given A	To find A given F	To find P given A	To find A given P	To find A given G	To find P given G
n	$F\|P\ i,n$	$P\|F\ i,n$	$F\|A\ i,n$	$A\|F\ i,n$	$P\|A\ i,n$	$A\|P\ i,n$	$A\|G\ i,n$	$P\|G\ i,n$
1	1.2500	0.8000	1.0000	1.0000	0.8000	1.2500	0.0000	0.0000
2	1.5625	0.6400	2.2500	0.4444	1.4400	0.6944	0.4444	0.6400
3	1.9531	0.5120	3.8125	0.2623	1.9520	0.5123	0.8525	1.6640
4	2.4414	0.4096	5.7656	0.1734	2.3616	0.4234	1.2249	2.8928
5	3.0518	0.3277	8.2070	0.1218	2.6893	0.3718	1.5631	4.2035
6	3.8147	0.2621	11.2588	0.0888	2.9514	0.3388	1.8683	5.5142
7	4.7684	0.2097	15.0735	0.0663	3.1611	0.3163	2.1424	6.7725
8	5.9605	0.1678	19.8419	0.0504	3.3289	0.3004	2.3872	7.9469
9	7.4506	0.1342	25.8023	0.0388	3.4631	0.2888	2.6048	9.0207
10	9.3132	0.1074	33.2529	0.0301	3.5705	0.2801	2.7971	9.9870
11	11.6415	0.0859	42.5661	0.0235	3.6564	0.2735	2.9663	10.8460
12	14.5519	0.0687	54.2077	0.0184	3.7251	0.2684	3.1145	11.6020
13	18.1899	0.0550	68.7596	0.0145	3.7801	0.2645	3.2437	12.2617
14	22.7374	0.0440	86.9495	0.0115	3.8241	0.2615	3.3559	12.8334
15	28.4217	0.0352	109.6868	0.0091	3.8593	0.2591	3.4530	13.3260
16	35.5271	0.0281	138.1085	0.0072	3.8874	0.2572	3.5366	13.7482
17	44.4089	0.0225	173.6357	0.0058	3.9099	0.2558	3.6084	14.1085
18	55.5112	0.0180	218.0446	0.0046	3.9279	0.2546	3.6698	14.4147
19	69.3889	0.0144	273.5558	0.0037	3.9424	0.2537	3.7222	14.6741
20	86.7362	0.0115	342.9447	0.0029	3.9539	0.2529	3.7667	14.8932
21	108.4202	0.0092	429.6809	0.0023	3.9631	0.2523	3.8045	15.0777
22	135.5253	0.0074	538.1011	0.0019	3.9705	0.2519	3.8365	15.2326
23	169.4066	0.0059	673.6264	0.0015	3.9764	0.2515	3.8634	15.3625
24	211.7582	0.0047	843.0329	0.0012	3.9811	0.2512	3.8861	15.4711
25	264.6978	0.0038	1054.7912	0.0009	3.9849	0.2509	3.9052	15.5618
26	330.8722	0.0030	1319.4890	0.0008	3.9879	0.2508	3.9212	15.6373
27	413.5903	0.0024	1650.3612	0.0006	3.9903	0.2506	3.9346	15.7002
28	516.9879	0.0019	2063.9515	0.0005	3.9923	0.2505	3.9457	15.7524
29	646.2349	0.0015	2580.9394	0.0004	3.9938	0.2504	3.9551	15.7957
30	807.7936	0.0012	3227.1743	0.0003	3.9950	0.2503	3.9628	15.8316
31	1009.7420	0.0010	4034.9678	0.0002	3.9960	0.2502	3.9693	15.8614
32	1262.1774	0.0008	5044.7098	0.0002	3.9968	0.2502	3.9746	15.8859
33	1577.7218	0.0006	6306.8872	0.0002	3.9975	0.2502	3.9791	15.9062
34	1972.1523	0.0005	7884.6091	0.0001	3.9980	0.2501	3.9828	15.9229
35	2465.1903	0.0004	9856.7613	0.0001	3.9984	0.2501	3.9858	15.9367
36	3081.4879	0.0003	12321.9516	0.0001	3.9987	0.2501	3.9883	15.9481
37	3851.8599	0.0003	15403.4396	0.0001	3.9990	0.2501	3.9904	15.9574
38	4814.8249	0.0002	19255.2994	0.0001	3.9992	0.2501	3.9921	15.9651
39	6018.5311	0.0002	24070.1243	0.0000	3.9993	0.2500	3.9935	15.9714
40	7523.1638	0.0001	30088.6554	0.0000	3.9995	0.2500	3.9947	15.9766
45	22958.8740	0.0000	91831.4962	0.0000	3.9998	0.2500	3.9980	15.9915
48	44841.5509	0.0000	179362.2034	0.0000	3.9999	0.2500	3.9989	15.9954

Table A.22 Discrete Compounding: $i = 30\%$

	Single payment		Uniform Series				Gradient series	
	Compound amount factor	Present worth factor	Compound amount factor	Sinking fund factor	Present worth factor	Capital recovery factor	Uniform series factor	Present worth factor
	To find F Given P	To find P given F	To find F given A	To find A given F	To find P given A	To find A given P	To find A given G	To find P given G
n	$F\|P\ i,n$	$P\|F\ i,n$	$F\|A\ i,n$	$A\|F\ i,n$	$P\|A\ i,n$	$A\|P\ i,n$	$A\|G\ i,n$	$P\|G\ i,n$
1	1.3000	0.7692	1.0000	1.0000	0.7692	1.3000	0.0000	0.0000
2	1.6900	0.5917	2.3000	0.4348	1.3609	0.7348	0.4348	0.5917
3	2.1970	0.4552	3.9900	0.2506	1.8161	0.5506	0.8271	1.5020
4	2.8561	0.3501	6.1870	0.1616	2.1662	0.4616	1.1783	2.5524
5	3.7129	0.2693	9.0431	0.1106	2.4356	0.4106	1.4903	3.6297
6	4.8268	0.2072	12.7560	0.0784	2.6427	0.3784	1.7654	4.6656
7	6.2749	0.1594	17.5828	0.0569	2.8021	0.3569	2.0063	5.6218
8	8.1573	0.1226	23.8577	0.0419	2.9247	0.3419	2.2156	6.4800
9	10.6045	0.0943	32.0150	0.0312	3.0190	0.3312	2.3963	7.2343
10	13.7858	0.0725	42.6195	0.0235	3.0915	0.3235	2.5512	7.8872
11	17.9216	0.0558	56.4053	0.0177	3.1473	0.3177	2.6833	8.4452
12	23.2981	0.0429	74.3270	0.0135	3.1903	0.3135	2.7952	8.9173
13	30.2875	0.0330	97.6250	0.0102	3.2233	0.3102	2.8895	9.3135
14	39.3738	0.0254	127.9125	0.0078	3.2487	0.3078	2.9685	9.6437
15	51.1859	0.0195	167.2863	0.0060	3.2682	0.3060	3.0344	9.9172
16	66.5417	0.0150	218.4722	0.0046	3.2832	0.3046	3.0892	10.1426
17	86.5042	0.0116	285.0139	0.0035	3.2948	0.3035	3.1345	10.3276
18	112.4554	0.0089	371.5180	0.0027	3.3037	0.3027	3.1718	10.4788
19	146.1920	0.0068	483.9734	0.0021	3.3105	0.3021	3.2025	10.6019
20	190.0496	0.0053	630.1655	0.0016	3.3158	0.3016	3.2275	10.7019
21	247.0645	0.0040	820.2151	0.0012	3.3198	0.3012	3.2480	10.7828
22	321.1839	0.0031	1067.2796	0.0009	3.3230	0.3009	3.2646	10.8482
23	417.5391	0.0024	1388.4635	0.0007	3.3254	0.3007	3.2781	10.9009
24	542.8008	0.0018	1806.0026	0.0006	3.3272	0.3006	3.2890	10.9433
25	705.6410	0.0014	2348.8033	0.0004	3.3286	0.3004	3.2979	10.9773
26	917.3333	0.0011	3054.4443	0.0003	3.3297	0.3003	3.3050	11.0045
27	1192.5333	0.0008	3971.7776	0.0003	3.3305	0.3003	3.3107	11.0263
28	1550.2933	0.0006	5164.3109	0.0002	3.3312	0.3002	3.3153	11.0437
29	2015.3813	0.0005	6714.6042	0.0001	3.3317	0.3001	3.3189	11.0576
30	2619.9956	0.0004	8729.9855	0.0001	3.3321	0.3001	3.3219	11.0687
31	3405.9943	0.0003	11349.9811	0.0001	3.3324	0.3001	3.3242	11.0775
32	4427.7926	0.0002	14755.9755	0.0001	3.3326	0.3001	3.3261	11.0845
33	5756.1304	0.0002	19183.7681	0.0001	3.3328	0.3001	3.3276	11.0901
34	7482.9696	0.0001	24939.8985	0.0000	3.3329	0.3000	3.3288	11.0945
35	9727.8604	0.0001	32422.8681	0.0000	3.3330	0.3000	3.3297	11.0980
36	12646.2186	0.0001	42150.7285	0.0000	3.3331	0.3000	3.3305	11.1007
37	16440.0841	0.0001	54796.9471	0.0000	3.3331	0.3000	3.3311	11.1029
38	21372.1094	0.0000	71237.0312	0.0000	3.3332	0.3000	3.3316	11.1047
39	27783.7422	0.0000	92609.1405	0.0000	3.3332	0.3000	3.3319	11.1060
40	36118.8648	0.0000	120392.8827	0.0000	3.3332	0.3000	3.3322	11.1071
45	134106.8167	0.0000	447019.3890	0.0000	3.3333	0.3000	3.3330	11.1099
48	294632.6763	0.0000	982105.5877	0.0000	3.3333	0.3000	3.3332	11.1105

Table A.23   Discrete Compounding: $i = 40\%$

	Single payment		Uniform Series				Gradient series	
	Compound amount factor	Present worth factor	Compound amount factor	Sinking fund factor	Present worth factor	Capital recovery factor	Uniform series factor	Present worth factor
	To find F Given P	To find P given F	To find F given A	To find A given F	To find P given A	To find A given P	To find A given G	To find P given G
n	$F\|P\ i,n$	$P\|F\ i,n$	$F\|A\ i,n$	$A\|F\ i,n$	$P\|A\ i,n$	$A\|P\ i,n$	$A\|G\ i,n$	$P\|G\ i,n$
1	1.4000	0.7143	1.0000	1.0000	0.7143	1.4000	0.0000	0.0000
2	1.9600	0.5102	2.4000	0.4167	1.2245	0.8167	0.4167	0.5102
3	2.7440	0.3644	4.3600	0.2294	1.5889	0.6294	0.7798	1.2391
4	3.8416	0.2603	7.1040	0.1408	1.8492	0.5408	1.0923	2.0200
5	5.3782	0.1859	10.9456	0.0914	2.0352	0.4914	1.3580	2.7637
6	7.5295	0.1328	16.3238	0.0613	2.1680	0.4613	1.5811	3.4278
7	10.5414	0.0949	23.8534	0.0419	2.2628	0.4419	1.7664	3.9970
8	14.7579	0.0678	34.3947	0.0291	2.3306	0.4291	1.9185	4.4713
9	20.6610	0.0484	49.1526	0.0203	2.3790	0.4203	2.0422	4.8585
10	28.9255	0.0346	69.8137	0.0143	2.4136	0.4143	2.1419	5.1696
11	40.4957	0.0247	98.7391	0.0101	2.4383	0.4101	2.2215	5.4166
12	56.6939	0.0176	139.2348	0.0072	2.4559	0.4072	2.2845	5.6106
13	79.3715	0.0126	195.9287	0.0051	2.4685	0.4051	2.3341	5.7618
14	111.1201	0.0090	275.3002	0.0036	2.4775	0.4036	2.3729	5.8788
15	155.5681	0.0064	386.4202	0.0026	2.4839	0.4026	2.4030	5.9688
16	217.7953	0.0046	541.9883	0.0018	2.4885	0.4018	2.4262	6.0376
17	304.9135	0.0033	759.7837	0.0013	2.4918	0.4013	2.4441	6.0901
18	426.8789	0.0023	1064.6971	0.0009	2.4941	0.4009	2.4577	6.1299
19	597.6304	0.0017	1491.5760	0.0007	2.4958	0.4007	2.4682	6.1601
20	836.6826	0.0012	2089.2064	0.0005	2.4970	0.4005	2.4761	6.1828
21	1171.3556	0.0009	2925.8889	0.0003	2.4979	0.4003	2.4821	6.1998
22	1639.8978	0.0006	4097.2445	0.0002	2.4985	0.4002	2.4866	6.2127
23	2295.8569	0.0004	5737.1423	0.0002	2.4989	0.4002	2.4900	6.2222
24	3214.1997	0.0003	8032.9993	0.0001	2.4992	0.4001	2.4925	6.2294
25	4499.8796	0.0002	11247.1990	0.0001	2.4994	0.4001	2.4944	6.2347
26	6299.8314	0.0002	15747.0785	0.0001	2.4996	0.4001	2.4959	6.2387
27	8819.7640	0.0001	22046.9099	0.0000	2.4997	0.4000	2.4969	6.2416
28	12347.6696	0.0001	30866.6739	0.0000	2.4998	0.4000	2.4977	6.2438
29	17286.7374	0.0001	43214.3435	0.0000	2.4999	0.4000	2.4983	6.2454
30	24201.4324	0.0000	60501.0809	0.0000	2.4999	0.4000	2.4988	6.2466
31	33882.0053	0.0000	84702.5132	0.0000	2.4999	0.4000	2.4991	6.2475
32	47434.8074	0.0000	118584.5185	0.0000	2.4999	0.4000	2.4993	6.2482
33	66408.7304	0.0000	166019.3260	0.0000	2.5000	0.4000	2.4995	6.2487
34	92972.2225	0.0000	232428.0563	0.0000	2.5000	0.4000	2.4996	6.2490
35	130161.1116	0.0000	325400.2789	0.0000	2.5000	0.4000	2.4997	6.2493

Table A.24 Discrete Compounding: $i = 50\%$

	Single payment		Uniform Series				Gradient series	
	Compound amount factor	Present worth factor	Compound amount factor	Sinking fund factor	Present worth factor	Capital recovery factor	Uniform series factor	Present worth factor
	To find F Given P $F\|P\ i,n$	To find P given F $P\|F\ i,n$	To find F given A $F\|A\ i,n$	To find A given F $A\|F\ i,n$	To find P given A $P\|A\ i,n$	To find A given P $A\|P\ i,n$	To find A given G $A\|G\ i,n$	To find P given G $P\|G\ i,n$
n								
1	1.5000	0.6667	1.0000	1.0000	0.6667	1.5000	0.0000	0.0000
2	2.2500	0.4444	2.5000	0.4000	1.1111	0.9000	0.4000	0.4444
3	3.3750	0.2963	4.7500	0.2105	1.4074	0.7105	0.7368	1.0370
4	5.0625	0.1975	8.1250	0.1231	1.6049	0.6231	1.0154	1.6296
5	7.5938	0.1317	13.1875	0.0758	1.7366	0.5758	1.2417	2.1564
6	11.3906	0.0878	20.7813	0.0481	1.8244	0.5481	1.4226	2.5953
7	17.0859	0.0585	32.1719	0.0311	1.8829	0.5311	1.5648	2.9465
8	25.6289	0.0390	49.2578	0.0203	1.9220	0.5203	1.6752	3.2196
9	38.4434	0.0260	74.8867	0.0134	1.9480	0.5134	1.7596	3.4277
10	57.6650	0.0173	113.3301	0.0088	1.9653	0.5088	1.8235	3.5838
11	86.4976	0.0116	170.9951	0.0058	1.9769	0.5058	1.8713	3.6994
12	129.7463	0.0077	257.4927	0.0039	1.9846	0.5039	1.9068	3.7842
13	194.6195	0.0051	387.2390	0.0026	1.9897	0.5026	1.9329	3.8459
14	291.9293	0.0034	581.8585	0.0017	1.9931	0.5017	1.9519	3.8904
15	437.8939	0.0023	873.7878	0.0011	1.9954	0.5011	1.9657	3.9224
16	656.8408	0.0015	1311.6817	0.0008	1.9970	0.5008	1.9756	3.9452
17	985.2613	0.0010	1968.5225	0.0005	1.9980	0.5005	1.9827	3.9614
18	1477.8919	0.0007	2953.7838	0.0003	1.9986	0.5003	1.9878	3.9729
19	2216.8378	0.0005	4431.6756	0.0002	1.9991	0.5002	1.9914	3.9811
20	3325.2567	0.0003	6648.5135	0.0002	1.9994	0.5002	1.9940	3.9868
21	4987.8851	0.0002	9973.7702	0.0001	1.9996	0.5001	1.9958	3.9908
22	7481.8276	0.0001	14961.6553	0.0001	1.9997	0.5001	1.9971	3.9936
23	11222.7415	0.0001	22443.4829	0.0000	1.9998	0.5000	1.9980	3.9955
24	16834.1122	0.0001	33666.2244	0.0000	1.9999	0.5000	1.9986	3.9969
25	25251.1683	0.0000	50500.3366	0.0000	1.9999	0.5000	1.9990	3.9979
26	37876.7524	0.0000	75751.5049	0.0000	1.9999	0.5000	1.9993	3.9985
27	56815.1287	0.0000	113628.2573	0.0000	2.0000	0.5000	1.9995	3.9990
28	85222.6930	0.0000	170443.3860	0.0000	2.0000	0.5000	1.9997	3.9993
29	127834.0395	0.0000	255666.0790	0.0000	2.0000	0.5000	1.9998	3.9995
30	191751.0592	0.0000	383500.1185	0.0000	2.0000	0.5000	1.9998	3.9997

SECTION II -- GEOMETRIC SERIES FACTORS

Table A.25 Discrete Compounding: $i=5\%$

	Geometric series present worth factor,				$(P\|A_1\ i,j,n)$
n	j = 4%	j = 6%	j = 8%	j = 10%	j = 15%
1	0.9524	0.9524	0.9524	0.9524	0.9524
2	1.8957	1.9138	1.9320	1.9501	1.9955
3	2.8300	2.8844	2.9396	2.9954	3.1379
4	3.7554	3.8643	3.9759	4.0904	4.3891
5	4.6721	4.8535	5.0419	5.2375	5.7595
6	5.5799	5.8521	6.1383	6.4393	7.2604
7	6.4792	6.8602	7.2661	7.6983	8.9043
8	7.3699	7.8779	8.4261	9.0173	10.7047
9	8.2521	8.9053	9.6192	10.3991	12.6765
10	9.1258	9.9425	10.8464	11.8467	14.8362
11	9.9913	10.9896	12.1087	13.3632	17.2016
12	10.8485	12.0466	13.4070	14.9519	19.7922
13	11.6976	13.1137	14.7425	16.6163	22.6295
14	12.5386	14.1910	16.1161	18.3599	25.7371
15	13.3715	15.2785	17.5289	20.1866	29.1407
16	14.1966	16.3764	18.9821	22.1002	32.8683
17	15.0137	17.4848	20.4769	24.1050	36.9510
18	15.8231	18.6037	22.0143	26.2052	41.4226
19	16.6248	19.7332	23.5956	28.4055	46.3200
20	17.4189	20.8736	25.2222	30.7105	51.6838
21	18.2054	22.0247	26.8952	33.1253	57.5584
22	18.9844	23.1869	28.6160	35.6550	63.9925
23	19.7559	24.3601	30.3860	38.3053	71.0394
24	20.5202	25.5445	32.2066	41.0817	78.7575
25	21.2771	26.7401	34.0791	43.9904	87.2106
26	22.0269	27.9472	36.0052	47.0375	96.4687
27	22.7695	29.1657	37.9863	50.2298	106.6086
28	23.5050	30.3959	40.0240	53.5741	117.7142
29	24.2335	31.6377	42.1199	57.0776	129.8774
30	24.9551	32.8914	44.2757	60.7480	143.1991
31	25.6698	34.1571	46.4931	64.5931	157.7895
32	26.3777	35.4348	48.7739	68.6213	173.7695
33	27.0789	36.7246	51.1198	72.8414	191.2713
34	27.7734	38.0267	53.5328	77.2624	210.4400
35	28.4612	39.3413	56.0146	81.8940	231.4343
36	29.1426	40.6683	58.5674	86.7461	254.4280
37	29.8174	42.0080	61.1932	91.8292	279.6116
38	30.4858	43.3605	63.8939	97.1544	307.1937
39	31.1478	44.7258	66.6719	102.7332	337.4026
40	31.8036	46.1042	69.5291	108.5776	370.4886
48	36.8296	57.6141	95.5310	166.5488	777.7891
50	38.0271	60.6306	102.9998	184.7384	934.9897
54	40.3544	66.8378	119.2613	226.6112	1349.7568
60	43.6826	76.6013	147.3617	306.0117	2336.9967
65	46.3138	85.1726	174.6930	391.3866	3688.7517
70	48.8221	94.1599	206.1585	499.1191	5819.0515
80	53.4925	113.4643	284.0874	806.6104	14467.1576
90	57.7367	134.6880	387.3737	1296.2390	35945.7801
100	61.5936	158.0219	524.2687	2075.8906	89290.5489
120	68.2835	211.8803	946.1871	5294.1808	550829.3461

Table A.26 Discrete Compounding: $i=5\%$

	Geometric series future worth factor,				$(F|A_1 \ i,j,n)$
n	j = 4%	j = 6%	j = 8%	j = 10%	j = 15%
1	1.0000	1.0000	1.0000	1.0000	1.0000
2	2.0900	2.1100	2.1300	2.1500	2.2000
3	3.2761	3.3391	3.4029	3.4675	3.6325
4	4.5648	4.6971	4.8328	4.9719	5.3350
5	5.9629	6.1944	6.4349	6.6846	7.3508
6	7.4777	7.8423	8.2260	8.6293	9.7297
7	9.1169	9.6530	10.2241	10.8323	12.5292
8	10.8886	11.6393	12.4492	13.3227	15.8157
9	12.8016	13.8151	14.9225	16.1324	19.6655
10	14.8650	16.1953	17.6677	19.2970	24.1666
11	17.0885	18.7959	20.7100	22.8556	29.4205
12	19.4824	21.6340	24.0771	26.8515	35.5439
13	22.0576	24.7279	27.7992	31.3325	42.6714
14	24.8255	28.0972	31.9087	36.3514	50.9577
15	27.7985	31.7630	36.4414	41.9664	60.5813
16	30.9893	35.7477	41.4356	48.2420	71.7475
17	34.4118	40.0754	46.9333	55.2491	84.6925
18	38.0803	44.7720	52.9800	63.0660	99.6883
19	42.0101	49.8649	59.6250	71.7792	117.0482
20	46.2175	55.3838	66.9220	81.4841	137.1324
21	50.7195	61.3601	74.9290	92.2858	160.3556
22	55.5342	67.8277	83.7093	104.3004	187.1948
23	60.6808	74.8226	93.3313	117.6557	218.1993
24	66.1796	82.3835	103.8694	132.4928	254.0008
25	72.0519	90.5516	115.4040	148.9672	295.3260
26	78.3203	99.3710	128.0227	167.2503	343.0112
27	85.0088	108.8890	141.8202	187.5310	398.0186
28	92.1426	119.1558	156.8992	210.0176	461.4548
29	99.7484	130.2252	173.3713	234.9395	534.5932
30	107.8545	142.1549	191.3572	262.5496	618.8983
31	116.4906	155.0061	210.9877	293.1266	716.0550
32	125.6883	168.8445	232.4047	326.9773	828.0013
33	135.4807	183.7401	255.7620	364.4400	956.9664
34	145.9032	199.7677	281.2262	405.8872	1105.5145
35	156.9926	217.0071	308.9776	451.7293	1276.5951
36	168.7884	235.5436	339.2119	502.4183	1473.6003
37	181.3317	255.4680	372.1406	558.4520	1700.4322
38	194.6664	276.8775	407.9933	620.3787	1961.5784
39	208.8385	299.8756	447.0182	688.8021	2262.2007
40	223.8968	324.5729	489.4844	764.3871	2608.2355
48	383.0741	599.2602	993.6434	1732.3243	8089.9942
50	436.0716	695.2754	1181.1404	2118.4755	10721.9001
54	562.4882	931.6324	1662.3477	3158.6756	18813.8487
60	815.9558	1430.8505	2752.5959	5716.0709	43653.1937
65	1104.1165	2030.5071	4164.6649	9330.6558	87939.4708
70	1485.4807	2864.9505	6272.6660	15186.4808	177052.9271
80	2651.1642	5623.4552	14079.7798	39976.9935	717013.1369
90	4661.1032	10873.4146	31272.8241	104646.5051	2901915.7503
100	8099.6310	20780.0826	68941.9999	272984.1812	11741818.5907
120	23824.9425	73927.5762	330136.0319	1847219.6697	192190942.7662

Table A.27 Discrete Compounding: $i=8\%$

	Geometric series present worth factor,			$(P\|A_1, i,j,n)$	
n	$j=4\%$	$j=6\%$	$j=8\%$	$j=10\%$	$j=15\%$
1	0.9259	0.9259	0.9259	0.9259	0.9259
2	1.8176	1.8347	1.8519	1.8690	1.9119
3	2.6762	2.7267	2.7778	2.8295	2.9617
4	3.5030	3.6021	3.7037	3.8079	4.0796
5	4.2992	4.4613	4.6296	4.8043	5.2699
6	5.0659	5.3046	5.5556	5.8192	6.5374
7	5.8042	6.1323	6.4815	6.8529	7.8871
8	6.5151	6.9447	7.4074	7.9057	9.3242
9	7.1997	7.7420	8.3333	8.9780	10.8545
10	7.8590	8.5246	9.2593	10.0702	12.4839
11	8.4939	9.2926	10.1852	11.1826	14.2190
12	9.1052	10.0465	11.1111	12.3157	16.0665
13	9.6939	10.7863	12.0370	13.4696	18.0338
14	10.2608	11.5125	12.9630	14.6450	20.1286
15	10.8067	12.2252	13.8889	15.8421	22.3592
16	11.3324	12.9248	14.8148	17.0614	24.7343
17	11.8386	13.6114	15.7407	18.3033	27.2634
18	12.3260	14.2852	16.6667	19.5682	29.9564
19	12.7954	14.9466	17.5926	20.8565	32.8239
20	13.2475	15.5957	18.5185	22.1687	35.8773
21	13.6827	16.2329	19.4444	23.5051	39.1286
22	14.1019	16.8582	20.3704	24.8663	42.5906
23	14.5055	17.4719	21.2963	26.2527	46.2771
24	14.8942	18.0743	22.2222	27.6648	50.2024
25	15.2685	18.6655	23.1482	29.1031	54.3822
26	15.6289	19.2458	24.0741	30.5679	58.8329
27	15.9760	19.8153	25.0000	32.0599	63.5721
28	16.3102	20.3743	25.9259	33.5796	68.6184
29	16.6321	20.9229	26.8519	35.1273	73.9919
30	16.9420	21.4614	27.7778	36.7038	79.7136
31	17.2404	21.9899	28.7037	38.3094	85.8061
32	17.5278	22.5086	29.6296	39.9447	92.2935
33	17.8046	23.0177	30.5556	41.6104	99.2015
34	18.0711	23.5173	31.4815	43.3069	106.5571
35	18.3277	24.0078	32.4074	45.0348	114.3895
36	18.5748	24.4891	33.3333	46.7947	122.7296
37	18.8128	24.9615	34.2593	48.5872	131.6102
38	19.0419	25.4252	35.1852	50.4129	141.0664
39	19.2626	25.8803	36.1111	52.2724	151.1355
40	19.4751	26.3269	37.0370	54.1663	161.8573
48	20.9149	29.6150	44.4444	70.6365	276.8249
50	21.2119	30.3630	46.2963	75.1459	315.7844
54	21.7427	31.7777	50.0000	84.6766	410.0431
60	22.4027	33.7109	55.5556	100.3508	604.2248
65	22.8494	35.1643	60.1852	114.7974	832.3907
70	23.2192	36.4881	64.8148	130.6322	1144.7262
80	23.7790	38.7917	74.0741	167.0123	2157.5589
90	24.1629	40.7026	83.3333	210.7196	4055.4818
100	24.4260	42.2878	92.5926	263.2297	7611.9539
120	24.7302	44.6933	111.1111	402.1073	26764.5779

Table A.28 Discrete Compounding: $i=8\%$

	Geometric series future worth factor,				$(F|A_1\ i,j,n)$
n	j=4%	j=6%	j=8%	j=10%	j=15%
1	1.0000	1.0000	1.0000	1.0000	1.0000
2	2.1200	2.1400	2.1600	2.1800	2.2300
3	3.3712	3.4348	3.4992	3.5644	3.7309
4	4.7658	4.9006	5.0388	5.1806	5.5502
5	6.3169	6.5551	6.8024	7.0591	7.7433
6	8.0389	8.4178	8.8160	9.2343	10.3741
7	9.9473	10.5097	11.1081	11.7446	13.5171
8	12.0590	12.8541	13.7106	14.6329	17.2585
9	14.3923	15.4763	16.6584	17.9472	21.6982
10	16.9670	18.4039	19.9900	21.7409	26.9519
11	19.8046	21.6670	23.7482	26.0739	33.1536
12	22.9284	25.2987	27.9797	31.0129	40.4583
13	26.3638	29.3348	32.7362	36.6324	49.0452
14	30.1379	33.8145	38.0747	43.0153	59.1216
15	34.2806	38.7805	44.0579	50.2540	70.9270
16	38.8240	44.2795	50.7547	58.4516	84.7383
17	43.8029	50.3623	58.2410	67.7227	100.8749
18	49.2551	57.0840	66.6003	78.1950	119.7062
19	55.2213	64.5051	75.9243	90.0105	141.6581
20	61.7459	72.6911	86.3139	103.3272	167.2226
21	68.8766	81.7135	97.8800	118.3209	196.9669
22	76.6655	91.6501	110.7442	135.1869	231.5458
23	85.1687	102.5857	125.0403	154.1421	271.7142
24	94.4469	114.6123	140.9150	175.4278	318.3428
25	104.5660	127.8302	158.5293	199.3118	372.4354
26	115.5971	142.3485	178.0601	226.0914	435.1492
27	127.6173	158.2858	199.7013	256.0970	507.8179
28	140.7101	175.7710	223.6654	289.6947	591.9786
29	154.9656	194.9443	250.1858	327.2914	689.4025
30	170.4815	215.9583	279.5179	369.3378	802.1302
31	187.3634	238.9784	311.9419	416.3343	932.5124
32	205.7256	264.1848	347.7649	468.8354	1083.2569
33	225.6917	291.7730	387.3232	527.4561	1257.4825
34	247.3954	321.9554	430.9850	592.8778	1458.7810
35	270.9814	354.9629	479.1539	665.8558	1691.2882
36	296.6060	391.0460	532.2715	747.2268	1959.7668
37	324.4384	430.4769	590.8214	837.9177	2269.7000
38	354.6616	473.5512	655.3327	938.9552	2627.4006
39	387.4733	520.5895	726.3845	1051.4760	3040.1360
40	423.0875	571.9402	804.6105	1176.7390	3516.2717
48	841.0011	1190.8351	1787.1327	2840.3400	11131.2874
50	994.8732	1424.0729	2171.3660	3524.4711	14810.7971
54	1387.3828	2027.7053	3190.4485	5403.1563	26164.4921
60	2268.4359	3413.4686	5625.3771	10161.2610	61182.0216
65	3399.5278	5231.7437	8954.3161	17079.6034	123842.9595
70	5075.8697	7976.5238	14168.8885	28557.1360	250244.4686
80	11222.6259	18307.9420	34959.4895	78822.6198	1018270.2942
90	24619.8939	41472.5289	84909.2410	214706.4761	4132191.3076
100	53731.4077	93022.9586	203680.6644	579045.9229	16744480.0582
120	253558.2595	458240.2597	1139215.1626	4122833.7485	274417003.9699

Table A.29 Discrete Compounding: $i = 10\%$

	Geometric series present worth factor,				$(P|A_1\ i,j,n)$
n	j = 4%	j = 6%	j = 8%	j = 10%	j = 15%
1	0.9091	0.9091	0.9091	0.9091	0.9091
2	1.7686	1.7851	1.8017	1.8182	1.8595
3	2.5812	2.6293	2.6780	2.7273	2.8531
4	3.3495	3.4428	3.5384	3.6364	3.8919
5	4.0759	4.2267	4.3831	4.5455	4.9779
6	4.7627	4.9821	5.2125	5.4545	6.1133
7	5.4120	5.7100	6.0269	6.3636	7.3002
8	6.0259	6.4115	6.8264	7.2727	8.5411
9	6.6063	7.0874	7.6113	8.1818	9.8385
10	7.1550	7.7388	8.3820	9.0909	11.1948
11	7.6738	8.3664	9.1387	10.0000	12.6127
12	8.1644	8.9713	9.8817	10.9091	14.0951
13	8.6281	9.5542	10.6111	11.8182	15.6449
14	9.0666	10.1158	11.3273	12.7273	17.2651
15	9.4811	10.6571	12.0304	13.6364	18.9590
16	9.8731	11.1786	12.7208	14.5455	20.7298
17	10.2436	11.6812	13.3986	15.4545	22.5812
18	10.5940	12.1656	14.0640	16.3636	24.5167
19	10.9252	12.6323	14.7174	17.2727	26.5402
20	11.2384	13.0820	15.3589	18.1818	28.6556
21	11.5345	13.5154	15.9888	19.0909	30.8672
22	11.8144	13.9330	16.6071	20.0000	33.1794
23	12.0791	14.3354	17.2143	20.9091	35.5966
24	12.3293	14.7232	17.8104	21.8182	38.1238
25	12.5659	15.0969	18.3957	22.7273	40.7658
26	12.7896	15.4570	18.9703	23.6364	43.5278
27	13.0011	15.8041	19.5345	24.5455	46.4155
28	13.2010	16.1385	20.0884	25.4545	49.4343
29	13.3900	16.4607	20.6322	26.3636	52.5905
30	13.5688	16.7712	21.1662	27.2727	55.8900
31	13.7377	17.0704	21.6904	28.1818	59.3396
32	13.8975	17.3588	22.2052	29.0909	62.9459
33	14.0485	17.6367	22.7105	30.0000	66.7162
34	14.1913	17.9044	23.2067	30.9091	70.6578
35	14.3264	18.1624	23.6938	31.8182	74.7786
36	14.4540	18.4111	24.1721	32.7273	79.0867
37	14.5747	18.6507	24.6417	33.6364	83.5907
38	14.6888	18.8816	25.1028	34.5455	88.2994
39	14.7967	19.1040	25.5555	35.4545	93.2221
40	14.8987	19.3184	25.9999	36.3636	98.3685
48	15.5379	20.7755	29.2766	43.6364	148.9186
50	15.6577	21.0772	30.0233	45.4545	164.6238
54	15.8605	21.6174	31.4370	49.0909	200.5507
60	16.0908	22.2915	33.3722	54.5455	267.9647
65	16.2317	22.7494	34.8299	59.0909	339.6376
70	16.3380	23.1299	36.1597	63.6364	429.1494
80	16.4791	23.7088	38.4799	72.7273	680.5553
90	16.5596	24.1085	40.4112	81.8182	1072.6824
100	16.6056	24.3845	42.0186	90.9091	1684.2979
120	16.6468	24.7066	44.4703	109.0909	4126.1845

Table A.30 Discrete Compounding: $i = 10\%$

	Geometric series future worth factor,				$(F|A_1 \ i,j,n)$
n	j=4%	j=6%	j=8%	j=10%	j=15%
1	1.0000	1.0000	1.0000	1.0000	1.0000
2	2.1400	2.1600	2.1800	2.2000	2.2500
3	3.4356	3.4996	3.5644	3.6300	3.7975
4	4.9040	5.0406	5.1806	5.3240	5.6981
5	6.5643	6.8071	7.0591	7.3205	8.0169
6	8.4374	8.8260	9.2343	9.6631	10.8300
7	10.5464	11.1272	11.7446	12.4009	14.2261
8	12.9170	13.7435	14.6329	15.5897	18.3087
9	15.5773	16.7117	17.9471	19.2923	23.1986
10	18.5583	20.0724	21.7409	23.5795	29.0363
11	21.8944	23.8705	26.0739	28.5312	35.9855
12	25.6233	28.1558	31.0129	34.2374	44.2364
13	29.7866	32.9836	36.6324	40.7996	54.0103
14	34.4304	38.4149	43.0152	48.3318	65.5641
15	39.6051	44.5172	50.2539	56.9625	79.1963
16	45.3665	51.3655	58.4515	66.8360	95.2530
17	51.7762	59.0424	67.7226	78.1146	114.1359
18	58.9017	67.6395	78.1948	90.9805	136.3107
19	66.8177	77.2577	90.0103	105.6385	162.3173
20	75.6063	88.0091	103.3271	122.3183	192.7807
21	85.3580	100.0172	118.3207	141.2776	228.4254
22	96.1726	113.4184	135.1866	162.8057	270.0894
23	108.1598	128.3638	154.1418	187.2265	318.7431
24	121.4405	145.0200	175.4274	214.9035	375.5089
25	136.1478	163.5709	199.3113	246.2436	441.6849
26	152.4284	184.2198	226.0909	281.7027	518.7724
27	170.4438	207.1912	256.0964	321.7911	608.5064
28	190.3715	232.7327	289.6940	367.0803	712.8924
29	212.4074	261.1176	327.2905	418.2093	834.2472
30	236.7667	292.6478	369.3368	475.8934	975.2474
31	263.6868	327.6560	416.3331	540.9322	1138.9839
32	293.4286	366.5098	468.8341	614.2198	1329.0258
33	326.2796	409.6141	527.4545	696.7556	1549.4935
34	362.5559	457.4161	592.8760	789.6564	1805.1426
35	402.6058	510.4088	665.8537	894.1698	2101.4617
36	446.8125	569.1357	747.2244	1011.6893	2444.7834
37	495.5976	634.1965	837.9149	1143.7710	2842.4136
38	549.4255	706.2523	938.9520	1292.1522	3302.7795
39	608.8069	786.0318	1051.4724	1458.7719	3835.6008
40	674.3039	874.3384	1176.7348	1645.7940	4452.0857
48	1507.4451	2015.5841	2840.3278	4233.4883	14447.6692
50	1838.0695	2474.2675	3524.4553	5335.9597	19325.3312
54	2725.9689	3715.4232	5403.1300	8437.3705	34469.0327
60	4899.3669	6787.3487	10161.2061	16608.1339	81590.3392
65	7959.5332	11155.6438	17079.5033	28976.5362	166548.3266
70	12902.9223	18266.7757	28556.9556	50256.7821	338919.4468
80	33755.8403	48565.1055	78822.0497	148975.0958	1394049.5118
90	87981.7213	128088.9525	214704.7272	434703.6079	5699185.7115
100	228835.1232	336032.7564	579040.6774	1252788.5754	23210655.1891
120	1543306.7710	2290522.0267	4122788.8759	10113771.2878	382534686.2027

Table A.31 Discrete Compounding: $i = 15\%$

	Geometric series present worth factor,				$(P|A_1 \ i,j,n)$
n	j=4%	j=6%	j=8%	j=10%	j=15%
1	0.8696	0.8696	0.8696	0.8696	0.8696
2	1.6560	1.6711	1.6862	1.7013	1.7391
3	2.3671	2.4099	2.4531	2.4969	2.6087
4	3.0103	3.0908	3.1734	3.2579	3.4783
5	3.5919	3.7185	3.8498	3.9858	4.3478
6	4.1179	4.2971	4.4850	4.6821	5.2174
7	4.5936	4.8303	5.0816	5.3481	6.0870
8	5.0237	5.3219	5.6418	5.9851	6.9565
9	5.4128	5.7749	6.1680	6.5945	7.8261
10	5.7646	6.1926	6.6621	7.1773	8.6957
11	6.0828	6.5775	7.1261	7.7348	9.5652
12	6.3705	6.9323	7.5619	8.2681	10.4348
13	6.6307	7.2593	7.9712	8.7782	11.3043
14	6.8660	7.5608	8.3556	9.2661	12.1739
15	7.0789	7.8386	8.7165	9.7328	13.0435
16	7.2713	8.0947	9.0555	10.1792	13.9130
17	7.4454	8.3308	9.3739	10.6062	14.7826
18	7.6028	8.5484	9.6729	11.0146	15.6522
19	7.7451	8.7489	9.9537	11.4053	16.5217
20	7.8738	8.9338	10.2173	11.7790	17.3913
21	7.9903	9.1042	10.4650	12.1364	18.2609
22	8.0955	9.2613	10.6976	12.4783	19.1304
23	8.1907	9.4060	10.9160	12.8053	20.0000
24	8.2768	9.5395	11.1211	13.1181	20.8696
25	8.3547	9.6625	11.3137	13.4173	21.7391
26	8.4251	9.7759	11.4946	13.7035	22.6087
27	8.4888	9.8803	11.6645	13.9773	23.4783
28	8.5464	9.9767	11.8241	14.2392	24.3478
29	8.5985	10.0655	11.9739	14.4896	25.2174
30	8.6456	10.1473	12.1146	14.7292	26.0870
31	8.6882	10.2227	12.2468	14.9584	26.9565
32	8.7267	10.2922	12.3709	15.1776	27.8261
33	8.7615	10.3563	12.4874	15.3873	28.6956
34	8.7930	10.4154	12.5969	15.5878	29.5652
35	8.8215	10.4698	12.6997	15.7796	30.4348
36	8.8473	10.5200	12.7962	15.9631	31.3043
37	8.8706	10.5663	12.8869	16.1386	32.1739
38	8.8917	10.6089	12.9720	16.3065	33.0435
39	8.9107	10.6482	13.0520	16.4671	33.9130
40	8.9280	10.6845	13.1271	16.6207	34.7826
48	9.0180	10.8888	13.5847	17.6320	41.7391
50	9.0313	10.9222	13.6674	17.8334	43.4783
54	9.0510	10.9748	13.8048	18.1864	46.9565
60	9.0691	11.0275	13.9558	18.6109	52.1739
65	9.0777	11.0555	14.0447	18.8878	56.5217
70	9.0829	11.0741	14.1096	19.1094	60.8696
80	9.0880	11.0947	14.1917	19.4290	69.5652
90	9.0898	11.1039	14.2356	19.6339	78.2609
100	9.0905	11.1079	14.2590	19.7653	86.9565
120	9.0909	11.1105	14.2781	19.9035	104.3478

Table A.32 Discrete Compounding: $i = 15\%$

	Geometric series future worth factor,				$(F|A_1 \ i,j,n)$
n	j = 4%	j = 6%	j = 8%	j = 10%	j = 15%
1	1.0000	1.0000	1.0000	1.0000	1.0000
2	2.1900	2.2100	2.2300	2.2500	2.3000
3	3.6001	3.6651	3.7309	3.7975	3.9675
4	5.2650	5.4059	5.5502	5.6981	6.0835
5	7.2246	7.4792	7.7433	8.0169	8.7450
6	9.5249	9.9394	10.3741	10.8300	12.0681
7	12.2190	12.8488	13.5171	14.2261	16.1914
8	15.3678	16.2797	17.2585	18.3087	21.2802
9	19.0415	20.3155	21.6982	23.1986	27.5312
10	23.3210	25.0523	26.9519	29.0363	35.1788
11	28.2994	30.6010	33.1536	35.9855	44.5011
12	34.0838	37.0895	40.4583	44.2365	55.8287
13	40.7974	44.6651	49.0452	54.0104	69.5532
14	48.5821	53.4978	59.1216	65.5642	86.1390
15	57.6011	63.7834	70.9270	79.1963	106.1356
16	68.0422	75.7474	84.7382	95.2530	130.1930
17	80.1215	89.6499	100.8749	114.1359	159.0795
18	94.0876	105.7902	119.7061	136.3108	193.7027
19	110.2266	124.5130	141.6581	162.3174	235.1336
20	128.8674	146.2156	167.2225	192.7809	284.6354
21	150.3886	171.3550	196.9684	228.4255	343.6972
22	175.2257	200.4579	231.5456	270.0896	414.0733
23	203.8795	234.1301	271.7140	318.7434	497.8291
24	236.9261	273.0694	318.3425	375.5092	597.3949
25	275.0283	318.0787	372.4351	441.6853	715.6293
26	318.9484	370.0824	435.1488	518.7729	855.8927
27	369.5631	430.1441	507.8175	608.5070	1022.1333
28	427.8810	499.4881	591.9781	712.8931	1218.9886
29	495.0618	579.5230	689.4019	834.2481	1451.9025
30	572.4398	671.8698	802.1295	975.2484	1727.2634
31	661.5491	778.3937	932.5115	1138.9851	2052.5646
32	764.1546	901.2409	1083.2559	1329.0273	2436.5929
33	882.2859	1042.8804	1257.4813	1549.4952	2889.6468
34	1018.2772	1206.1531	1458.7795	1805.1447	3423.7937
35	1174.8130	1394.3271	1691.2865	2101.4641	4053.1675
36	1354.9811	1611.1622	1959.7648	2444.7863	4794.3182
37	1562.3322	1860.9838	2269.6977	2842.4170	5666.6177
38	1800.9501	2148.7675	2627.3979	3302.7836	6692.7350
39	2075.5314	2480.2368	3040.1328	3835.6057	7899.1885
40	2391.4775	2861.9759	3516.2679	4452.0914	9316.9916
48	7389.3653	8922.2982	11131.2743	14447.6900	34201.0678
50	9786.8251	11835.9699	14810.7795	19325.3598	47115.5338
54	17154.6348	20800.7624	26164.4602	34469.0859	88997.7921
60	39758.9011	48344.5673	61181.9447	81590.4721	228730.3333
65	80045.3514	97484.9157	123842.8007	166548.6084	498396.5994
70	161092.2584	196407.1568	250244.1432	338920.0390	1079565.3949
80	652071.1782	796056.4823	1018268.9462	1394052.0739	4991364.6825
90	2638529.1443	3223142.8966	4132185.7913	5699196.5809	22716960.7653
100	10675117.6887	13044157.2069	16744457.6481	23210700.6830	102114195.2907
120	174721221.2670	213537297.9353	274416638.3034	382535463.5055	2005506937.5163

APPENDIX B

CONTINUOUS COMPOUNDING

SECTION I -- CONTINUOUS COMPOUNDING INTEREST FACTORS

SECTION II -- CONTINUOUS COMPOUNDING, CONTINUOUS
FLOW INTEREST FACTORS

SECTION III-- CONTINUOUS COMPOUNDING GEOMETRIC SERIES FACTORS

521

Table B.1 Continuous Compounding: $r = 1/2\%$

	Single payment		Uniform Series				Gradient series	
	Compound amount factor	Present worth factor	Compound amount factor	Sinking fund factor	Present worth factor	Capital recovery factor	Uniform series factor	Present worth factor
n	To find F given P $(F\|P\ r,n)_\infty$	To find P given F $(P\|F\ r,n)_\infty$	To find F given A $(F\|A\ r,n)_\infty$	To find A given F $(A\|F\ r,n)_\infty$	To find P given A $(P\|A\ r,n)_\infty$	To find A given P $(A\|P\ r,n)_\infty$	To find A given G $(A\|G\ r,n)_\infty$	To find P given G $(P\|G\ r,n)_\infty$
1	1.0050	0.9950	1.0000	1.0000	0.9950	1.0050	0.0000	0.0000
2	1.0101	0.9900	2.0050	0.4988	1.9851	0.5038	0.4988	0.9900
3	1.0151	0.9851	3.0151	0.3317	2.9702	0.3367	0.9967	2.9603
4	1.0202	0.9802	4.0302	0.2481	3.9504	0.2531	1.4938	5.9009
5	1.0253	0.9753	5.0504	0.1980	4.9257	0.2030	1.9900	9.8021
6	1.0305	0.9704	6.0757	0.1646	5.8961	0.1696	2.4854	14.6543
7	1.0356	0.9656	7.1061	0.1407	6.8617	0.1457	2.9800	20.4480
8	1.0408	0.9608	8.1418	0.1228	7.8225	0.1278	3.4738	27.1735
9	1.0460	0.9560	9.1826	0.1089	8.7785	0.1139	3.9667	34.8215
10	1.0513	0.9512	10.2286	0.0978	9.7298	0.1028	4.4588	43.3825
11	1.0565	0.9465	11.2799	0.0887	10.6762	0.0937	4.9500	52.8474
12	1.0618	0.9418	12.3364	0.0811	11.6180	0.0861	5.4404	63.2068
13	1.0672	0.9371	13.3983	0.0746	12.5551	0.0796	5.9300	74.4516
14	1.0725	0.9324	14.4654	0.0691	13.4875	0.0741	6.4188	86.5727
15	1.0779	0.9277	15.5379	0.0644	14.4152	0.0694	6.9067	99.5611
16	1.0833	0.9231	16.6158	0.0602	15.3383	0.0652	7.3938	113.4079
17	1.0887	0.9185	17.6991	0.0565	16.2568	0.0615	7.8800	128.1041
18	1.0942	0.9139	18.7878	0.0532	17.1708	0.0582	8.3654	143.6409
19	1.0997	0.9094	19.8820	0.0503	18.0801	0.0553	8.8500	160.0096
20	1.1052	0.9048	20.9816	0.0477	18.9850	0.0527	9.3338	177.2015
21	1.1107	0.9003	22.0868	0.0453	19.8853	0.0503	9.8167	195.2080
22	1.1163	0.8958	23.1975	0.0431	20.7811	0.0481	10.2988	214.0205
23	1.1219	0.8914	24.3138	0.0411	21.6725	0.0461	10.7800	233.6306
24	1.1275	0.8869	25.4357	0.0393	22.5594	0.0443	11.2605	254.0298
25	1.1331	0.8825	26.5632	0.0376	23.4419	0.0427	11.7401	275.2097
26	1.1388	0.8781	27.6963	0.0361	24.3200	0.0411	12.2188	297.1621
27	1.1445	0.8737	28.8351	0.0347	25.1937	0.0397	12.6968	319.8787
28	1.1503	0.8694	29.9797	0.0334	26.0631	0.0384	13.1739	343.3514
29	1.1560	0.8650	31.1300	0.0321	26.9281	0.0371	13.6501	367.5720
30	1.1618	0.8607	32.2860	0.0310	27.7888	0.0360	14.1256	392.5325
31	1.1677	0.8564	33.4478	0.0299	28.6452	0.0349	14.6002	418.2250
32	1.1735	0.8521	34.6155	0.0289	29.4974	0.0339	15.0739	444.6414
33	1.1794	0.8479	35.7890	0.0279	30.3453	0.0330	15.5469	471.7740
34	1.1853	0.8437	36.9684	0.0271	31.1889	0.0321	16.0190	499.6150
35	1.1912	0.8395	38.1537	0.0262	32.0284	0.0312	16.4903	528.1565
36	1.1972	0.8353	39.3449	0.0254	32.8637	0.0304	16.9607	557.3910
37	1.2032	0.8311	40.5422	0.0247	33.6948	0.0297	17.4303	587.3107
38	1.2092	0.8270	41.7454	0.0240	34.5217	0.0290	17.8991	617.9082
39	1.2153	0.8228	42.9546	0.0233	35.3446	0.0283	18.3671	649.1759
40	1.2214	0.8187	44.1699	0.0226	36.1633	0.0277	18.8342	681.1064
45	1.2523	0.7985	50.3385	0.0199	40.1961	0.0249	21.1574	850.4440
48	1.2712	0.7866	54.1143	0.0185	42.5678	0.0235	22.5413	959.5358
50	1.2840	0.7788	56.6632	0.0176	44.1293	0.0227	23.4598	1035.2669
54	1.3100	0.7634	61.8380	0.0162	47.2059	0.0212	25.2869	1193.6902
60	1.3499	0.7408	69.7970	0.0143	51.7069	0.0193	28.0027	1447.9301
65	1.3840	0.7225	76.6143	0.0131	55.3559	0.0181	30.2431	1674.1338
70	1.4191	0.7047	83.6042	0.0120	58.9148	0.0170	32.4629	1912.5472
75	1.4550	0.6873	90.7710	0.0110	62.3859	0.0160	34.6621	2162.4296
80	1.4918	0.6703	98.1192	0.0102	65.7713	0.0152	36.8408	2423.0691
90	1.5683	0.6376	113.3785	0.0088	72.2933	0.0138	41.1368	2973.9130
100	1.6487	0.6065	129.4202	0.0077	78.4973	0.0127	45.3510	3559.9316
120	1.8221	0.5488	164.0130	0.0061	90.0123	0.0111	53.5361	4818.9067

Table B.2 Continuous Compounding: $r = 3/4\%$

	Single payment		Uniform Series				Gradient series	
	Compound amount factor	Present worth factor	Compound amount factor	Sinking fund factor	Present worth factor	Capital recovery factor	Uniform series factor	Present worth factor
	To find F given P	To find P given F	To find F given A	To find A given F	To find P given A	To find A given P	To find A given G	To find P given G
n	$(F\|P\ r,n)_\infty$	$(P\|F\ r,n)_\infty$	$(F\|A\ r,n)_\infty$	$(A\|F\ r,n)_\infty$	$(P\|A\ r,n)_\infty$	$(A\|P\ r,n)_\infty$	$(A\|G\ r,n)_\infty$	$(P\|G\ r,n)_\infty$
1	1.0075	0.9925	1.0000	1.0000	0.9925	1.0075	0.0000	0.0000
2	1.0151	0.9851	2.0075	0.4981	1.9776	0.5057	0.4981	0.9851
3	1.0228	0.9778	3.0226	0.3308	2.9554	0.3384	0.9950	2.9406
4	1.0305	0.9704	4.0454	0.2472	3.9258	0.2547	1.4906	5.8520
5	1.0382	0.9632	5.0759	0.1970	4.8890	0.2045	1.9850	9.7047
6	1.0460	0.9560	6.1141	0.1636	5.8450	0.1711	2.4781	14.4847
7	1.0539	0.9489	7.1601	0.1397	6.7939	0.1472	2.9700	20.1778
8	1.0618	0.9418	8.2140	0.1217	7.7356	0.1293	3.4606	26.7702
9	1.0698	0.9347	9.2758	0.1078	8.6704	0.1153	3.9500	34.2480
10	1.0779	0.9277	10.3457	0.0967	9.5981	0.1042	4.4381	42.5977
11	1.0860	0.9208	11.4235	0.0875	10.5189	0.0951	4.9250	51.8058
12	1.0942	0.9139	12.5095	0.0799	11.4329	0.0875	5.4106	61.8591
13	1.1024	0.9071	13.6037	0.0735	12.3400	0.0810	5.8950	72.7443
14	1.1107	0.9003	14.7061	0.0680	13.2403	0.0755	6.3781	84.4485
15	1.1191	0.8936	15.8168	0.0632	14.1339	0.0708	6.8600	96.9589
16	1.1275	0.8869	16.9359	0.0590	15.0208	0.0666	7.3407	110.2627
17	1.1360	0.8803	18.0634	0.0554	15.9011	0.0629	7.8200	124.3474
18	1.1445	0.8737	19.1994	0.0521	16.7748	0.0596	8.2982	139.2005
19	1.1532	0.8672	20.3439	0.0492	17.6420	0.0567	8.7751	154.8099
20	1.1618	0.8607	21.4971	0.0465	18.5027	0.0540	9.2507	171.1634
21	1.1706	0.8543	22.6589	0.0441	19.3570	0.0517	9.7251	188.2489
22	1.1794	0.8479	23.8295	0.0420	20.2049	0.0495	10.1983	206.0547
23	1.1883	0.8416	25.0089	0.0400	21.0464	0.0475	10.6702	224.5690
24	1.1972	0.8353	26.1972	0.0382	21.8817	0.0457	11.1408	243.7802
25	1.2062	0.8290	27.3944	0.0365	22.7107	0.0440	11.6102	263.6769
26	1.2153	0.8228	28.6006	0.0350	23.5336	0.0425	12.0784	284.2477
27	1.2245	0.8167	29.8159	0.0335	24.3503	0.0411	12.5453	305.4816
28	1.2337	0.8106	31.0404	0.0322	25.1608	0.0397	13.0110	327.3674
29	1.2430	0.8045	32.2741	0.0310	25.9654	0.0385	13.4754	349.8941
30	1.2523	0.7985	33.5170	0.0298	26.7639	0.0374	13.9386	373.0511
31	1.2618	0.7925	34.7693	0.0288	27.5564	0.0363	14.4005	396.8276
32	1.2712	0.7866	36.0311	0.0278	28.3431	0.0353	14.8612	421.2131
33	1.2808	0.7808	37.3023	0.0268	29.1238	0.0343	15.3207	446.1971
34	1.2905	0.7749	38.5832	0.0259	29.8987	0.0334	15.7789	471.7693
35	1.3002	0.7691	39.8736	0.0251	30.6679	0.0326	16.2359	497.9196
36	1.3100	0.7634	41.1738	0.0243	31.4312	0.0318	16.6916	524.6379
37	1.3198	0.7577	42.4838	0.0235	32.1889	0.0311	17.1461	551.9142
38	1.3298	0.7520	43.8036	0.0228	32.9409	0.0304	17.5993	579.7387
39	1.3398	0.7464	45.1334	0.0222	33.6873	0.0297	18.0514	608.1018
40	1.3499	0.7408	46.4731	0.0215	34.4281	0.0290	18.5021	636.9937
45	1.4014	0.7136	53.3248	0.0188	38.0500	0.0263	20.7374	789.0584
48	1.4333	0.6977	57.5609	0.0174	40.1589	0.0249	22.0637	886.0539
50	1.4550	0.6873	60.4383	0.0165	41.5386	0.0241	22.9418	952.9693
54	1.4993	0.6670	66.3243	0.0151	44.2368	0.0226	24.6831	1091.9005
60	1.5683	0.6376	75.4912	0.0132	48.1353	0.0208	27.2582	1312.0804
65	1.6282	0.6142	83.4517	0.0120	51.2527	0.0195	29.3704	1505.3115
70	1.6905	0.5916	91.7164	0.0109	54.2553	0.0184	31.4521	1706.4438
75	1.7551	0.5698	100.2969	0.0100	57.1475	0.0175	33.5034	1914.6339
80	1.8221	0.5488	109.2053	0.0092	59.9331	0.0167	35.5244	2129.0900
90	1.9640	0.5092	128.0563	0.0078	65.2007	0.0153	39.4762	2573.8728
100	2.1170	0.4724	148.3755	0.0067	70.0876	0.0143	43.3084	3035.3867
120	2.4596	0.4066	193.8849	0.0052	78.8277	0.0127	50.6198	3990.2447

Table B.3 Continuous Compounding: r = 1%

	Single payment		Uniform Series				Gradient series	
	Compound amount factor	Present worth factor	Compound amount factor	Sinking fund factor	Present worth factor	Capital recovery factor	Uniform series factor	Present worth factor
	To find F given P	To find P given F	To find F given A	To find A given F	To find P given A	To find A given P	To find A given G	To find P given G
n	$(F\|P\ r,n)_\infty$	$(P\|F\ r,n)_\infty$	$(F\|A\ r,n)_\infty$	$(A\|F\ r,n)_\infty$	$(P\|A\ r,n)_\infty$	$(A\|P\ r,n)_\infty$	$(A\|G\ r,n)_\infty$	$(P\|G\ r,n)_\infty$
1	1.0101	0.9900	1.0000	1.0000	0.9900	1.0101	0.0000	0.0000
2	1.0202	0.9802	2.0101	0.4975	1.9702	0.5076	0.4975	0.9802
3	1.0305	0.9704	3.0303	0.3300	2.9407	0.3401	0.9933	2.9211
4	1.0408	0.9608	4.0607	0.2463	3.9015	0.2563	1.4875	5.8035
5	1.0513	0.9512	5.1015	0.1960	4.8527	0.2061	1.9800	9.6084
6	1.0618	0.9418	6.1528	0.1625	5.7945	0.1726	2.4708	14.3172
7	1.0725	0.9324	7.2146	0.1386	6.7269	0.1487	2.9600	19.9116
8	1.0833	0.9231	8.2871	0.1207	7.6500	0.1307	3.4475	26.3734
9	1.0942	0.9139	9.3704	0.1067	8.5639	0.1168	3.9333	33.6848
10	1.1052	0.9048	10.4646	0.0956	9.4688	0.1056	4.4175	41.8284
11	1.1163	0.8958	11.5698	0.0864	10.3646	0.0965	4.9000	50.7867
12	1.1275	0.8869	12.6860	0.0788	11.2515	0.0889	5.3809	60.5428
13	1.1388	0.8781	13.8135	0.0724	12.1296	0.0824	5.8600	71.0800
14	1.1503	0.8694	14.9524	0.0669	12.9990	0.0769	6.3376	82.3816
15	1.1618	0.8607	16.1026	0.0621	13.8597	0.0722	6.8134	94.4315
16	1.1735	0.8521	17.2645	0.0579	14.7118	0.0680	7.2876	107.2137
17	1.1853	0.8437	18.4380	0.0542	15.5555	0.0643	7.7601	120.7123
18	1.1972	0.8353	19.6233	0.0510	16.3908	0.0610	8.2310	134.9119
19	1.2092	0.8270	20.8205	0.0480	17.2177	0.0581	8.7002	149.7972
20	1.2214	0.8187	22.0298	0.0454	18.0364	0.0554	9.1677	165.3531
21	1.2337	0.8106	23.2512	0.0430	18.8470	0.0531	9.6336	181.5648
22	1.2461	0.8025	24.4848	0.0408	19.6495	0.0509	10.0978	198.4177
23	1.2586	0.7945	25.7309	0.0389	20.4441	0.0489	10.5604	215.8974
24	1.2712	0.7866	26.9895	0.0371	21.2307	0.0471	11.0213	233.9898
25	1.2840	0.7788	28.2608	0.0354	22.0095	0.0454	11.4805	252.6811
26	1.2969	0.7711	29.5448	0.0338	22.7806	0.0439	11.9381	271.9573
27	1.3100	0.7634	30.8417	0.0324	23.5439	0.0425	12.3941	291.8052
28	1.3231	0.7558	32.1517	0.0311	24.2997	0.0412	12.8484	312.2114
29	1.3364	0.7483	33.4748	0.0299	25.0480	0.0399	13.3010	333.1628
30	1.3499	0.7408	34.8112	0.0287	25.7888	0.0388	13.7520	354.6465
31	1.3634	0.7334	36.1611	0.0277	26.5222	0.0377	14.2013	376.6499
32	1.3771	0.7261	37.5245	0.0266	27.2484	0.0367	14.6490	399.1605
33	1.3910	0.7189	38.9017	0.0257	27.9673	0.0358	15.0950	422.1661
34	1.4049	0.7118	40.2926	0.0248	28.6791	0.0349	15.5394	445.6545
35	1.4191	0.7047	41.6976	0.0240	29.3838	0.0340	15.9821	469.6139
36	1.4333	0.6977	43.1166	0.0232	30.0815	0.0332	16.4232	494.0326
37	1.4477	0.6907	44.5500	0.0224	30.7722	0.0325	16.8626	518.8990
38	1.4623	0.6839	45.9977	0.0217	31.4561	0.0318	17.3004	544.2019
39	1.4770	0.6771	47.4600	0.0211	32.1331	0.0311	17.7365	569.9300
40	1.4918	0.6703	48.9370	0.0204	32.8034	0.0305	18.1710	596.0725
45	1.5683	0.6376	56.5475	0.0177	36.0563	0.0277	20.3190	732.6280
48	1.6161	0.6188	61.2999	0.0163	37.9314	0.0264	21.5882	818.8687
50	1.6487	0.6065	64.5483	0.0155	39.1505	0.0255	22.4261	877.9948
54	1.7160	0.5827	71.2433	0.0140	41.5169	0.0241	24.0826	999.8332
60	1.8221	0.5488	81.8015	0.0122	44.8936	0.0223	26.5187	1190.5195
65	1.9155	0.5220	91.0971	0.0110	47.5568	0.0210	28.5045	1355.5862
70	2.0138	0.4966	100.8692	0.0099	50.0902	0.0200	30.4505	1525.2692
75	2.1170	0.4724	111.1424	0.0090	52.5000	0.0190	32.3567	1698.7256
80	2.2255	0.4493	121.9423	0.0082	54.7922	0.0183	34.2235	1875.1837
90	2.4596	0.4066	145.2317	0.0069	59.0468	0.0169	37.8402	2234.3453
100	2.7183	0.3679	170.9705	0.0058	62.8965	0.0159	41.3032	2597.8253
120	3.3201	0.3012	230.8536	0.0043	69.5318	0.0144	47.7793	3322.1789

Table B.4 Continuous Compounding: $r = 3/2\%$

	Single payment		Uniform Series				Gradient series	
	Compound amount factor	Present worth factor	Compound amount factor	Sinking fund factor	Present worth factor	Capital recovery factor	Uniform series factor	Present worth factor
n	To find F given P $(F\|P\ r,n)_\infty$	To find P given F $(P\|F\ r,n)_\infty$	To find F given A $(F\|A\ r,n)_\infty$	To find A given F $(A\|F\ r,n)_\infty$	To find P given A $(P\|A\ r,n)_\infty$	To find A given P $(A\|P\ r,n)_\infty$	To find A given G $(A\|G\ r,n)_\infty$	To find P given G $(P\|G\ r,n)_\infty$
1	1.0151	0.9851	1.0000	1.0000	0.9851	1.0151	0.0000	0.0000
2	1.0305	0.9704	2.0151	0.4963	1.9556	0.5114	0.4963	0.9704
3	1.0460	0.9560	3.0456	0.3283	2.9116	0.3435	0.9900	2.8824
4	1.0618	0.9418	4.0916	0.2444	3.8533	0.2595	1.4813	5.7077
5	1.0779	0.9277	5.1534	0.1940	4.7811	0.2092	1.9700	9.4187
6	1.0942	0.9139	6.2313	0.1605	5.6950	0.1756	2.4563	13.9884
7	1.1107	0.9003	7.3255	0.1365	6.5953	0.1516	2.9400	19.3903
8	1.1275	0.8869	8.4362	0.1185	7.4822	0.1336	3.4213	25.5988
9	1.1445	0.8737	9.5637	0.1046	8.3560	0.1197	3.9000	32.5885
10	1.1618	0.8607	10.7082	0.0934	9.2167	0.1085	4.3763	40.3349
11	1.1794	0.8479	11.8701	0.0842	10.0646	0.0994	4.8501	48.8138
12	1.1972	0.8353	13.0495	0.0766	10.8998	0.0917	5.3213	58.0018
13	1.2153	0.8228	14.2467	0.0702	11.7227	0.0853	5.7901	67.8758
14	1.2337	0.8106	15.4620	0.0647	12.5332	0.0798	6.2564	78.4134
15	1.2523	0.7985	16.6957	0.0599	13.3318	0.0750	6.7202	89.5926
16	1.2712	0.7866	17.9480	0.0557	14.1184	0.0708	7.1816	101.3920
17	1.2905	0.7749	19.2192	0.0520	14.8933	0.0671	7.6404	113.7907
18	1.3100	0.7634	20.5097	0.0488	15.6567	0.0639	8.0967	126.7681
19	1.3298	0.7520	21.8197	0.0458	16.4087	0.0609	8.5506	140.3044
20	1.3499	0.7408	23.1494	0.0432	17.1495	0.0583	9.0020	154.3799
21	1.3703	0.7298	24.4993	0.0408	17.8793	0.0559	9.4509	168.9757
22	1.3910	0.7189	25.8695	0.0387	18.5982	0.0538	9.8973	184.0731
23	1.4120	0.7082	27.2605	0.0367	19.3065	0.0518	10.3413	199.6540
24	1.4333	0.6977	28.6725	0.0349	20.0041	0.0500	10.7828	215.7005
25	1.4550	0.6873	30.1058	0.0332	20.6914	0.0483	11.2218	232.1955
26	1.4770	0.6771	31.5608	0.0317	21.3685	0.0468	11.6584	249.1219
27	1.4993	0.6670	33.0378	0.0303	22.0355	0.0454	12.0925	266.4633
28	1.5220	0.6570	34.5371	0.0290	22.6925	0.0441	12.5241	284.2035
29	1.5450	0.6473	36.0591	0.0277	23.3398	0.0428	12.9533	302.3270
30	1.5683	0.6376	37.6040	0.0266	23.9774	0.0417	13.3800	320.8182
31	1.5920	0.6281	39.1723	0.0255	24.6055	0.0406	13.8043	339.6622
32	1.6161	0.6188	40.7644	0.0245	25.2243	0.0396	14.2261	358.8445
33	1.6405	0.6096	42.3804	0.0236	25.8339	0.0387	14.6455	378.3508
34	1.6653	0.6005	44.0209	0.0227	26.4344	0.0378	15.0625	398.1671
35	1.6905	0.5916	45.6862	0.0219	27.0259	0.0370	15.4770	418.2800
36	1.7160	0.5827	47.3767	0.0211	27.6087	0.0362	15.8891	438.6762
37	1.7419	0.5741	49.0927	0.0204	28.1828	0.0355	16.2987	459.3428
38	1.7683	0.5655	50.8346	0.0197	28.7483	0.0348	16.7059	480.2672
39	1.7950	0.5571	52.6029	0.0190	29.3054	0.0341	17.1108	501.4373
40	1.8221	0.5488	54.3979	0.0184	29.8542	0.0335	17.5131	522.8409
45	1.9640	0.5092	63.7881	0.0157	32.4781	0.0308	19.4890	632.9661
48	2.0544	0.4868	69.7696	0.0143	33.9605	0.0294	20.6458	701.1434
50	2.1170	0.4724	73.9096	0.0135	34.9124	0.0286	21.4052	747.3056
54	2.2479	0.4449	82.5715	0.0121	36.7326	0.0272	22.8955	841.0107
60	2.4596	0.4066	96.5789	0.0104	39.2660	0.0255	25.0609	984.0406
65	2.6512	0.3772	109.2543	0.0092	41.2099	0.0243	26.8018	1104.5001
70	2.8577	0.3499	122.9169	0.0081	43.0133	0.0232	28.4859	1225.2725
75	3.0802	0.3247	137.6436	0.0073	44.6863	0.0224	30.1140	1345.6837
80	3.3201	0.3012	153.5173	0.0065	46.2385	0.0216	31.6869	1465.1553
90	3.8574	0.2592	189.0699	0.0053	49.0145	0.0204	34.6710	1699.3843
100	4.4817	0.2231	230.3761	0.0043	51.4039	0.0195	37.4462	1924.8805
120	6.0496	0.1653	334.1247	0.0030	55.2304	0.0181	42.4039	2341.9848

Table B.5 Continuous Compounding: $r = 2\%$

	Single payment		Uniform Series				Gradient series	
	Compound amount factor	Present worth factor	Compound amount factor	Sinking fund factor	Present worth factor	Capital recovery factor	Uniform series	Present worth
n	To find F given P $(F\|P\ r,n)_\infty$	To find P given F $(P\|F\ r,n)_\infty$	To find F given A $(F\|A\ r,n)_\infty$	To find A given F $(A\|F\ r,n)_\infty$	To find P given A $(P\|A\ r,n)_\infty$	To find A given P $(A\|P\ r,n)_\infty$	To find A given G $(A\|G\ r,n)_\infty$	To find P given G $(P\|G\ r,n)_\infty$
1	1.0202	0.9802	1.0000	1.0000	0.9802	1.0202	0.0000	0.0000
2	1.0408	0.9608	2.0202	0.4950	1.9410	0.5152	0.4950	0.9608
3	1.0618	0.9418	3.0610	0.3267	2.8828	0.3469	0.9867	2.8443
4	1.0833	0.9231	4.1228	0.2426	3.8059	0.2628	1.4750	5.6137
5	1.1052	0.9048	5.2061	0.1921	4.7107	0.2123	1.9600	9.2330
6	1.1275	0.8869	6.3113	0.1584	5.5976	0.1786	2.4417	13.6676
7	1.1503	0.8694	7.4388	0.1344	6.4670	0.1546	2.9200	18.8838
8	1.1735	0.8521	8.5891	0.1164	7.3191	0.1366	3.3950	24.8488
9	1.1972	0.8353	9.7626	0.1024	8.1544	0.1226	3.8667	31.5309
10	1.2214	0.8187	10.9598	0.0912	8.9731	0.1114	4.3351	38.8995
11	1.2461	0.8025	12.1812	0.0821	9.7756	0.1023	4.8002	46.9247
12	1.2712	0.7866	13.4273	0.0745	10.5623	0.0947	5.2619	55.5776
13	1.2969	0.7711	14.6985	0.0680	11.3333	0.0882	5.7203	64.8302
14	1.3231	0.7558	15.9955	0.0625	12.0891	0.0827	6.1754	74.6554
15	1.3499	0.7408	17.3186	0.0577	12.8299	0.0779	6.6272	85.0269
16	1.3771	0.7261	18.6685	0.0536	13.5561	0.0738	7.0757	95.9191
17	1.4049	0.7118	20.0456	0.0499	14.2678	0.0701	7.5209	107.3074
18	1.4333	0.6977	21.4505	0.0466	14.9655	0.0668	7.9628	119.1679
19	1.4623	0.6839	22.8839	0.0437	15.6494	0.0639	8.4014	131.4774
20	1.4918	0.6703	24.3461	0.0411	16.3197	0.0613	8.8368	144.2135
21	1.5220	0.6570	25.8380	0.0387	16.9768	0.0589	9.2688	157.3545
22	1.5527	0.6440	27.3599	0.0365	17.6208	0.0568	9.6976	170.8792
23	1.5841	0.6313	28.9126	0.0346	18.2521	0.0548	10.1231	184.7675
24	1.6161	0.6188	30.4967	0.0328	18.8709	0.0530	10.5453	198.9995
25	1.6487	0.6065	32.1128	0.0311	19.4774	0.0513	10.9643	213.5562
26	1.6820	0.5945	33.7615	0.0296	20.0719	0.0498	11.3800	228.4192
27	1.7160	0.5827	35.4435	0.0282	20.6547	0.0484	11.7925	243.5707
28	1.7507	0.5712	37.1595	0.0269	21.2259	0.0471	12.2018	258.9933
29	1.7860	0.5599	38.9102	0.0257	21.7858	0.0459	12.6078	274.6705
30	1.8221	0.5488	40.6963	0.0246	22.3346	0.0448	13.0106	290.5860
31	1.8589	0.5379	42.5184	0.0235	22.8725	0.0437	13.4102	306.7243
32	1.8965	0.5273	44.3773	0.0225	23.3998	0.0427	13.8065	323.0704
33	1.9348	0.5169	46.2738	0.0216	23.9167	0.0418	14.1997	339.6097
34	1.9739	0.5066	48.2086	0.0207	24.4233	0.0409	14.5897	356.3280
35	2.0138	0.4966	50.1824	0.0199	24.9199	0.0401	14.9765	373.2119
36	2.0544	0.4868	52.1962	0.0192	25.4066	0.0394	15.3601	390.2482
37	2.0959	0.4771	54.2506	0.0184	25.8837	0.0386	15.7406	407.4243
38	2.1383	0.4677	56.3466	0.0177	26.3514	0.0379	16.1179	424.7280
39	2.1815	0.4584	58.4848	0.0171	26.8098	0.0373	16.4920	442.1474
40	2.2255	0.4493	60.6663	0.0165	27.2591	0.0367	16.8630	459.6713
45	2.4596	0.4066	72.2528	0.0138	29.3758	0.0340	18.6714	548.4862
48	2.6117	0.3829	79.7817	0.0125	30.5478	0.0327	19.7194	602.3844
50	2.7183	0.3679	85.0578	0.0118	31.2910	0.0320	20.4028	638.4254
54	2.9447	0.3396	96.2649	0.0104	32.6911	0.0306	21.7336	710.4956
60	3.3201	0.3012	114.8497	0.0087	34.5921	0.0289	23.6409	817.7873
65	3.6693	0.2725	132.1346	0.0076	36.0109	0.0278	25.1507	905.6984
70	4.0552	0.2466	151.2375	0.0066	37.2947	0.0268	26.5899	991.6629
75	4.4817	0.2231	172.3494	0.0058	38.4564	0.0260	27.9604	1075.2549
80	4.9530	0.2019	195.6817	0.0051	39.5075	0.0253	29.2640	1156.1476
90	6.0496	0.1653	249.9660	0.0040	41.3191	0.0242	31.6786	1308.9328
100	7.3891	0.1353	316.2689	0.0032	42.8023	0.0234	33.8499	1448.8552
120	11.0232	0.0907	496.1639	0.0020	45.0110	0.0222	37.5294	1689.2356

Table B.6 Continuous Compounding: $r = 3\%$

	Single payment		Uniform Series				Gradient series	
	Compound amount factor	Present worth factor	Compound amount factor	Sinking fund factor	Present worth factor	Capital recovery factor	Uniform series factor	Present worth factor
n	To find F given P $(F\|P\ r,n)_\infty$	To find P given F $(P\|F\ r,n)_\infty$	To find F given A $(F\|A\ r,n)_\infty$	To find A given F $(A\|F\ r,n)_\infty$	To find P given A $(P\|A\ r,n)_\infty$	To find A given P $(A\|P\ r,n)_\infty$	To find A given G $(A\|G\ r,n)_\infty$	To find P given G $(P\|G\ r,n)_\infty$
1	1.0305	0.9704	1.0000	1.0000	0.9704	1.0305	0.0000	0.0000
2	1.0618	0.9418	2.0305	0.4925	1.9122	0.5230	0.4925	0.9418
3	1.0942	0.9139	3.0923	0.3234	2.8261	0.3538	0.9800	2.7696
4	1.1275	0.8869	4.1865	0.2389	3.7131	0.2693	1.4625	5.4304
5	1.1618	0.8607	5.3140	0.1882	4.5738	0.2186	1.9400	8.8732
6	1.1972	0.8353	6.4758	0.1544	5.4090	0.1849	2.4125	13.0496
7	1.2337	0.8106	7.6730	0.1303	6.2196	0.1608	2.8801	17.9131
8	1.2712	0.7866	8.9067	0.1123	7.0063	0.1427	3.3427	23.4195
9	1.3100	0.7634	10.1779	0.0983	7.7696	0.1287	3.8002	29.5265
10	1.3499	0.7408	11.4879	0.0870	8.5104	0.1175	4.2529	36.1939
11	1.3910	0.7189	12.8378	0.0779	9.2294	0.1083	4.7005	43.3831
12	1.4333	0.6977	14.2287	0.0703	9.9270	0.1007	5.1433	51.0575
13	1.4770	0.6771	15.6621	0.0638	10.6041	0.0943	5.5811	59.1822
14	1.5220	0.6570	17.1390	0.0583	11.2612	0.0888	6.0139	67.7238
15	1.5683	0.6376	18.6610	0.0536	11.8988	0.0840	6.4419	76.6506
16	1.6161	0.6188	20.2293	0.0494	12.5176	0.0799	6.8649	85.9324
17	1.6653	0.6005	21.8454	0.0458	13.1181	0.0762	7.2831	95.5403
18	1.7160	0.5827	23.5107	0.0425	13.7008	0.0730	7.6964	105.4470
19	1.7683	0.5655	25.2267	0.0396	14.2663	0.0701	8.1048	115.6265
20	1.8221	0.5488	26.9950	0.0370	14.8151	0.0675	8.5084	126.0539
21	1.8776	0.5326	28.8171	0.0347	15.3477	0.0652	8.9072	136.7057
22	1.9348	0.5169	30.6947	0.0326	15.8646	0.0630	9.3012	147.5596
23	1.9937	0.5016	32.6295	0.0306	16.3662	0.0611	9.6904	158.5943
24	2.0544	0.4868	34.6232	0.0289	16.8529	0.0593	10.0748	169.7896
25	2.1170	0.4724	36.6776	0.0273	17.3253	0.0577	10.4545	181.1264
26	2.1815	0.4584	38.7946	0.0258	17.7837	0.0562	10.8294	192.5866
27	2.2479	0.4449	40.9761	0.0244	18.2285	0.0549	11.1996	204.1529
28	2.3164	0.4317	43.2240	0.0231	18.6603	0.0536	11.5652	215.8090
29	2.3869	0.4190	45.5404	0.0220	19.0792	0.0524	11.9261	227.5397
30	2.4596	0.4066	47.9273	0.0209	19.4858	0.0513	12.2823	239.3302
31	2.5345	0.3946	50.3869	0.0198	19.8803	0.0503	12.6339	251.1668
32	2.6117	0.3829	52.9214	0.0189	20.2632	0.0494	12.9810	263.0365
33	2.6912	0.3716	55.5331	0.0180	20.6348	0.0485	13.3235	274.9270
34	2.7732	0.3606	58.2243	0.0172	20.9954	0.0476	13.6614	286.8266
35	2.8577	0.3499	60.9975	0.0164	21.3453	0.0468	13.9948	298.7245
36	2.9447	0.3396	63.8552	0.0157	21.6849	0.0461	14.3238	310.6103
37	3.0344	0.3296	66.7999	0.0150	22.0145	0.0454	14.6483	322.4744
38	3.1268	0.3198	69.8342	0.0143	22.3343	0.0448	14.9683	334.3077
39	3.2220	0.3104	72.9610	0.0137	22.6447	0.0442	15.2840	346.1017
40	3.3201	0.3012	76.1830	0.0131	22.9459	0.0436	15.5953	357.8483
45	3.8574	0.2592	93.8259	0.0107	24.3235	0.0411	17.0874	415.6245
48	4.2207	0.2369	105.7542	0.0095	25.0561	0.0399	17.9322	449.3118
50	4.4817	0.2231	114.3242	0.0087	25.5092	0.0392	18.4750	471.2816
54	5.0531	0.1979	133.0866	0.0075	26.3377	0.0380	19.5127	513.9181
60	6.0496	0.1653	165.8094	0.0060	27.4081	0.0365	20.9538	574.3044
65	7.0287	0.1423	197.9570	0.0051	28.1641	0.0355	22.0541	621.1335
70	8.1662	0.1225	235.3072	0.0042	28.8149	0.0347	23.0677	664.6933
75	9.4877	0.1054	278.7019	0.0036	29.3750	0.0340	23.9996	704.9860
80	11.0232	0.0907	329.1193	0.0030	29.8570	0.0335	24.8543	742.0766
90	14.8797	0.0672	455.7526	0.0022	30.6291	0.0326	26.3516	807.1241
100	20.0855	0.0498	626.6895	0.0016	31.2010	0.0321	27.5963	861.0319
120	36.5982	0.0273	1168.8977	0.0009	31.9386	0.0313	29.4649	941.0681

Table B.7 Continuous Compounding: $r = 4\%$

	Single payment		Uniform Series				Gradient series	
	Compound amount factor	Present worth factor	Compound amount factor	Sinking fund factor	Present worth factor	Capital recovery factor	Uniform series factor	Present worth factor
	To find F given P $(F\|P\ r,n)_\infty$	To find P given F $(P\|F\ r,n)_\infty$	To find F given A $(F\|A\ r,n)_\infty$	To find A given F $(A\|F\ r,n)_\infty$	To find P given A $(P\|A\ r,n)_\infty$	To find A given P $(A\|P\ r,n)_\infty$	To find A given G $(A\|G\ r,n)_\infty$	To find P given G $(P\|G\ r,n)_\infty$
n								
1	1.0408	0.9608	1.0000	1.0000	0.9608	1.0408	0.0000	0.0000
2	1.0833	0.9231	2.0408	0.4900	1.8839	0.5308	0.4900	0.9231
3	1.1275	0.8869	3.1241	0.3201	2.7708	0.3609	0.9733	2.6970
4	1.1735	0.8521	4.2516	0.2352	3.6230	0.2760	1.4500	5.2534
5	1.2214	0.8187	5.4251	0.1843	4.4417	0.2251	1.9201	8.5283
6	1.2712	0.7866	6.6465	0.1505	5.2283	0.1913	2.3834	12.4615
7	1.3231	0.7558	7.9178	0.1263	5.9841	0.1671	2.8402	16.9962
8	1.3771	0.7261	9.2409	0.1082	6.7103	0.1490	3.2904	22.0792
9	1.4333	0.6977	10.6180	0.0942	7.4079	0.1350	3.7339	27.6606
10	1.4918	0.6703	12.0513	0.0830	8.0783	0.1238	4.1709	33.6935
11	1.5527	0.6440	13.5432	0.0738	8.7223	0.1146	4.6013	40.1339
12	1.6161	0.6188	15.0959	0.0662	9.3411	0.1071	5.0252	46.9405
13	1.6820	0.5945	16.7120	0.0598	9.9356	0.1006	5.4425	54.0747
14	1.7507	0.5712	18.3940	0.0544	10.5068	0.0952	5.8534	61.5004
15	1.8221	0.5488	20.1447	0.0496	11.0556	0.0905	6.2578	69.1838
16	1.8965	0.5273	21.9668	0.0455	11.5829	0.0863	6.6558	77.0932
17	1.9739	0.5066	23.8633	0.0419	12.0895	0.0827	7.0473	85.1991
18	2.0544	0.4868	25.8371	0.0387	12.5763	0.0795	7.4326	93.4738
19	2.1383	0.4677	27.8916	0.0359	13.0439	0.0767	7.8114	101.8918
20	2.2255	0.4493	30.0298	0.0333	13.4933	0.0741	8.1840	110.4291
21	2.3164	0.4317	32.2554	0.0310	13.9250	0.0718	8.5503	119.0633
22	2.4109	0.4148	34.5717	0.0289	14.3398	0.0697	8.9104	127.7737
23	2.5093	0.3985	36.9826	0.0270	14.7383	0.0679	9.2644	136.5412
24	2.6117	0.3829	39.4919	0.0253	15.1212	0.0661	9.6122	145.3477
25	2.7183	0.3679	42.1036	0.0238	15.4891	0.0646	9.9539	154.1768
26	2.8292	0.3535	44.8219	0.0223	15.8425	0.0631	10.2896	163.0132
27	2.9447	0.3396	47.6511	0.0210	16.1821	0.0618	10.6193	171.8427
28	3.0649	0.3263	50.5958	0.0198	16.5084	0.0606	10.9431	180.6522
29	3.1899	0.3135	53.6607	0.0186	16.8219	0.0594	11.2609	189.4298
30	3.3201	0.3012	56.8506	0.0176	17.1231	0.0584	11.5730	198.1645
31	3.4556	0.2894	60.1707	0.0166	17.4125	0.0574	11.8792	206.8460
32	3.5966	0.2780	63.6263	0.0157	17.6905	0.0565	12.1797	215.4651
33	3.7434	0.2671	67.2230	0.0149	17.9576	0.0557	12.4746	224.0135
34	3.8962	0.2567	70.9664	0.0141	18.2143	0.0549	12.7638	232.4833
35	4.0552	0.2466	74.8626	0.0134	18.4609	0.0542	13.0475	240.8676
36	4.2207	0.2369	78.9178	0.0127	18.6978	0.0535	13.3256	249.1600
37	4.3929	0.2276	83.1385	0.0120	18.9255	0.0528	13.5984	257.3550
38	4.5722	0.2187	87.5314	0.0114	19.1442	0.0522	13.8657	265.4473
39	4.7588	0.2101	92.1036	0.0109	19.3543	0.0517	14.1277	273.4325
40	4.9530	0.2019	96.8625	0.0103	19.5562	0.0511	14.3845	281.3065
45	6.0496	0.1653	123.7332	0.0081	20.4530	0.0489	15.5918	318.8989
48	6.8210	0.1466	142.6329	0.0070	20.9110	0.0478	16.2573	339.9553
50	7.3891	0.1353	156.5532	0.0064	21.1872	0.0472	16.6775	353.3480
54	8.6711	0.1153	187.9684	0.0053	21.6775	0.0461	17.4640	378.5747
60	11.0232	0.0907	245.6012	0.0041	22.2804	0.0449	18.5172	412.5715
65	13.4637	0.0743	305.4031	0.0033	22.6834	0.0441	19.2882	437.5217
70	16.4446	0.0608	378.4453	0.0026	23.0133	0.0435	19.9710	459.5987
75	20.0855	0.0498	467.6593	0.0021	23.2834	0.0429	20.5737	479.0243
80	24.5325	0.0408	576.6254	0.0017	23.5045	0.0425	21.1038	496.0344
90	36.5982	0.0273	872.2754	0.0011	23.8338	0.0420	21.9751	523.7508
100	54.5982	0.0183	1313.3333	0.0008	24.0545	0.0416	22.6376	544.5370
120	121.5104	0.0082	2952.9069	0.0003	24.3017	0.0411	23.5076	571.2733

Table B.8 Continuous Compounding: r = 5%

	Single payment		Uniform Series				Gradient series	
	Compound amount factor	Present worth factor	Compound amount factor	Sinking fund factor	Present worth factor	Capital recovery factor	Uniform series factor	Present worth factor
	To find F given P $(F \vert P\ r, n)_\infty$	To find P given F $(P \vert F\ r, n)_\infty$	To find F given A $(F \vert A\ r, n)_\infty$	To find A given F $(A \vert F\ r, n)_\infty$	To find P given A $(P \vert A\ r, n)_\infty$	To find A given P $(A \vert P\ r, n)_\infty$	To find A given G $(A \vert G\ r, n)_\infty$	To find P given G $(P \vert G\ r, n)_\infty$
n								
1	1.0513	0.9512	1.0000	1.0000	0.9512	1.0513	0.0000	0.0000
2	1.1052	0.9048	2.0513	0.4875	1.8561	0.5388	0.4875	0.9048
3	1.1618	0.8607	3.1564	0.3168	2.7168	0.3681	0.9667	2.6263
4	1.2214	0.8187	4.3183	0.2316	3.5355	0.2828	1.4375	5.0824
5	1.2840	0.7788	5.5397	0.1805	4.3143	0.2318	1.9001	8.1976
6	1.3499	0.7408	6.8237	0.1465	5.0551	0.1978	2.3544	11.9017
7	1.4191	0.7047	8.1736	0.1223	5.7598	0.1736	2.8004	16.1299
8	1.4918	0.6703	9.5926	0.1042	6.4301	0.1555	3.2382	20.8221
9	1.5683	0.6376	11.0845	0.0902	7.0678	0.1415	3.6678	25.9231
10	1.6487	0.6065	12.6528	0.0790	7.6743	0.1303	4.0892	31.3819
11	1.7333	0.5769	14.3015	0.0699	8.2512	0.1212	4.5025	37.1514
12	1.8221	0.5488	16.0347	0.0624	8.8001	0.1136	4.9077	43.1883
13	1.9155	0.5220	17.8569	0.0560	9.3221	0.1073	5.3049	49.4529
14	2.0138	0.4966	19.7724	0.0506	9.8187	0.1018	5.6941	55.9085
15	2.1170	0.4724	21.7862	0.0459	10.2911	0.0972	6.0753	62.5216
16	2.2255	0.4493	23.9032	0.0418	10.7404	0.0931	6.4487	69.2616
17	2.3396	0.4274	26.1287	0.0383	11.1678	0.0895	6.8143	76.1002
18	2.4596	0.4066	28.4683	0.0351	11.5744	0.0864	7.1720	83.0119
19	2.5857	0.3867	30.9279	0.0323	11.9611	0.0836	7.5221	89.9732
20	2.7183	0.3679	33.5137	0.0298	12.3290	0.0811	7.8646	96.9629
21	2.8577	0.3499	36.2319	0.0276	12.6789	0.0789	8.1996	103.9617
22	3.0042	0.3329	39.0896	0.0256	13.0118	0.0769	8.5270	110.9520
23	3.1582	0.3166	42.0938	0.0238	13.3284	0.0750	8.8471	117.9180
24	3.3201	0.3012	45.2519	0.0221	13.6296	0.0734	9.1599	124.8455
25	3.4903	0.2865	48.5721	0.0206	13.9161	0.0719	9.4654	131.7216
26	3.6693	0.2725	52.0624	0.0192	14.1887	0.0705	9.7638	138.5349
27	3.8574	0.2592	55.7317	0.0179	14.4479	0.0692	10.0551	145.2751
28	4.0552	0.2466	59.5891	0.0168	14.6945	0.0681	10.3395	151.9332
29	4.2631	0.2346	63.6443	0.0157	14.9291	0.0670	10.6170	158.5012
30	4.4817	0.2231	67.9074	0.0147	15.1522	0.0660	10.8877	164.9720
31	4.7115	0.2122	72.3891	0.0138	15.3644	0.0651	11.1517	171.3394
32	4.9530	0.2019	77.1006	0.0130	15.5663	0.0642	11.4091	177.5982
33	5.2070	0.1920	82.0536	0.0122	15.7584	0.0635	11.6601	183.7438
34	5.4739	0.1827	87.2606	0.0115	15.9411	0.0627	11.9046	189.7724
35	5.7546	0.1738	92.7346	0.0108	16.1149	0.0621	12.1429	195.6807
36	6.0496	0.1653	98.4892	0.0102	16.2801	0.0614	12.3750	201.4661
37	6.3598	0.1572	104.5388	0.0096	16.4374	0.0608	12.6009	207.1267
38	6.6859	0.1496	110.8986	0.0090	16.5870	0.0603	12.8210	212.6607
39	7.0287	0.1423	117.5845	0.0085	16.7292	0.0598	13.0351	218.0671
40	7.3891	0.1353	124.6132	0.0080	16.8646	0.0593	13.2435	223.3452
45	9.4877	0.1054	165.5462	0.0060	17.4484	0.0573	14.2024	247.8097
48	11.0232	0.0907	195.4937	0.0051	17.7348	0.0564	14.7153	260.9721
50	12.1825	0.0821	218.1052	0.0046	17.9032	0.0559	15.0329	269.1364
54	14.8797	0.0672	270.7126	0.0037	18.1934	0.0550	15.6136	284.0642
60	20.0855	0.0498	372.2475	0.0027	18.5331	0.0540	16.3604	303.2096
65	25.7903	0.0388	483.5149	0.0021	18.7479	0.0533	16.8822	316.5055
70	33.1155	0.0302	626.3851	0.0016	18.9152	0.0529	17.3245	327.6968
75	42.5211	0.0235	809.8341	0.0012	19.0455	0.0525	17.6979	337.0640
80	54.5982	0.0183	1045.3872	0.0010	19.1469	0.0522	18.0116	344.8665
90	90.0171	0.0111	1736.2049	0.0006	19.2875	0.0518	18.4931	356.6861
100	148.4132	0.0067	2875.1708	0.0003	19.3727	0.0516	18.8258	364.7075
120	403.4288	0.0025	7849.0382	0.0001	19.4558	0.0514	19.2060	373.6680

Table B.9 Continuous Compounding: $r = 6\%$

	Single payment		Uniform Series				Gradient series	
	Compound amount factor	Present worth factor	Compound amount factor	Sinking fund factor	Present worth factor	Capital recovery factor	Uniform series factor	Present worth factor
n	To find F given P $(F\|P\ r,n)_{\infty}$	To find P given F $(P\|F\ r,n)_{\infty}$	To find F given A $(F\|A\ r,n)_{\infty}$	To find A given F $(A\|F\ r,n)_{\infty}$	To find P given A $(P\|A\ r,n)_{\infty}$	To find A given P $(A\|P\ r,n)_{\infty}$	To find A given G $(A\|G\ r,n)_{\infty}$	To find P given G $(P\|G\ r,n)_{\infty}$
1	1.0618	0.9418	1.0000	1.0000	0.9418	1.0618	0.0000	0.0000
2	1.1275	0.8869	2.0618	0.4850	1.8287	0.5468	0.4850	0.8869
3	1.1972	0.8353	3.1893	0.3135	2.6640	0.3754	0.9600	2.5575
4	1.2712	0.7866	4.3866	0.2280	3.4506	0.2898	1.4251	4.9173
5	1.3499	0.7408	5.6578	0.1767	4.1914	0.2386	1.8802	7.8806
6	1.4333	0.6977	7.0077	0.1427	4.8891	0.2045	2.3254	11.3690
7	1.5220	0.6570	8.4410	0.1185	5.5461	0.1803	2.7607	15.3113
8	1.6161	0.6188	9.9629	0.1004	6.1649	0.1622	3.1862	19.6428
9	1.7160	0.5827	11.5790	0.0864	6.7477	0.1482	3.6020	24.3047
10	1.8221	0.5488	13.2950	0.0752	7.2965	0.1371	4.0080	29.2441
11	1.9348	0.5169	15.1171	0.0662	7.8133	0.1280	4.4043	34.4126
12	2.0544	0.4868	17.0519	0.0586	8.3001	0.1205	4.7911	39.7668
13	2.1815	0.4584	19.1064	0.0523	8.7585	0.1142	5.1684	45.2677
14	2.3164	0.4317	21.2878	0.0470	9.1902	0.1088	5.5363	50.8800
15	2.4596	0.4066	23.6042	0.0424	9.5968	0.1042	5.8949	56.5719
16	2.6117	0.3829	26.0638	0.0384	9.9797	0.1002	6.2442	62.3153
17	2.7732	0.3606	28.6755	0.0349	10.3402	0.0967	6.5845	68.0848
18	2.9447	0.3396	31.4487	0.0318	10.6798	0.0936	6.9156	73.8580
19	3.1268	0.3198	34.3934	0.0291	10.9997	0.0909	7.2379	79.6147
20	3.3201	0.3012	37.5202	0.0267	11.3009	0.0885	7.5514	85.3374
21	3.5254	0.2837	40.8403	0.0245	11.5845	0.0863	7.8562	91.0105
22	3.7434	0.2671	44.3657	0.0225	11.8516	0.0844	8.1525	96.6203
23	3.9749	0.2516	48.1091	0.0208	12.1032	0.0826	8.4403	102.1550
24	4.2207	0.2369	52.0840	0.0192	12.3401	0.0810	8.7199	107.6044
25	4.4817	0.2231	56.3047	0.0178	12.5633	0.0796	8.9912	112.9595
26	4.7588	0.2101	60.7864	0.0165	12.7734	0.0783	9.2546	118.2129
27	5.0531	0.1979	65.5452	0.0153	12.9713	0.0771	9.5101	123.3583
28	5.3656	0.1864	70.5983	0.0142	13.1577	0.0760	9.7578	128.3904
29	5.6973	0.1755	75.9639	0.0132	13.3332	0.0750	9.9980	133.3049
30	6.0496	0.1653	81.6612	0.0122	13.4985	0.0741	10.2307	138.0986
31	6.4237	0.1557	87.7109	0.0114	13.6542	0.0732	10.4560	142.7688
32	6.8210	0.1466	94.1346	0.0106	13.8008	0.0725	10.6743	147.3136
33	7.2427	0.1381	100.9556	0.0099	13.9389	0.0717	10.8855	151.7318
34	7.6906	0.1300	108.1983	0.0092	14.0689	0.0711	11.0899	156.0228
35	8.1662	0.1225	115.8889	0.0086	14.1913	0.0705	11.2876	160.1863
36	8.6711	0.1153	124.0551	0.0081	14.3067	0.0699	11.4788	164.2227
37	9.2073	0.1086	132.7262	0.0075	14.4153	0.0694	11.6635	168.1326
38	9.7767	0.1023	141.9335	0.0070	14.5176	0.0689	11.8420	171.9171
39	10.3812	0.0963	151.7102	0.0066	14.6139	0.0684	12.0144	175.5776
40	11.0232	0.0907	162.0915	0.0062	14.7046	0.0680	12.1809	179.1156
45	14.8797	0.0672	224.4584	0.0045	15.0848	0.0663	12.9295	195.0399
48	17.8143	0.0561	271.9148	0.0037	15.2639	0.0655	13.3169	203.2682
50	20.0855	0.0498	308.6449	0.0032	15.3665	0.0651	13.5519	208.2453
54	25.5337	0.0392	396.7512	0.0025	15.5383	0.0644	13.9706	217.0799
60	36.5982	0.0273	575.6828	0.0017	15.7298	0.0636	14.4862	227.8648
65	49.4024	0.0202	782.7483	0.0013	15.8443	0.0631	14.8288	234.9516
70	66.6863	0.0150	1062.2574	0.0009	15.9292	0.0628	15.1060	240.6259
75	90.0171	0.0111	1439.5553	0.0007	15.9920	0.0625	15.3291	245.1437
80	121.5104	0.0082	1948.8543	0.0005	16.0386	0.0623	15.5078	248.7234
90	221.4064	0.0045	3564.3390	0.0003	16.0986	0.0621	15.7633	253.7679
100	403.4288	0.0025	6507.9442	0.0002	16.1316	0.0620	15.9232	256.8660
120	1339.4308	0.0007	21644.6558	0.0000	16.1596	0.0619	16.0820	259.8787

Table B.10 Continuous Compounding: r = 7%

	Single payment		Uniform Series				Gradient series	
	Compound amount factor	Present worth factor	Compound amount factor	Sinking fund factor	Present worth factor	Capital recovery factor	Uniform series factor	Present worth factor
	To find F given P $(F|P\ r,n)_\infty$	To find P given F $(P|F\ r,n)_\infty$	To find F given A $(F|A\ r,n)_\infty$	To find A given F $(A|F\ r,n)_\infty$	To find P given A $(P|A\ r,n)_\infty$	To find A given P $(A|P\ r,n)_\infty$	To find A given G $(A|G\ r,n)_\infty$	To find P given G $(P|G\ r,n)_\infty$
n								
1	1.0725	0.9324	1.0000	1.0000	0.9324	1.0725	0.0000	0.0000
2	1.1503	0.8694	2.0725	0.4825	1.8018	0.5550	0.4825	0.8694
3	1.2337	0.8106	3.2228	0.3103	2.6123	0.3828	0.9534	2.4905
4	1.3231	0.7558	4.4565	0.2244	3.3681	0.2969	1.4126	4.7579
5	1.4191	0.7047	5.7796	0.1730	4.0728	0.2455	1.8603	7.5766
6	1.5220	0.6570	7.1987	0.1389	4.7299	0.2114	2.2964	10.8619
7	1.6323	0.6126	8.7206	0.1147	5.3425	0.1872	2.7211	14.5376
8	1.7507	0.5712	10.3529	0.0966	5.9137	0.1691	3.1344	18.5361
9	1.8776	0.5326	12.1036	0.0826	6.4463	0.1551	3.5364	22.7968
10	2.0138	0.4966	13.9812	0.0715	6.9429	0.1440	3.9272	27.2661
11	2.1598	0.4630	15.9950	0.0625	7.4059	0.1350	4.3069	31.8962
12	2.3164	0.4317	18.1547	0.0551	7.8376	0.1276	4.6755	36.6450
13	2.4843	0.4025	20.4711	0.0488	8.2401	0.1214	5.0333	41.4753
14	2.6645	0.3753	22.9554	0.0436	8.6154	0.1161	5.3804	46.3544
15	2.8577	0.3499	25.6199	0.0390	8.9654	0.1115	5.7168	51.2535
16	3.0649	0.3263	28.4775	0.0351	9.2916	0.1076	6.0428	56.1477
17	3.2871	0.3042	31.5424	0.0317	9.5959	0.1042	6.3585	61.0152
18	3.5254	0.2837	34.8295	0.0287	9.8795	0.1012	6.6640	65.8374
19	3.7810	0.2645	38.3549	0.0261	10.1440	0.0986	6.9596	70.5979
20	4.0552	0.2466	42.1359	0.0237	10.3906	0.0962	7.2453	75.2833
21	4.3492	0.2299	46.1911	0.0216	10.6205	0.0942	7.5215	79.8818
22	4.6646	0.2144	50.5404	0.0198	10.8349	0.0923	7.7881	84.3838
23	5.0028	0.1999	55.2050	0.0181	11.0348	0.0906	8.0456	88.7813
24	5.3656	0.1864	60.2078	0.0166	11.2212	0.0891	8.2940	93.0679
25	5.7546	0.1738	65.5733	0.0153	11.3949	0.0878	8.5335	97.2385
26	6.1719	0.1620	71.3279	0.0140	11.5570	0.0865	8.7643	101.2891
27	6.6194	0.1511	77.4998	0.0129	11.7080	0.0854	8.9867	105.2170
28	7.0993	0.1409	84.1192	0.0119	11.8489	0.0844	9.2009	109.0202
29	7.6141	0.1313	91.2185	0.0110	11.9802	0.0835	9.4070	112.6976
30	8.1662	0.1225	98.8326	0.0101	12.1027	0.0826	9.6052	116.2488
31	8.7583	0.1142	106.9987	0.0093	12.2169	0.0819	9.7958	119.6742
32	9.3933	0.1065	115.7570	0.0086	12.3233	0.0811	9.9790	122.9744
33	10.0744	0.0993	125.1504	0.0080	12.4226	0.0805	10.1550	126.1507
34	10.8049	0.0926	135.2248	0.0074	12.5151	0.0799	10.3239	129.2049
35	11.5883	0.0863	146.0297	0.0068	12.6014	0.0794	10.4860	132.1389
36	12.4286	0.0805	157.6180	0.0063	12.6819	0.0789	10.6416	134.9550
37	13.3298	0.0750	170.0466	0.0059	12.7569	0.0784	10.7907	137.8557
38	14.2963	0.0699	183.3764	0.0055	12.8269	0.0780	10.9336	140.2438
39	15.3329	0.0652	197.6727	0.0051	12.8921	0.0776	11.0705	142.7221
40	16.4446	0.0608	213.0056	0.0047	12.9529	0.0772	11.2017	145.0937
45	23.3361	0.0429	308.0489	0.0032	13.2006	0.0758	11.7769	155.4611
48	28.7892	0.0347	383.2559	0.0026	13.3125	0.0751	12.0643	160.6053
50	33.1155	0.0302	442.9218	0.0023	13.3751	0.0748	12.2347	163.6396
54	43.8160	0.0228	590.4995	0.0017	13.4768	0.0742	12.5303	168.8687
60	66.6863	0.0150	905.9161	0.0011	13.5847	0.0736	12.8781	174.9458
65	94.6324	0.0106	1291.3358	0.0008	13.6458	0.0733	13.0973	178.7238
70	134.2898	0.0074	1838.2723	0.0005	13.6888	0.0731	13.2664	181.6014
75	190.5663	0.0052	2614.4121	0.0004	13.7192	0.0729	13.3959	183.7808
80	270.4264	0.0037	3715.8070	0.0003	13.7405	0.0728	13.4946	185.4235
90	544.5719	0.0018	7496.6976	0.0001	13.7662	0.0726	13.6260	187.5782
100	1096.6332	0.0009	15110.4764	0.0001	13.7790	0.0726	13.7003	188.7757
120	4447.0667	0.0002	61318.1392	0.0000	13.7884	0.0725	13.7646	189.7918

Table B.11 Continuous Compounding: $r = 8\%$

	Single payment		Uniform Series				Gradient series	
	Compound amount factor	Present worth factor	Compound amount factor	Sinking fund factor	Present worth factor	Capital recovery factor	Uniform series factor	Present worth factor
n	To find F given P $(F\|P\ r,n)_\infty$	To find P given F $(P\|F\ r,n)_\infty$	To find F given A $(F\|A\ r,n)_\infty$	To find A given F $(A\|F\ r,n)_\infty$	To find P given A $(P\|A\ r,n)_\infty$	To find A given P $(A\|P\ r,n)_\infty$	To find A given G $(A\|G\ r,n)_\infty$	To find P given G $(P\|G\ r,n)_\infty$
1	1.0833	0.9231	1.0000	1.0000	0.9231	1.0833	0.0000	0.0000
2	1.1735	0.8521	2.0833	0.4800	1.7753	0.5633	0.4800	0.8521
3	1.2712	0.7866	3.2568	0.3071	2.5619	0.3903	0.9467	2.4254
4	1.3771	0.7261	4.5280	0.2208	3.2880	0.3041	1.4002	4.6038
5	1.4918	0.6703	5.9052	0.1693	3.9584	0.2526	1.8404	7.2851
6	1.6161	0.6188	7.3970	0.1352	4.5771	0.2185	2.2676	10.3790
7	1.7507	0.5712	9.0131	0.1109	5.1483	0.1942	2.6817	13.8063
8	1.8965	0.5273	10.7637	0.0929	5.6756	0.1762	3.0829	17.4973
9	2.0544	0.4868	12.6602	0.0790	6.1624	0.1623	3.4713	21.3914
10	2.2255	0.4493	14.7147	0.0680	6.6117	0.1512	3.8470	25.4353
11	2.4109	0.4148	16.9402	0.0590	7.0265	0.1423	4.2102	29.5832
12	2.6117	0.3829	19.3511	0.0517	7.4094	0.1350	4.5611	33.7950
13	2.8292	0.3535	21.9628	0.0455	7.7629	0.1288	4.8998	38.0364
14	3.0649	0.3263	24.7920	0.0403	8.0891	0.1236	5.2265	42.2781
15	3.3201	0.3012	27.8569	0.0359	8.3903	0.1192	5.5415	46.4948
16	3.5966	0.2780	31.1770	0.0321	8.6684	0.1154	5.8449	50.6653
17	3.8962	0.2567	34.7736	0.0288	8.9250	0.1120	6.1369	54.7719
18	4.2207	0.2369	38.6698	0.0259	9.1620	0.1091	6.4178	58.7997
19	4.5722	0.2187	42.8905	0.0233	9.3807	0.1066	6.6879	62.7365
20	4.9530	0.2019	47.4627	0.0211	9.5826	0.1044	6.9473	66.5725
21	5.3656	0.1864	52.4158	0.0191	9.7689	0.1024	7.1963	70.3000
22	5.8124	0.1720	57.7813	0.0173	9.9410	0.1006	7.4352	73.9100
23	6.2965	0.1588	63.5938	0.0157	10.0998	0.0990	7.6642	77.4069
24	6.8210	0.1466	69.8903	0.0143	10.2464	0.0976	7.8836	80.7789
25	7.3891	0.1353	76.7113	0.0130	10.3817	0.0963	8.0937	84.0270
26	8.0045	0.1249	84.1003	0.0119	10.5067	0.0952	8.2947	87.1502
27	8.6711	0.1153	92.1048	0.0109	10.6220	0.0941	8.4870	90.1487
28	9.3933	0.1065	100.7759	0.0099	10.7285	0.0932	8.6707	93.0230
29	10.1757	0.0983	110.1693	0.0091	10.8267	0.0924	8.8461	95.7747
30	11.0232	0.0907	120.3449	0.0083	10.9174	0.0916	9.0136	98.4055
31	11.9413	0.0837	131.3681	0.0076	11.0012	0.0909	9.1734	100.9178
32	12.9358	0.0773	143.3094	0.0070	11.0785	0.0903	9.3257	103.3143
33	14.0132	0.0714	156.2452	0.0064	11.1499	0.0897	9.4708	105.5978
34	15.1803	0.0659	170.2584	0.0059	11.2157	0.0892	9.6090	107.7717
35	16.4446	0.0608	185.4387	0.0054	11.2765	0.0887	9.7405	109.8392
36	17.8143	0.0561	201.8834	0.0050	11.3327	0.0882	9.8656	111.8039
37	19.2980	0.0518	219.6976	0.0046	11.3845	0.0878	9.9846	113.6694
38	20.9052	0.0478	238.9956	0.0042	11.4323	0.0875	10.0976	115.4393
39	22.6464	0.0442	259.9008	0.0038	11.4765	0.0871	10.2050	117.1173
40	24.5325	0.0408	282.5472	0.0035	11.5172	0.0868	10.3069	118.7070
45	36.5982	0.0273	427.4161	0.0023	11.6786	0.0856	10.7426	125.4580
48	46.5255	0.0215	546.6092	0.0018	11.7486	0.0851	10.9523	128.6743
50	54.5982	0.0183	643.5351	0.0016	11.7868	0.0848	11.0738	130.5242
54	75.1886	0.0133	890.7581	0.0011	11.8470	0.0844	11.2788	133.6196
60	121.5104	0.0082	1446.9283	0.0007	11.9079	0.0840	11.5088	137.0449
65	181.2722	0.0055	2164.4686	0.0005	11.9404	0.0837	11.6461	139.0594
70	270.4264	0.0037	3234.9129	0.0003	11.9623	0.0836	11.7469	140.5190
75	403.4288	0.0025	4831.8281	0.0002	11.9769	0.0835	11.8203	141.5706
80	601.8450	0.0017	7214.1457	0.0001	11.9867	0.0834	11.8735	142.3245
90	1339.4308	0.0007	16070.0911	0.0001	11.9977	0.0833	11.9394	143.2456
100	2980.9580	0.0003	35779.3601	0.0000	12.0026	0.0833	11.9731	143.7089
120	14764.7816	0.0001	177263.7935	0.0000	12.0059	0.0833	11.9985	144.0527

Table B.12 Continuous Compounding: $r = 9\%$

	Single payment		Uniform Series				Gradient series	
	Compound amount factor	Present worth factor	Compound amount factor	Sinking fund factor	Present worth factor	Capital recovery factor	Uniform series factor	Present worth factor
n	To find F given P $(F\|P\ r,n)_\infty$	To find P given F $(P\|F\ r,n)_\infty$	To find F given A $(F\|A\ r,n)_\infty$	To find A given F $(A\|F\ r,n)_\infty$	To find P given A $(P\|A\ r,n)_\infty$	To find A given P $(A\|P\ r,n)_\infty$	To find A given G $(A\|G\ r,n)_\infty$	To find P given G $(P\|G\ r,n)_\infty$
1	1.0942	0.9139	1.0000	1.0000	0.9139	1.0942	0.0000	0.0000
2	1.1972	0.8353	2.0942	0.4775	1.7492	0.5717	0.4775	0.8353
3	1.3100	0.7634	3.2914	0.3038	2.5126	0.3980	0.9401	2.3620
4	1.4333	0.6977	4.6014	0.2173	3.2103	0.3115	1.3878	4.4551
5	1.5683	0.6376	6.0347	0.1657	3.8479	0.2599	1.8206	7.0056
6	1.7160	0.5827	7.6030	0.1315	4.4306	0.2257	2.2388	9.9193
7	1.8776	0.5326	9.3190	0.1073	4.9632	0.2015	2.6424	13.1149
8	2.0544	0.4868	11.1966	0.0893	5.4500	0.1835	3.0316	16.5221
9	2.2479	0.4449	13.2510	0.0755	5.8948	0.1696	3.4065	20.0810
10	2.4596	0.4066	15.4990	0.0645	6.3014	0.1587	3.7674	23.7401
11	2.6912	0.3716	17.9586	0.0557	6.6730	0.1499	4.1145	27.4559
12	2.9447	0.3396	20.6498	0.0484	7.0126	0.1426	4.4479	31.1914
13	3.2220	0.3104	23.5945	0.0424	7.3229	0.1366	4.7680	34.9158
14	3.5254	0.2837	26.8165	0.0373	7.6066	0.1315	5.0750	38.6033
15	3.8574	0.2592	30.3419	0.0330	7.8658	0.1271	5.3691	42.2327
16	4.2207	0.2369	34.1993	0.0292	8.1028	0.1234	5.6507	45.7866
17	4.6182	0.2165	38.4200	0.0260	8.3193	0.1202	5.9201	49.2512
18	5.0531	0.1979	43.0382	0.0232	8.5172	0.1174	6.1776	52.6155
19	5.5290	0.1809	48.0913	0.0208	8.6981	0.1150	6.4234	55.8711
20	6.0496	0.1653	53.6202	0.0186	8.8634	0.1128	6.6579	59.0117
21	6.6194	0.1511	59.6699	0.0168	9.0144	0.1109	6.8815	62.0332
22	7.2427	0.1381	66.2893	0.0151	9.1525	0.1093	7.0945	64.9326
23	7.9248	0.1262	73.5320	0.0136	9.2787	0.1078	7.2972	67.7087
24	8.6711	0.1153	81.4568	0.0123	9.3940	0.1065	7.4900	70.3612
25	9.4877	0.1054	90.1280	0.0111	9.4994	0.1053	7.6732	72.8908
26	10.3812	0.0963	99.6157	0.0100	9.5957	0.1042	7.8471	75.2990
27	11.3589	0.0880	109.9969	0.0091	9.6838	0.1033	8.0122	77.5879
28	12.4286	0.0805	121.3558	0.0082	9.7642	0.1024	8.1686	79.7603
29	13.5991	0.0735	133.7844	0.0075	9.8378	0.1016	8.3168	81.8193
30	14.8797	0.0672	147.3835	0.0068	9.9050	0.1010	8.4572	83.7683
31	16.2810	0.0614	162.2632	0.0062	9.9664	0.1003	8.5899	85.6109
32	17.8143	0.0561	178.5442	0.0056	10.0225	0.0998	8.7155	87.3511
33	19.4919	0.0513	196.3585	0.0051	10.0738	0.0993	8.8340	88.9928
34	21.3276	0.0469	215.8504	0.0046	10.1207	0.0988	8.9460	90.5401
35	23.3361	0.0429	237.1780	0.0042	10.1636	0.0984	9.0516	91.9970
36	25.5337	0.0392	260.5140	0.0038	10.2027	0.0980	9.1512	93.3678
37	27.9383	0.0358	286.0477	0.0035	10.2385	0.0977	9.2451	94.6563
38	30.5694	0.0327	313.9861	0.0032	10.2712	0.0974	9.3335	95.8667
39	33.4483	0.0299	344.5555	0.0029	10.3011	0.0971	9.4167	97.0028
40	36.5982	0.0273	378.0038	0.0026	10.3285	0.0968	9.4950	98.0684
45	57.3975	0.0174	598.8626	0.0017	10.4336	0.0958	9.8207	102.4654
48	75.1886	0.0133	787.7801	0.0013	10.4774	0.0954	9.9716	104.4764
50	90.0171	0.0111	945.2382	0.0011	10.5006	0.0952	10.0569	105.6042
54	129.0242	0.0078	1359.4391	0.0007	10.5363	0.0949	10.1968	107.4368
60	221.4064	0.0045	2340.4098	0.0004	10.5707	0.0946	10.3464	109.3680
65	347.2344	0.0029	3676.5279	0.0003	10.5880	0.0944	10.4309	110.4424
70	544.5719	0.0018	5771.9782	0.0002	10.5991	0.0943	10.4898	111.1829
75	854.0588	0.0012	9058.2984	0.0001	10.6062	0.0943	10.5307	111.6904
80	1339.4308	0.0007	14212.2744	0.0001	10.6107	0.0942	10.5588	112.0365
90	3294.4681	0.0003	34972.0534	0.0000	10.6154	0.0942	10.5913	112.4306
100	8103.0839	0.0001	86032.8702	0.0000	10.6173	0.0942	10.6063	112.6099
120	49020.8011	0.0000	520522.1554	0.0000	10.6184	0.0942	10.6162	112.7266

Table B.13 Continuous Compounding: $r = 10\%$

	Single payment		Uniform Series				Gradient series	
	Compound amount factor	Present worth factor	Compound amount factor	Sinking fund factor	Present worth factor	Capital recovery factor	Uniform series factor	Present worth factor
n	To find F given P $(F\|P\ r,n)_\infty$	To find P given F $(P\|F\ r,n)_\infty$	To find F given A $(F\|A\ r,n)_\infty$	To find A given F $(A\|F\ r,n)_\infty$	To find P given A $(P\|A\ r,n)_\infty$	To find A given P $(A\|P\ r,n)_\infty$	To find A given G $(A\|G\ r,n)_\infty$	To find P given G $(P\|G\ r,n)_\infty$
1	1.1052	0.9048	1.0000	1.0000	0.9048	1.1052	0.0000	0.0000
2	1.2214	0.8187	2.1052	0.4750	1.7236	0.5802	0.4750	0.8187
3	1.3499	0.7408	3.3266	0.3006	2.4644	0.4058	0.9334	2.3004
4	1.4918	0.6703	4.6764	0.2138	3.1347	0.3190	1.3754	4.3113
5	1.6487	0.6065	6.1683	0.1621	3.7412	0.2673	1.8009	6.7374
6	1.8221	0.5488	7.8170	0.1279	4.2900	0.2331	2.2101	9.4815
7	2.0138	0.4966	9.6391	0.1037	4.7866	0.2089	2.6033	12.4610
8	2.2255	0.4493	11.6528	0.0858	5.2360	0.1910	2.9806	15.6063
9	2.4596	0.4066	13.8784	0.0721	5.6425	0.1772	3.3423	18.8589
10	2.7183	0.3679	16.3380	0.0612	6.0104	0.1664	3.6886	22.1698
11	3.0042	0.3329	19.0563	0.0525	6.3433	0.1576	4.0198	25.4985
12	3.3201	0.3012	22.0604	0.0453	6.6445	0.1505	4.3362	28.8116
13	3.6693	0.2725	25.3806	0.0394	6.9170	0.1446	4.6381	32.0820
14	4.0552	0.2466	29.0499	0.0344	7.1636	0.1396	4.9260	35.2878
15	4.4817	0.2231	33.1051	0.0302	7.3867	0.1354	5.2001	38.4116
16	4.9530	0.2019	37.5867	0.0266	7.5886	0.1318	5.4608	41.4401
17	5.4739	0.1827	42.5398	0.0235	7.7713	0.1287	5.7086	44.3630
18	6.0496	0.1653	48.0137	0.0208	7.9366	0.1260	5.9437	47.1731
19	6.6859	0.1496	54.0634	0.0185	8.0862	0.1237	6.1667	49.8653
20	7.3891	0.1353	60.7493	0.0165	8.2215	0.1216	6.3780	52.4367
21	8.1662	0.1225	68.1383	0.0147	8.3440	0.1198	6.5779	54.8858
22	9.0250	0.1108	76.3045	0.0131	8.4548	0.1183	6.7669	57.2127
23	9.9742	0.1003	85.3295	0.0117	8.5550	0.1169	6.9454	59.4184
24	11.0232	0.0907	95.3037	0.0105	8.6458	0.1157	7.1139	61.5049
25	12.1825	0.0821	106.3269	0.0094	8.7278	0.1146	7.2727	63.4749
26	13.4637	0.0743	118.5094	0.0084	8.8021	0.1136	7.4223	65.3318
27	14.8797	0.0672	131.9731	0.0076	8.8693	0.1127	7.5630	67.0791
28	16.4446	0.0608	146.8528	0.0068	8.9301	0.1120	7.6954	68.7210
29	18.1741	0.0550	163.2975	0.0061	8.9852	0.1113	7.8197	70.2616
30	20.0855	0.0498	181.4716	0.0055	9.0349	0.1107	7.9365	71.7054
31	22.1980	0.0450	201.5572	0.0050	9.0800	0.1101	8.0459	73.0569
32	24.5325	0.0408	223.7551	0.0045	9.1208	0.1096	8.1485	74.3206
33	27.1126	0.0369	248.2876	0.0040	9.1576	0.1092	8.2446	75.5008
34	29.9641	0.0334	275.4003	0.0036	9.1910	0.1088	8.3345	76.6021
35	33.1155	0.0302	305.3644	0.0033	9.2212	0.1084	8.4185	77.6288
36	36.5982	0.0273	338.4798	0.0030	9.2485	0.1081	8.4970	78.5852
37	40.4473	0.0247	375.0781	0.0027	9.2733	0.1078	8.5704	79.4752
38	44.7012	0.0224	415.5254	0.0024	9.2956	0.1076	8.6388	80.3029
39	49.4024	0.0202	460.2266	0.0022	9.3159	0.1073	8.7026	81.0721
40	54.5982	0.0183	509.6290	0.0020	9.3342	0.1071	8.7620	81.7864
45	90.0171	0.0111	846.4044	0.0012	9.4027	0.1064	9.0028	84.6508
48	121.5104	0.0082	1145.8531	0.0009	9.4301	0.1060	9.1100	85.9083
50	148.4132	0.0067	1401.6532	0.0007	9.4443	0.1059	9.1691	86.5959
54	221.4064	0.0045	2095.8974	0.0005	9.4654	0.1056	9.2633	87.8810
60	403.4288	0.0025	3826.4266	0.0003	9.4848	0.1054	9.3592	88.7701
65	665.1416	0.0015	6314.8791	0.0002	9.4940	0.1053	9.4105	89.3433
70	1096.6332	0.0009	10417.6438	0.0001	9.4997	0.1053	9.4444	89.7190
75	1808.0424	0.0006	17181.9591	0.0001	9.5031	0.1052	9.4668	89.9640
80	2980.9580	0.0003	28334.4297	0.0000	9.5051	0.1052	9.4815	90.1229
90	8103.0839	0.0001	77037.3034	0.0000	9.5072	0.1052	9.4972	90.2916
100	22026.4658	0.0000	209425.4400	0.0000	9.5079	0.1052	9.5038	90.3611
120	162754.7914	0.0000	1547517.0741	0.0000	9.5083	0.1052	9.5076	90.4008

Table B.14 Continuous Compounding: *r* = 12%

	Single payment		Uniform Series				Gradient series	
	Compound amount factor	Present worth factor	Compound amount factor	Sinking fund factor	Present worth factor	Capital recovery factor	Uniform series factor	Present worth factor
	To find F given P	To find P given F	To find F given A	To find A given F	To find P given A	To find A given P	To find A given G	To find P given G
n	(F|P *r,n*)∞	(P|F *r,n*)∞	(F|A *r,n*)∞	(A|F *r,n*)∞	(P|A *r,n*)∞	(A|P *r,n*)∞	(A|G *r,n*)∞	(P|G *r,n*)∞
1	1.1275	0.8869	1.0000	1.0000	0.8869	1.1275	0.0000	0.0000
2	1.2712	0.7866	2.1275	0.4700	1.6735	0.5975	0.4700	0.7866
3	1.4333	0.6977	3.3987	0.2942	2.3712	0.4217	0.9202	2.1820
4	1.6161	0.6188	4.8321	0.2070	2.9900	0.3344	1.3506	4.0383
5	1.8221	0.5488	6.4481	0.1551	3.5388	0.2826	1.7615	6.2336
6	2.0544	0.4868	8.2703	0.1209	4.0256	0.2484	2.1531	8.6673
7	2.3164	0.4317	10.3247	0.0969	4.4573	0.2244	2.5257	11.2576
8	2.6117	0.3829	12.6411	0.0791	4.8402	0.2066	2.8796	13.9379
9	2.9447	0.3396	15.2528	0.0656	5.1798	0.1931	3.2153	16.6546
10	3.3201	0.3012	18.1974	0.0550	5.4810	0.1824	3.5332	19.3654
11	3.7434	0.2671	21.5176	0.0465	5.7481	0.1740	3.8337	22.0367
12	4.2207	0.2369	25.2610	0.0396	5.9850	0.1671	4.1174	24.6429
13	4.7588	0.2101	29.4817	0.0339	6.1952	0.1614	4.3848	27.1646
14	5.3656	0.1864	34.2405	0.0292	6.3815	0.1567	4.6364	29.5874
15	6.0496	0.1653	39.6061	0.0252	6.5468	0.1527	4.8728	31.9016
16	6.8210	0.1466	45.6557	0.0219	6.6934	0.1494	5.0946	34.1007
17	7.6906	0.1300	52.4767	0.0191	6.8235	0.1466	5.3025	36.1812
18	8.6711	0.1153	60.1673	0.0166	6.9388	0.1441	5.4969	38.1417
19	9.7767	0.1023	68.8384	0.0145	7.0411	0.1420	5.6785	39.9828
20	11.0232	0.0907	78.6151	0.0127	7.1318	0.1402	5.8480	41.7064
21	12.4286	0.0805	89.6383	0.0112	7.2123	0.1387	6.0058	43.3156
22	14.0132	0.0714	102.0669	0.0098	7.2836	0.1373	6.1527	44.8142
23	15.7998	0.0633	116.0801	0.0086	7.3469	0.1361	6.2893	46.2066
24	17.8143	0.0561	131.8799	0.0076	7.4030	0.1351	6.4160	47.4977
25	20.0855	0.0498	149.6942	0.0067	7.4528	0.1342	6.5334	48.6926
26	22.6464	0.0442	169.7797	0.0059	7.4970	0.1334	6.6422	49.7966
27	25.5337	0.0392	192.4261	0.0052	7.5362	0.1327	6.7428	50.8148
28	28.7892	0.0347	217.9598	0.0046	7.5709	0.1321	6.8357	51.7527
29	32.4597	0.0308	246.7490	0.0041	7.6017	0.1315	6.9215	52.6153
30	36.5982	0.0273	279.2087	0.0036	7.6290	0.1311	7.0006	53.4077
31	41.2644	0.0242	315.8070	0.0032	7.6533	0.1307	7.0734	54.1347
32	46.5255	0.0215	357.0714	0.0028	7.6747	0.1303	7.1404	54.8010
33	52.4573	0.0191	403.5968	0.0025	7.6938	0.1300	7.2020	55.4110
34	59.1455	0.0169	456.0542	0.0022	7.7107	0.1297	7.2586	55.9690
35	66.6863	0.0150	515.1996	0.0019	7.7257	0.1294	7.3105	56.4788
36	75.1886	0.0133	581.8860	0.0017	7.7390	0.1292	7.3581	56.9443
37	84.7749	0.0118	657.0746	0.0015	7.7508	0.1290	7.4017	57.3690
38	95.5835	0.0105	741.8495	0.0013	7.7613	0.1288	7.4416	57.7561
39	107.7701	0.0093	837.4330	0.0012	7.7706	0.1287	7.4781	58.1087
40	121.5104	0.0082	945.2031	0.0011	7.7788	0.1286	7.5114	58.4296
45	221.4064	0.0045	1728.7205	0.0006	7.8079	0.1281	7.6392	59.6459
48	317.3483	0.0032	2481.2246	0.0004	7.8186	0.1279	7.6916	60.1377
50	403.4288	0.0025	3156.3822	0.0003	7.8239	0.1278	7.7191	60.3933
54	651.9709	0.0015	5105.7806	0.0002	7.8313	0.1277	7.7604	60.7739
60	1339.4308	0.0007	10497.7554	0.0001	7.8375	0.1276	7.7985	61.1206

Table B.15 Continuous Compounding: $r = 15\%$

	Single payment		Uniform Series				Gradient series	
	Compound amount factor	Present worth factor	Compound amount factor	Sinking fund factor	Present worth factor	Capital recovery factor	Uniform series factor	Present worth factor
	To find F given P $(F\|P\ r,n)_\infty$	To find P given F $(P\|F\ r,n)_\infty$	To find F given A $(F\|A\ r,n)_\infty$	To find A given F $(A\|F\ r,n)_\infty$	To find P given A $(P\|A\ r,n)_\infty$	To find A given P $(A\|P\ r,n)_\infty$	To find A given G $(A\|G\ r,n)_\infty$	To find P given G $(P\|G\ r,n)_\infty$
n								
1	1.1618	0.8607	1.0000	1.0000	0.8607	1.1618	0.0000	0.0000
2	1.3499	0.7408	2.1618	0.4626	1.6015	0.6244	0.4626	0.7408
3	1.5683	0.6376	3.5117	0.2848	2.2392	0.4466	0.9004	2.0161
4	1.8221	0.5488	5.0800	0.1969	2.7880	0.3587	1.3137	3.6625
5	2.1170	0.4724	6.9021	0.1449	3.2603	0.3067	1.7029	5.5520
6	2.4596	0.4066	9.0191	0.1109	3.6669	0.2727	2.0685	7.5848
7	2.8577	0.3499	11.4787	0.0871	4.0168	0.2490	2.4110	9.6845
8	3.3201	0.3012	14.3364	0.0698	4.3180	0.2316	2.7311	11.7928
9	3.8574	0.2592	17.6565	0.0566	4.5773	0.2185	3.0295	13.8667
10	4.4817	0.2231	21.5139	0.0465	4.8004	0.2083	3.3070	15.8749
11	5.2070	0.1920	25.9956	0.0385	4.9925	0.2003	3.5645	17.7954
12	6.0496	0.1653	31.2026	0.0320	5.1578	0.1939	3.8028	19.6137
13	7.0287	0.1423	37.2522	0.0268	5.3000	0.1887	4.0228	21.3210
14	8.1662	0.1225	44.2809	0.0226	5.4225	0.1844	4.2255	22.9129
15	9.4877	0.1054	52.4471	0.0191	5.5279	0.1809	4.4119	24.3885
16	11.0232	0.0907	61.9348	0.0161	5.6186	0.1780	4.5829	25.7493
17	12.8071	0.0781	72.9580	0.0137	5.6967	0.1755	4.7394	26.9986
18	14.8797	0.0672	85.7651	0.0117	5.7639	0.1735	4.8823	28.1411
19	17.2878	0.0578	100.6448	0.0099	5.8217	0.1718	5.0126	29.1823
20	20.0855	0.0498	117.9326	0.0085	5.8715	0.1703	5.1312	30.1282
21	23.3361	0.0429	138.0182	0.0072	5.9144	0.1691	5.2390	30.9853
22	27.1126	0.0369	161.3542	0.0062	5.9513	0.1680	5.3367	31.7598
23	31.5004	0.0317	188.4669	0.0053	5.9830	0.1671	5.4251	32.4582
24	36.5982	0.0273	219.9673	0.0045	6.0103	0.1664	5.5050	33.0867
25	42.5211	0.0235	256.5655	0.0039	6.0338	0.1657	5.5771	33.6511
26	49.4024	0.0202	299.0866	0.0033	6.0541	0.1652	5.6420	34.1571
27	57.3975	0.0174	348.4890	0.0029	6.0715	0.1647	5.7004	34.6101
28	66.6863	0.0150	405.8865	0.0025	6.0865	0.1643	5.7529	35.0150
29	77.4785	0.0129	472.5728	0.0021	6.0994	0.1640	5.8000	35.3764
30	90.0171	0.0111	550.0513	0.0018	6.1105	0.1637	5.8421	35.6985
31	104.5850	0.0096	640.0684	0.0016	6.1201	0.1634	5.8799	35.9854
32	121.5104	0.0082	744.6534	0.0013	6.1283	0.1632	5.9136	36.2405
33	141.1750	0.0071	866.1638	0.0012	6.1354	0.1630	5.9437	36.4672
34	164.0219	0.0061	1007.3388	0.0010	6.1415	0.1628	5.9706	36.6684
35	190.5663	0.0052	1171.3607	0.0009	6.1467	0.1627	5.9945	36.8468
36	221.4064	0.0045	1361.9269	0.0007	6.1513	0.1626	6.0158	37.0049
37	257.2376	0.0039	1583.3334	0.0006	6.1551	0.1625	6.0348	37.1448
38	298.8674	0.0033	1840.5709	0.0005	6.1585	0.1624	6.0516	37.2686
39	347.2344	0.0029	2139.4383	0.0005	6.1614	0.1623	6.0665	37.3781
40	403.4288	0.0025	2486.6723	0.0004	6.1638	0.1622	6.0798	37.4747
45	854.0588	0.0012	5271.1883	0.0002	6.1719	0.1620	6.1264	37.8118
48	1339.4308	0.0007	8270.3805	0.0001	6.1745	0.1620	6.1433	37.9321
50	1808.0424	0.0006	11166.0078	0.0001	6.1757	0.1619	6.1515	37.9900
54	3294.4681	0.0003	20350.8727	0.0000	6.1773	0.1619	6.1628	38.0692
60	8103.0839	0.0001	50064.0890	0.0000	6.1784	0.1619	6.1718	38.1316

Table B.16 Continuous Compounding: r = 18%

	Single payment		Uniform Series				Gradient series	
	Compound amount factor	Present worth factor	Compound amount factor	Sinking fund factor	Present worth factor	Capital recovery factor	Uniform series factor	Present worth factor
	To find F given P $(F\|P\ r,n)_\infty$	To find P given F $(P\|F\ r,n)_\infty$	To find F given A $(F\|A\ r,n)_\infty$	To find A given F $(A\|F\ r,n)_\infty$	To find P given A $(P\|A\ r,n)_\infty$	To find A given P $(A\|P\ r,n)_\infty$	To find A given G $(A\|G\ r,n)_\infty$	To find P given G $(P\|G\ r,n)_\infty$
n								
1	1.1972	0.8353	1.0000	1.0000	0.8353	1.1972	0.0000	0.0000
2	1.4333	0.6977	2.1972	0.4551	1.5329	0.6523	0.4551	0.6977
3	1.7160	0.5827	3.6305	0.2754	2.1157	0.4727	0.8806	1.8632
4	2.0544	0.4868	5.3466	0.1870	2.6024	0.3843	1.2770	3.3234
5	2.4596	0.4066	7.4010	0.1351	3.0090	0.3323	1.6450	4.9497
6	2.9447	0.3396	9.8606	0.1014	3.3486	0.2986	1.9852	6.6477
7	3.5254	0.2837	12.8053	0.0781	3.6323	0.2753	2.2987	8.3496
8	4.2207	0.2369	16.3307	0.0612	3.8692	0.2585	2.5866	10.0081
9	5.0531	0.1979	20.5514	0.0487	4.0671	0.2459	2.8500	11.5913
10	6.0496	0.1653	25.6045	0.0391	4.2324	0.2363	3.0902	13.0790
11	7.2427	0.1381	31.6541	0.0316	4.3705	0.2288	3.3085	14.4597
12	8.6711	0.1153	38.8969	0.0257	4.4858	0.2229	3.5062	15.7283
13	10.3812	0.0963	47.5680	0.0210	4.5821	0.2182	3.6848	16.8842
14	12.4286	0.0805	57.9492	0.0173	4.6626	0.2145	3.8456	17.9302
15	14.8797	0.0672	70.3778	0.0142	4.7298	0.2114	3.9898	18.8710
16	17.8143	0.0561	85.2576	0.0117	4.7859	0.2089	4.1190	19.7131
17	21.3276	0.0469	103.0718	0.0097	4.8328	0.2069	4.2342	20.4633
18	25.5337	0.0392	124.3994	0.0080	4.8720	0.2053	4.3369	21.1290
19	30.5694	0.0327	149.9331	0.0067	4.9047	0.2039	4.4280	21.7179
20	36.5982	0.0273	180.5025	0.0055	4.9320	0.2028	4.5087	22.2370
21	43.8160	0.0228	217.1008	0.0046	4.9548	0.2018	4.5801	22.6935
22	52.4573	0.0191	260.9168	0.0038	4.9739	0.2011	4.6430	23.0938
23	62.8028	0.0159	313.3741	0.0032	4.9898	0.2004	4.6984	23.4441
24	75.1886	0.0133	376.1770	0.0027	5.0031	0.1999	4.7470	23.7500
25	90.0171	0.0111	451.3656	0.0022	5.0142	0.1994	4.7897	24.0166
26	107.7701	0.0093	541.3827	0.0018	5.0235	0.1991	4.8270	24.2486
27	129.0242	0.0078	649.1528	0.0015	5.0312	0.1988	4.8596	24.4501
28	154.4700	0.0065	778.1770	0.0013	5.0377	0.1985	4.8881	24.6249
29	184.9342	0.0054	932.6470	0.0011	5.0431	0.1983	4.9129	24.7763
30	221.4064	0.0045	1117.5812	0.0009	5.0476	0.1981	4.9344	24.9073
31	265.0716	0.0038	1338.9876	0.0007	5.0514	0.1980	4.9532	25.0205
32	317.3483	0.0032	1604.0592	0.0006	5.0546	0.1978	4.9694	25.1181
33	379.9349	0.0026	1921.4075	0.0005	5.0572	0.1977	4.9835	25.2024
34	454.8647	0.0022	2301.3425	0.0004	5.0594	0.1977	4.9956	25.2749
35	544.5719	0.0018	2756.2072	0.0004	5.0612	0.1976	5.0062	25.3374
36	651.9709	0.0015	3300.7791	0.0003	5.0628	0.1975	5.0152	25.3910
37	780.5509	0.0013	3952.7500	0.0003	5.0641	0.1975	5.0231	25.4372
38	934.4891	0.0011	4733.3010	0.0002	5.0651	0.1974	5.0298	25.4768
39	1118.7866	0.0009	5667.7901	0.0002	5.0660	0.1974	5.0357	25.5107
40	1339.4308	0.0007	6786.5767	0.0001	5.0668	0.1974	5.0407	25.5398
45	3294.4681	0.0003	16699.6862	0.0001	5.0690	0.1973	5.0569	25.6334
48	5653.3298	0.0002	28660.4066	0.0000	5.0697	0.1973	5.0621	25.6629
50	8103.0839	0.0001	41082.0011	0.0000	5.0699	0.1972	5.0644	25.6760
54	16647.2447	0.0001	84405.5740	0.0000	5.0702	0.1972	5.0673	25.6925
60	49020.8011	0.0000	248557.2282	0.0000	5.0704	0.1972	5.0693	25.7037

Table B.17 Continuous Compounding: r = 20%

	Single payment		Uniform Series				Gradient series	
	Compound amount factor	Present worth factor	Compound amount factor	Sinking fund factor	Present worth factor	Capital recovery factor	Uniform series factor	Present worth factor
	To find F given P $(F\|P\ r,n)_\infty$	To find P given F $(P\|F\ r,n)_\infty$	To find F given A $(F\|A\ r,n)_\infty$	To find A given F $(A\|F\ r,n)_\infty$	To find P given A $(P\|A\ r,n)_\infty$	To find A given P $(A\|P\ r,n)_\infty$	To find A given G $(A\|G\ r,n)_\infty$	To find P given G $(P\|G\ r,n)_\infty$
n								
1	1.2214	0.8187	1.0000	1.0000	0.8187	1.2214	0.0000	0.0000
2	1.4918	0.6703	2.2214	0.4502	1.4891	0.6716	0.4502	0.6703
3	1.8221	0.5488	3.7132	0.2693	2.0379	0.4907	0.8675	1.7679
4	2.2255	0.4493	5.5353	0.1807	2.4872	0.4021	1.2528	3.1159
5	2.7183	0.3679	7.7609	0.1289	2.8551	0.3503	1.6068	4.5874
6	3.3201	0.3012	10.4792	0.0954	3.1563	0.3168	1.9306	6.0934
7	4.0552	0.2466	13.7993	0.0725	3.4029	0.2939	2.2255	7.5730
8	4.9530	0.2019	17.8545	0.0560	3.6048	0.2774	2.4929	8.9863
9	6.0496	0.1653	22.8075	0.0438	3.7701	0.2652	2.7344	10.3087
10	7.3891	0.1353	28.8572	0.0347	3.9054	0.2561	2.9515	11.5267
11	9.0250	0.1108	36.2462	0.0276	4.0162	0.2490	3.1459	12.6347
12	11.0232	0.0907	45.2712	0.0221	4.1069	0.2435	3.3194	13.6326
13	13.4637	0.0743	56.2944	0.0178	4.1812	0.2392	3.4736	14.5239
14	16.4446	0.0608	69.7581	0.0143	4.2420	0.2357	3.6102	15.3144
15	20.0855	0.0498	86.2028	0.0116	4.2918	0.2330	3.7307	16.0114
16	24.5325	0.0408	106.2883	0.0094	4.3325	0.2308	3.8367	16.6229
17	29.9641	0.0334	130.8209	0.0076	4.3659	0.2290	3.9297	17.1569
18	36.5982	0.0273	160.7850	0.0062	4.3932	0.2276	4.0110	17.6214
19	44.7012	0.0224	197.3832	0.0051	4.4156	0.2265	4.0819	18.0240
20	54.5982	0.0183	242.0844	0.0041	4.4339	0.2255	4.1435	18.3720
21	66.6863	0.0150	296.6825	0.0034	4.4489	0.2248	4.1970	18.6719
22	81.4509	0.0123	363.3689	0.0028	4.4612	0.2242	4.2432	18.9298
23	99.4843	0.0101	444.8197	0.0022	4.4713	0.2237	4.2831	19.1509
24	121.5104	0.0082	544.3040	0.0018	4.4795	0.2232	4.3175	19.3402
25	148.4132	0.0067	665.8145	0.0015	4.4862	0.2229	4.3471	19.5019
26	181.2722	0.0055	814.2276	0.0012	4.4917	0.2226	4.3724	19.6398
27	221.4064	0.0045	995.4999	0.0010	4.4963	0.2224	4.3942	19.7572
28	270.4264	0.0037	1216.9063	0.0008	4.5000	0.2222	4.4127	19.8571
29	330.2996	0.0030	1487.3327	0.0007	4.5030	0.2221	4.4286	19.9419
30	403.4288	0.0025	1817.6323	0.0006	4.5055	0.2220	4.4421	20.0137
31	492.7490	0.0020	2221.0610	0.0005	4.5075	0.2219	4.4536	20.0746
32	601.8450	0.0017	2713.8101	0.0004	4.5092	0.2218	4.4634	20.1261
33	735.0952	0.0014	3315.6551	0.0003	4.5105	0.2217	4.4717	20.1697
34	897.8473	0.0011	4050.7503	0.0002	4.5116	0.2216	4.4787	20.2064
35	1096.6332	0.0009	4948.5976	0.0002	4.5125	0.2216	4.4847	20.2374
36	1339.4308	0.0007	6045.2308	0.0002	4.5133	0.2216	4.4898	20.2636
37	1635.9844	0.0006	7384.6615	0.0001	4.5139	0.2215	4.4940	20.2856
38	1998.1959	0.0005	9020.6460	0.0001	4.5144	0.2215	4.4976	20.3041
39	2440.6020	0.0004	11018.8419	0.0001	4.5148	0.2215	4.5007	20.3196
40	2980.9580	0.0003	13459.4438	0.0001	4.5151	0.2215	4.5032	20.3327
45	8103.0839	0.0001	36594.3225	0.0000	4.5161	0.2214	4.5111	20.3726
48	14764.7816	0.0001	66682.9162	0.0000	4.5163	0.2214	4.5134	20.3841
50	22026.4658	0.0000	99481.4427	0.0000	4.5165	0.2214	4.5144	20.3890
54	49020.8011	0.0000	221405.5577	0.0000	4.5166	0.2214	4.5156	20.3948
60	162754.7914	0.0000	735102.8179	0.0000	4.5166	0.2214	4.5163	20.3984

Table B.18 Continuous Compounding: $r = 25\%$

	Single payment		Uniform Series				Gradient series	
	Compound amount factor	Present worth factor	Compound amount factor	Sinking fund factor	Present worth factor	Capital recovery factor	Uniform series factor	Present worth factor
n	To find F given P $(F\|P\ r,n)_\infty$	To find P given F $(P\|F\ r,n)_\infty$	To find F given A $(F\|A\ r,n)_\infty$	To find A given F $(A\|F\ r,n)_\infty$	To find P given A $(P\|A\ r,n)_\infty$	To find A given P $(A\|P\ r,n)_\infty$	To find A given G $(A\|G\ r,n)_\infty$	To find P given G $(P\|G\ r,n)_\infty$
1	1.2840	0.7788	1.0000	1.0000	0.7788	1.2840	0.0000	0.0000
2	1.6487	0.6065	2.2840	0.4378	1.3853	0.7218	0.4378	0.6065
3	2.1170	0.4724	3.9327	0.2543	1.8577	0.5383	0.8350	1.5513
4	2.7183	0.3679	6.0497	0.1653	2.2256	0.4493	1.1929	2.6549
5	3.4903	0.2865	8.7680	0.1141	2.5121	0.3981	1.5131	3.8009
6	4.4817	0.2231	12.2584	0.0816	2.7352	0.3656	1.7975	4.9166
7	5.7546	0.1738	16.7401	0.0597	2.9090	0.3438	2.0486	5.9592
8	7.3891	0.1353	22.4947	0.0445	3.0443	0.3285	2.2687	6.9066
9	9.4877	0.1054	29.8837	0.0335	3.1497	0.3175	2.4605	7.7498
10	12.1825	0.0821	39.3715	0.0254	3.2318	0.3094	2.6266	8.4885
11	15.6426	0.0639	51.5539	0.0194	3.2957	0.3034	2.7696	9.1278
12	20.0855	0.0498	67.1966	0.0149	3.3455	0.2989	2.8921	9.6755
13	25.7903	0.0388	87.2821	0.0115	3.3843	0.2955	2.9964	10.1407
14	33.1155	0.0302	113.0725	0.0088	3.4145	0.2929	3.0849	10.5333
15	42.5211	0.0235	146.1879	0.0068	3.4380	0.2909	3.1595	10.8626
16	54.5982	0.0183	188.7090	0.0053	3.4563	0.2893	3.2223	11.1373
17	70.1054	0.0143	243.3071	0.0041	3.4706	0.2881	3.2748	11.3655
18	90.0171	0.0111	313.4126	0.0032	3.4817	0.2872	3.3186	11.5544
19	115.5843	0.0087	403.4297	0.0025	3.4904	0.2865	3.3550	11.7101
20	148.4132	0.0067	519.0140	0.0019	3.4971	0.2860	3.3851	11.8381
21	190.5663	0.0052	667.4271	0.0015	3.5023	0.2855	3.4100	11.9431
22	244.6919	0.0041	857.9934	0.0012	3.5064	0.2852	3.4305	12.0289
23	314.1907	0.0032	1102.6853	0.0009	3.5096	0.2849	3.4474	12.0989
24	403.4288	0.0025	1416.8760	0.0007	3.5121	0.2847	3.4612	12.1559
25	518.0128	0.0019	1820.3048	0.0005	3.5140	0.2846	3.4725	12.2023
26	665.1416	0.0015	2338.3176	0.0004	3.5155	0.2845	3.4817	12.2399
27	854.0588	0.0012	3003.4592	0.0003	3.5167	0.2844	3.4892	12.2703
28	1096.6332	0.0009	3857.5180	0.0003	3.5176	0.2843	3.4953	12.2949
29	1408.1048	0.0007	4954.1512	0.0002	3.5183	0.2842	3.5002	12.3148
30	1808.0424	0.0006	6362.2560	0.0002	3.5189	0.2842	3.5042	12.3308
31	2321.5724	0.0004	8170.2984	0.0001	3.5193	0.2841	3.5075	12.3438
32	2980.9580	0.0003	10491.8708	0.0001	3.5196	0.2841	3.5101	12.3542
33	3827.6258	0.0003	13472.8288	0.0001	3.5199	0.2841	3.5122	12.3625
34	4914.7688	0.0002	17300.4546	0.0001	3.5201	0.2841	3.5139	12.3692
35	6310.6881	0.0002	22215.2235	0.0000	3.5203	0.2841	3.5153	12.3746
36	8103.0839	0.0001	28525.9116	0.0000	3.5204	0.2841	3.5164	12.3789
37	10404.5657	0.0001	36628.9955	0.0000	3.5205	0.2841	3.5173	12.3824
38	13359.7268	0.0001	47033.5612	0.0000	3.5205	0.2840	3.5180	12.3852
39	17154.2288	0.0001	60393.2881	0.0000	3.5206	0.2840	3.5185	12.3874
40	22026.4658	0.0000	77547.5169	0.0000	3.5207	0.2840	3.5190	12.3892
45	76879.9198	0.0000	270676.1974	0.0000	3.5208	0.2840	3.5202	12.3939
48	162754.7914	0.0000	573025.4472	0.0000	3.5208	0.2840	3.5205	12.3950

Table B.19 Continuous Compounding: $r = 30\%$

	Single payment		Uniform Series				Gradient series	
	Compound amount factor	Present worth factor	Compound amount factor	Sinking fund factor	Present worth factor	Capital recovery factor	Uniform series factor	Present worth factor
n	To find F given P $(F\|P\ r,n)_\infty$	To find P given F $(P\|F\ r,n)_\infty$	To find F given A $(F\|A\ r,n)_\infty$	To find A given F $(A\|F\ r,n)_\infty$	To find P given A $(P\|A\ r,n)_\infty$	To find A given P $(A\|P\ r,n)_\infty$	To find A given G $(A\|G\ r,n)_\infty$	To find P given G $(P\|G\ r,n)_\infty$
1	1.3499	0.7408	1.0000	1.0000	0.7408	1.3499	0.0000	0.0000
2	1.8221	0.5488	2.3499	0.4256	1.2896	0.7754	0.4256	0.5488
3	2.4596	0.4066	4.1720	0.2397	1.6962	0.5896	0.8029	1.3620
4	3.3201	0.3012	6.6316	0.1508	1.9974	0.5007	1.1342	2.2655
5	4.4817	0.2231	9.9517	0.1005	2.2205	0.4503	1.4222	3.1581
6	6.0496	0.1653	14.4334	0.0693	2.3858	0.4191	1.6701	3.9845
7	8.1662	0.1225	20.4830	0.0488	2.5083	0.3987	1.8815	4.7193
8	11.0232	0.0907	28.6492	0.0349	2.5990	0.3848	2.0601	5.3543
9	14.8797	0.0672	39.6724	0.0252	2.6662	0.3751	2.2099	5.8920
10	20.0855	0.0498	54.5521	0.0183	2.7160	0.3682	2.3343	6.3400
11	27.1126	0.0369	74.6376	0.0134	2.7529	0.3633	2.4370	6.7089
12	36.5982	0.0273	101.7503	0.0098	2.7802	0.3597	2.5212	7.0094
13	49.4024	0.0202	138.3485	0.0072	2.8004	0.3571	2.5897	7.2523
14	66.6863	0.0150	187.7510	0.0053	2.8154	0.3552	2.6452	7.4473
15	90.0171	0.0111	254.4373	0.0039	2.8265	0.3538	2.6898	7.6028
16	121.5104	0.0082	344.4544	0.0029	2.8348	0.3528	2.7255	7.7263
17	164.0219	0.0061	465.9648	0.0021	2.8409	0.3520	2.7540	7.8238
18	221.4064	0.0045	629.9868	0.0016	2.8454	0.3514	2.7766	7.9006
19	298.8674	0.0033	851.3932	0.0012	2.8487	0.3510	2.7945	7.9608
20	403.4288	0.0025	1150.2606	0.0009	2.8512	0.3507	2.8086	8.0079
21	544.5719	0.0018	1553.6894	0.0006	2.8530	0.3505	2.8197	8.0446
22	735.0952	0.0014	2098.2613	0.0005	2.8544	0.3503	2.8283	8.0732
23	992.2747	0.0010	2833.3565	0.0004	2.8554	0.3502	2.8351	8.0954
24	1339.4308	0.0007	3825.6312	0.0003	2.8562	0.3501	2.8404	8.1125
25	1808.0424	0.0006	5165.0619	0.0002	2.8567	0.3501	2.8445	8.1258
26	2440.6020	0.0004	6973.1044	0.0001	2.8571	0.3500	2.8476	8.1361
27	3294.4681	0.0003	9413.7063	0.0001	2.8574	0.3500	2.8501	8.1440
28	4447.0667	0.0002	12708.1744	0.0001	2.8577	0.3499	2.8520	8.1500
29	6002.9122	0.0002	17155.2412	0.0001	2.8578	0.3499	2.8535	8.1547
30	8103.0839	0.0001	23158.1534	0.0000	2.8579	0.3499	2.8546	8.1583
31	10938.0192	0.0001	31261.2373	0.0000	2.8580	0.3499	2.8555	8.1610
32	14764.7816	0.0001	42199.2565	0.0000	2.8581	0.3499	2.8561	8.1631
33	19930.3704	0.0001	56964.0381	0.0000	2.8582	0.3499	2.8566	8.1647
34	26903.1861	0.0000	76894.4085	0.0000	2.8582	0.3499	2.8570	8.1659
35	36315.5027	0.0000	103797.5946	0.0000	2.8582	0.3499	2.8573	8.1669
36	49020.8011	0.0000	140113.0973	0.0000	2.8582	0.3499	2.8576	8.1676
37	66171.1602	0.0000	189133.8984	0.0000	2.8583	0.3499	2.8577	8.1681
38	89321.7234	0.0000	255305.0586	0.0000	2.8583	0.3499	2.8579	8.1685
39	120571.7150	0.0000	344626.7819	0.0000	2.8583	0.3499	2.8580	8.1689
40	162754.7914	0.0000	465198.4969	0.0000	2.8583	0.3499	2.8581	8.1691

Table B.20 Continuous Compounding: $r = 40\%$

	Single payment		Uniform Series				Gradient series	
	Compound amount factor	Present worth factor	Compound amount factor	Sinking fund factor	Present worth factor	Capital recovery factor	Uniform series factor	Present worth factor
n	To find F given P $(F\|P\ r,n)_{\infty}$	To find P given F $(P\|F\ r,n)_{\infty}$	To find F given A $(F\|A\ r,n)_{\infty}$	To find A given F $(A\|F\ r,n)_{\infty}$	To find P given A $(P\|A\ r,n)_{\infty}$	To find A given P $(A\|P\ r,n)_{\infty}$	To find A given G $(A\|G\ r,n)_{\infty}$	To find P given G $(P\|G\ r,n)_{\infty}$
1	1.4918	0.6703	1.0000	1.0000	0.6703	1.4918	0.0000	0.0000
2	2.2255	0.4493	2.4918	0.4013	1.1196	0.8931	0.4013	0.4493
3	3.3201	0.3012	4.7174	0.2120	1.4208	0.7038	0.7402	1.0517
4	4.9530	0.2019	8.0375	0.1244	1.6227	0.6162	1.0214	1.6574
5	7.3891	0.1353	12.9905	0.0770	1.7581	0.5688	1.2507	2.1987
6	11.0232	0.0907	20.3796	0.0491	1.8488	0.5409	1.4346	2.6523
7	16.4446	0.0608	31.4027	0.0318	1.9096	0.5237	1.5800	3.0172
8	24.5325	0.0408	47.8474	0.0209	1.9504	0.5127	1.6933	3.3025
9	36.5982	0.0273	72.3799	0.0138	1.9777	0.5056	1.7804	3.5211
10	54.5982	0.0183	108.9782	0.0092	1.9960	0.5010	1.8467	3.6860
11	81.4509	0.0123	163.5763	0.0061	2.0083	0.4979	1.8965	3.8087
12	121.5104	0.0082	245.0272	0.0041	2.0165	0.4959	1.9337	3.8993
13	181.2722	0.0055	366.5376	0.0027	2.0220	0.4946	1.9611	3.9655
14	270.4264	0.0037	547.8098	0.0018	2.0257	0.4937	1.9813	4.0135
15	403.4288	0.0025	818.2362	0.0012	2.0282	0.4930	1.9960	4.0482
16	601.8450	0.0017	1221.6650	0.0008	2.0299	0.4926	2.0066	4.0732
17	897.8473	0.0011	1823.5101	0.0005	2.0310	0.4924	2.0143	4.0910
18	1339.4308	0.0007	2721.3574	0.0004	2.0317	0.4922	2.0198	4.1037
19	1998.1959	0.0005	4060.7881	0.0002	2.0322	0.4921	2.0237	4.1127
20	2980.9580	0.0003	6058.9840	0.0002	2.0326	0.4920	2.0265	4.1191
21	4447.0667	0.0002	9039.9420	0.0001	2.0328	0.4919	2.0285	4.1236
22	6634.2440	0.0002	13487.0088	0.0001	2.0329	0.4919	2.0299	4.1267
23	9897.1291	0.0001	20121.2528	0.0000	2.0330	0.4919	2.0309	4.1289
24	14764.7816	0.0001	30018.3818	0.0000	2.0331	0.4919	2.0316	4.1305
25	22026.4658	0.0000	44783.1634	0.0000	2.0332	0.4918	2.0321	4.1316
26	32859.6257	0.0000	66809.6292	0.0000	2.0332	0.4918	2.0325	4.1323
27	49020.8011	0.0000	99669.2549	0.0000	2.0332	0.4918	2.0327	4.1329
28	73130.4418	0.0000	148690.0560	0.0000	2.0332	0.4918	2.0329	4.1332
29	109097.7993	0.0000	221820.4978	0.0000	2.0332	0.4918	2.0330	4.1335
30	162754.7914	0.0000	330918.2971	0.0000	2.0332	0.4918	2.0331	4.1337

Table B.21 Continuous Compounding: $r = 50\%$

	Single payment		Uniform Series				Gradient series	
	Compound amount factor	Present worth factor	Compound amount factor	Sinking fund factor	Present worth factor	Capital recovery factor	Uniform series factor	Present worth factor
	To find F given P $(F\|P\ r,n)_\infty$	To find P given F $(P\|F\ r,n)_\infty$	To find F given A $(F\|A\ r,n)_\infty$	To find A given F $(A\|F\ r,n)_\infty$	To find P given A $(P\|A\ r,n)_\infty$	To find A given P $(A\|P\ r,n)_\infty$	To find A given G $(A\|G\ r,n)_\infty$	To find P given G $(P\|G\ r,n)_\infty$
n								
1	1.6487	0.6065	1.0000	1.0000	0.6065	1.6487	0.0000	0.0000
2	2.7183	0.3679	2.6487	0.3775	0.9744	1.0263	0.3775	0.3679
3	4.4817	0.2231	5.3670	0.1863	1.1975	0.8350	0.6798	0.8141
4	7.3891	0.1353	9.8487	0.1015	1.3329	0.7503	0.9154	1.2201
5	12.1825	0.0821	17.2377	0.0580	1.4150	0.7067	1.0944	1.5485
6	20.0855	0.0498	29.4202	0.0340	1.4647	0.6827	1.2271	1.7974
7	33.1155	0.0302	49.5058	0.0202	1.4949	0.6689	1.3235	1.9786
8	54.5982	0.0183	82.6212	0.0121	1.5133	0.6608	1.3922	2.1068
9	90.0171	0.0111	137.2194	0.0073	1.5244	0.6560	1.4404	2.1957
10	148.4132	0.0067	227.2365	0.0044	1.5311	0.6531	1.4737	2.2563
11	244.6919	0.0041	375.6497	0.0027	1.5352	0.6514	1.4964	2.2972
12	403.4288	0.0025	620.3416	0.0016	1.5377	0.6503	1.5117	2.3245
13	665.1416	0.0015	1023.7704	0.0010	1.5392	0.6497	1.5219	2.3425
14	1096.6332	0.0009	1688.9120	0.0006	1.5401	0.6493	1.5287	2.3544
15	1808.0424	0.0006	2785.5452	0.0004	1.5406	0.6491	1.5332	2.3621
16	2980.9580	0.0003	4593.5876	0.0002	1.5410	0.6489	1.5361	2.3671
17	4914.7688	0.0002	7574.5456	0.0001	1.5412	0.6489	1.5380	2.3704
18	8103.0839	0.0001	12489.3144	0.0001	1.5413	0.6488	1.5393	2.3725
19	13359.7268	0.0001	20592.3984	0.0000	1.5414	0.6488	1.5401	2.3738
20	22026.4658	0.0000	33952.1252	0.0000	1.5414	0.6488	1.5406	2.3747
21	36315.5027	0.0000	55978.5910	0.0000	1.5415	0.6487	1.5409	2.3752
22	59874.1417	0.0000	92294.0937	0.0000	1.5415	0.6487	1.5411	2.3756
23	98715.7710	0.0000	152168.2354	0.0000	1.5415	0.6487	1.5413	2.3758
24	162754.7914	0.0000	250884.0064	0.0000	1.5415	0.6487	1.5413	2.3760
25	268337.2865	0.0000	413638.7978	0.0000	1.5415	0.6487	1.5414	2.3761

Table B.22 Continuous Compounding: $r=$ 4%

	Continuous flow, Uniform series			
	Present worth factor	Capital recovery factor	Compound amount factor	Sinking fund factor
	To find P given $\overline{A}$	To find $\overline{A}$ given P	To find F given $\overline{A}$	To find $\overline{A}$ given F
n	$P\vert\overline{A}\ r,n$	$\overline{A}\vert P\ r,n$	$F\vert\overline{A}\ r,n$	$\overline{A}\vert F\ r,n$
1	0.9803	1.0201	1.0203	0.9801
2	1.9221	0.5203	2.0822	0.4803
3	2.8270	0.3537	3.1874	0.3137
4	3.6964	0.2705	4.3378	0.2305
5	4.5317	0.2207	5.5351	0.1807
6	5.3343	0.1875	6.7812	0.1475
7	6.1054	0.1638	8.0782	0.1238
8	6.8463	0.1461	9.4282	0.1061
9	7.5581	0.1323	10.8332	0.0923
10	8.2420	0.1213	12.2956	0.0813
11	8.8991	0.1124	13.8177	0.0724
12	9.5304	0.1049	15.4019	0.0649
13	10.1370	0.0986	17.0507	0.0586
14	10.7198	0.0933	18.7668	0.0533
15	11.2797	0.0887	20.5530	0.0487
16	11.8177	0.0846	22.4120	0.0446
17	12.3346	0.0811	24.3469	0.0411
18	12.8312	0.0779	26.3608	0.0379
19	13.3083	0.0751	28.4569	0.0351
20	13.7668	0.0726	30.6385	0.0326
21	14.2072	0.0704	32.9092	0.0304
22	14.6304	0.0684	35.2725	0.0284
23	15.0370	0.0665	37.7323	0.0265
24	15.4277	0.0648	40.2924	0.0248
25	15.8030	0.0633	42.9570	0.0233
26	16.1636	0.0619	45.7304	0.0219
27	16.5101	0.0606	48.6170	0.0206
28	16.8430	0.0594	51.6214	0.0194
29	17.1628	0.0583	54.7483	0.0183
30	17.4701	0.0572	58.0029	0.0172
31	17.7654	0.0563	61.3903	0.0163
32	18.0491	0.0554	64.9160	0.0154
33	18.3216	0.0546	68.5855	0.0146
34	18.5835	0.0538	72.4048	0.0138
35	18.8351	0.0531	76.3800	0.0131
36	19.0768	0.0524	80.5174	0.0124
37	19.3091	0.0518	84.8236	0.0118
38	19.5322	0.0512	89.3056	0.0112
39	19.7466	0.0506	93.9705	0.0106
40	19.9526	0.0501	98.8258	0.0101
45	20.8675	0.0479	126.2412	0.0079
48	21.3348	0.0469	145.5240	0.0069
50	21.6166	0.0463	159.7264	0.0063
54	22.1169	0.0452	191.7784	0.0052
60	22.7321	0.0440	250.5794	0.0040
65	23.1432	0.0432	311.5935	0.0032
70	23.4797	0.0426	386.1162	0.0026
75	23.7553	0.0421	477.1384	0.0021
80	23.9809	0.0417	588.3133	0.0017
90	24.3169	0.0411	889.9559	0.0011
100	24.5421	0.0407	1339.9538	0.0007
120	24.7943	0.0403	3012.7604	0.0003

Table B.23 Continuous Compounding: $r = 5\%$

	Continuous flow, Uniform series			
	Present worth factor	Capital recovery factor	Compound amount factor	Sinking fund factor
	To find P given $\bar{A}$	To find $\bar{A}$ given P	To find F given $\bar{A}$	To find $\bar{A}$ given F
n	$P\|\bar{A}$ $_{r,n}$	$\bar{A}\|P$ $_{r,n}$	$F\|\bar{A}$ $_{r,n}$	$\bar{A}\|F$ $_{r,n}$
1	0.9754	1.0252	1.0254	0.9752
2	1.9033	0.5254	2.1034	0.4754
3	2.7858	0.3590	3.2367	0.3090
4	3.6254	0.2758	4.4281	0.2258
5	4.4240	0.2260	5.6805	0.1760
6	5.1836	0.1929	6.9972	0.1429
7	5.9062	0.1693	8.3814	0.1193
8	6.5936	0.1517	9.8365	0.1017
9	7.2474	0.1380	11.3662	0.0880
10	7.8694	0.1271	12.9744	0.0771
11	8.4610	0.1182	14.6651	0.0682
12	9.0238	0.1108	16.4424	0.0608
13	9.5591	0.1046	18.3108	0.0546
14	10.0683	0.0993	20.2751	0.0493
15	10.5527	0.0948	22.3400	0.0448
16	11.0134	0.0908	24.5108	0.0408
17	11.4517	0.0873	26.7929	0.0373
18	11.8686	0.0843	29.1921	0.0343
19	12.2652	0.0815	31.7142	0.0315
20	12.6424	0.0791	34.3656	0.0291
21	13.0012	0.0769	37.1530	0.0269
22	13.3426	0.0749	40.0833	0.0249
23	13.6673	0.0732	43.1639	0.0232
24	13.9761	0.0716	46.4023	0.0216
25	14.2699	0.0701	49.8069	0.0201
26	14.5494	0.0687	53.3859	0.0187
27	14.8152	0.0675	57.1485	0.0175
28	15.0681	0.0664	61.1040	0.0164
29	15.3086	0.0653	65.2623	0.0153
30	15.5374	0.0644	69.6338	0.0144
31	15.7550	0.0635	74.2294	0.0135
32	15.9621	0.0626	79.0606	0.0126
33	16.1590	0.0619	84.1396	0.0119
34	16.3463	0.0612	89.4789	0.0112
35	16.5245	0.0605	95.0921	0.0105
36	16.6940	0.0599	100.9929	0.0099
37	16.8553	0.0593	107.1964	0.0093
38	17.0086	0.0588	113.7179	0.0088
39	17.1545	0.0583	120.5738	0.0083
40	17.2933	0.0578	127.7811	0.0078
45	17.8920	0.0559	169.7547	0.0059
48	18.1856	0.0550	200.4635	0.0050
50	18.3583	0.0545	223.6499	0.0045
54	18.6559	0.0536	277.5946	0.0036
60	19.0043	0.0526	381.7107	0.0026
65	19.2245	0.0520	495.8068	0.0020
70	19.3961	0.0516	642.3090	0.0016
75	19.5296	0.0512	830.4216	0.0012
80	19.6337	0.0509	1071.9630	0.0009
90	19.7778	0.0506	1780.3426	0.0006
100	19.8652	0.0503	2948.2632	0.0003
120	19.9504	0.0501	8048.5759	0.0001

Table B.24 Continuous Compounding: $r = 6\%$

	Continuous flow, Uniform series			
	Present worth factor	Capital recovery factor	Compound amount factor	Sinking fund factor
	To find P given $\bar{A}$	To find $\bar{A}$ given P	To find F given $\bar{A}$	To find $\bar{A}$ given F
n	$P\|\bar{A}\ r,n$	$\bar{A}\|P\ r,n$	$F\|\bar{A}\ r,n$	$\bar{A}\|F\ r,n$
1	0.9706	1.0303	1.0306	0.9703
2	1.8847	0.5306	2.1249	0.4706
3	2.7455	0.3642	3.2870	0.3042
4	3.5562	0.2812	4.5208	0.2212
5	4.3197	0.2315	5.8310	0.1715
6	5.0387	0.1985	7.2222	0.1385
7	5.7159	0.1750	8.6994	0.1150
8	6.3536	0.1574	10.2679	0.0974
9	6.9542	0.1438	11.9334	0.0838
10	7.5198	0.1330	13.7020	0.0730
11	8.0525	0.1242	15.5799	0.0642
12	8.5541	0.1169	17.5739	0.0569
13	9.0266	0.1108	19.6912	0.0508
14	9.4715	0.1056	21.9394	0.0456
15	9.8905	0.1011	24.3267	0.0411
16	10.2851	0.0972	26.8616	0.0372
17	10.6568	0.0938	29.5532	0.0338
18	11.0067	0.0909	32.4113	0.0309
19	11.3363	0.0882	35.4461	0.0282
20	11.6468	0.0859	38.6686	0.0259
21	11.9391	0.0838	42.0904	0.0238
22	12.2144	0.0819	45.7237	0.0219
23	12.4737	0.0802	49.5817	0.0202
24	12.7179	0.0786	53.6783	0.0186
25	12.9478	0.0772	58.0282	0.0172
26	13.1644	0.0760	62.6470	0.0160
27	13.3684	0.0748	67.5515	0.0148
28	13.5604	0.0737	72.7593	0.0137
29	13.7413	0.0728	78.2891	0.0128
30	13.9117	0.0719	84.1608	0.0119
31	14.0721	0.0711	90.3956	0.0111
32	14.2232	0.0703	97.0160	0.0103
33	14.3655	0.0696	104.0457	0.0096
34	14.4995	0.0690	111.5102	0.0090
35	14.6257	0.0684	119.4362	0.0084
36	14.7446	0.0678	127.8523	0.0078
37	14.8565	0.0673	136.7888	0.0073
38	14.9619	0.0668	146.2780	0.0068
39	15.0612	0.0664	156.3539	0.0064
40	15.1547	0.0660	167.0529	0.0060
45	15.5466	0.0643	231.3289	0.0043
48	15.7311	0.0636	280.2379	0.0036
50	15.8369	0.0631	318.0923	0.0031
54	16.0139	0.0624	408.8954	0.0024
60	16.2113	0.0617	593.3039	0.0017
65	16.3293	0.0612	806.7075	0.0012
70	16.4167	0.0609	1094.7722	0.0009
75	16.4815	0.0607	1483.6189	0.0007
80	16.5295	0.0605	2008.5070	0.0005
90	16.5914	0.0603	3673.4403	0.0003
100	16.6254	0.0601	6707.1466	0.0001
120	16.6542	0.0600	22307.1794	0.0000

Table B.25 Continuous Compounding: $r=8\%$

	Continuous flow, Uniform series			
	Present worth factor	Capital recovery factor	Compound amount factor	Sinking fund factor
	To find P given $\bar{A}$	To find $\bar{A}$ given P	To find F given $\bar{A}$	To find $\bar{A}$ given F
n	$P\vert\bar{A}\ r,n$	$\bar{A}\vert P\ r,n$	$F\vert\bar{A}\ r,n$	$\bar{A}\vert F\ r,n$
1	0.9610	1.0405	1.0411	0.9605
2	1.8482	0.5411	2.1689	0.4611
3	2.6672	0.3749	3.3906	0.2949
4	3.4231	0.2921	4.7141	0.2121
5	4.1210	0.2427	6.1478	0.1627
6	4.7652	0.2099	7.7009	0.1299
7	5.3599	0.1866	9.3834	0.1066
8	5.9088	0.1692	11.2060	0.0892
9	6.4156	0.1559	13.1804	0.0759
10	6.8834	0.1453	15.3193	0.0653
11	7.3152	0.1367	17.6362	0.0567
12	7.7138	0.1296	20.1462	0.0496
13	8.0818	0.1237	22.8652	0.0437
14	8.4215	0.1187	25.8107	0.0387
15	8.7351	0.1145	29.0015	0.0345
16	9.0245	0.1108	32.4580	0.0308
17	9.2917	0.1076	36.2024	0.0276
18	9.5384	0.1048	40.2587	0.0248
19	9.7661	0.1024	44.6528	0.0224
20	9.9763	0.1002	49.4129	0.0202
21	10.1703	0.0983	54.5694	0.0183
22	10.3494	0.0966	60.1555	0.0166
23	10.5148	0.0951	66.2067	0.0151
24	10.6674	0.0937	72.7620	0.0137
25	10.8083	0.0925	79.8632	0.0125
26	10.9384	0.0914	87.5559	0.0114
27	11.0584	0.0904	95.8892	0.0104
28	11.1693	0.0895	104.9166	0.0095
29	11.2716	0.0887	114.6959	0.0087
30	11.3660	0.0880	125.2897	0.0080
31	11.4532	0.0873	136.7658	0.0073
32	11.5337	0.0867	149.1977	0.0067
33	11.6080	0.0861	162.6650	0.0061
34	11.6766	0.0856	177.2540	0.0056
35	11.7399	0.0852	193.0581	0.0052
36	11.7983	0.0848	210.1784	0.0048
37	11.8523	0.0844	228.7246	0.0044
38	11.9021	0.0840	248.8155	0.0040
39	11.9480	0.0837	270.5797	0.0037
40	11.9905	0.0834	294.1566	0.0034
45	12.1585	0.0822	444.9779	0.0022
48	12.2313	0.0818	569.0684	0.0018
50	12.2711	0.0815	669.9769	0.0015
54	12.3338	0.0811	927.3579	0.0011
60	12.3971	0.0807	1506.3802	0.0007
65	12.4310	0.0804	2253.4030	0.0004
70	12.4538	0.0803	3367.8301	0.0003
75	12.4690	0.0802	5030.3599	0.0002
80	12.4792	0.0801	7510.5630	0.0001
90	12.4907	0.0801	16730.3846	0.0001
100	12.4958	0.0800	37249.4748	0.0000
120	12.4992	0.0800	184547.2696	0.0000

Table B.26 Continuous Compounding: $r = 10\%$

	Continuous flow, Uniform series			
	Present worth factor	Capital recovery factor	Compound amount factor	Sinking fund factor
	To find P given $\bar{A}$	To find $\bar{A}$ given P	To find F given $\bar{A}$	To find $\bar{A}$ given F
n	$P|\bar{A}\ r,n$	$\bar{A}|P\ r,n$	$F|\bar{A}\ r,n$	$\bar{A}|F\ r,n$
1	0.9516	1.0508	1.0517	0.9508
2	1.8127	0.5517	2.2140	0.4517
3	2.5918	0.3858	3.4986	0.2858
4	3.2968	0.3033	4.9182	0.2033
5	3.9347	0.2541	6.4872	0.1541
6	4.5119	0.2216	8.2212	0.1216
7	5.0341	0.1986	10.1375	0.0986
8	5.5067	0.1816	12.2554	0.0816
9	5.9343	0.1685	14.5960	0.0685
10	6.3212	0.1582	17.1828	0.0582
11	6.6713	0.1499	20.0417	0.0499
12	6.9881	0.1431	23.2012	0.0431
13	7.2747	0.1375	26.6930	0.0375
14	7.5340	0.1327	30.5520	0.0327
15	7.7687	0.1287	34.8169	0.0287
16	7.9810	0.1253	39.5303	0.0253
17	8.1732	0.1224	44.7395	0.0224
18	8.3470	0.1198	50.4965	0.0198
19	8.5043	0.1176	56.8589	0.0176
20	8.6466	0.1157	63.8906	0.0157
21	8.7754	0.1140	71.6617	0.0140
22	8.8920	0.1125	80.2501	0.0125
23	8.9974	0.1111	89.7418	0.0111
24	9.0928	0.1100	100.2318	0.0100
25	9.1792	0.1089	111.8249	0.0089
26	9.2573	0.1080	124.6374	0.0080
27	9.3279	0.1072	138.7973	0.0072
28	9.3919	0.1065	154.4465	0.0065
29	9.4498	0.1058	171.7415	0.0058
30	9.5021	0.1052	190.8554	0.0052
31	9.5495	0.1047	211.9795	0.0047
32	9.5924	0.1042	235.3253	0.0042
33	9.6312	0.1038	261.1264	0.0038
34	9.6663	0.1035	289.6410	0.0035
35	9.6980	0.1031	321.1545	0.0031
36	9.7268	0.1028	355.9823	0.0028
37	9.7528	0.1025	394.4730	0.0025
38	9.7763	0.1023	437.0118	0.0023
39	9.7976	0.1021	484.0245	0.0021
40	9.8168	0.1019	535.9815	0.0019
45	9.8889	0.1011	890.1713	0.0011
48	9.9177	0.1008	1205.1042	0.0008
50	9.9326	0.1007	1474.1316	0.0007
54	9.9548	0.1005	2204.0642	0.0005
60	9.9752	0.1002	4024.2879	0.0002
65	9.9850	0.1002	6641.4163	0.0002
70	9.9909	0.1001	10956.3316	0.0001
75	9.9945	0.1001	18070.4241	0.0001
80	9.9966	0.1000	29799.5799	0.0000
90	9.9988	0.1000	81020.8393	0.0000
100	9.9995	0.1000	220254.6579	0.0000
120	9.9999	0.1000	1627537.9142	0.0000

Table B.27 Continuous Compounding: *r*= 15%

	Continuous flow, Uniform series			
	Present worth factor	Capital recovery factor	Compound amount factor	Sinking fund factor
	To find P given Ā	To find Ā given P	To find F given Ā	To find Ā given F
n	P|Ā *r,n*	Ā|P *r,n*	F|Ā *r,n*	Ā|F *r,n*
1	0.9286	1.0769	1.0789	0.9269
2	1.7279	0.5787	2.3324	0.4287
3	2.4158	0.4139	3.7887	0.2639
4	3.0079	0.3325	5.4808	0.1825
5	3.5176	0.2843	7.4467	0.1343
6	3.9562	0.2528	9.7307	0.1028
7	4.3337	0.2307	12.3843	0.0807
8	4.6587	0.2147	15.4674	0.0647
9	4.9384	0.2025	19.0495	0.0525
10	5.1791	0.1931	23.2113	0.0431
11	5.3863	0.1857	28.0465	0.0357
12	5.5647	0.1797	33.6643	0.0297
13	5.7182	0.1749	40.1913	0.0249
14	5.8503	0.1709	47.7745	0.0209
15	5.9640	0.1677	56.5849	0.0177
16	6.0619	0.1650	66.8212	0.0150
17	6.1461	0.1627	78.7140	0.0127
18	6.2186	0.1608	92.5315	0.0108
19	6.2810	0.1592	108.5852	0.0092
20	6.3348	0.1579	127.2369	0.0079
21	6.3810	0.1567	148.9071	0.0067
22	6.4208	0.1557	174.0843	0.0057
23	6.4550	0.1549	203.3359	0.0049
24	6.4845	0.1542	237.3216	0.0042
25	6.5099	0.1536	276.8072	0.0036
26	6.5317	0.1531	322.6830	0.0031
27	6.5505	0.1527	375.9830	0.0027
28	6.5667	0.1523	437.9089	0.0023
29	6.5806	0.1520	509.8564	0.0020
30	6.5926	0.1517	593.4475	0.0017
31	6.6029	0.1514	690.5666	0.0014
32	6.6118	0.1512	803.4028	0.0012
33	6.6194	0.1511	934.4998	0.0011
34	6.6260	0.1509	1086.8127	0.0009
35	6.6317	0.1508	1263.7751	0.0008
36	6.6366	0.1507	1469.3761	0.0007
37	6.6408	0.1506	1708.2504	0.0006
38	6.6444	0.1505	1985.7827	0.0005
39	6.6475	0.1504	2308.2292	0.0004
40	6.6501	0.1504	2682.8586	0.0004
45	6.6589	0.1502	5687.0584	0.0002
48	6.6617	0.1501	8922.8718	0.0001
50	6.6630	0.1501	12046.9494	0.0001
54	6.6646	0.1500	21956.4538	0.0000
60	6.6658	0.1500	54013.8929	0.0000
65	6.6663	0.1500	114354.8587	0.0000
70	6.6665	0.1500	242096.6845	0.0000
75	6.6666	0.1500	512526.1318	0.0000
80	6.6666	0.1500	1085025.2761	0.0000
90	6.6667	0.1500	4862769.1323	0.0000
100	6.6667	0.1500	21793442.4831	0.0000
120	6.6667	0.1500	437733120.9155	0.0000

Table B.28 Continuous Compounding: $r = 20\%$

	Continuous flow, Uniform series			
	Present worth factor	Capital recovery factor	Compound amount factor	Sinking fund factor
	To find P given $\bar{A}$	To find $\bar{A}$ given P	To find F given $\bar{A}$	To find $\bar{A}$ given F
n	$P\|\bar{A}$ r,n	$\bar{A}\|P$ r,n	$F\|\bar{A}$ r,n	$\bar{A}\|F$ r,n
1	0.9063	1.1033	1.1070	0.9033
2	1.6484	0.6066	2.4591	0.4066
3	2.2559	0.4433	4.1106	0.2433
4	2.7534	0.3632	6.1277	0.1632
5	3.1606	0.3164	8.5914	0.1164
6	3.4940	0.2862	11.6006	0.0862
7	3.7670	0.2655	15.2760	0.0655
8	3.9905	0.2506	19.7652	0.0506
9	4.1735	0.2396	25.2482	0.0396
10	4.3233	0.2313	31.9453	0.0313
11	4.4460	0.2249	40.1251	0.0249
12	4.5464	0.2200	50.1159	0.0200
13	4.6286	0.2160	62.3187	0.0160
14	4.6959	0.2129	77.2232	0.0129
15	4.7511	0.2105	95.4277	0.0105
16	4.7962	0.2085	117.6627	0.0085
17	4.8331	0.2069	144.8205	0.0069
18	4.8634	0.2056	177.9912	0.0056
19	4.8881	0.2046	218.5059	0.0046
20	4.9084	0.2037	267.9908	0.0037
21	4.9250	0.2030	328.4317	0.0030
22	4.9386	0.2025	402.2543	0.0025
23	4.9497	0.2020	492.4216	0.0020
24	4.9589	0.2017	602.5521	0.0017
25	4.9663	0.2014	737.0658	0.0014
26	4.9724	0.2011	901.3612	0.0011
27	4.9774	0.2009	1102.0321	0.0009
28	4.9815	0.2007	1347.1320	0.0007
29	4.9849	0.2006	1646.4978	0.0006
30	4.9876	0.2005	2012.1440	0.0005
31	4.9899	0.2004	2458.7452	0.0004
32	4.9917	0.2003	3004.2252	0.0003
33	4.9932	0.2003	3670.4759	0.0003
34	4.9944	0.2002	4484.2365	0.0002
35	4.9954	0.2002	5478.1658	0.0002
36	4.9963	0.2001	6692.1538	0.0001
37	4.9969	0.2001	8174.9221	0.0001
38	4.9975	0.2001	9985.9795	0.0001
39	4.9980	0.2001	12198.0099	0.0001
40	4.9983	0.2001	14899.7899	0.0001
45	4.9994	0.2000	40510.4196	0.0000
48	4.9997	0.2000	73818.9078	0.0000
50	4.9998	0.2000	110127.3290	0.0000
54	4.9999	0.2000	245099.0057	0.0000
60	5.0000	0.2000	813768.9571	0.0000
65	5.0000	0.2000	2212061.9600	0.0000
70	5.0000	0.2000	6013016.4208	0.0000
75	5.0000	0.2000	16345081.8624	0.0000
80	5.0000	0.2000	44430547.6025	0.0000
90	5.0000	0.2000	328299840.6867	0.0000

Table B.29 Continuous Compounding: r= 25%

	Continuous flow, Uniform series			
	Present worth factor	Capital recovery factor	Compound amount factor	Sinking fund factor
	To find P given Ā	To find Ā given P	To find F given Ā	To find Ā given F
n	P|Ā r,n	Ā|P r,n	F|Ā r,n	Ā|F r,n
1	0.8848	1.1302	1.1361	0.8802
2	1.5739	0.6354	2.5949	0.3854
3	2.1105	0.4738	4.4680	0.2238
4	2.5285	0.3955	6.8731	0.1455
5	2.8540	0.3504	9.9614	0.1004
6	3.1075	0.3218	13.9268	0.0718
7	3.3049	0.3026	19.0184	0.0526
8	3.4587	0.2891	25.5562	0.0391
9	3.5784	0.2795	33.9509	0.0295
10	3.6717	0.2724	44.7300	0.0224
11	3.7443	0.2671	58.5705	0.0171
12	3.8009	0.2631	76.3421	0.0131
13	3.8449	0.2601	99.1614	0.0101
14	3.8792	0.2578	128.4618	0.0078
15	3.9059	0.2560	166.0843	0.0060
16	3.9267	0.2547	214.3926	0.0047
17	3.9429	0.2536	276.4216	0.0036
18	3.9556	0.2528	356.0685	0.0028
19	3.9654	0.2522	458.3371	0.0022
20	3.9730	0.2517	589.6526	0.0017
21	3.9790	0.2513	758.2651	0.0013
22	3.9837	0.2510	974.7677	0.0010
23	3.9873	0.2508	1252.7626	0.0008
24	3.9901	0.2506	1609.7152	0.0006
25	3.9923	0.2505	2068.0513	0.0005
26	3.9940	0.2504	2656.5665	0.0004
27	3.9953	0.2503	3412.2351	0.0003
28	3.9964	0.2502	4382.5326	0.0002
29	3.9972	0.2502	5628.4194	0.0002
30	3.9978	0.2501	7228.1697	0.0001
31	3.9983	0.2501	9282.2897	0.0001
32	3.9987	0.2501	11919.8319	0.0001
33	3.9990	0.2501	15306.5033	0.0001
34	3.9992	0.2501	19655.0754	0.0001
35	3.9994	0.2500	25238.7524	0.0000
36	3.9995	0.2500	32408.3357	0.0000
37	3.9996	0.2500	41614.2629	0.0000
38	3.9997	0.2500	53434.9073	0.0000
39	3.9998	0.2500	68612.9152	0.0000
40	3.9998	0.2500	88101.8632	0.0000
45	3.9999	0.2500	307515.8791	0.0000
48	4.0000	0.2500	651015.1657	0.0000
50	4.0000	0.2500	1073345.1461	0.0000
54	4.0000	0.2500	2917661.4794	0.0000
60	4.0000	0.2500	13076065.4899	0.0000

Table B.30 Continuous Compounding: $r=5\%$

	Geometric series present worth factor,				$(P \vert A_1 \quad r,c,n)_\infty$
n	c = 4%	c = 6%	c = 8%	c = 10%	c = 15%
1	0.9512	0.9512	0.9512	0.9512	0.9512
2	1.8930	1.9120	1.9313	1.9512	2.0025
3	2.8254	2.8825	2.9412	3.0025	3.1643
4	3.7485	3.8627	3.9817	4.1077	4.4484
5	4.6624	4.8527	5.0538	5.2695	5.8674
6	5.5673	5.8527	6.1584	6.4909	7.4357
7	6.4631	6.8628	7.2965	7.7749	9.1690
8	7.3500	7.8830	8.4692	9.1248	11.0845
9	8.2281	8.9134	9.6775	10.5439	13.2015
10	9.0975	9.9542	10.9224	12.0357	15.5412
11	9.9582	11.0055	12.2052	13.6040	18.1269
12	10.8103	12.0673	13.5269	15.2527	20.9845
13	11.6540	13.1398	14.8887	16.9860	24.1427
14	12.4893	14.2231	16.2918	18.8081	27.6331
15	13.3162	15.3173	17.7375	20.7236	31.4905
16	14.1350	16.4225	19.2271	22.7374	35.7536
17	14.9455	17.5388	20.7619	24.8544	40.4651
18	15.7481	18.6663	22.3433	27.0800	45.6721
19	16.5426	19.8051	23.9726	29.4196	51.4267
20	17.3292	20.9554	25.6515	31.8792	57.7865
21	18.1080	22.1172	27.3813	34.4649	64.8152
22	18.8791	23.2907	29.1636	37.1832	72.5831
23	19.6425	24.4760	30.9999	40.0409	81.1679
24	20.3982	25.6732	32.8921	43.0450	90.6557
25	21.1465	26.8825	34.8416	46.2032	101.1412
26	21.8873	28.1039	36.8503	49.5234	112.7296
27	22.6208	29.3376	38.9200	53.0137	125.5367
28	23.3469	30.5836	41.0525	56.6830	139.6907
29	24.0658	31.8422	43.2498	60.5405	155.3334
30	24.7776	33.1135	45.5137	64.5957	172.6211
31	25.4823	34.3975	47.8463	68.8588	191.7271
32	26.1800	35.6944	50.2498	73.3405	212.8424
33	26.8707	37.0044	52.7262	78.0520	236.1785
34	27.5546	38.3275	55.2777	83.0050	261.9688
35	28.2316	39.6640	57.9067	88.2120	290.4716
36	28.9019	41.0138	60.6155	93.6860	321.9720
37	29.5656	42.3772	63.4065	99.4406	356.7853
38	30.2226	43.7544	66.2822	105.4903	395.2599
39	30.8732	45.1453	69.2452	111.8501	437.7810
40	31.5172	46.5503	72.2981	118.5361	484.7741
48	36.4441	58.3103	100.3046	185.9600	1089.9691
50	37.6154	61.4002	108.4172	207.4688	1333.2938
54	39.8890	67.7687	126.1728	257.5108	1993.4889
60	43.1333	77.8120	157.1240	354.0943	3639.8093
65	45.6921	86.6542	187.5146	459.9358	6006.8985
70	48.1261	95.9498	222.8057	595.8390	9909.5687
80	52.6438	115.9951	311.3781	994.4092	26952.5414
90	56.7316	138.1487	430.8188	1651.5409	73280.1440
100	60.4303	162.6321	591.8858	2734.9690	199211.6230
120	66.8054	219.5947	1101.9828	7466.3106	1472043.6146

Table B.31 Continuous Compounding: $r=5\%$

	Geometric series future worth factor,				$(P|A_1 \ i,j,n)_{\infty}$
n	c = 4%	c = 6%	c = 8%	c = 10%	c = 15%
1	1.0000	1.0000	1.0000	1.0000	1.0000
2	2.0921	2.1131	2.1346	2.1564	2.2131
3	3.2826	3.3489	3.4175	3.4884	3.6764
4	4.5784	4.7179	4.8640	5.0171	5.4332
5	5.9867	6.2310	6.4905	6.7662	7.5339
6	7.5150	7.9003	8.3151	8.7618	10.0372
7	9.1716	9.7387	10.3575	11.0332	13.0114
8	10.9650	11.7600	12.6392	13.6126	16.5362
9	12.9043	13.9790	15.1837	16.5361	20.7041
10	14.9992	16.4118	18.0166	19.8435	25.6231
11	17.2601	19.0753	21.1659	23.5792	31.4185
12	19.6977	21.9881	24.6620	27.7923	38.2363
13	22.3237	25.1699	28.5381	32.5373	46.2464
14	25.1503	28.6419	32.8305	37.8749	55.6462
15	28.1905	32.4267	37.5786	43.8720	66.6654
16	31.4579	36.5489	42.8254	50.6030	79.5711
17	34.9673	41.0345	48.6178	58.1505	94.6740
18	38.7340	45.9116	55.0066	66.6059	112.3352
19	42.7743	51.2102	62.0476	76.0705	132.9744
20	47.1057	56.9626	69.8010	86.6567	157.0800
21	51.7464	63.2032	78.3328	98.4887	185.2192
22	56.7159	69.9691	87.7146	111.7045	218.0516
23	62.0347	77.2999	98.0242	126.4568	256.3440
24	67.7245	85.2381	109.3466	142.9146	300.9874
25	73.8085	93.8290	121.7738	161.2652	353.0176
26	80.3111	103.1215	135.4064	181.7159	413.6383
27	87.2579	113.1674	150.3532	204.4965	484.2484
28	94.6764	124.0227	166.7331	229.8610	566.4738
29	102.5954	135.7471	184.6750	258.0909	662.2039
30	111.0455	148.4043	204.3192	289.4977	773.6343
31	120.0591	162.0628	225.8180	324.4262	903.3165
32	129.6703	176.7957	249.3371	363.2579	1054.2155
33	139.9152	192.6812	275.0567	406.4151	1229.7767
34	150.8323	209.8029	303.1723	454.3652	1434.0037
35	162.4618	228.2503	333.8966	507.6252	1671.5485
36	174.8466	248.1191	367.4604	566.7673	1947.8169
37	188.0319	269.5116	404.1147	632.4244	2269.0900
38	202.0654	292.5371	444.1320	705.2970	2642.6663
39	216.9977	317.3125	487.8083	786.1597	3077.0261
40	232.8823	343.9627	535.4651	875.8696	3582.0229
48	401.7293	642.7645	1108.8901	2049.8694	12014.9216
50	458.2495	748.0082	1324.8240	2527.4875	16242.8432
54	593.5376	1008.3801	1883.7065	3831.6909	29662.5802
60	866.3558	1562.8958	3167.9343	7112.1736	73107.5250
65	1178.4151	2234.8418	4856.3652	11861.9002	154919.9539
70	1593.7181	3177.4206	7412.2338	19731.4790	328159.8458
80	2874.2537	6333.1204	17092.8416	54292.9010	1471558.8994
90	5106.8118	12435.7489	39024.4542	148666.9722	8596468.3423
100	8968.6539	24136.7505	88472.2371	405905.3880	29565626.2924
120	26951.2135	88590.8273	448564.3760	3012124.6728	593864779.4096

Table B.32 Continuous Compounding: $r=8\%$

	Geometric series present worth factor,				$(P\|A_1 \ r,c,n)_{oo}$
n	c = 4%	c = 6%	c = 8%	c = 10%	c = 15%
1	0.9231	0.9231	0.9231	0.9231	0.9231
2	1.8100	1.8280	1.8461	1.8649	1.9132
3	2.6622	2.7149	2.7691	2.8257	2.9750
4	3.4809	3.5842	3.6919	3.8059	4.1138
5	4.2675	4.4364	4.6147	4.8059	5.3352
6	5.0233	5.2716	5.5373	5.8261	6.6452
7	5.7495	6.0904	6.4599	6.8669	8.0501
8	6.4471	6.8929	7.3823	7.9287	9.5570
9	7.1175	7.6795	8.3047	9.0120	11.1730
10	7.7615	8.4506	9.2270	10.1172	12.9063
11	8.3803	9.2064	10.1492	11.2447	14.7652
12	8.9748	9.9472	11.0713	12.3950	16.7589
13	9.5460	10.6733	11.9933	13.5685	18.8972
14	10.0948	11.3851	12.9152	14.7657	21.1905
15	10.6221	12.0828	13.8371	15.9871	23.6501
16	11.1287	12.7666	14.7588	17.2332	26.2881
17	11.6155	13.4370	15.6804	18.5044	29.1173
18	12.0832	14.0940	16.6020	19.8014	32.1517
19	12.5325	14.7380	17.5234	21.1245	35.4060
20	12.9642	15.3693	18.4448	22.4743	38.8964
21	13.3790	15.9881	19.3661	23.8515	42.6398
22	13.7775	16.5946	20.2873	25.2564	46.6546
23	14.1604	17.1892	21.2083	26.6898	50.9606
24	14.5283	17.7719	22.1293	28.1520	55.5788
25	14.8817	18.3431	23.0502	29.6439	60.5318
26	15.2213	18.9030	23.9710	31.1658	65.8440
27	15.5476	19.4518	24.8918	32.7186	71.5413
28	15.8611	19.9898	25.8124	34.3026	77.6518
29	16.1623	20.5171	26.7329	35.9187	84.2053
30	16.4517	21.0339	27.6534	37.5674	91.2340
31	16.7297	21.5405	28.5737	39.2495	98.7723
32	16.9968	22.0371	29.4940	40.9655	106.8572
33	17.2535	22.5239	30.4142	42.7162	115.5283
34	17.5001	23.0010	31.3342	44.5022	124.8282
35	17.7370	23.4686	32.2542	46.3243	134.8024
36	17.9647	23.9271	33.1741	48.1833	145.4998
37	18.1834	24.3764	34.0939	50.0797	156.9728
38	18.3935	24.8168	35.0136	52.0145	169.2777
39	18.5954	25.2485	35.9332	53.9884	182.4749
40	18.7894	25.6717	36.8527	56.0022	196.6289
48	20.0910	28.7689	44.2056	73.6480	353.7898
50	20.3564	29.4688	46.0429	78.5185	408.8683
54	20.8275	30.7873	49.7164	88.8640	545.0997
60	21.4068	32.5776	55.2239	106.0200	836.2660
65	21.7940	33.9138	59.8110	121.9761	1192.0531
70	22.1109	35.1228	64.3957	139.6104	1696.9391
80	22.5829	37.2067	73.5584	180.6379	3430.1220
90	22.8993	38.9129	82.7119	230.7490	6920.3236
100	23.1113	40.3097	91.8562	291.9549	13948.7265
120	23.3488	42.3897	110.1175	458.0208	56603.7706

Table B.33 Continuous Compounding: $r=8\%$

	Geometric series future worth factor,				$(P\|A_1 \; i,j,n)_{\infty}$
n	c = 4%	c = 6%	c = 8%	c = 10%	c = 15%
1	1.0000	1.0000	1.0000	1.0000	1.0000
2	2.1241	2.1451	2.1666	2.1885	2.2451
3	3.3843	3.4513	3.5205	3.5921	3.7820
4	4.7937	4.9359	5.0850	5.2412	5.6853
5	6.3664	6.6183	6.8856	7.1695	7.9592
6	8.1181	8.5194	8.9509	9.4154	10.7391
7	10.0654	10.6623	11.3125	12.0217	14.0932
8	12.2269	13.0722	14.0054	15.0367	18.1246
9	14.6224	15.7771	17.0683	18.5146	22.9543
10	17.2735	18.8071	20.5443	22.5162	28.7235
11	20.2040	22.1956	24.4809	27.1098	35.5975
12	23.4395	25.9790	28.9308	32.3719	43.7693
13	27.0078	30.1972	33.9520	38.3881	53.4643
14	30.9392	34.8937	39.6090	45.2547	64.9459
15	35.2667	40.1162	45.9728	53.0790	78.5212
16	40.0261	45.9170	53.1218	61.9815	94.5487
17	45.2562	52.3530	61.1428	72.0968	113.4466
18	50.9993	59.4865	70.1314	83.5755	135.7023
19	57.3014	67.3856	80.1931	96.5859	161.8843
20	64.2121	76.1247	91.4444	111.3162	192.6550
21	71.7857	85.7851	104.0136	127.9764	228.7862
22	80.0809	96.4553	118.0421	146.8014	271.1772
23	89.1614	108.2322	133.6859	168.0531	320.8754
24	99.0967	121.2214	151.1167	192.0239	379.1005
25	109.9619	135.5383	170.5238	219.0403	447.2729
26	121.8386	151.3086	192.1152	249.4660	527.0481
27	134.8154	168.6694	216.1204	283.7071	620.3446
28	148.9884	187.7705	242.7915	322.2160	729.4088
29	164.4621	208.7749	272.4062	365.4971	856.8454
30	181.3496	231.8605	305.2698	414.1125	1005.6880
31	199.7738	257.2211	341.7180	468.6883	1179.4860
32	219.8680	285.0681	382.1199	529.9220	1382.2852
33	241.7768	315.6315	426.8813	598.5902	1618.9221
34	265.6571	349.1623	476.4481	675.5578	1894.9323
35	291.6791	385.9336	531.3104	761.7872	2216.7776
36	320.0274	426.2430	592.0062	858.3498	2591.9728
37	350.9022	470.4147	659.1269	966.4376	3029.2570
38	384.5208	518.8015	733.3216	1087.3768	3538.7925
39	421.1186	571.7877	815.3029	1222.6426	4132.3955
40	460.9512	629.7914	905.8534	1373.8755	4823.8050
48	934.7453	1338.4850	2061.5196	3426.5061	16460.2386
50	1111.4222	1608.9405	2520.0161	4286.9639	22323.4534
54	1565.9909	2314.8578	3748.0141	6681.5593	40985.2972
60	2601.1507	3958.5138	6730.0753	12882.5306	101615.0248
65	3950.6391	6147.6265	10876.7642	22110.8837	216086.1379
70	5979.3767	9498.1411	17474.3923	37754.3366	458897.1434
80	13591.4050	22392.6784	44445.6638	108716.0047	2064401.8789
90	30671.9919	52121.0809	111280.0438	309072.2980	9269294.3280
100	68893.9547	120161.6118	275175.7425	870305.3412	41580567.7757
120	344739.8525	625875.2745	1635543.6143	6762576.6413	835742309.0391

Table B.34 Continuous Compounding: $r = 10\%$

	Geometric series present worth factor,				$(P\|A_1 \; r,c,n)_{oo}$
n	c = 4%	c = 6%	c = 8%	c = 10%	c = 15%
1	0.9048	0.9048	0.9048	0.9048	0.9048
2	1.7570	1.7742	1.7917	1.8097	1.8561
3	2.5595	2.6095	2.6609	2.7145	2.8561
4	3.3153	3.4120	3.5127	3.6194	3.9073
5	4.0271	4.1830	4.3477	4.5242	5.0125
6	4.6974	4.9239	5.1660	5.4290	6.1743
7	5.3287	5.6356	5.9680	6.3339	7.3957
8	5.9232	6.3195	6.7541	7.2387	8.6798
9	6.4831	6.9765	7.5245	8.1435	10.0296
10	7.0104	7.6078	8.2797	9.0484	11.4487
11	7.5070	8.2143	9.0197	9.9532	12.9405
12	7.9746	8.7971	9.7451	10.8581	14.5088
13	8.4151	9.3570	10.4560	11.7629	16.1576
14	8.8298	9.8949	11.1528	12.6677	17.8908
15	9.2205	10.4118	11.8357	13.5726	19.7129
16	9.5883	10.9084	12.5050	14.4774	21.6285
17	9.9348	11.3855	13.1610	15.3822	23.6422
18	10.2611	11.8439	13.8039	16.2871	25.7592
19	10.5684	12.2843	14.4341	17.1919	27.9848
20	10.8577	12.7075	15.0517	18.0968	30.3244
21	11.1303	13.1141	15.6570	19.0016	32.7840
22	11.3869	13.5047	16.2503	19.9064	35.3697
23	11.6286	13.8800	16.8317	20.8113	38.0880
24	11.8563	14.2406	17.4016	21.7161	40.9457
25	12.0707	14.5870	17.9602	22.6210	43.9498
26	12.2726	14.9199	18.5076	23.5258	47.1080
27	12.4627	15.2397	19.0442	24.4306	50.4281
28	12.6418	15.5470	19.5701	25.3355	53.9185
29	12.8104	15.8422	20.0855	26.2403	57.5878
30	12.9692	16.1259	20.5906	27.1452	61.4452
31	13.1188	16.3984	21.0857	28.0500	65.5004
32	13.2597	16.6603	21.5710	28.9548	69.7635
33	13.3923	16.9119	22.0465	29.8597	74.2452
34	13.5172	17.1536	22.5127	30.7645	78.9567
35	13.6349	17.3858	22.9695	31.6694	83.9097
36	13.7457	17.6089	23.4173	32.5742	89.1167
37	13.8500	17.8233	23.8561	33.4790	94.5906
38	13.9483	18.0293	24.2862	34.3839	100.3452
39	14.0409	18.2272	24.7078	35.2887	106.3949
40	14.1280	18.4173	25.1210	36.1936	112.7547
48	14.6654	19.6932	28.1437	43.4323	176.8900
50	14.7640	19.9533	28.8264	45.2420	197.3498
54	14.9291	20.4151	30.1122	48.8614	244.9509
60	15.1130	20.9829	31.8571	54.2904	336.8234
65	15.2231	21.3624	33.1588	58.8146	437.5024
70	15.3046	21.6731	34.3360	63.3388	566.7767
80	15.4097	22.1357	36.3635	72.3873	945.9054
90	15.4674	22.4458	38.0218	81.4357	1570.9831
100	15.4991	22.6537	39.3782	90.4842	2601.5619
120	15.5260	22.8865	41.3949	108.5811	7102.1027

Table B.35 Continuous Compounding: $r = 10\%$

	Geometric series future worth factor,				$(P\|A_1 \ i,j,n)_{\infty}$
n	c = 4%	c = 6%	c = 8%	c = 10%	c = 15%
1	1.0000	1.0000	1.0000	1.0000	1.0000
2	2.1460	2.1670	2.1885	2.2103	2.2670
3	3.4550	3.5224	3.5921	3.6642	3.8553
4	4.9458	5.0901	5.2412	5.3994	5.8291
5	6.6395	6.8967	7.1695	7.4591	8.2642
6	8.5592	8.9718	9.4154	9.8923	11.2504
7	10.7306	11.3488	12.0217	12.7548	14.8932
8	13.1823	14.0643	15.0367	16.1100	19.3172
9	15.9458	17.1595	18.5146	20.0299	24.6689
10	19.0562	20.6802	22.5162	24.5960	31.1208
11	22.5521	24.6773	27.1098	29.9011	38.8755
12	26.4767	29.2074	32.3718	36.0500	48.1710
13	30.8773	34.3336	38.3881	43.1615	59.2869
14	35.8067	40.1260	45.2546	51.3702	72.5508
15	41.3232	46.6624	53.0789	60.8280	88.3472
16	47.4914	54.0295	61.9814	71.7071	107.1265
17	54.3826	62.3236	72.0967	84.2016	129.4163
18	62.0759	71.6514	83.5753	98.5311	155.8342
19	70.6589	82.1317	96.5857	114.9434	187.1032
20	80.2285	93.8963	111.3159	133.7180	224.0688
21	90.8917	107.0916	127.9762	155.1703	267.7198
22	102.7672	121.8800	146.8011	179.6559	319.2122
23	115.9863	138.4416	168.0527	207.5755	379.8967
24	130.6939	156.9766	192.0235	239.3807	451.3512
25	147.0508	177.7066	219.0397	275.5797	535.4184
26	165.2346	200.8779	249.4654	316.7452	634.2499
27	185.4417	226.7632	283.7063	363.5214	750.3570
28	207.8894	255.6652	322.2151	416.6331	886.6702
29	232.8182	287.9193	365.4961	476.8954	1046.6085
30	260.4938	323.8974	414.1113	545.2252	1234.1597
31	291.2103	364.0116	468.6869	622.6526	1453.9746
32	325.2928	408.7188	529.9203	710.3355	1711.4754
33	363.1008	458.5251	598.5883	809.5748	2012.9832
34	405.0318	513.9913	675.5555	921.8312	2365.8655
35	451.5256	575.7389	761.7846	1048.7453	2778.7076
36	503.0682	644.4560	858.3468	1192.1584	3261.5131
37	560.1970	720.9052	966.4341	1354.1371	3825.9359
38	623.5064	805.9308	1087.3728	1537.0004	4485.5506
39	693.6533	900.4679	1222.6379	1743.3495	5256.1675
40	771.3643	1005.5522	1373.8702	1976.1018	6156.1978
48	1781.9951	2392.9306	3426.4900	5277.4767	21493.9790
50	2191.1713	2961.3357	4286.9429	6714.5054	29289.3015
54	3305.3885	4520.0308	6681.5239	10818.2164	54233.6964
60	6097.0283	8465.1178	12882.4546	21902.3127	135884.2753
65	10125.4885	14209.0254	22110.7422	39120.0527	291001.0399
70	16783.5006	23767.4343	37754.0761	69459.4698	621546.0867
80	45935.6564	65985.6707	108715.1458	215783.4387	2819704.3620
90	125333.5559	181880.5081	309069.5473	659880.5555	12729807.8516
100	341389.3494	498981.2068	870296.7252	1993046.9108	57303214.4912

Table B.36 Continuous Compounding: $r = 15\%$

	Geometric series present worth factor,				$(P \mid A_1 \quad r,c,n)_{\infty}$
n	c = 4%	c = 6%	c = 8%	c = 10%	c = 15%
1	0.8607	0.8607	0.8607	0.8607	0.8607
2	1.6318	1.6473	1.6631	1.6794	1.7214
3	2.3225	2.3663	2.4113	2.4582	2.5821
4	2.9413	3.0233	3.1087	3.1991	3.4428
5	3.4956	3.6238	3.7590	3.9037	4.3035
6	3.9922	4.1726	4.3652	4.5741	5.1642
7	4.4370	4.6742	4.9304	5.2117	6.0250
8	4.8356	5.1326	5.4573	5.8182	6.8857
9	5.1926	5.5515	5.9486	6.3952	7.7464
10	5.5124	5.9344	6.4066	6.9440	8.6071
11	5.7989	6.2844	6.8335	7.4660	9.4678
12	6.0556	6.6042	7.2316	7.9626	10.3285
13	6.2855	6.8965	7.6028	8.4350	11.1892
14	6.4915	7.1636	7.9488	8.8843	12.0499
15	6.6760	7.4078	8.2713	9.3117	12.9106
16	6.8413	7.6309	8.5721	9.7183	13.7713
17	6.9894	7.8348	8.8525	10.1050	14.6320
18	7.1220	8.0212	9.1139	10.4729	15.4927
19	7.2409	8.1915	9.3576	10.8229	16.3535
20	7.3473	8.3472	9.5848	11.1557	17.2142
21	7.4427	8.4895	9.7966	11.4724	18.0749
22	7.5281	8.6195	9.9941	11.7736	18.9356
23	7.6046	8.7383	10.1782	12.0601	19.7963
24	7.6732	8.8470	10.3499	12.3326	20.6570
25	7.7346	8.9462	10.5099	12.5918	21.5177
26	7.7897	9.0369	10.6591	12.8384	22.3784
27	7.8389	9.1198	10.7982	13.0730	23.2391
28	7.8831	9.1956	10.9278	13.2961	24.0998
29	7.9227	9.2649	11.0487	13.5084	24.9605
30	7.9581	9.3282	11.1615	13.7103	25.8212
31	7.9898	9.3860	11.2665	13.9023	26.6819
32	8.0183	9.4389	11.3645	14.0850	27.5427
33	8.0438	9.4872	11.4559	14.2588	28.4034
34	8.0666	9.5313	11.5410	14.4241	29.2641
35	8.0870	9.5717	11.6204	14.5813	30.1248
36	8.1053	9.6086	11.6944	14.7309	30.9855
37	8.1218	9.6423	11.7634	14.8732	31.8462
38	8.1365	9.6731	11.8277	15.0085	32.7069
39	8.1498	9.7013	11.8877	15.1373	33.5676
40	8.1614	9.7270	11.9436	15.2597	34.4283
48	8.2208	9.8672	12.2742	16.0471	41.3140
50	8.2291	9.8891	12.3317	16.1995	43.0354
54	8.2411	9.9227	12.4251	16.4621	46.4782
60	8.2516	9.9551	12.5242	16.7695	51.6425
65	8.2564	9.9714	12.5802	16.9638	55.9460
70	8.2591	9.9819	12.6197	17.1152	60.2496
80	8.2616	9.9928	12.6670	17.3249	68.8566
90	8.2624	9.9972	12.6905	17.4521	77.4637
100	8.2627	9.9990	12.7022	17.5292	86.0708
120	8.2628	10.0000	12.7108	17.6044	103.2850

Table B.37 Continuous Compounding: $r = 15\%$

	Geometric series future worth factor,				$(P\|A_1 \; i,j,n)_{oo}$
n	$c = 4\%$	$c = 6\%$	$c = 8\%$	$c = 10\%$	$c = 15\%$
1	1.0000	1.0000	1.0000	1.0000	1.0000
2	2.2026	2.2237	2.2451	2.2670	2.3237
3	3.6424	3.7110	3.7820	3.8553	4.0496
4	5.3594	5.5088	5.6653	5.8291	6.2732
5	7.4002	7.6716	7.9592	8.2642	9.1106
6	9.8192	10.2630	10.7391	11.2504	12.7020
7	12.6795	13.3572	14.0932	14.8932	17.2172
8	16.0546	17.0408	18.1246	19.3172	22.8612
9	20.0300	21.4147	22.9543	24.6689	29.8811
10	24.7048	26.5963	28.7235	31.1208	38.5743
11	30.1947	32.7226	35.5974	38.8755	49.2986
12	36.6340	39.9531	43.7692	48.1711	62.4838
13	44.1787	48.4733	53.4643	59.2869	78.6454
14	53.0104	58.4994	64.9458	72.5508	98.4016
15	63.3399	70.2829	78.5212	88.3473	122.4926
16	75.4126	84.1167	94.5487	107.1266	151.8038
17	89.5134	100.3414	113.4465	129.4164	187.3940
18	105.9736	119.3533	135.7023	155.8343	230.5279
19	125.1782	141.6134	161.8842	187.1033	282.7149
20	147.5746	167.6581	192.6548	224.0689	345.7556
21	173.6828	198.1110	228.7860	267.7200	421.7963
22	204.1070	233.6976	271.1770	319.2125	513.3934
23	239.5494	275.2612	320.8751	379.8970	623.5907
24	280.8260	323.7828	379.1002	451.3516	756.0094
25	328.8850	380.4027	447.2726	535.4189	914.9559
26	384.8281	446.4466	527.0456	634.2506	1105.5482
27	449.9357	523.4557	620.3441	750.3578	1333.8662
28	525.6954	613.2219	729.4081	886.6712	1607.1289
29	613.8357	717.8277	856.8446	1046.6096	1933.9036
30	716.3653	839.6942	1005.6871	1234.1611	2324.3539
31	835.6179	981.6351	1179.4648	1453.9762	2790.5312
32	974.3051	1146.9210	1382.2839	1711.4774	3346.7197
33	1135.5776	1339.3531	1618.9205	2012.9856	4009.8439
34	1323.0964	1563.3490	1894.9304	2365.8684	4799.9489
35	1541.1149	1824.0430	2216.7753	2778.7111	5740.7670
36	1794.5752	2127.4018	2591.9701	3261.5172	6860.3858
37	2089.2197	2480.3594	3029.2538	3825.9408	8192.0378
38	2431.7199	2890.9738	3538.7887	4485.5565	9775.0276
39	2829.8277	3368.6090	4132.3910	5256.1745	11655.8289
40	3292.5495	3924.1466	4823.7997	6156.2061	13889.3755
48	11011.1685	13216.4694	16460.2193	21494.0113	55337.2208
50	14878.5469	17879.9807	22323.4269	29289.3465	77809.8292
54	27150.0807	32690.0967	40985.2475	54233.7831	153121.0486
60	66863.5666	80666.7153	101614.8986	135884.5034	418463.3444
65	141631.7453	171052.2043	216085.8659	291001.5449	959710.8181
70	299934.0414	362496.5147	458896.5602	621547.1956	2187993.0460
80	1344617.3776	1626370.2786	2064399.2283	2819709.8127	11206747.8079
90	6026763.0769	7292117.6314	9269282.3964	12729832.2906	56503304.4551
100	27010982.3646	32686892.4830	41580514.3560	57303326.9635	281366935.9595

Answers to Even-Numbered Problems

Because of space limitations, a few even-numbered problems are not answered.

CHAPTER 1

2. One person. Total cost = $138.00

4. (a) Type A. Total cost = $656.67/month
   (b) $0.43/mile

6. Ceramic, 247.75 fpm

8. (a) Machine A for 100 units ($810/order); Machine B for 500 units ($3793.50/order)
   (b) 280 units

10. 100 inch long ingots ($49,535.60/month)

12. 32-inch × 32-inch × 6-inch pallet ($0.60/carton)

CHAPTER 2

2. (a) M1 for $0 < X \le 50$ units, M2 for $50 \le X \le 125$ units, M3 for ≥ 125 units
   (b) M2, $875

4. Operate, TR > TC

6. (a) −$25,000
   (b) 220,000 pallets/year
   (c) $125,000
   (d) 132,000 pallets/year

8. (a) 300 hours/year
   (b) $225
   (c) $5.13/hour

10. (b) Book value = $31,100
    (c) Depreciation reserve = $18,900

12. Loss of $5125

14. (a) 22,360 barrels/week;
       $24.67/barrel
    (b) 74,000 barrels; $497,600

16. (a) 224 units/order
    (b) $418.33
    (c) 9
    (d) 158 units/order

18. (a) 3.0 square inches
    (b) $200
    (c) $200.19 for 2-inch; $418.48
        for 4-inch

20. (a) A—28%,   B—55.56%,
       C—60%,   D—42.86%
    (b) A—24.14%, B—33.33%,
       C—40%,   D—33.80%
    (c) A—$2.10/hour,
       B—$2.50/hour,
       C—$3.6/hour,
       D—$1.71/hour

CHAPTER 3

2. $1041.94
4. $4632.00
6. (a) $4626.00
   (b) $3683.40
8. $3900.00
10. $402.10
12. $2030.50
14. $26,399.13
16. $5336.06
18. $2484.35
20. $5308.58
22. $4025.23
24. $292,647.50
26. $8977.60
28. $433.95
30. $3618.94
32. $11,323.90
34. $18,412.89

36. 12.68%
38. $13,903.98
40. 9.54%
42. (a) $10,817.80
    (b) $573.34
44. $2300.10
46. $1557.35
48. (a) $18,372.64
    (b) $47,653.12
    (c) $2989.23
50. $68,496.00
52. (a) $41,763.22
    (b) $113,524.96
54. Pay cash.
56. $3767.16
58. −$815.16
60. $1083.46
62. 11.80%
64. $1755.94
66. $4357.22
68. $1058.70
70. −$1412.19
72. −$1212.34
74. $25,037.50
76. $X = 908.52
    $Y = 768.12
78. 12.36%
80. $2308
82. $9276.43
84. $1347.02
86. $13,107.30
88. $1073.35
90. $20,474.11
92. $7795.55
94. $8869.00
96. $T_1 = $474,600$
    $C_1 = $435,413$
    $T_2 = $536,298$
    $C_2 = $451,391$

$T_3 = \$606,017$
$C_3 = \$467,956$
$T_4 = \$684,799$
$C_4 = \$485,129$

98. 12.32%

100. (a) $49,296.54
    (b) $49,296.54

102. (a) $66,147.65
    (b) $66,147.65

104. $1722,407

106. $1742,751.40

108. $5081,095

110. $350.45

112. $1014.53

114. $2213.15

116. $7012.04

118. 11.34%

120. $2234.23 ·

CHAPTER 4

2. (a) $5,652.50
   (b) $1161.02
   (c) $216,703.28
   (d) 22.46%
   (e) 20.47%
   (f) 1.0808
   (g) 3.75 years

4. (a) −$211,350.62
   (b) −$37,409.06
   (c) −$656,412.75

6. (a) $1478.18
   (b) $444.00
   (c) $8809.78
   (d) 27.7%
   (e) 26.12%
   (f) 1.074

All measures of economic worth
indicate the condenser should be
purchased.

8. (a) $39,946.40
   (b) $7070.51

(c) $124,064.30
(d) 23.77%
(e) 18.77%
(f) 1.7989

All measures of economic worth
indicate the filter should be purchased.

10. (a) $1003.35
    (b) $210.30
    (c) $3529.68
    (d) 17.31%
    (e) 16.44%
    (f) 1.0907
    (g) 6 years and 7 months

12. $8333.33

14. $50,240.19

16. $28,145.87

18. −$36,300

20. (a) −3.175%
    (b) 36.93%

22. (a) 0%
    (b) 100%
    (c) 200%

24. $U = -\$4000$
    $V = \$25,000$
    $W = \$2000$
    $X = 2$
    $Y = A|F$

26. $CR = \$43,560/\text{year}$

28. $12,630.10

30. 9 years

CHAPTER 5

2. {A}, {B}, {D}, {A,C}, {B,C},
   {B,D}, {B,C,D}

4. {Do nothing.}, {A,B}, {A,B,C},
   {A,C}, {C}

14. (a) 16.64%
    (b) 10.98%

16. (a) 12.55%
    (b) 8.28%

18. Yes, 10.53%

20. 9.46%

24. (a) 1: Do nothing.
2: B only.
3: A only.
(b) $PW_1(25\%) = \$0$
$PW_2(25\%) = \$392,227$
$PW_3(25\%) = \$371,553$
Select Proposal B.
(c) $PW_{3-1}(25\%) = \$371,553$
$PW_{2-3}(25\%) = \$20,674$
Select Proposal B.
(d) $AW_1(25\%) = \$0$
$AW_2(25\%) = \$109,863$
$AW_3(25\%) = \$104,072$
Select Proposal B.
(e) $AW_{3-1}(25\%) = \$104,072$
$AW_{2-3}(25\%) = \$5791$
Select Proposal B.
(f) $FW_1(25\%) = \$0$
$FW_2(25\%) = \$3,652,888$
$FW_3(25\%) = \$3,460,347$
Select Proposal B.
(g) $FW_{3-1}(25\%) = \$3,460,347$
$FW_{2-3}(25\%) = \$192,541$
Select Proposal B.
(h) $i_{3-1} = 42.8\%$
$i_{2-3} = 27.9\%$
Select Proposal B.
(i) $i'_{3-1} = 31.2\%$
$i'_{2-3} = 26.2\%$
Select Proposal B.
(j) $SIR_{3-1}(25\%) = 1.619$
$SIR_{2-3}(25\%) = 1.103$
Select Proposal B.

26. (a) 1: Do nothing.
2: C only.
3: B only.
4: B and D.
5: A only.
Alternative 5 dominates alternative 3, due to same investment but greater return
(b–j) Alternative 4 (Proposals B and D) is the economic choice.

28. (a) 1: No insulation.
2: 1-inch insulation.
3: 2-inch insulation.
(b—j) Select 2-inch insulation.

30. Alternative 1.

32. (a) $PW_1(20\%) = \$2813.53$
$PW_2(20\%) = \$722.00$
$PW_3(20\%) = \$319.50$
Select Alternative 1.
(b) $PBP_1 = 3.125$ years (or 4 years)
$PBP_2 = 1$ year
$PBP_3 = 2$ years
Select Alternative 2.

34. Concrete bridge design.

36. (a)

EOY	NCF(U)	NCF(V)
0	−$1360.00	−$ 300.00
1–8	− 4376.53	− 4689.14

(b) $PW_U(20\%) = -\$18,153.62$
(select)
$PW_V(20\%) = -\$18,293.17$
(c) $AW_U(20\%) = -\$4,730.83$
(select)
$AW_V(20\%) = -\$4,767.20$
(d) $FW_U(20\%) = -\$78,056.94$
(select)
$FW_V(20\%) = -\$78,656.97$
(e) $i_{U-V} = 29.19\%$ (select U)
(f) $i'_{U-V} = 21.9\%$ (select U)
(g) $SIR_{U-V}(20\%) = 1.132$ (select U)

38. (a)

EOY	NCF(I)	NCF(II)
0	−$325,000	−$620,000
1–9	− 125,000	− 110,000
10	− 250,000	− 470,000
11–20	− 125,000	− 110,000

(b) $PW_I(12\%) = -\$1,271,637$
(select)
$PW_{II}(12\%) = -\$1,478,966$

(c) $AW_I(12\%) = -\$170,272$
(select)
$AW_{II}(12\%) = -\$198,034$
(d) $FW_I(12\%) = -\$12,266,597$
(select)
$FW_{II}(12\%) = -\$14,266,550$
(e) $i_{II-I} < 0\%$ (select I)
(f) $i'_{II-I} = 4.51\%$ (select I)
(g) $SIR_{II-I}(12\%) = 0.351$ (select I)

40. (b) $PW_1(12\%) = -\$13,826$
$PW_2(12\%) = -\$13,620$
(select)
(c) $AW_1(12\%) = -\$3,362$
$AW_2(12\%) = -\$3,312$
(select)
(d) $FW_1(12\%) = -\$27,289$
$FW_2(12\%) = -\$26,883$
(select)
(e) $i_{2-1} = 15.73\%$ (select 2)
(f) $i'_{2-1} = 13.85\%$ (select 2)
(g) $SIR_{2-1}(12\%) = 1.103$
(select 2)

42. (a)

EOY	NCF(1)	NCF(2)
0	$-\$\quad 0$	$-\$53,000$
1–10	$-\ 32,000$	$-\ 15,000$

(b) $PW_1(20\%) = -\$134,160$
$PW_2(20\%) = -\$115,888$
(select)
(c) $AW_1(20\%) = -\$31,997$
$AW_2(20\%) = -\$27,639$
(select)
(d) $FW_1(20\%) = -\$830,678$
$FW_2(20\%) = -\$717,541$
(select)
(e) $i_{2-1} = 29.61\%$ (select 2)
(f) $i'_{2-1} = 23.61\%$ (select 2)
(g) $SIR_{2-1}(20\%) = 1.345$ (select 2)

44. (b) $PW_1(15\%) = -\$21,505$
$PW_2(15\%) = -\$19,273$
(select)

(c) $AW_1(15\%) = -\$4286$
$AW_2(15\%) = -\$3841$
(select)
(d) $FW_1(15\%) = -\$87,000$
$FW_2(15\%) = -\$77,973$
(e) $i_{2-1} = 23.90\%$ (select 2)
(f) $i'_{2-1} = 19.30\%$ (select 2)
(g) $SIR_{2-1}(15\%) = 1.4463$
(select 2)

46. (a) $AW_1(15\%) = -\$67,898$
$AW_2(15\%) = -\$58,978$
(select)
(b) $AW_1(15\%) = -\$64,916$
$AW_2(15\%) = -\$58,978$
(select)

48. (a) $PW_1(12\%) = -\$41,114$
(select)
$PW_2(12\%) = -\$66,117$
(b) Negative rate of return, since the sum of the "net savings" is less than the investment. Do not purchase!

50. $166,667/year

52. $76,790/year (approximately; both methods obtain same answer for all problems).

54. 50-inch culvert.

56. Keep old pump.

58. Keep old crane.

60. Buy new reverberatory furnace.

62. Keep old machine.

64. Trade in old equipment and buy new equipment.

66. Sell old mixer and buy new larger mixer.

68. Keep old machine and buy machine X.

70. Keep old tractor and scraper.

72. Purchase new machine immediately.

74. (a) Replace every 10 years.
    (b) Replace every 12 years.

CHAPTER 6

2. (a) 5-year property; 200%
       DBSLH.
   (b) 27.5-year residential rental
       property; SLM.
   (c) Not MACRS property; pre-
       1981 straight-line.
   (d) 7-year property; 200%
       DBSLH.
   (e) 31.5-year nonresidential
       real property; SLM.
   (f) 5-year property; 200%
       DBSLH.
   (g) 5-year property; 200%
       DBSLH.
   (h) 15-year property; 150%
       DBSLH.
   (i) 5-year property; 200%
       DBSLH.
   (j) 15-year property; 150%
       DBSLH.

4. (a) $D_2 = \$3840.00$
       $D_4 = 1382.40$
       $D_6 = 691.20$
       $B_2 = \$5760.00$
       $B_4 = 2073.60$
       $B_6 = 0.00$
   (b) $D_2 = \$2400.00$
       $D_4 = 2400.00$
       $D_6 = 1200.00$
       $B_2 = \$8400.00$
       $B_4 = 3600.00$
       $B_6 = 0.00$

6. (a) $D_1 = \$8000.00$
       $D_4 = 4608.00$
       $D_7 = 0.00$
   (b) $D_1 = \$4000.00$
       $D_4 = 8000.00$
       $D_7 = 0.00$

8. (a) $D_2 = \$45,000.00$
       $D_4 = 28,800.00$
       $D_6 = 9,212.50$

   (b) $D_2 = \$14,705.88$
       $D_4 = 14,705.88$
       $D_6 = 7,352.94$

10. (a) $D_1 = \$9126.00$
        $D_{10} = 9525.00$
    (b) $D_1 = \$7187.50$
        $D_{10} = 7500.00$

12. (a) $D_1 = \$1586.40$
        $D_3 = 7620.00$
        $D_5 = 4762.50$
        $B_1 = \$238,413.60$
        $B_3 = 223,173.60$
        $B_5 = 210,791.10$
    (b) $D_1 = \$1250.00$
        $D_3 = 6000.00$
        $D_5 = 3750.00$
        $B_1 = \$238,750.00$
        $B_3 = 226,750.00$
        $B_5 = 217,000.00$

14. (a) 3-year property; Section
        1245 class property.
    (c) Not recovery property.
    (e) 15-, 18-, or 19-year real
        property; Section 1250
        class property.
    (g) 5-year property; Section
        1245 class property.
    (i) 5-year property; Section
        1245 class property.

16. (a) $D_2 = \$4560.00$
        $D_4 = 0.00$
        $D_6 = 0.00$
        $B_2 = \$4440.00$
        $B_4 = 0.00$
        $B_6 = 0.00$
    (b) $D_2 = \$4000.00$
        $D_4 = 2000.00$
        $D_6 = 0.00$
        $B_2 = \$6000.00$
        $B_4 = 0.00$
        $B_6 = 0.00$

18. (a) $D_2 = \$55,000.00$
        $D_4 = 52,500.00$
        $D_6 = 0.00$
    (b) $D_2 = \$50,000.00$

$D_4 = $ 50,000.00
$D_6 = $ 25,000.00

20. (a) $D_2 = $ \$18,000.00
    $D_4 = $ 14,400.00
    $D_6 = $ 1,800.00
    (b) $D_2 = $ \$12,000.00
    $D_4 = $ 12,000.00
    $D_6 = $ 2,000.00

22. (a) $D_2 = $ \$1583.33
    $D_4 = $ 1583.33
    $D_6 = $ 1583.33
    $B_2 = $ \$8833.33
    $B_4 = $ 5666.67
    $B_6 = $ 2500.00
    (b) $D_2 = $ \$2261.90
    $D_4 = $ 1357.14
    $D_6 = $ 452.38
    $B_2 = $ \$7023.81
    $B_4 = $ 3857.14
    $B_6 = $ 2500.00
    (c) $D_2 = $ \$2666.67
    $D_4 = $ 1055.56
    $D_6 = $ 0.00
    $B_2 = $ \$5333.33
    $B_4 = $ 2500.00
    $B_6 = $ 2500.00

24. (a) $D_1 = $ \$1999.80
    $D_3 = $ 888.60
    $B_1 = $ \$4000.20
    $B_3 = $ 444.60
    $R_1 = $ \$900.00
    $R_3 = $ 199.98
    $CR + R_1 = $ \$2899.80
    $CR + R_3 = $ 1088.58
    $AECR + R = $ \$2101.59/
    year
    (b) $D_1 = $ \$1500.00
    $D_3 = $ 2220.00
    $B_1 = $ \$4500.00
    $B_3 = $ 0.00
    $R_1 = $ \$900.00
    $R_3 = $ 333.00
    $CR + R_1 = $ \$2400.00
    $CR + R_3 = $ 2553.00
    $AECR + R = $ \$2101.59/
    year

(c) $D_1 = $ \$1500.00
    $D_3 = $ 1500.00
    $B_1 = $ \$4500.00
    $B_3 = $ 1500.00
    $R_1 = $ \$900.00
    $R_3 = $ 450.00
    $CR + R_1 = $ \$2400.00
    $CR + R_3 = $ 1950.00
    $AECR + R = $ \$2101.59/
    year

26. $D_1 = $ \$4000.00
    $D_2 = $ 2800.00
    $D_3 = $ 6000.00
    $D_4 = $ 1200.00

28. (a) \$2250.00
    (b) \$11,500.00
    (c) \$18,170.00
    (d) \$53,450.00
    (e) \$136,000.00

30. (a) $ATCF_0 = -\$50,000.00$
    $ATCF_1 = $ 12,640.00
    $ATCF_5 = $ 11,198.40
    $ATCF_7 = $ 9,240.00
    (b) $ATCF_0 = -\$50,000.00$
    $ATCF_1 = $ 10,656.67
    $ATCF_5 = $ 12,073.33
    $ATCF_7 = $ 10,656.67
    (c) $ATCF_0 = -\$50,000.00$
    $ATCF_1 = $ 11,790.00
    $ATCF_5 = $ 12,810.00
    $ATCF_7 = $ 9,240.00
    (d) $ATCF_0 = -\$50,000.00$
    $ATCF_1 = $ 11,365.00
    $ATCF_5 = $ 11,365.00
    $ATCF_7 = $ 11,365.00

32. (a) \$6522.60/year
    (b) \$6577.99/year
    (c) \$6621.01/year

34. \$23,192.24

36. (a) \$55,987.14/year
    (b) \$3.50/unit

38. (a) 1. \$80,000.00   2. \$80,000.00
    (b) 1. 55,213.19   2. 39,647.94
    (c) 1. 61,200.00   2. 61,200.00
    (d) 1. 25,593.43   2. 30,885.61

(e) Regardless of depreciation method, the undiscounted total depreciation and total taxes paid will be the same.

(f) Accelerated depreciation methods increase the present worth of depreciation allowed and decrease the present worth of taxes paid.

40. (a) a. $BTCF_1 = \$150,000.00$
$BTCF_3 = 150,000.00$
$BTCF_5 = 150,000.00$
b. $LPP_1 = \$57,595.39$
$LPP_3 = 76,169.91$
$LPP_5 = 0.00$
c. $LIP_1 = \$30,000.00$
$LIP_3 = 11,425.49$
$LIP_5 = 0.00$
d. $D_1 = \$100,000.00$
$D_3 = 96,000.00$
$D_5 = 57,600.00$
e. $TI_1 = \$20,000.00$
$TI_3 = 42,574.51$
$TI_5 = 92,400.00$
f. $TAX_1 = \$ 6,800.00$
$TAX_3 = 14,475.33$
$TAX_5 = 31,416.00$
g. $ATCF_1 = \$ 55,604.61$
$ATCF_3 = 47,929.27$
$ATCF_5 = 118,584.00$

(b) a. $BTCF_1 = \$150,000.00$
$BTCF_3 = 150,000.00$
$BTCF_5 = 150,000.00$
b. $LPP_1 = \$66,666.67$
$LPP_3 = 66,666.67$
$LPP_5 = 0.00$
c. $LIP_1 = \$30,000.00$
$LIP_3 = 10,000.00$
$LIP_5 = 0.00$
d. $D_1 = \$100,000.00$
$D_3 = 96,000.00$
$D_5 = 57,600.00$
e. $TI_1 = \$20,000.00$
$TI_3 = 44,000.00$
$TI_5 = 92,400.00$

f. $TAX_1 = \$ 6,800.00$
$TAX_3 = 14,960.00$
$TAX_5 = 31,416.00$
g. $ATCF_1 = \$ 46,533.33$
$ATCF_3 = 58,373.33$
$ATCF_5 = 118,584.00$

42. $B4TLCF_3 = \$30,000.00$
$B4TLCF_4 = \$23,000.00$
$B4TLCF_7 = \$32,000.00$
$B4TLCF_8 = \$40,000.00$
$LIP_3 = \$4800.00$
$LIP_4 = \$4800.00$
$D_4 = \$12,490.00$
$D_5 = \$8930.00$
$D_6 = \$8920.00$
$D_7 = \$8930.00$
$D_8 = \$4460.00$
$D_9 = \$0.00$
$D_{10} = \$0.00$
$TI_1 = \$910.00$
$TI_2 = -\$9290.00$
$TI_3 = \$7710.00$
$TI_4 = \$5710.00$
$TI_5 = \$26,270.00$
$TI_6 = \$6280.00$
$TI_9 = \$40,200.00$
$TI_{10} = \$25,200.00$
$TAX_1 = \$309.40$
$TAX_2 = -\$3158.60$
$TAX_6 = \$2135.20$
$TAX_7 = \$6211.80$
$TAX_8 = \$10,451.60$
$TAX_9 = \$13,668.00$
$TAX_{10} = \$8568.00$
$ATCF_1 = \$14,890.60$
$ATCF_2 = \$18,358.60$
$ATCF_3 = \$22,578.60$
$ATCF_4 = \$16,258.60$
$ATCF_5 = \$26,268.20$
$ATCF_7 = \$20,988.20$
$ATCF_8 = \$24,748.40$
$ATCF_9 = \$26,532.00$
$ATCF_{10} = -\$23,368.00$

44. \$3636.64/year

46. (a) \$39,410.00
(b) \$34,330.00

48. (a) $23,200.00
    (b) $3200.00
    (c) Net capital loss of $40,000.00. May be carried back or over.
    (d) $3080.00
    (e) Net capital loss of $9000.00. May be carried back or over.
    (f) Net capital loss of $11,000.00. May be carried back or over.

50. (a) $8040.00 Section 1231 loss, resulting in −$2733.60 tax.
    (b) $51,666.67 Section 1231 loss, resulting in −$17,566.67 tax.

52. (a) $33,333.33 Depreciation recapture plus $20,000.00 Section 1231 gain, resulting in tax of $18,133.33.
    (b) $6666.67 Section 1231 loss, resulting in tax of −$2266.67.
    (c) $156,666.67 Section 1231 loss, resulting in tax of −$53,266.67

54. (a) $32,000.00 Section 1231 gain, resulting in tax of $10,880.00.
    (b) $58,000.00 Section 1231 loss, resulting in tax of −$19,720.00.

56. (a) $170,416.67 Section 1231 gain, resulting in tax of $57,941.67.
    (b) $70,416.67 Section 1231 gain, resulting in tax of $23,941.67.
    (c) $359,583.33 Section 1231 loss, resulting in tax of −$122,258.33.

58. $15,416.01

60. −$1504.65/year

62. (a) $PW = \$45,053.36$; Invest
    (b) $PW = \$45,053.36$; Invest

64. (a) $114.03/year
    (b) $114.03/year

66. $ATCF_0 = -\$30,000,000.00$
    $ATCF_1 = 3,225,000.00$
    $ATCF_2 = 4,300,000.00$

CHAPTER 7

2. $X = \$0.1014$/minute (Stop-start costs/year = $2600; waiting costs/year = $121,333.33x$)

4. $B/C = 0.7412$ ($B = \$76,200,127$ and $C = \$102,804,571$)

6. (a) For $i = 8\%$, Route A, with $B/C = 1.6567$
   (b) For $i = 0\%$, Route A, with $B/C = 2.9126$

8. Enlarge zoo, with $B - C = \$37,071.90$/year

10. For $i = 20\%$ and $75,000 budget, projects A and B should be undertaken ($B/C_A = 1.4043$ and $B/C_B = 0.9998$). Opportunity cost slightly less than 20%

12. An opinion question

14. $B/C = 1.1356$ and $B - C = \$15,250$ year. Build.

16. $B - C = -\$428.10$. Do not build.

18. (a) $B/C = 20$
    (b) $B/C_{\text{error}} = 1.1765$
    (c) $B - C = \$285,000$
    (d) $B - C$ preferred.

20. Design 1 has both lowest cost ($PW = -\$48,347,558$) and highest payload (180 units). Eliminate other designs.

CHAPTER 8

2. (a) 83,333 units/year
   (b) 115,385 units/year
   (c) 187,500 units/year

4. Pump B is not justified for any value of n; select Pump A.

6. (a) 66.15 days/year
   (b) 78.69 days/year
   (c) 110.85 days/year

8. $10,688/year

10. $650.40/year

12. One-year horizon assumed.
    (a) $\geq$ 25 units
    (b) < 25 units

14. 2376.5 units, 12623.5 units

20. (b) $Pr\,[PW(25\%) \geq 0] = 0.15$

22. Analytical method:
    $Pr\,[PW(25\%) \geq 0] = 0.15$

24. $Pr\,[PW \geq 0 | I = \$9000] = 0.304$, $Pr\,[PW \geq 0 | I = \$11000] = 0.068$, and $Pr\,[PW \geq 0] = 0.186$

26. $Pr\,(PW < 0) = 0.2667$
    Average $PW = \$569.87$

28. Average $PW = \$384.46$

30. (a) $E[PW\,(0\%)] = \$14000$;
    Var $[PW\,(0\%)] = 896 \times 10^4$;
    $Pr\,[PW\,(0\%) \geq 0] = Pr\,[AW\,(0\%) \geq 0] = 1.0$.
    (b) $E[PW\,(15\%)] = -\$1925.20$;
    Var $[PW\,(15\%)] = 713.93 \times 10^4$; $Pr\,[PW\,(15\%) \geq 0] = Pr\,[AW\,(15\%) \geq 0] = 0.2358$.

32. $Pr\,[PW_A\,(25\%) \geq PW_B\,(25\%)] = Pr\,[PW_{A-B}\,(25\%) \geq 0] = 0$.
    NOTE: $E[PW_{A-B}\,(25\%) = -\$4342.15; \mathrm{Var}\,[PW_{A-B}(25\%)] = 142.174 \times 10^4$

34. $E(EUAC) = \$3,795.19$

CHAPTER 9

2. A_3 dominates A_1.
   Laplace: A_4, with $E(A_4) = \$5$
   Maximin: Either A_3 or A_4
   Maximax: A_2
   Hurwicz: A_4, with $H_4 = \$3.75$
   Savage: A_4

4. If $0 \leq \alpha < \frac{1}{2}$, choose A_1. If $\alpha = \frac{1}{2}$, choose either A_1 or A_2. If $\frac{1}{2} < \alpha \leq 1.0$, choose A_2.

6. (a) 1. $\alpha = 0.7098$
       2. A_2
   (b) A_2, with $E(A_2) = \$218$ and $\mathrm{Var}(A_2) = 0$.

8. (a) Minimax: Choose a_2, with value = $60
   (b) Minimin: Choose a_4, with value = $-\$50$
   (c) Minimax regret: Choose either a_1 or a_3 with value = 100
   (d) Expectation: Choose a_2, with $E(a_2) = \$60$ and Var $(a_2) = 0$.
   (e) Hurwicz: Choose a_4, with $H_4 = \$10$

10. A_4 dominates A_3.
    Expectation: A_2, with $E(A_2) = \$273$
    Most probable future: A_2
    Maximax: A_2, with value = $500
    Maximin: A_2, with value = $-\$20$
    Minimax regret: Either A_1 or A_2, with value = 300.

12. (a) A_2, with $P(\text{loss}) = 0$.
    (b) A_5, with $P(\text{annual profit} \geq \$9000) = 0.60$.
    (c) A_3, with probability = 0.70.
    (d) No, because A_4 is choice and has large loss possible.

14. (a) $P(S_1) = 0.216$; $P(S_2) = 0.432$; $P(S_3) = 0.288$; $P(S_4) = 0.064$
    (b) $E(\text{Guess } S_1) = -\$0.136$; $E(\text{Guess } S_2) = \$0.296$; $E(\text{Guess } S_3) = \$0.152$; $E(\text{Guess } S_4) = -\$0.680$. Both principles choose "Guess S_2".

16. (a) Event A, with $P(A|h) = 0.5$
    (b) White die, with $E(\text{guess white die}) = \$1.50$

18. (a) $P(H) = 0.62$
    (b) $P(C|L) = 0.434$

20. (a) Choose a_1 with $PEP = \$6,100,000$.
    (b) $EVPI = \$1,000,000$
    (c) $EVSI = \$89,637$
    (d) $-\$160,363$.

22. (b) Keep the present truck (by dominance principle). For interest, $E(\text{Keep truck}) = -\$21,700$, and $E(\text{Buy new}) = -\$30,500$ at D_1.

24. Choose A_1, by dominance

Index